Adipose Tissue and Adipokines in Health and Disease

Nutrition ◊ and ◊ Health

Adrianne Bendich, Series Editor

Handbook of Nutrition and Ophthalmology, by **Richard D. Semba,** 2007
Adipose Tissue and Adipokines in Health and Disease, edited by **Giamila Fantuzzi and Theodore Mazzone,** 2007
Nutritional Health: *Strategies for Disease Prevention, Second Edition,* edited by **Norman J. Temple**, **Ted Wilson, and David R. Jacobs, Jr.,** 2006
Nutrients, Stress, and Medical Disorders, edited by **Shlomo Yehuda and David I. Mostofsky,** 2006
Calcium in Human Health, edited by **Connie M. Weaver and Robert P. Heaney,** 2006
Preventive Nutrition: *The Comprehensive Guide for Health Professionals, Third Edition,* edited by **Adrianne Bendich and Richard J. Deckelbaum,** 2005
The Management of Eating Disorders and Obesity, *Second Edition,* edited by **David J. Goldstein,** 2005
Nutrition and Oral Medicine, edited by **Riva Touger-Decker, David A. Sirois, and Connie C. Mobley,** 2005
IGF and Nutrition in Health and Disease, edited by **M. Sue Houston, Jeffrey M. P. Holly, and Eva L. Feldman,** 2005
Epilepsy and the Ketogenic Diet, edited by **Carl E. Stafstrom and Jong M. Rho,** 2004
Handbook of Drug–Nutrient Interactions, edited by **Joseph I. Boullata and Vincent T. Armenti,** 2004
Nutrition and Bone Health, edited by **Michael F. Holick and Bess Dawson-Hughes,** 2004
Diet and Human Immune Function, edited by **David A. Hughes, L. Gail Darlington, and Adrianne Bendich,** 2004
Beverages in Nutrition and Health, edited by **Ted Wilson and Norman J. Temple,** 2004
Handbook of Clinical Nutrition and Aging, edited by **Connie Watkins Bales and Christine Seel Ritchie,** 2004
Fatty Acids: *Physiological and Behavioral Functions,* edited by **David I. Mostofsky, Shlomo Yehuda, and Norman Salem, Jr.,** 2001
Nutrition and Health in Developing Countries, edited by **Richard D. Semba and Martin W. Bloem,** 2001
Preventive Nutrition: *The Comprehensive Guide for Health Professionals, Second Edition,* edited
by **Adrianne Bendich and Richard J. Deckelbaum,** 2001
Nutritional Health: *Strategies for Disease Prevention,* edited by **Ted Wilson and Norman J. Temple**, 2001
Clinical Nutrition of the Essential Trace Elements and Minerals: *The Guide for Health Professionals,* edited by **John D. Bogden and Leslie M. Klevey**, 2000
Primary and Secondary Preventive Nutrition, edited by **Adrianne Bendich and Richard J. Deckelbaum**, 2000
The Management of Eating Disorders and Obesity, edited by **David J. Goldstein**, 1999
Vitamin D: *Physiology, Molecular Biology, and Clinical Applications,* edited by **Michael F. Holick**, 1999
Preventive Nutrition: *The Comprehensive Guide for Health Professionals,* edited by **Adrianne Bendich and Richard J. Deckelbaum**, 1997

ADIPOSE TISSUE AND ADIPOKINES IN HEALTH AND DISEASE

Edited by

GIAMILA FANTUZZI, PhD

Department of Human Nutrition
University of Illinois at Chicago
Chicago, IL

THEODORE MAZZONE, MD

Section of Endocrinology, Diabetes, and Metabolism
Department of Medicine
University of Illinois at Chicago
Chicago, IL

Foreword by

ANDREW P. GOLDBERG, MD AND SUSAN K. FRIED, PhD

Department of Medicine, University of Maryland School of Medicine
Baltimore Veteran's Affairs Hospital, Gastric Research, Education and Clinical Center
Baltimore, MD

HUMANA PRESS
TOTOWA, NEW JERSEY

999 Riverview Drive, Suite 208
Totowa, New Jersey 07512

Cover design by Nancy Fallatt.
Production Editor: Amy Thau

For additional copies, pricing for bulk purchases, and/or information about other Humana titles, contact Humana at the above address or at any of the following numbers: Tel.: 973-256-1699; Fax: 973-256-8341; E-mail: orders@humanapr.com or visit our website at http://humanapress.com

This publication is printed on acid-free paper. ∞
ANSI Z39.48-1984 (American National Standards Institute) Permanence of Paper for Printed Library Materials.

Printed in the United States of America. 10 9 8 7 6 5 4 3 2 1
eISBN: 1-59745-370-6
13-digit eISBN: 978-1-59745-370-7
13-digit ISBN: 978-1-58829-721-1

Library of Congress Cataloging-in-Publication Data
Adipose tissue and adipokines in health and disease / edited by
Giamila Fantuzzi, Theodore Mazzone; foreword by Andrew P. Goldberg,
Susan K. Fried.
p. ; cm. -- (Nutrition and health)
Includes bibliographical references and index.
ISBN 1-58829-721-7 (alk. paper)
1. Adipose tissues. 2. Fat cells. I. Fantuzzi, Giamila. II.
Mazzone, Theodore. III. Series: Nutrition and health (Totowa, N.J.)
[DNLM: 1. Adipose Tissue--physiology. 2. Adipocytes--physiology.
3. Obesity--complications. 4. Obesity--etiology. QS 532.5.A3
A23475 2007]
QP88.23.A35 2007
571.5'7--dc22

2006017678

Dedication

GF would like to dedicate this book to Sergio, with love.

Series Editor's Introduction

The Nutrition and Health™ series of books has as an overriding mission to provide health professionals with texts that are considered essential because each includes: (1) a synthesis of the state of the science; (2) timely, in-depth reviews by the leading researchers in their respective fields; (3) extensive, up-to-date, fully annotated reference lists; (4) a detailed index; (5) relevant tables and figures; (6) identification of paradigm shifts and the consequences; (7) virtually no overlap of information between chapters, but targeted, interchapter referrals; (8) suggestions of areas for future research; and (9) balanced, data-driven answers to patient/health professional's questions, which are based on the totality of evidence rather than the findings of any single study.

The series volumes are not the outcome of a symposium. Rather, each editor has the potential to examine a chosen area with a broad perspective, both in subject matter as well as in the choice of chapter authors. The international perspective, especially with regard to public health initiatives, is emphasized where appropriate. The editors, whose trainings are both research- and practice-oriented, have the opportunity to develop a primary objective for their book, define the scope and focus, and then invite the leading authorities from around the world to be a part of their initiative. The authors are encouraged to provide an overview of the field, discuss their own research, and relate the research findings to potential human health consequences. Because each book is developed de novo, the chapters are coordinated so that the resulting volume imparts greater knowledge than the sum of the information contained in the individual chapters.

Adipose Tissue and Adipokines in Health and Disease, edited by Drs. Giamila Fantuzzi and Theodore Mazzone, is a very welcome addition to the Nutrition and Health Series and fully exemplifies the Series' goals. This volume is especially timely because the obesity epidemic continues to increase around the world and the comorbidities, such as type 2 diabetes, are seen even in very young children. It is only recently that scientists have begun to think about adipose tissue as more than a fat storage site. The last decade has seen an explosion of identification and characterization of the many bioactive molecules that are synthesized and secreted by adipose cells (adipokines). Adipokines have been associated with the development of obesity and its comorbidities as well as many, often thought of as unrelated, consequences including insulin resistance, cardiovascular complications, lipid disorders, hypertension, hormonal imbalances, endometriosis, asthma, kidney disease, certain autoimmune diseases, brain and peripheral nervous system disorders, inflammatory bowel disease, as well as cancer. Thus, the relevance of adipocyte pathophysiology to the clinical setting is of great interest to not only academic researchers, but also to health care providers. This text is the first to synthesize the knowledge base concerning adipose tissue, adipokines, and obesity for the practicing health professional as well as those professionals who have an interest in the latest, up-to-date information on fat cell function and its implications for human health and disease.

Adipose Tissue and Adipokines in Health and Disease serves a dual purpose of providing in-depth focus on the biological functions of adipocytes, as well as examining the current clinical findings associated with the consequences of obesity and putting these into historic perspective, as well as pointing the way to future research opportunities. Both of the editors are internationally recognized leaders in the field of adipocyte research as well as clinical outcomes. Both are excellent communicators and they have worked tirelessly to develop a book that is destined to be the benchmark in the field because of its extensive, in-depth chapters covering the most important aspects of the complex interactions between cellular functions, diet and obesity, and its impact on disease states. The introductory chapters provide readers with the basics so that the more clinically related chapters can be easily understood. The editors have chosen 55 of the most well-recognized and respected authors from around the world to contribute the 27 informative chapters in the volume. Hallmarks of all of the chapters include complete definitions of terms with the abbreviations fully defined for the reader and consistent use of terms between chapters. Key features of this comprehensive volume include the informative abstract and key words that are at the beginning of each chapter, appendices that include a detailed list of abbreviations as well as extensive lists of relevant books, journals, and websites, more than 80 detailed tables and informative figures, an extensive, detailed index and more than 2500 up-to-date references that provide the reader with excellent sources of worthwhile information about adiposity and health.

Drs. Fantuzzi and Mazzone have chosen chapter authors who are internationally distinguished researchers, clinicians, and epidemiologists who provide a comprehensive foundation for understanding the role of adipose tissue and its secretions in the maintenance of human health as well as its role in obesity and related comorbidities. Outstanding chapters have been contributed by the actual discoverers of potent adipokines, including Dr. Matsuzawa, the discoverer of adiponectin, who provides an historical perspective on his research since his discovery (in Chapter 9). There are more than a dozen chapters that cover the most important aspects of the complex interactions between adipose cell functions including the synthesis of more than 20 adipokines, receptors for these active molecules, and consequent activation of other cells and tissues in the central nervous system as well as peripherally. The volume includes separate chapters on leptin and adiponectin, as well as several chapters that include significant analyses of the roles of factors including interleukin-6, tumor necrosis factor, adipsin, resistin, visfatin, and prostaglandins as examples. Likewise, the different cells that comprise fat tissue have the capacity to produce signals that lead to the production of other active biomolecules. One example that is discussed in detail is the interaction between interleukin-6 from adipose tissue that triggers the production of C-reactive protein in the liver. Thus, the chapter authors have integrated the newest research findings so the reader can better understand the complex interactions that can result from adipocyte function. Another important chapter highlights the novel interactions between adipose tissue and lymph nodes that are physically in close proximity and influence the functioning of both tissues (Chapter 11). Moreover, wherever there is sufficient evidence, the authors have highlighted the potential impact of these interactions on disease states and health conditions that may increase or decrease the risk of chronic disease.

Given the growing concern with the increase in adult as well as childhood obesity, it is not surprising to find 15 chapters in this book that are devoted to the clinical aspects of obesity, weight control, diabetes, and other chronic diseases associated with obesity. Unique to *Adipose Tissue and Adipokines in Health and Disease* are chapters that examine the biological plausibility of the "thrifty gene" hypothesis; the potential functions of bone marrow adipose tissue and bone formation; the cultural aspects of weight gain; three in-depth chapters on the effects of weight loss that is either deliberate, genetically inherited, or acquired because of cultural factors or as a consequence of obesity-related diseases. The growing awareness that obesity is associated with a low-grade inflammatory state is examined extensively in each of the chapters on insulin resistance, non-alcoholic liver disease, kidney disease, and joint disease, as well as in the chapters on heart disease, asthma, and cancer. The editors and authors have integrated the information within these chapters so that the health care practitioner can provide guidance to the patient about the potential consequences of chronic obesity. The inclusion of both the earlier chapters on the complexity of adipose tissue and the chapters that contain clinical discussions help the reader to have a broader basis of understanding of fat cells, adipose tissue, obesity, and disease.

In conclusion, *Adipose Tissue and Adipokines in Health and Disease* provides health professionals in many areas of research and practice with the most up-to-date, well-referenced volume on the importance of adipose tissue and its secretions in determining the potential for obesity-related chronic diseases that can affect human health. This volume will serve the reader as the benchmark in this complex area of interrelationships between fat cells, adipose tissue, the adipose organ, and the functioning of all other organ systems in the human body. Moreover, the interactions between obesity, genetic factors, and the numerous comorbidities are clearly delineated so that students as well as practitioners can better understand the complexities of these interactions. The editors are applauded for their efforts to develop the most authoritative resource in the field to date and this excellent text is a very welcome addition to the Nutrition and Health Series.

Adrianne Bendich, *PhD, FACN*

Foreword

Nearly 35% of the American population is now considered obese, with a projected increase expected to reach nearly 50% by the year 2025 unless there are major changes in the obesifying environment that promotes a sedentary lifestyle and overconsumption of calories. In addition, effective treatments must be developed for the most severely and morbidly obese. Understanding the pleiotropic effects of the adipose organ on health and disease is crucial to achieving the goal of preventing the comorbidities associated with obesity and abdominal fat distribution.

Adipose tissue is a highly specialized tissue that stores excess energy in the form of triglyceride and releases it when energy is needed by other tissues. As pointed out in Chapter 1 of *Adipose Tissue and Adipokines in Health and Disease*, the adipose organ includes multiple adipose depots in distinct anatomical locations, and it can comprise up to 70% of body mass in massively obese individuals. Too little or too much adipose tissue has deleterious consequences for the function of numerous physiological systems. In addition, excess fat deposited in the abdomen (visceral depots) increases morbidity independent of overall adiposity. If the adipose tissue storage capacity is reached—for example, if maximal fat cell size is reached and reservoirs of preadipocytes are depleted—ectopic fat deposition in non-adipose organs is thought to impair cellular function. Thus, the ability of the adipose organ to expand through hypertrophy and hyperplasia (recruitment and differentiation of adipose precursors) is critical to the prevention of obesity-related comorbidities. As highlighted in *Adipose Tissue and Adipokines in Health and Disease*, understanding the basic mechanisms regulating adipose tissue growth is critical to understanding the links between obesity and disease.

Different adipose tissue depots grow at different rates in males compared with females, and depots preferentially expand (visceral) or shrink (peripheral subcutaneous) with aging. The classic studies of Vague and others showed that body shape, largely determined by body fat distribution in the trunk and periphery, has a large impact on risk for metabolic diseases and complications associated with obesity. It remains unclear whether visceral compared with subcutaneous adipocytes are truly genetically distinct or develop different metabolic profiles as a result of their anatomy, cellular composition, innervation, and blood flow. The cellular and molecular mechanisms that contribute to the metabolic heterogeneity of adipose tissue depots are active and evolving topics of research. We still do not understand why males and females preferentially deposit fat in different depots and why upper body depots (both abdominal subcutaneous and visceral) increase risk for insulin resistance, type 2 diabetes, and dyslipidemia, while lower body (subcutaneous gluteal–femoral) fat depots may be protective.

The discovery of leptin in 1994 and adiponectin in 1995 ushered in a new era in the study of the endocrine functions of the adipose organ and mechanisms that link adiposity to the maintenance of reproductive, immunological, and endocrine function. The effects of genetic and environmental factors on the size and distribution of the adipose organ has an impact on the function of virtually all physiological systems. As is clear from *Adipose Tissue and Adipokines in Health and Disease*, we are only beginning to understand the crosstalk among myriad paracrine mediators that modulate the growth, remodeling, and functions of each adipose depot. Furthermore, the impacts of circulating adipose hormones (adipokines) on the pathophysiological events in the cardiovascular, immune, and endocrine systems in obesity are beginning to be understood. It is now known that obesity is linked to the dysfunction of many physiological systems. As reviewed in this volume, adipose function is mechanistically linked to diseases such as arthritis, asthma, and inflammatory bowel disease. An exceptionally valuable feature of this book is that it merges the medical literature with basic research related to the pathobiology of adipose tissue and adipokines.

Adipose Tissue and Adipokines in Health and Disease highlights emerging and important new research on the interrelationships of adipocytes with the immune system. It is now clear that macrophages take up residence in adipose tissue in obesity and appear to drive the local chronic inflammation and, to a large extent, the altered adipocyte function. For example, macrophage production of tumor necrosis factor and interleukin-6, as well as perhaps monocyte chemotactic protein-1, appears to contribute to adipocyte resistance and elevated lipolysis through their paracrine effects. In addition, adipokines produced by the adipocyte itself (leptin and adiponectin) act in an autocrine fashion via their receptors to alter adipocyte function. Furthermore, understanding the interrelationships of lymphoid and other immune tissues with surrounding adipocytes, as well as the function of resident macrophages, mast and mononuclear cells in adipose tissue are critical areas of research in this field. This work provides insights into mechanisms linking adiposity to inflammation, insulin resistance, atherogenesis, and cancer. Research on adipocyte-immunocyte relationships will undoubtedly provide insights into new therapeutic approaches to prevent or treat adipose tissue-associated diseases.

A major issue in the field of obesity, however, that remains unresolved is: "Why do so few obese people maintain their weight after successful weight loss?" No particular diets effectively promote successful long-term weight maintenance, as most weight-reduced obese people regain lost weight within 3 years, with the re-emergence of the metabolic, cardiovascular, and cancer-related consequences of obesity. However, some can succeed, as evidenced by the weight control registry and the success of obesity surgeries. Thus, it will be of great interest to determine how the altered function of adipocytes and adipose tissues, and in particular altered adipokine action, contributes to the biological variability in response to weight loss.

The underlying molecular mechanisms by which the expanded adipose tissue organ and circulating adipose hormones (adipokines) contribute to the pathophysiological events in the cardiovascular, neuroendocrine, immune, gastrointestinal, renal, hepatobiliary, and musculoskeletal systems in obesity are now beginning to be understood. An exceptionally valuable component of *Adipose Tissue and Adipokines in Health and Disease* is that it brings together the diverse medical literature on the roles of adipose tissue and adipokines in the pathogenesis of disease. We are confident that *Adipose Tissue and Adipokines in Health and Disease* will be of great value in supporting the multidisciplinary efforts to understand the many functions of the adipose tissue that will undoubtedly lead to the discovery of new pharmacological targets and therapeutic tools for the prevention and treatment of the inflammatory, endocrine, immune, pro-carcinogenic, and cardiovascular consequences of obesity.

Andrew P. Goldberg, MD
Susan K. Fried, PhD

Preface

Adipose tissue has traditionally been considered as a virtually inert tissue, mainly devoted to energy storage. As a consequence, research on the role of adipose tissue in physiology and pathology has been relatively neglected and "out-of-fashion" until the mid 1990s, when the discovery of leptin (in 1994) and adiponectin (in 1995) led to a major renaissance in this field. Concomitant with the discovery of important mediators produced by adipocytes has been the awareness of a dramatically increasing prevalence of obesity and associated pathologies worldwide in both children and adults. High rates of overweight and obesity in the population have awakened the interest of both clinicians and researchers in understanding the role of adipose tissue in inducing and/or participating in the array of comorbidities associated with excess adiposity, which range from insulin resistance and diabetes to cardiovascular disease, osteoarthritis, and other chronic pathologies. As a consequence of the increased attention devoted to the study of adipose tissue in the last 12–15 yr, our current understanding of the biology of adipocytes and their role in physiology and pathology has dramatically improved.

The aim of *Adipose Tissue and Adipokines in Health and Disease* is to provide comprehensive information regarding adipose tissue, its physiological functions, and its role in disease. We have strived to collect information spanning the entire range of adipose tissue studies, from basic anatomical and physiological research to epidemiology and clinical aspects, in one place. *Adipose Tissue and Adipokines in Health and Disease* is addressed to both basic researchers and clinicians interested in the fields of obesity, metabolic diseases, inflammation, and immunity, and specialists in each of the pathologies associated with obesity.

In Part I (Adipose Tissue: Structure and Function), Chapter 1 provides a comprehensive review of adipose tissue anatomy with the new concept of the adipose organ, and Chapter 2 summarizes fundamental physiological aspects of adipose tissue metabolism. Each chapter does an excellent job of incorporating late-breaking studies into what can only be called a "classical" description of adipose tissue properties. Part I also includes chapters on two of the best-known and best-characterized adipokines, leptin (Chapter 3) and adiponectin (Chapter 4). These two chapters, in addition to providing a comprehensive overview of the biology of leptin and adiponectin, introduce the reader to a "novel" view of adipose tissue, as a view that has led to a revolution in our understanding of many metabolic and inflammatory conditions. This novel view pictures adipose tissue as a dynamic organ regulating not only systemic substrate availability and metabolism, but also a variety of other functions, spanning from immune responses to bone structure and susceptibility to cancer.

Parts II (Adipokines as Regulators of Immunity and Inflammation) and III (Interactions Between Adipocytes and Immune Cells) expand on the novel concept of adipose tissue as a critical component in maintaining body homeostasis. Adipose tissue and the immune system are engaged in a constant dialogue that leads to the modulation of both inflammatory and metabolic responses. Part II describes how the discovery of adipocytes as a source of mediators that modulate immune and inflammatory responses has opened up a whole new field of investigation. We are now able to include non-conventional soluble mediators derived from adipose tissue, such as leptin, adiponectin, and other adipokines (or adipocytokines), into the vast array of molecules that participate in the control of immunity and inflammation. This section introduces the different vasoactive and inflammatory molecules produced by adipose tissue (Chapter 5) and presents a detailed description of the role of leptin (Chapters 6–8), and adiponectin (Chapter 9) in modulating immunity and inflammation. In Part III, Chapter 10 describes how macrophages are normal components of adipose tissue and become activated in obesity, possibly mediating some of the metabolic alterations associated with excess adiposity. Chapters 11 and 12 deal with interactions between adipocytes and other components of the immune system, specifically lymphocytes and mast cells, whereas Chapter 13 discusses the interaction between adipocytes and hematopoietic cells in bone marrow.

The authors of the chapters presented in Part IV (Weight Gain and Weight Loss), present an overview of the worldwide trends in obesity, providing an evolutionary perspective and discussing possible causes. A discussion presenting updated and detailed epidemiological data on the obesity epidemics (Chapter 14) opens this section. Chapter 15 deals with environmental contributions to the ever-increasing incidence of obesity, including a discussion of food supply and physical activity trends, fast-food consumption, portion size, and others. Two very interesting chapters discuss the genetics of obesity from an evolutionary perspective. Chapter 16 introduces the reader to the concept of the famine hypothesis and discusses possible problems with this notion, whereas Chapter 17 deals with the role of intrauterine development patterns in shaping the postnatal propensity to develop obesity and insulin resistance. Chapter 18 analyzes the metabolic disturbances associated with disorders of adipose tissue development and distribution, such as the different forms of congenital or acquired lipodystrophies. Chapter 19 describes the mechanisms leading to cachexia, still an important risk factor in chronic kidney disease and chronic pathologies. This section closes with a discussion of different approaches to the difficult problem of inducing and sustaining weight loss in obese subjects and the health effects of such treatments (Chapter 20).

In Part V (Adipose Tissue and Disease), researchers and clinicians describe the association and potential role of adipose tissue in mediating disease. Pathologies covered range from metabolic conditions such as insulin resistance (Chapter 21) and fatty liver disease (Chapter 22) to cancer (discussed in Chapter 23) and cardiovascular disease (Chapter 24). Epidemiological evidence on the association between obesity and asthma and possible underlying mechanisms are discussed in Chapter 25, whereas Chapters 26 and 27 deal with the effect of obesity in kidney and joint disease, respectively.

The amount of information regarding the role of adipose tissue in health and disease continues to grow dramatically. Integration of epidemiological, physiological, and pathophysiological information will be critical for optimizing use of this information for improving human health. By bringing together a group of distinguished researchers, clinicians and epidemiologists as contributors to this volume, we hope to provide a comprehensive foundation for understanding new developments in adipose tissue biology.

We thank Paul Dolgert and the staff at Humana Press for their confidence in our vision and their excellent assistance. We also wish to thank Dr. Adrianne Bendich, the Series Editor, for her guidance and helpful suggestions, as well as Drs. Andrew Goldberg and Susan Fried for their eloquent introduction. Finally, a special thanks to each of the authors who contributed their insightful chapters.

Giamila Fantuzzi
Theodore Mazzone

Contents

Contributors

ANIL K. AGARWAL, PhD • *Division of Nutrition and Metabolic Diseases, Department of Internal Medicine and Center for Human Nutrition, The University of Texas Southwestern Medical Center at Dallas Dallas, TX*

REXFORD S. AHIMA, MD, PhD • *Division of Endocrinology, Diabetes, and Metabolism, Department of Medicine, University of Pennsylvania School of Medicine, Philadelphia, PA*

LUIGI ALOE, PhD • *NGF Section, Institute of Neurobiology and Molecular Medicine, National Research Council-European Brain Research Institute, Rome, Italy*

PEPA ATANASSOVA, MD, PhD • *Department of Anatomy and Histology, Medical University, Plovdiv, Bulgaria*

ANCHA BARANOVA, PhD • *Center for Liver Diseases, Inova Fairfax Hospital; Center for the Study of Genomics in Liver Diseases; Molecular and Microbiology Department, George Mason University; and George Mason-Inova Health System's Translational Research Centers, Falls Church, VA*

SERGIO JOSÉ BARDARO, MD • *Division of Minimally Invasive Surgery, Legacy Health System, Portland, OR*

ARVIND BATRA, PhD • *Charité, Campus Benjamin Franklin, Medizinische Klinik I, Berlin, Germany*

SRINIVASAN BEDDHU, MD • *Medical Service, Veterans Affairs Salt Lake City Healthcare System; and Division of Nephrology and Hypertension, University of Utah, Salt Lake City, UT*

MICHEL BEYLOT, MD, PhD • *INSERM U499, Metabolic and Renal Physiopathology, Faculte RTH Laennec, Lyon, France*

STEPHEN E. BORST, PhD • *Geriatric Research, Education and Clinical Center, Malcom Randal Veterans Administration Medical Center and University of Florida, Gainesville, FL*

CAROL A. BRAUNSCHWEIG, PhD, RD • *Department of Human Nutrition, University of Illinois at Chicago, Chicago, IL*

EUGENIA E. CALLE, PhD • *Department of Epidemiology and Surveillance Research at the American Cancer Society, Atlanta, GA*

LOUIS CASTEILLA, PhD • *UMR CNRS 5018, Institut Louis Bugnard, Toulouse, France*

GEORGE N. CHALDAKOV, MD, PhD • *Division of Cell Biology, Medical University, Varna, Bulgaria*

WAI W. CHEUNG, PhD • *Division of Pediatric Nephrology, Oregon Health and Science University, Portland, OR*

BONNIE CHING-HA KWAN, MB BS • *Division of Nephrology and Hypertension, University of Utah, Salt Lake City, UT; and Department of Medicine and Therapeutics, Prince of Wales Hospital, The Chinese University of Hong Kong, Hong Kong*

FLAVIA M. CICUTTINI, MB BS, FRACP, MSc, PhD • *Department of Epidemiology and Preventive Medicine, Monash University, Melbourne, Victoria, Australia*

SAVERIO CINTI, MD • *Institute of Normal Human Morphology–Anatomy, School of Medicine, University of Ancona, Ancona, Italy*
LISA DIEWALD, MS, RD, LDN • *Weight and Eating Disorders Program, University of Pennsylvania School of Medicine, Philadelphia, PA*
MEREDITH S. DOLAN, MS, RD, LDN • *The Children's Hospital of Philadelphia, Philadelphia, PA*
ROBERT H. ECKEL, MD • *Division of Endocrinology, Metabolism, and Diabetes, University of Colorado Health Sciences Center, Aurora, CO*
MYLES S. FAITH, PhD • *Weight and Eating Disorders Program, University of Pennsylvania School of Medicine and the Children's Hospital of Philadelphia, Philadelphia, PA*
GIAMILA FANTUZZI, PhD • *Department of Human Nutrition, University of Illinois at Chicago, Chicago, IL*
PATRICIA FERNÁNDEZ-RIEJOS, PhD • *Department of Medical Biochemistry and Molecular Biology, School of Medicine, Virgen Macarena University Hospital, Seville, Spain*
ANTHONY W. FERRANTE, JR., MD • *Naomi Berrie Diabetes Center, Columbia University, New York, NY*
SUSAN K. FRIED, PhD • *Department of Medicine, University of Maryland School of Medicine, Baltimore Veteran's Affairs Hospital, Gastric Research, Education and Clinical Center, Baltimore, MD*
GEMA FRÜHBECK, R Nutr, MD, PhD • *Department of Endocrinology, Clínica Universitaria de Navarra, Pamplona, Spain*
PETER D. GLUCKMAN, CNZM, MBChB, MMedSc, DSc • *Centre for Human Evolution, Adaptation, and Disease, Liggins Institute, University of Auckland, and National Research Centre for Growth and Development, Auckland, New Zealand*
ANDREW P. GOLDBERG, MD • *Department of Medicine, University of Maryland School of Medicine, Baltimore Veteran's Affairs Hospital, Gastric Research, Education and Clinical Center, Baltimore, MD*
CARMEN GONZÁLEZ-YANES, PhD • *Department of Medical Biochemistry and Molecular Biology, School of Medicine, Virgen Macarena University Hospital, Seville, Spain*
MARK A. HANSON, MA, Dphil,CertEd, FRCOG • *Centre for Developmental Origins of Health and Disease, University of Southampton, Southampton, UK*
DENNIS HONG, MD, FRCSC • *Oregon Weight Loss Surgery, Legacy Health System, Portland, OR*
MALAKA B. JACKSON, MD • *Division of Endocrinology, University of Pennsylvania School of Medicine, Children's Hospital of Philadelphia, Philadelphia, PA*
PATRICK LAHARRAGUE, MD, PhD • *Laboratoire d'Hématologie, Hôpital Rangueil, CHU Toulouse, France*
KAREN S. L. LAM, MD • *Department of Medicine, University of Hong Kong, Hong Kong, China*
ELISABETH LUDER, PhD • *Mount Sinai School of Medicine, New York, NY*
CHRISTOPHER W. KUZAWA, PhD, MsPH • *Department of Anthropology, Northwestern University, Evanston, IL*
ROBERT H. MAK, MD, PhD • *Division of Pediatric Nephrology, Oregon Health and Science University, Portland, OR*
CONSUELO MARTÍN-ROMERO, PhD • *Department of Medical Biochemistry and Molecular Biology, School of Medicine, Virgen Macarena University Hospital, Seville, Spain*

GIUSEPPE MATARESE, MD, PhD • *Gruppo di ImmunoEndocrinologia, Istituto di Endocrinologia e Oncologia Sperimentale, Consiglio Nazionale delle Ricerche c/o Dipartimento di Biologia e Patologia Cellulare e Molecolare, Università di Napoli "Federico," Napoli, Italy*

YUJI MATSUZAWA, MD • *Metabolic Syndrome Institute, Sumitomo Hospital, Osaka, Japan*

THEODORE MAZZONE, MD • *Section of Endocrinology, Diabetes, and Metabolism, Department of Medicine, University of Illinois at Chicago,Chicago, IL*

ALISON M. MORRIS, PhD • *Nutritional Physiology Research Centre, School of Health Sciences, University of South Australia, Adelaide, South Australia, Australia*

SOUAD NAJIB, PhD • *Department of Medical Biochemistry and Molecular Biology, School of Medicine, Virgen Macarena University Hospital, Seville, Spain*

PAUL POIRIER, MD, PhD • *Quebec Heart Institute/Laval Hospital, Sainte-Foy, Quebec, Canada*

CAROLINE M. POND, PhD • *Department of Biological Sciences, The Open University, Milton Keynes, UK*

VÍCTOR SÁNCHEZ-MARGALET, MD, PhD • *Department of Medical Biochemistry and Molecular Biology, School of Medicine, Virgen Macarena University Hospital, Seville, Spain*

JOSÉ SANTOS-ALVAREZ, PhD • *Department of Medical Biochemistry and Molecular Biology, School of Medicine, Virgen Macarena University Hospital, Seville, Spain*

BRITTA SIEGMUND, MD • *Charité, Campus Benjamin Franklin, Medizinische Klinik I, Berlin, Germany*

VINAYA SIMHA, MD • *Division of Nutrition and Metabolic Diseases, Department of Internal Medicine and Center for Human Nutrition, The University of Texas Southwestern Medical Center at Dallas, Dallas, TX*

JOHN R. SPEAKMAN, DSc • *School of Biological Sciences, University of Aberdeen, Aberdeen, Scotland, UK*

LEE SWANSTRÖM, MD, FACS • *Oregon Health Sciences University, Division of Minimally Invasive Surgery, Legacy Health System, Portland, OR*

ANDREW J. TEICHTAHL, BS • *Department of Epidemiology and Preventive Medicine, Monash University, Melbourne, Victoria, Australia*

ANTON B. TONCHEV, MD PhD • *Division of Cell Biology, Medical University, Varna, Bulgaria*

NESE TUNCEL, MD PhD • *Department of Physiology, Osmangazi University, Eskisehir, Turkey*

YU WANG, PhD • *Genome Research Center, University of Hong Kong, Hong Kong, China*

ANITA E. WLUKA, MB BS, FRACP, PhD • *Department of Epidemiology and Preventive Medicine, Monash University, Melbourne, Victoria, Australia*

AIMIN XU, PhD • *Department of Medicine, University of Hong Kong, Hong Kong, China*

ZOBAIR M. YOUNOSSI, MD, MPH • *Center for Liver Diseases, Inova Fairfax Hospital; and George Mason-Inova Health System's Translational Research Centers, Falls Church, VA*

I

Adipose Tissue: Structure and Function

1 The Adipose Organ

Saverio Cinti

Abstract

Mammals are provided with an organ that has been neglected by scientists in the past: the adipose organ. This organ is formed by a series of well-defined depots mainly located at two corporal levels: superficial (subcutaneous depots) and deep (visceral depots). In adult rodents, two main depots are the anterior and posterior subcutaneous depots. The first consists of a central body located in the area between the scapulae and several elongated projections abutting toward the cervical region and the axillae. The second is extended from the dorsolumbar area to the gluteal, with an intermediate region located in the inguinal area.

The main visceral depots are tightly connected with viscera. In adult rodents, the main visceral depots are mediastinic, perirenal, perigonadal, mesenteric, and retroperitoneal. The weight of the adipose organ is about 20% of the body weight and therefore it is one of the biggest organs in the body. Its color is mainly white but some areas are brown. In young-adult rodents, maintained in standard conditions, the interscapular region and parts of the cervical and axillary projections of the anterior subcutaneous depot, as well as parts of the mediastinic and perirenal depots, are brown. These two colors correspond to the two tissues: white and brown adipose tissues. The relative amount of the two tissues varies with age, strain, environmental and metabolic conditions, and subsequently, the distribution of the two colors is also variable and implies the ability of reversible transdifferentiation of the two types of adipocytes. During pregnancy and lactation, the subcutaneous depots are transformed into mammary glands.

Each depot of the organ receives its own neurovascular peduncle that is specific for the subcutaneous depots and is usually dependent on the peduncle related to the connected organ in the case of visceral depots. The vascular and nerve supply is much more dense in the brown areas than in the white areas. Their density changes in conjunction with the number of brown adipocytes.

It has been shown that the white adipose tissue of obese mice and humans is infiltrated by macrophages and that the level of infiltration correlates with body mass index and mean size of adipocytes.

This infiltration seems to be an important cause for the insulin resistance associated with obesity. We recently observed that macrophages are mainly located at the level of dead adipocytes in white adipose tissue of obese mice, obese humans, and in transgenic mice, which are lean but have hypertrophic adipocytes (HSL knockout mice). The suggested function of these macrophages is mainly to reabsorb the lipid droplet from dead adipocytes.

Key Words: Adipose tissue; obesity; metabolism; anatomy; adipocytes.

1. INTRODUCTION

Adipose tissues have been neglected by scientists until recently. However, adipose tissues are now emerging as collaborative tissues into an active organ, the adipose organ, that significantly contributes to the regulation of body's homeostasis. This chapter will

From: *Nutrition and Health: Adipose Tissue and Adipokines in Health and Disease*
Edited by: G. Fantuzzi and T. Mazzone

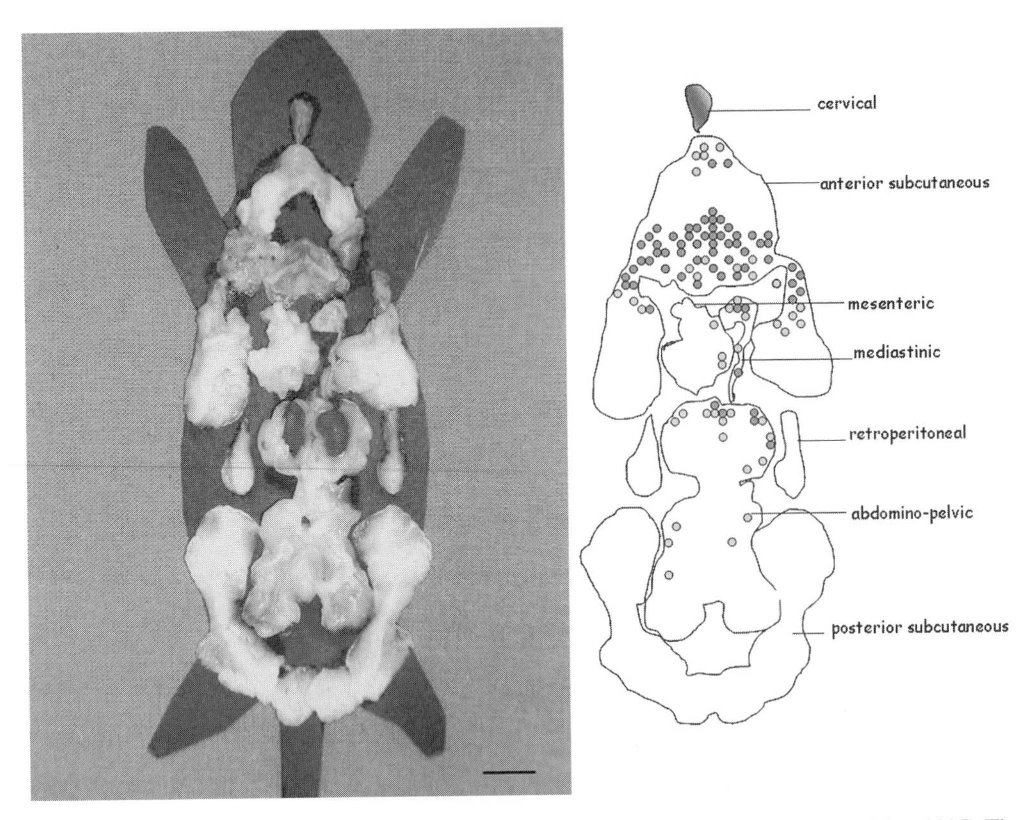

Fig. 1. Gross anatomy of the adipose organ. Lean Sv 129 female mouse maintained at 28 to 29°C. The organ is composed of two subcutaneous depots and several visceral depots. The most representative of these visceral depots are visible. Kidneys and ovaries were dissected together with the depots. White areas are made mainly by white adipose tissue and brown areas are made mainly by brown adipose tissue. Both are visible and indicated in the scheme. Circles indicate brown areas. Bar = 2 cm.

discuss the anatomy and physiology of white and brown adipose tissues in rodents and humans, with an emphasis on two novel concepts. The first concerns the anatomy of the adipose organ. The second is related to a developmental property of adipocytes: transdifferentiation.

2. GROSS ANATOMY

The adipose organ is a multidepot organ with a complex shape (Fig. 1) *(1–4)*.

In small mammals there are two main subcutaneous depots (anterior and posterior) and several visceral depots located inside the thorax (mediastinic) and abdomen (omental, mesenteric, perirenal, retroperitoneal, parametrial, periovaric, epididymal, perivescical). Discrete subcutaneous depots are also dissectable at the level of the major joints in the limbs.

The colors of the organ are white and brown. The white parts are made mainly by white adipocytes. The brown parts are made mainly by brown adipocytes. The relative amounts of white and brown parts are genetically determined and depend on several factors (mainly age, sex, environmental temperature, and nutritional status).

Brown adipocytes are present in all the aforementioned subcutaneous (including the limbic ones) and visceral depots of the adipose organ, but the areas where they are most constantly found in young/adult mice maintained at standard conditions are the interscapular, subscapular, axillary, and cervical areas of the subcutaneous anterior depot. Also, brown adipocytes are found in the inguinal part of the subcutaneous posterior depot and the periaortal part of the mediastinic depot, and the interrenal part (near the hilus of both kidneys) of the perirenal depot. It must be outlined that clear anatomical boundaries between brown and white adipose tissues do not exist. Many other parts of the adipose organ are mixed with brown adipocytes widespread within the white depot. We have recently described quantitative data of the adipose organ of Sv129 adult mice maintained at different environmental temperatures *(5)*.

Most of the adipocytes are located in the depots of the adipose organ described above, but white adipocytes are also found in the skin, thymus, lymph nodes, bone marrow, parotid, parathyroid, pancreas, and other tissues.

3. LIGHT AND ELECTRON MICROSCOPY

3.1. White Adipose Tissue

As described under the previous heading, in the areas where the adipose organ is white (or pale) the parenchymal element is the white adipocyte. These spherical cells have a diameter from a minimum of 30 to 40 μm to a maximum of 150 to 160 μm (lean, mammary subcutaneous) and from a minimum of about 20 to 30 μm to a maximum of about 90 to 100 μm (lean, visceral perirenal) (by light microscopy of fixed, but not embedded, human white adipose tissue [WAT]). In white adipocytes, most of the cytoplasm is occupied by the lipid droplet and only a thin rim of cytoplasm is visible (Fig. 2). Here, elongated mitochondria (Fig. 3), Golgi complex, rough and smooth endoplasmic reticulum, vesicles, and other organelles are usually visible by transmission electron microscopy.

Many pinocytotic vesicles are present in the proximity of the plasma membrane and an external lamina surrounds the cell.

3.2. Brown Adipose Tissue

Although these cells share the name "adipocyte," they differ greatly in their anatomy and, consequently, in their physiology. The common part of the name is due to the fact that they both accumulate lipids (triglycerides) into the cytoplasm. However, white adipocytes form only one big vacuole (unilocular cell), whereas brown adipocytes form numerous small vacuoles (multilocular cell; Fig. 4). The shape of brown adipocytes is polygonal or ellipsoid, with a maximum diameter between a minimum of 15 to 20 μm and a maximum of 40 to 50 μm. The most important organelles are the mitochondria. They are numerous, big, and rich in transverse cristae (Fig. 5). Peroxisomes, Golgi complex, rough and smooth endoplasmic reticulum, vesicles, and other organelles are also visible by transmission electron microscopy. Pinocytotic vesicles and external lamina are also present in this cell. Brown adipocytes are joined by gap junctions *(6)*.

4. VASCULAR SUPPLY

The adipose organ is diffuse into the organism. Most of its depots receive vascular supply by regional visceral or parietal nerve–vascular bundles. Specific bundles are present in the two main subcutaneous depots (murine adipose organ). The best studied is the anterior subcutaneous depot: two symmetrical bundles reach the depot at the lateral extremities. In the superior lateral bundles, four big and two small nerves are

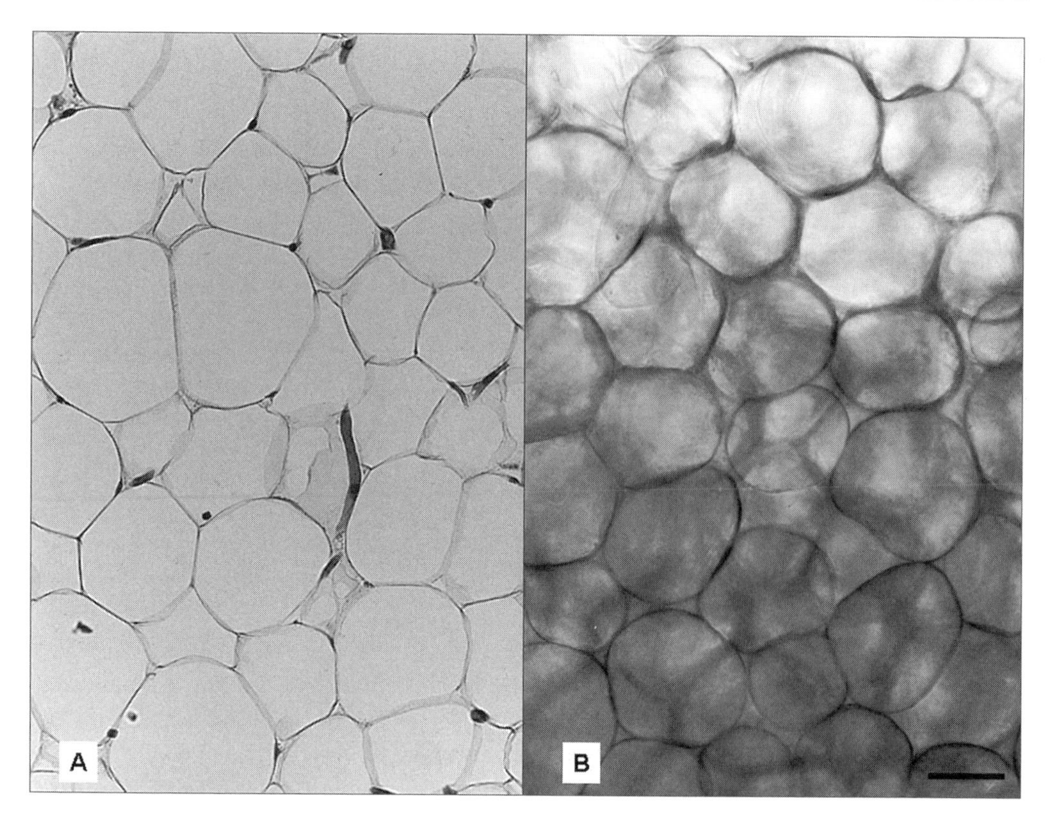

Fig. 2. Light microscopy of human white adipose tissue. **(A)** subcutaneous adipose tissue (formaldehyde-fixed and paraffin-embedded). **(B)** subcutaneous adipose tissue (formaldehyde-fixed and not embedded). Bar = 30 μm.

present. The same bundle also contains an artery and a vein. The lateral inferior bundle does not contain nerves, but only has an artery and a vein. The main vein of the depot is located at the apex of the deep part of the interscapular region and is directly connected to the azygos vein.

The posterior subcutaneous depot is reached by two main nerve–vascular bundles. The first is collateral of the femoral nerve–vascular tract and reaches the depot in the inguinal part; the second is a parietal bundle peduncle and reaches the depot in the dorsolumbar part.

The extension of the capillary network is quite different in the white and brown parts of the organ. In the brown areas the density of the capillaries is much higher than in the white areas.

5. NERVE SUPPLY

The nerve supply to the adipose organ is different in the brown and white areas. The former are more innervated than the latter *(7,8)*. In brown areas, numerous noradrenergic fibers are also found in fat lobules (parenchymal nerves), running with blood vessels (until the level of precapillary and postcapillary structures) and directly in contact with adipocytes.

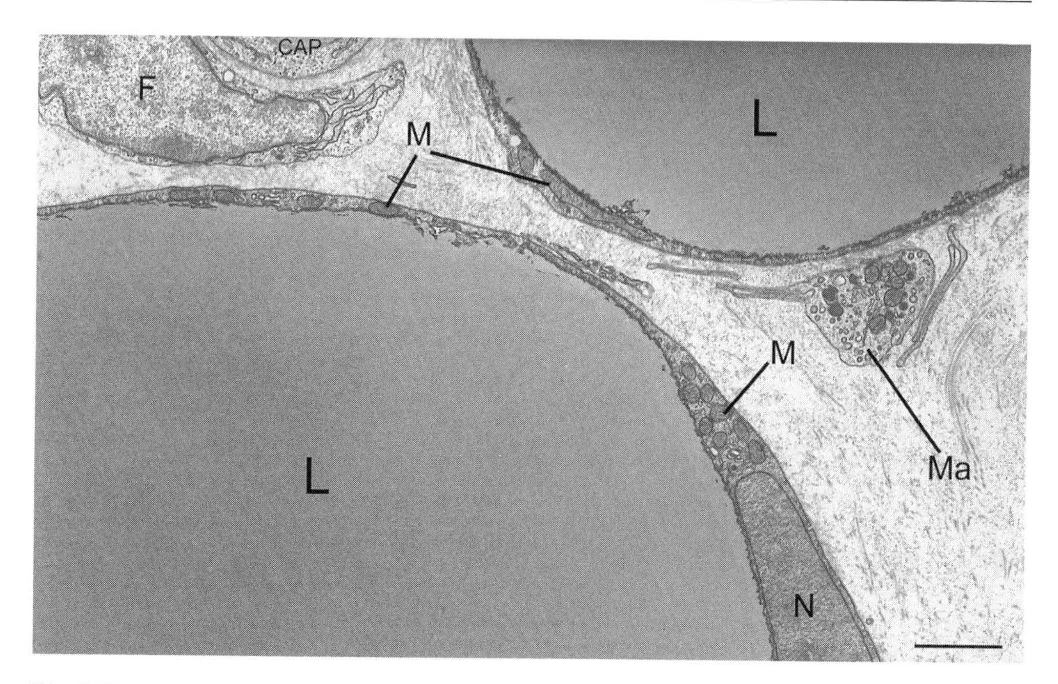

Fig. 3. Electron microscopy of murine white adipose tissue. Note the small and elongated mitochondria in the perinuclear area and in the thin rim of cytoplasm surrounding the large unilocular lipid droplet. F, fibroblast; CAP, capillary lumen; Ma, macrophage; M, mitochondria; N, nucleus; L, liquid droplet. Bar = 2 μm.

Adrenergic receptors (α1, 2 and β1, 2, and 3) are present in the adipose organ.

The density of parenchymal fibers varies according to the functional status of the organ. During cold exposure the noradrenergic parenchymal fibers increase their number in the brown part of the organ *(8)*. During fasting, the noradrenergic parenchymal fibers increase their number in the white part of the organ *(9)*.

Vascular noradrenergic fibers are also immunoreactive for neuropeptide Y. The vast majority of these nerves also contain noradrenaline *(9,10)*, suggesting that they belong to the sympathetic nerve supply to WAT blood vessels.

Brown and white areas also have a provision of sensory nerves *(11)* that are capsaicin-sensitive and are immunoreactive for calcitonin gene-related peptide and substance P. The functional significance of these sensory nerves is not precisely known, although in the rat periovarian adipose depot they affect the recruitment of brown adipocytes during cold acclimation *(12)*.

6. HISTOPHYSIOLOGY

White adipocytes' main purpose is to be a depot of highly energetic molecules (fatty acids) that can supply fuel to the organism during intervals between meals. When the interval is prolonged for weeks, WAT represents the survival tissue.

Brown adipocytes use the same highly energetic molecules to produce heat (non-shivering thermogenesis). This function is fulfilled by the activity of a unique protein, uncoupling protein 1 (UCP1), exclusively expressed by brown adipocytes in mitochondria (and therefore representing a molecular marker of brown adipocytes) *(13–19)*.

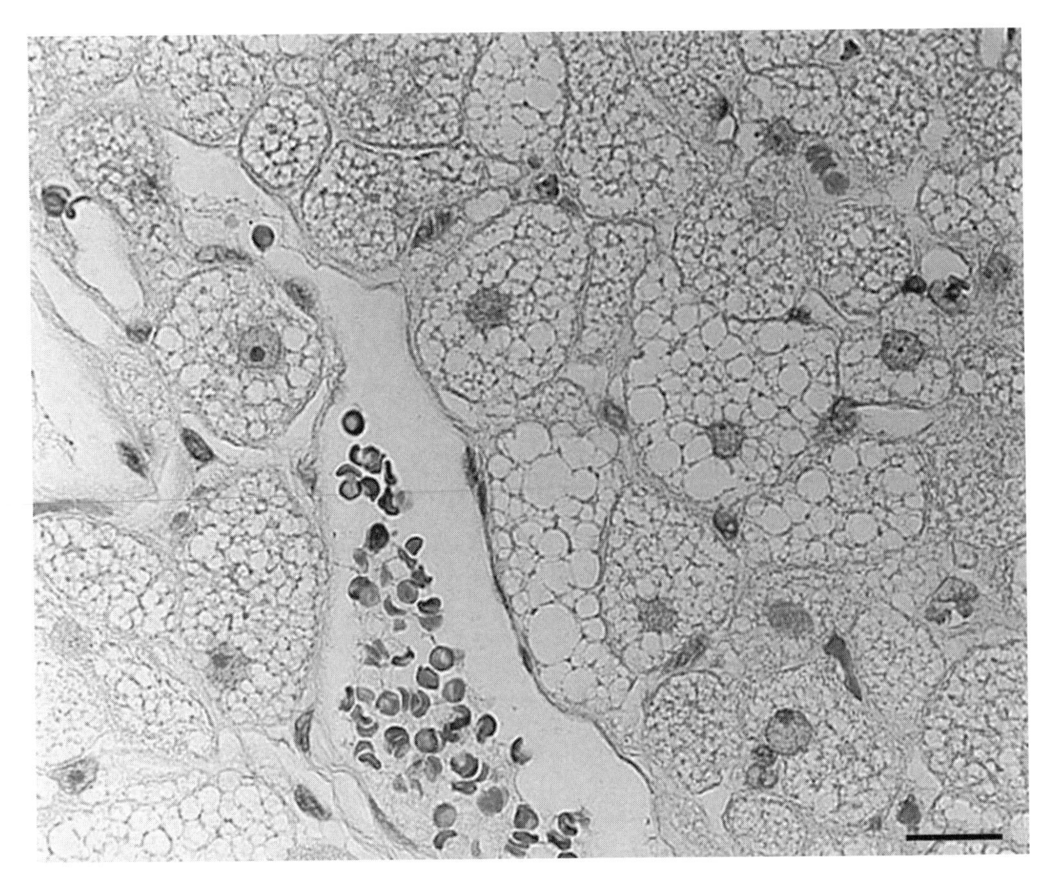

Fig. 4. Light microscopy of human brown adipose tissue. Note the characteristic multilocular lipid organization of the cytoplasm of adipocytes. Hibernoma removed from the skin of a 16-yr-old male patient. Bar = 15 μm.

The signal for brown adipocyte activation is a temperature below thermoneutrality (34°C for mice, 28°C for rats, 20–22°C for humans), that induces activation of the sympathetic nervous system. These neurons of the sympathetic chain directly reach brown adipocytes in the adipose organ *(14)*.

These two functions of the two tissues of the adipose organ (WAT and BAT) are therefore balanced between them, because the intrinsic energy of lipids can be accumulated (WAT) or dissipated (BAT). The total volume of the adipose organ is also dependent on the equilibrium between WAT and BAT activities. Of note, genetic ablation of BAT induces obesity in mice *(20)*, although mice lacking UCP1 are cold-sensitive but not obese *(21)*.

In 1994, another primary function of white adipocytes was discovered: production of leptin, a hormone able to influence animal behavior concerning food intake *(22)*. This hormone also induces energy dispersion (via BAT activation) and has gonadotrophic properties. Leptinemia is positively correlated to fat mass; therefore most obese patients are leptin-"resistant," but rare human cases of leptin or leptin receptor congenital absence have been found. Recombinant leptin administration has reversed cases of human massive obesity caused by congenital leptin absence *(23)*. Brown adipocytes in

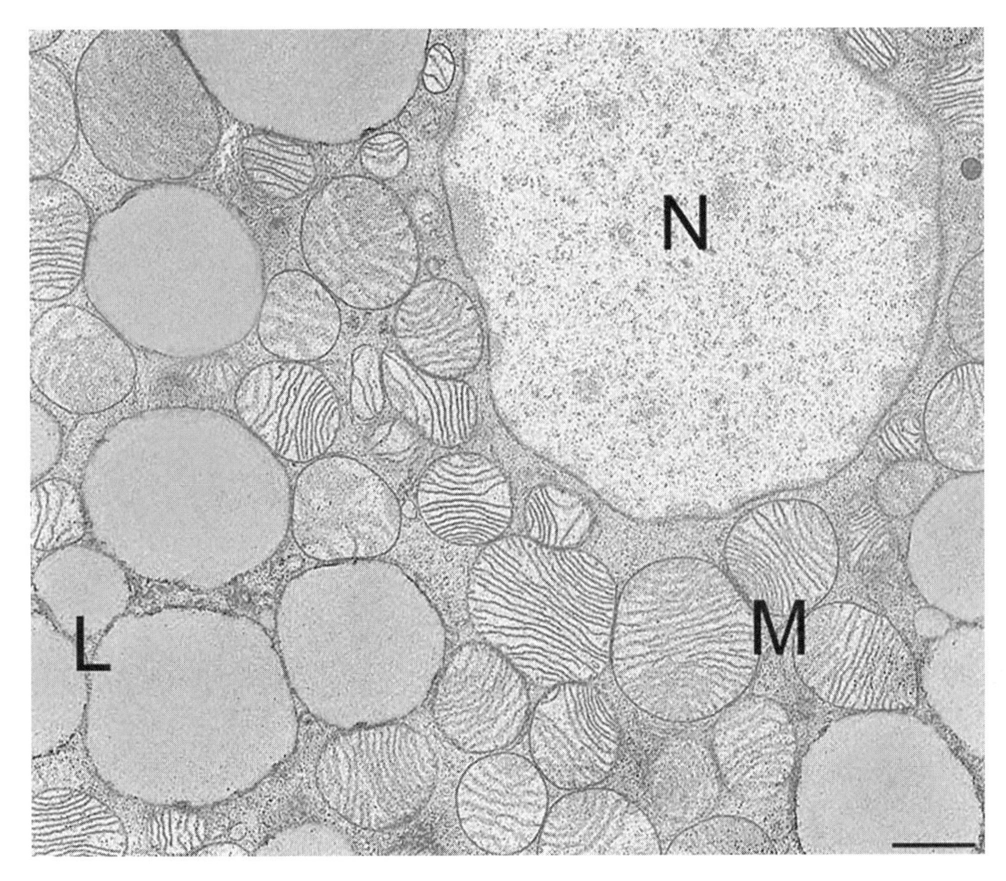

Fig. 5. Electron microscopy of mouse brown adipose tissue. Note the typical mitochondria (abundant, large, and rich in cristae). L, lipid droplets; N, nucleus of the adipocyte; M, mitochondria. Bar = 1 μm.

their classic multilocular configuration (i.e., during thermogenic activity) are not immunoreactive for leptin *(24,25)*.

In addition to these primary functions, the two cell types have many other "secondary" functions. Among them we should remember the activity as a thermo-insulator of subcutaneous white adipose tissue and the regulation of hydric compartments of the organism by production of receptors C and ANP and angiotensinogen II *(26)*. Many other functions have been recognized for WAT, as a growing body of evidence suggests production of several factors known as adipokines, controlling several important functions such as glucose and lipid metabolism, blood coagulation, blood pressure, and steroid hormone modulation. Brown adipocytes produce and secrete many substances, such as autocrine, paracrine, and endocrine factors. The production of all these adipokines raised the recent concept of the adipose organ as an endocrine organ *(27,28)*. In this context, it must be outlined that adipocytes are not the only cell type present in the adipose organ. It has been calculated that only about 50% of its cells are adipocytes *(29)*. Vascular elements, preadipocytes, fibroblasts, mast cells, macrophages, nervous elements, and mesenchymal cells with unknown functions *(1)* are usually found in all depots of the adipose organ.

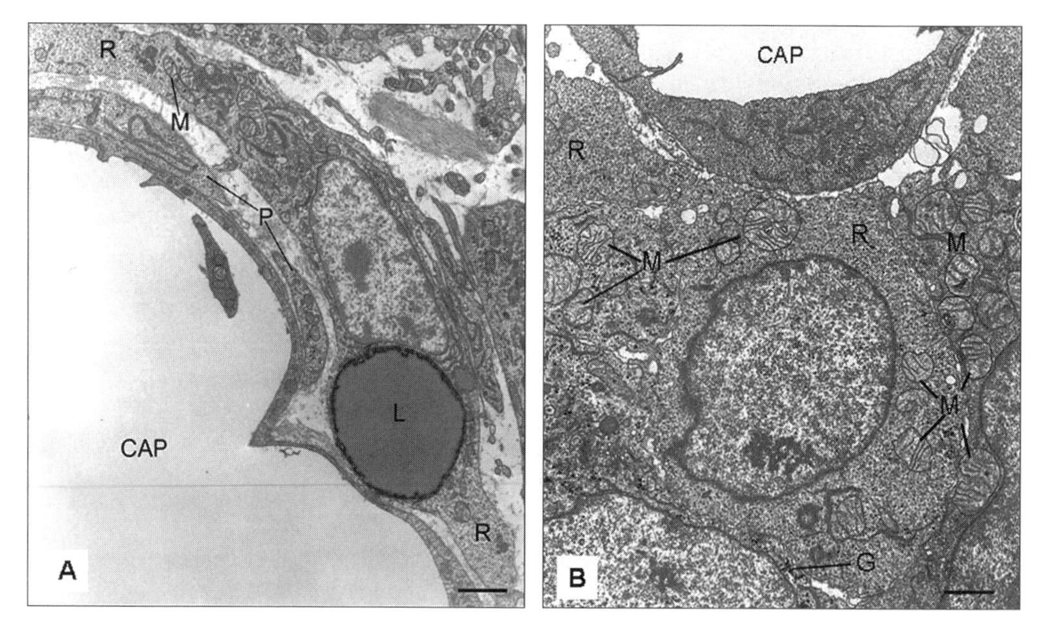

Fig. 6. Electron microscopy of white **(A)** and brown **(B)** adipocyte precursors. Note the different morphology of mitochondria (M). At this early stage of differentiation (note areas indicated by R, occupied only by ribosomes and polyribosomes, typical of poorly differentiated cells), small lipid droplets are visible in the white precursor. Glycogen (dark particles and "G)" and numerous pretypical M (compare with Fig. 5) are visible in the brown precursor. P, pericytes (probably an earlier stage of adipocyte precursor development); CAP, capillary lumen; L, liquid droplets; G, glycogen. Bar in **A** = 2 μm; bar in **B** = 1 μm.

7. DEVELOPMENT AND PLASTICITY

The origin of adipocytes is still unknown, but the in vivo and in vitro steps of their development have been described.

7.1. White Adipocytes

In the first week of postnatal development, the most "white" depot of murine adipose organ (epididymal depot) shows a high number of poorly differentiated cells with minimum adipose differentiation; these are usually referred to as white preadipocytes.

These cells are always tightly connected to the wall of capillaries. Electron microscopy can easily detect in these "blast-like cells" minimal signs of adipose differentiation—i.e., clusters of glycogen and small lipid droplets (Fig. 6A).

Preadipocytes can be distinguished from other cell types in the tissue because they are surrounded by a distinct basal membrane (external lamina).

Often there is a predominant lipid droplet surrounded by numerous small lipid droplets. Very soon the growing white adipocyte assumes the characteristic aspect of a unilocular cell. In adult human adipocytes we showed that the external lamina is immunoreactive for laminin, collagen IV, and heparan sulfate, but not for fibronectin, which is present in the external lamina of adipocyte precursors *(30)*. In vitro studies by electron microscopy confirmed these steps and showed that precursors of adipocytes in adult rats do not reach a complete differentiation *(31,32)*.

7.2. Brown Adipocytes

Brown adipocyte precursors—similarly to white precursors—are always tightly associated with the walls of capillaries. The early steps of development are characterized by pretypical mitochondria, a morphological marker that appears earlier than the molecular marker UCP1 (Fig. 6B). The second step of development is characterized by mitochondrial proliferation and lipid accumulation. Lipid droplets are mainly small and similar in size; therefore lipid accumulation is quite different from that of white precursors and these droplets tend to form multilocular cells. Pretypical mitochondria gradually assume the morphology of the mature adipocyte; this coincides with expression of the molecular marker UCP1. In vitro studies have confirmed the ultrastructure of the developmental steps described in vivo and demonstrated the importance of noradrenaline for mitochondriogenesis *(33–36)*.

7.3. Reversible Transdifferentiation of White Adipocytes Into Brown Adipocytes

Transdifferentiation is a process of direct transformation of a differentiated cell into another cell type with different morphology and physiology. We are convinced that a significant amount of adipocytes in the adipose organ can reciprocally transdifferentiate.

The concept of adipose organ discussed here implies that all depots have variable amount of both types of adipocytes, whose proportion depends on several factors (species, age, strain, environmental temperature, etc.) (Fig. 7). Therefore, in theory, all depots should have both types of precursors, whose morphological characteristics are quite different (*see* Subheadings 6.1. and 6.2.). The description of the developmental steps of the two types of precursors made above refers to the ontogenetic development into two depots that are quite characteristic for WAT (epididymal) and BAT (interscapular part of the anterior subcutaneous). In other words, the depot developed from adipocyte precursors of that specific area in the adipose organ became predominantly WAT (epididymal) or BAT (interscapular) in adult animals. Detailed morphological studies of ontogenesis in depots becoming mixed in adult animals are lacking, but, in our experience, both types of adipocyte precursors with the morphological characteristics of those reported in Figs. 1 and 2 are found during the ontogenesis of these depots *(1)*.

Furthermore, it is well-known that white adipocytes develop in brown areas of the organ (i.e., interscapular area) in genetic or diet-induced obese animals, as well as that brown adipocytes develop in white areas of the organ during cold exposure or treatment with β3-adreneregic (AR) agonists.

We have studied mainly the phenomenon of brown adipocyte development in white areas of the adipose organ, because this phenomenon is associated with amelioration of obesity and diabetes *(37–40)*.

After exposure to cold, the increase in the number of brown adipocytes in white areas of the adipose organ is accompanied by the appearance of brown adipocyte precursors *(41)*. Treatment with β3AR agonists induces development of brown adipocytes in white areas of the adipose organ. This is accompanied by the appearance of cells with morphological characteristics intermediate between white and brown adipocytes and quite different from those of brown adipocyte precursors. These cells show a multilocular lipid depot, usually with a predominant central vacuole and numerous small ones at the periphery of the cell. Their mitochondria are more numerous than in white adipocytes, but less numerous than in brown adipocytes. The mitochondrial morphology is intermediate

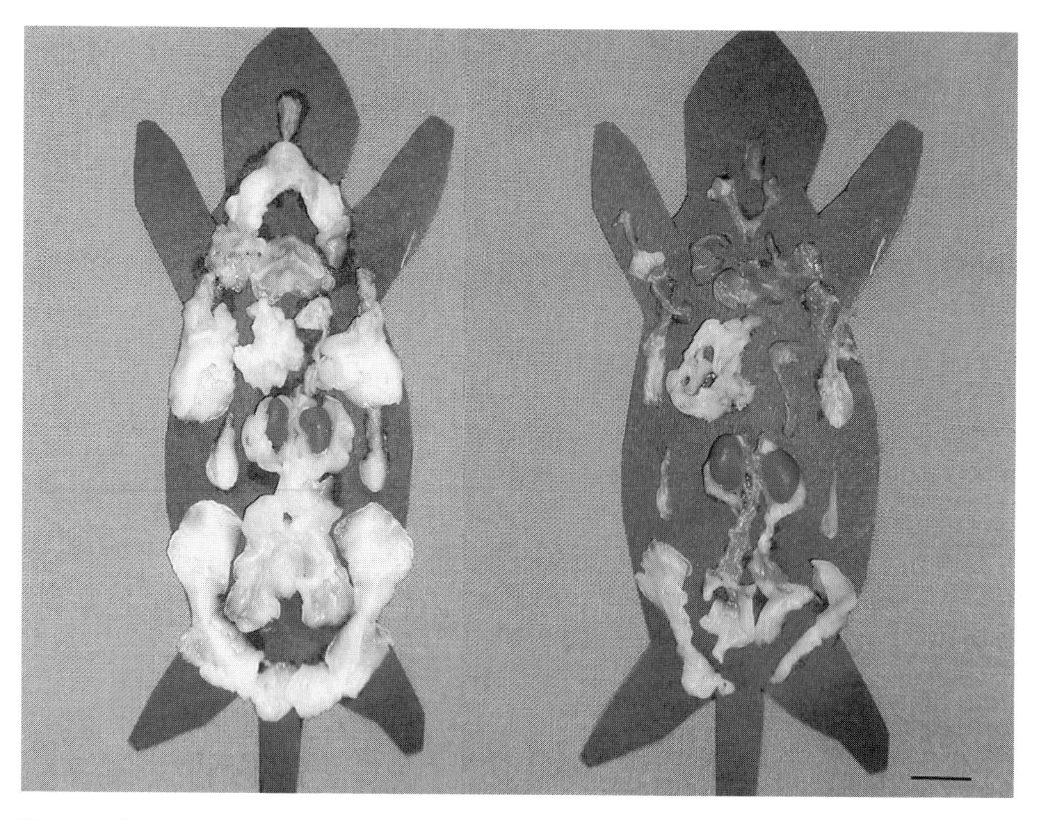

Fig. 7. Adipose organs of adult Sv129 mice maintained at 28 to 29°C (left) or 6°C (right) for 10 d. Note the evident reduction in size of the organ and the increased brown areas in the cold-acclimated mice. Bar = 2 cm.

between that of "white" mitochondria and that of "brown" mitochondria. We think that this type of multilocular adipocyte rich in mitochondria is the morphological equivalent of a white adipocyte transdifferentiating into a brown adipocyte. Some of these cells, also found in white areas of the adipose organ of cold-exposed animals, are immunoreactive for UCP1. Of note, 80 to 95% of these cells are BrdU negative, suggesting that their development is independent from mitotic processes *(42,43)*.

Recently, it has been shown that Sv129 mice are quite resistant to obesity and diabetes in comparison with B6 mice *(44)*. A recent morphometric investigation of the cellular composition of the adipose organ of Sv129 mice showed that the most numerous cells in the organ were multilocular (brown) adipocytes (60% of all adipocytes in the adipose organ of controls and 80% in cold-acclimated mice). In addition, in line with the transdifferentiation concept, cold acclimation did not significantly affect overall adipocyte number, but induced a significant increase in the number of brown adipocytes and an equivalent, significant reduction in the number of white adipocytes *(45)*.

White-into-brown transdifferentiation is also suggested by in vitro studies using primary cultures from human subcutaneous adipose tissue; in these studies, UCP1 expression was induced by treatment with peroxisome proliferator-activated receptor (PPAR)γ agonists *(46)* or PGC1 transfection *(47)*.

All together, these data suggest a possible role for β3AR agonists in the treatment of human obesity and diabetes. Human WAT is immunoreactive to β3AR monoclonal highly specific antibodies *(48)*, but a β3AR agonist producing curative effects for human obesity has not yet been identified *(49)*.

Energy expenditure via the sympathetic system seems to be essential for energy balance; mice lacking all three subtypes of β-adrenoceptors develop massive obesity in the absence of alterations in food intake and locomotion. These mice show a precocious and massive transformation (we think transdifferentiation) of brown into white adipose tissue *(50)*. This is in line with the obese phenotype of mice lacking brown adipose tissue *(20)* and with the obesity resistance of transgenic mice that express UCP1 ectopically *(51)*, but it must be remembered that mice lacking UCP1 do not develop obesity *(21)*.

Studies of genetically manipulated models seem to suggest plasticity of the adipose organ, with white-into-brown adipocyte transdifferentiation. In this context, it is interesting to note that mice lacking the subunit RIIβ (one of the regulatory subunits of cAMP-dependent proteinphosphokinase A, abundant in adipose tissues) have a compensatory hyperexpression of the RIα subunit, with an increase in phosphokinase A sensitivity to cAMP in white adipose tissue and activation of UCP1. These mice have a brown phenotype of abdominal fat and resistance to obesity *(52)*.

FOXC2 is a gene for a transcription factor that is exclusively expressed in adipose tissue. Its transgenic expression in the adipose tissue in mice results in a lean, obesity-resistant, and insulin-sensitive phenotype. The adipose organ of these mice has a browner phenotype than that of controls *(53)*. Of note, humans with insulin resistance have a reduction of *FOXC2* expression in biopsies from the abdominal subcutaneous adipose tissue, together with a reduction of other genes of brown adipocyte phenotype *(54)*.

WAT has high expression of 4E-BP1, a protein important for the posttranscriptional regulation of protein synthesis. Mice lacking 4E-BP1show a reduction of the total fat mass and a brown phenotype of their adipose organ, suggesting that a posttranscriptionally regulated protein can be responsible for the white phenotype. The author suggests that PGC1 (a cofactor of PPARγ) is the protein whose posttranscriptional synthesis is blocked by 4E-BP1 *(55)*.

We recently reported a completely new example of adipose organ plasticity, with reversible transdifferentiation of adipocytes into epithelial cells. The adipose tissue of the mammary gland is a subcutaneous adipose tissue. In female mice, all the subcutaneous part of the adipose organ (anterior and posterior subcutaneous depots; *see* Heading 1) belong to the five symmetrical mammary glands. During pregnancy and lactation almost all adipocytes in these depots disappear, in association with development of the mammary gland (mainly the milk-producing and secreting lobuloalveolar parts). This phenomenon was previously viewed as caused by a hiding of adipocytes, which undergo a delipidation process and apparently disappear. In the postlactation period the milk-secreting epithelial part of the gland disappears by massive apoptosis and the slimmed adipocytes refill and reconstruct the prepregnancy anatomy of the gland. We have brought evidence for a direct reversible transformation of adipocytes into the milk-secreting alveolar cells during pregnancy *(56)*. Of course, this example of extreme plasticity of adipocytes needs further demonstration.

7.4. Hypertrophy and Hyperplasia (Positive Energy Balance: Overweight and Obesity)

When the energy balance becomes positive, the adipose organ increases its white part. White adipocytes undergo hypertrophy followed by hyperplasia.

In fact, it has been proposed that adipocytes have a maximum volume and cannot be further expanded. This maximum volume, also referred to as "critical cell size," is genetically determined and specific for each depot *(57)*. Adipocytes with the critical cell size trigger an increase in cell numbers *(58,59)*. Not all depots have the same tendency to hypertrophy and hyperplasia: the former seems to be more characteristic of epididymal and mesenteric depots, the latter of inguinal and perirenal depots *(57)*.

Hausman et al., in a recent review *(60)*, after considering evidence supporting this theory as well as the conflicting data, conclude that not only paracrine factors but also circulating factors are involved; neural influences can also be important to regulate adipose tissue development and growth. In any case, paracrine factors seem to play a pivotal role. Adipose tissue expresses numerous factors that could be implicated in modulation of adipogenesis: insulin-like growth factor-1, transforming growth factor-β, tumor necrosis factor-α, macrophage colony-stimulating factor, angiotensin-2, autotaxin–lysophosphatidic acid, leptin, resistin, and the like *(61)*. Interestingly, it has been shown in mice that obesity induced by a high-fat diet is hypertrophic, whereas obesity induced by hypothalamic lesions caused by administration of monosodium glutamate is hyperplastic *(62)*.

It has been shown that the WAT of obese mice and humans is infiltrated by macrophages and that the level of infiltration correlates with body mass index and mean size of adipocytes *(63–65)*. This infiltration seems to be an important cause for the insulin resistance associated with obesity. We recently observed that macrophages are located mainly at the level of dead adipocytes in white adipose tissue of obese mice, obese humans, and transgenic mice that are lean but with hypertrophic adipocytes (hormone-sensitive lipase knockout mice). The suggested function of these macrophages is mainly to reabsorb by phagocytosis the lipid droplets from dead adipocytes *(66)*.

Also, the brown part of the organ is modified under this condition of positive energy balance. In obese mice, the rate of apoptosis of brown adipocytes increases; this is strongly attenuated in mice lacking tumor necrosis factor-α receptors *(67)*. The morphology of brown adipocytes gradually changes into a morphology similar to that of white adipocytes, including transformation of the multilocular lipid depot into a unilocular one. This is accompanied by activation of the leptin gene; these cells are also immunoreactive for leptin *(24,25)*.

7.5. Hypoplasia (Negative Energy Balance: Caloric Restriction, Fasting)

The morphology of the adipose organ during fasting is quite characteristic, because a variable amount of slimmed cells are present in WAT. The slimmed cell is barely visible at light microscopy but is easily recognized by electron microscopy—i.e., it has a specific ultrastructural morphology: cytoplasmatic irregular and thin projections with numerous invaginations rich in pinocytotic vesicles. These projections enlarge in proximity of the nucleus and of the residual lipid droplet. In acute fasting, completely delipidized adipocytes can be found near apparently unaffected unilocular cells (Fig. 8).

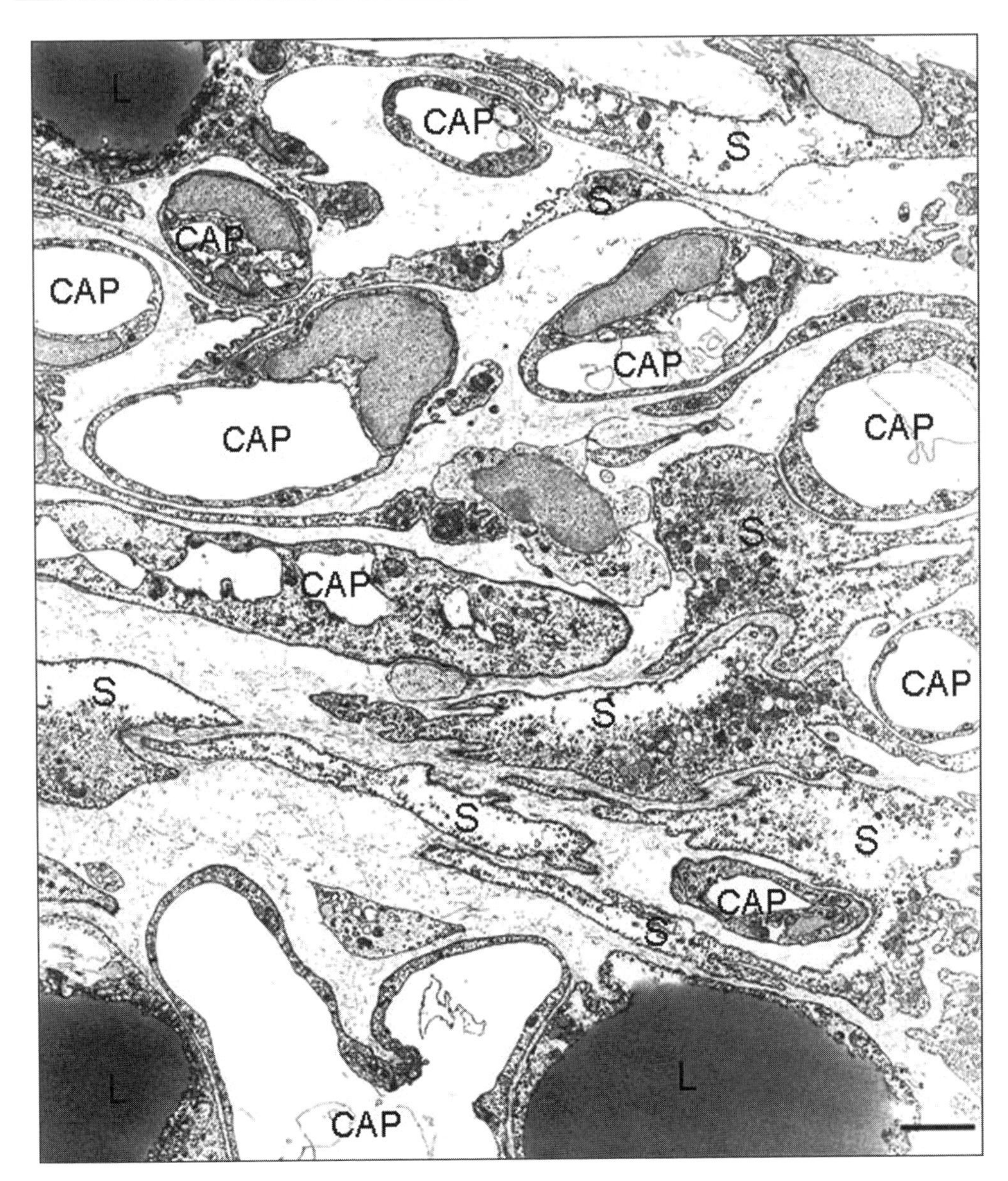

Fig. 8. Electron microscopy of white adipose tissue of fasted mouse. Numerous slimmed adipocytes are visible (S), showing a characteristic morphology different from that of other cell types found in the tissue. Note the dense vascular supply. CAP, capillary lumen. Bar = 4 μm.

Vasculogenesis and neurogenesis is also found in white adipose tissue of fasted animals. Neurogenesis is mainly supported by an increase of noradrenergic fibers *(9)*.

In chronic caloric restriction, the reduction in size of adipocytes is homogeneously distributed *(68)*.

8. THE ADIPOSE ORGAN OF HUMANS

The basic concepts of the adipose organs of small mammals reported above are applicable also to the adipose organ of humans. In fact, white, brown, and mixed adipose tissues are also present in the adipose organ of humans, with all the morphological and physiological characteristics described for the murine adipose organ.

Although a detailed description of the gross anatomy of the human adipose organ has never been performed, it is well-known that it is composed of subcutaneous and visceral depots. In humans, the subcutaneous adipose tissue is in continuity with the dermal adipose tissue (in rodents, dermal adipose tissue is separated from subcutaneous adipose tissue by a smooth muscle layer) and it is not limited to defined areas but is present as a continuous layer beneath the skin. Mammary and gluteofemoral subcutaneous adipose tissue is more developed in females than in males.

Visceral depots correspond to those described previously for the rodent adipose organ, but the omental depot is particularly well developed in humans.

The weight of the human adipose organ of lean adults is about 8 to 18% of body weight in males and 14 to 28% in females (and about 5% in monkeys) *(69)*.

Light and electron microscopy of human adipose tissues is identical to that of murine adipose tissues, but the size of adipocytes is about 30 to 40% bigger than that of mice and rats.

Development of the human adipose organ extends for a long period, until puberty, mainly through proliferation *(70)*. During the first year of age there is mainly an increase in size. In line with these data, the number of adipocytes, total fat mass, and the percentage of body fat correlate positively with age in both sexes. Instead, adipocyte size does not seem to positively correlate with age, but it seems to be correlated to the amount of fat mass and percentage in both sexes *(71)*. In massively obese humans, the adipose organ can increase four times and reach 60 to 70% of body weight *(46,60)*.

In case of negative energy balance, the adipose organ reduces its volume and the size of adipocytes. The reduction in size of adipocytes is important because the size of adipocytes correlates with insulin sensitivity *(72)*. Not all depots react in the same way to negative energy balance. Subcutaneous adipose tissue from the gluteofemoral region of premenopausal women is more resistant to slimming than subcutaneous abdominal adipose tissue, but after menopause the slimming process is similar. This seems to be due to a combination of increased lipoproteinlipase activity and reduced lipolytic activity in the gluteofemoral adipose tissue. The reduced lipolytic activity seems to be due to a relative preponderance of antilipolytic activity of α2-adrenoceptors over the lipolytic β-adrenoceptors *(73)*. In general, α2-adrenoceptors are more abundant in human adipose tissue than in murine adipose tissue. In genetically modified mice lacking β3- and expressing human α2-adrenoceptors, obesity induces hyperplasia (but not hypertrophy) of adipose tissue and mice are insulin-sensitive. These experiments are in line with the importance of α2 AR for adipocyte hyperplasia and with the relationship of their size with insulin sensitivity *(74)*.

Like the murine adipose organ, the human adipose organ contains brown adipose tissue. It is easy to understand that the relationship between surface and volume of the human body is quite different from that of small mammals; therefore, human thermodispersion is much lower than that in rodents. This alone justifies a reduced need for brown adipose tissue in adult humans. Newborns have a different surface/volume relationship and a considerable amount of brown adipose tissue is present at that age. Nevertheless, brown adipocytes dispersed among white adipocytes have been described in several histological studies (including studies showing the presence of UCP1) *(75,76)* and in our own studies of a case series of 100 consecutive perirenal biopsies of adult humans, brown adipocytes were found in 24% of cases (all ages) and in 50% of cases after exclusion of patients over 50 yr old (unpublished data).

BAT in human newborns has been described in almost all the same sites described for rodents, and *UCP1* gene expression was found in biopsies from visceral adipose tissue of adult lean and obese patients. In the same paper, the authors calculated the presence of one brown adipocyte every 100 to 200 white adipocytes in the visceral adipose tissue of adult lean humans *(77)*.

BAT has also been described as increased in outdoor workers in northern Europe *(78)* and in patients with feocromocitoma (a noradrenalin-secreting tumor). Furthermore, rare cases of hibernoma—BAT tumors occurring in several anatomical sites, including subcutaneous and visceral fat—have been described (about 100 cases have been described in the literature *[79]* and we recently observed a case [Fig. 4] in which brown adipocytes expressed UCP1 and had the classic electron microscopy with typical mitochondria).

The physiological role of BAT in humans is debated, but the possibility to artificially increase it in order to treat obesity and related disorders cannot be excluded. On this matter, it is interesting to note that human adults with reduced brown phenotype of abdominal subcutaneous adipose tissue have reduced insulin sensitivity and that human white adipocyte precursors can be induced in vitro to express UCP1 by administration of drugs *(46)*.

9. CONCLUSIONS

This chapter described in detail the anatomy of adipose tissue and introduced the novel concept of the adipose organ. Both white and brown adipose tissues are organised into a real organ, with a complex multi-depot organisation. Each depot has its own discrete vascular and nerve supply. The characteristics of the organ can be adapted to functional requirements in relation to the energy balance of the organism. The two tissues seem to derive from precursors with different morphological and functional characteristics, but with possibilities of reciprocal conversion, with an important role played by the nervous system. Both white and brown adipocytes produce factors that can influence the tissue pattern, adapting it to the functional needs.

REFERENCES

1. Cinti S. *The Adipose Organ*. Kurtis, Milan, 1999.
2. Cinti S. Proc Nutr Soc 2001;60:319–328.
3. Cinti S. J Endocrinol Invest 2002;25:823–835.
4. Cinti S. Prostaglandins Leukot Essent Fatty Acids 2005;73:9–15.
5. Murano I, Zingaretti MC, Cinti S. Adipocytes 2005;1:121–130.
6. Barbatelli G, Heinzelmann M, Ferrara P, et al. Tissue Cell 1994;26:667–676.
7. Yasuda T, Masaki T, Kakuma T, et al. Endocrinology 2005;146:2744–2748.
8. De Matteis R, Ricquier D, Cinti S. L. J Neurocytol 1998;27(12):877–886.
9. Giordano, A, Frontini A, Murano I, et al. J Histochem Cytochem 2005;53:679–687.
10. Cannon B, Nedergaard J. Biochem Soc Trans 1986;14:233–236.
11. Cinti S. Ital J Anat Embryol 1999;104:37–51.
12. Giordano A, Morroni M, Carle F, et al. J Cell Sci 1998;111(Pt 17):2587–2594.
13. Cannon B, Hedin A, Nedergaard J. FEBS Lett 1982;150:129–132.
14. Trayhurn P, Nicholls DG, *Brown Adipose Tissue*. Edward Arnold, London, 1986.

15. Cinti S, Zancanaro C, Sbarbati A, et al. Biol Cell 1989;67:359–362.
16. Klaus S, Casteilla L, Bouillaud F, et al. Int J Biochem 1991;23:791–801.
17. Ricquier D, Casteilla L, Bouillaud F. FASEB J 1991;5:2237–2242.
18. Klaus, S. *Adipose Tissue* Eureka.com, Austin, TX, 2001.
19. Cannon B, Nedergaard J. Physiol Rev 2004;84:277–359.
20. Lowell BB, S-Susulic V, Hamann A, Lawitts JA, et al. Nature 1993;366:740–742.
21. Enerback S, Jacobsson A, Simpson EM, et al. Nature 1997;387:90–94.
22. Zhang Y, Proenca R, Maffei M, et al. Nature 1994;372:425–432.
23. O'Rahilly S. Nutr Rev 2002;60:S30–34; discussion S68–84, 85–87.
24. Cinti S, Frederich RC, Zingaretti MC, et al. Endocrinology 1997;138:797–804.
25. Cancello R, Zingaretti MC, Sarzani R, et al. Endocrinology 1998;139:4747–4750.
26. Sarzani R, Paci VM, Zingaretti CM, et al. J Hypertens 1995;13:1241–1246.
27. Kershaw EE, Flier JS. J Clin Endocrinol Metab 2004;89:2548–2556.
28. Trayhurn P, Wood IS. Br J Nutr 2004;92:347–355.
29. Bukowiecki L, Collet AJ, Follea N, et al. Am J Physiol 1982;242:E353–359.
30. Pierleoni C, Verdenelli F, Castellucci M, et al. Eur J Histochem 1998;42:183–188.
31. Cinti S, Cigolini M, Bosello O, et al. J Submicrosc Cytol 1984;16:243–251.
32. Cinti S, Cigolini M, Gazzanelli G, et al. J Submicrosc Cytol 1985;17:631–636.
33. Sbarbati A, Zancanaro C, Cigolini M, et al. Acta Anat (Basel) 1987;128:84–88.
34. Cinti S, Cigolini M, Sbarbati A, et al. Tissue Cell 1987;19:809–815.
35. Cigolini M, Cinti S, Bosello O, et al. J Anat 1986;145:207–216.
36. Cigolini M, Cinti S, Brunetti L, et al. Exp Cell Res 1985;159:261–266.
37. Himms-Hagen J, Cui J, Danforth E Jr, et al. Am J Physiol 1994;266:R1371–R1382.
38. Collins S, Daniel KW, Petro AE, et al. Endocrinology 1997;138:405–413.
39. Ghorbani M, Claus TH, Himms-Hagen J. Biochem Pharmacol 1997;54:121–131.
40. Ghorbani M, Himms-Hagen J. Int J Obes Relat Metab Disord 1997;21:465–475.
41. Cousin B, Cinti S, Morroni M, et al. J Cell Sci 1992;103(Pt 4):931–942.
42. Himms-Hagen J, Melnyk A, Zingaretti MC et al, Am J Physiol Cell Physiol 2000;279:C670–C681.
43. Granneman JG, Li P, Zhu Z, et al. Am J Physiol Endocrinol Metab 2005;289:E608–E616.
44. Almind K, Kahn CR. Diabetes2004;53:3274–3785.
45. Murano I, Zingaretti CM, Cinti S. Adipocytes 2005;1:121–130.
46. Prins JB, O'Rahilly S. Clin Sci 1997;92:3–11.
47. Tiraby C, Tavernier G, Lefort C, et al. J Biol Chem 2003;278:33,370–33,376.
48. De Matteis R, Arch JR, Petroni ML, et al. Int J Obes Relat Metab Disord 2002;26:1442–1450.
49. Larsen TM, Toubro S, van Baak MA, et al. Am J Clin Nutr 2002;76:780–788.
50. Bachman ES, Dhillon H, Zhang CY, et al. Science 2002;297:843–845.
51. Kopecky J, Clarke G, Enerback S, et al. J Clin Invest 1995;96:2914–2923.
52. Cummings DE, Brandon EP, Planas JV, et al. Nature 1996;382:622–626.
53. Cederberg A, Gronning LM, Ahren B, et al. Cell 2001;106:563–573.
54. Yang X, Enerback S, Smith U. Obes Res 2003;11:1182–1191.
55. Tsukiyama-Kohara K, Poulin F, Kohara M, et al. Nat Med 2001;7:1128–1132.
56. Morroni M, Giordano A, Zingaretti MC, et al. Proc Natl Acad Sci USA 2004;101:16,801–16,806.
57. DiGirolamo M, Fine JB, Tagra K, et al. Am J Physiol 1998;274: R1460–R1467.
58. Faust IM, Johnson PR, Stern JS, et al. Am J Physiol 1978;235: E279–E286.
59. Bjorntorp P.Int J Obes 1991;15Suppl2:67–81.
60. Hausman DB, DiGirolamo M, Bartness TJ, et al. Obes Rev 2001;2:239–254.
61. Ahima RS, Flier JS. Trends Endocrinol Metab 2000;11:327–332.
62. Imai T, Jiang M, Kastner P, et al. Proc Natl Acad Sci USA 2001;98:4581–4586.
63. Weisberg SP, McCann D, Desai M, et al. J Clin Invest 2003;112:1796–1808.
64. Xu H, Barnes GT, Yang Q, et al. J Clin Invest 2003;112:1821–1830.
65. Cancello R, Henegar C, Viguerie N, et al. Diabetes 2005;54:2277–2286.
66. Cinti S, Mitchell G, Barbatelli G, et al. J Lipid Res 2005;46:2347–2355.
67. Nisoli E, Briscini L, Giordano A, et al. Proc Natl Acad Sci USA 2000;97:8033–8038.
68. Napolitano L. J Cell Biol 1963;18:663–679.

69. Pond CM, Mattacks CA. Folia Primatol (Basel) 1987;48:164–185.
70. Hager A, Sjostrom L, Arvidsson B, et al. Metabolism 1977;26:607–614.
71. Chumlea WC, Knittle JL, Roche AF, et al. Am J Clin Nutr 1981;34:1791–1797.
72. Stern JS, Batchelor BR, Hollander, N., et al. Lancet 1972;2:948–951.
73. Rebuffe-Scrive M, Eldh J, Hafstrom LO, et al. Metabolism 1986;35:792–797.
74. Valet P, Grujic D, Wade J, et al. J Biol Chem 2000;275:34,797–34,802.
75. Kortelainen ML, Pelletier G, Ricquier D, et al. J Histochem Cytochem 1993;41: 759–764.
76. Garruti G, Ricquier D. Int J Obes Relat Metab Disord 1992;16:383–390.
77. Oberkofler H, Dallinger G, Liu YM, et al. J Lipid Res 1997;38:2125–2133.
78. Huttunen P, Hirvonen J, Kinnula V. Eur J Appl Physiol Occup Physiol 1981;46:339–345.
79. Baldi A, Santini M, Mellone P, et al. J Clin Pathol 2004;57:993–994.

2 Metabolism of White Adipose Tissue

Michel Beylot

Abstract

Triacylglycerols (TAGs) stored in adipose tissue are by far the largest site of energy storage. Adipocytes continuously synthesize and break down these TAGs depending on the body energy status and its hormonal environment. They act as a "buffer" for plasma lipids and also for lipids stored in other tissues. This chapter presents the metabolic pathways used for adipose tissue TAG synthesis and breakdown and the way they are controlled. It points out important recent findings that have modified our conception of these pathways and of their regulation—particularly the role of glyceroneogensis in TAG synthesis—of proteins associated to lipids droplets—particularly perilipins—and of lipases other than the classic hormone-sensitive lipase in TAG hydrolysis.

Key Words: Triacylglycerols; lipolysis; lipogenesis; adipocyte; glyceroneogenesis; insulin; hormone-sensitive lipase; PAT proteins.

1. INTRODUCTION

The contribution of white adipose tissue (WAT) to whole-body oxygen consumption and energy production is limited, as it represents about 5% of whole-body energy expenditure *(1)*. This is, on a per-kilogram basis, much less than organs such as the liver, kidney, or brain. WAT is, however, continuously synthesizing and breaking down lipids and has important metabolic functions that play a significant role in the regulation of lipids and also in glucose metabolism. WAT is by far the largest site of energy storage of the body. This energy, stored as triacylglycerols (TAG) within intracellular lipid droplets, represents in a young healthy nonobese adult around 12 to 15 kg—i.e., 110,000 to 135,000 calories. This energy is stored during the postprandial periods; most of it comes from the ingested TAG. Some of the fatty acids used for the synthesis and storage of TAG may be synthesized from carbohydrates through the pathway of *de novo* lipogenesis, but this contribution is minor in humans except in situations of massive and prolonged carbohydrate overfeeding. This energy is released from WAT between meals and in situations of caloric restriction and of exercise to meet the energy needs of other organs. This release, through the process of lipolysis—i.e. the hydrolysis of intracellular TAG—provides glycerol and fatty acids. Glycerol will be used mostly by gluconeogenic tissues to synthesize new molecules of glucose. Fatty acids appear in the circulation as albumin-bound nonesterified fatty acids (NEFA) that will be used mainly by muscles (mostly for oxidation), liver (oxidation, complete to CO_2 or incomplete to

From: *Nutrition and Health: Adipose Tissue and Adipokines in Health and Disease*
Edited by: G. Fantuzzi and T. Mazzone © Humana Press Inc., Totowa, NJ

ketone bodies, but also TAG synthesis, storage, and secretion as very-low-density lipoprotein [VLDL]-TAG), and adipose tissue (re-esterification).

These processes of TAG storage and hydrolysis in WAT are highly regulated by hormonal (mainly insulin and catecholamines), metabolic (glucose, NEFA), and nutritional (energy intake, contribution of carbohydrates and lipids to this intake) factors. These regulations are essential to maintain body weight homeostasis; alterations in these processes, resulting in imbalance between storage and mobilization of TAG in WAT, may result in obesity. Moreover, the way plasma lipids are cleared from plasma by adipose tissue and fatty acids released during lipolysis plays an important role in the everyday regulation of plasma lipids concentrations, and alterations of WAT metabolism may result in increased plasma lipids levels and therefore increased risks of cardiovascular disease. Last, WAT, through the secretion of various hormones such as leptin or adiponectin and the way it stores or release lipids, controls in part the amount of lipids stored in other tissues and the sensitivity to insulin of tissues such as muscles or liver. Therefore, modifications of WAT metabolism may be implicated in the excessive accumulation of lipid substrates in other tissues, their unfavorable consequences (lipotoxicity *(2)*), and the development of insulin resistance and diabetes.

2. TAG SYNTHESIS AND STORAGE

TAG stored in adipocytes are synthesized within these cells from fatty acids and glycerol-3-phosphate (G3P). Most of the fatty acids used for this synthesis are provided by circulating plasma lipids, whereas G3P has two main possible origins, glycolysis and glyceroneogenesis. The exact intracellular site of TAG synthesis and the way new TAG molecules are directed toward lipids droplets is still debated.

2.1. Sources of Fatty Acids

2.1.1. Fatty Acids From Circulating Lipids

These fatty acids are provided either by the albumin-bound NEFA pool or by the TAG incorporated in TAG-rich lipoproteins, mainly VLDL in the postabsorptive state and chylomicrons in the postprandial state. These lipoproteins–TAG must first be hydrolyzed by the enzyme lipoprotein lipase (LPL) bound to the wall of capillaries in adipose tissue *(3)* in order to release their fatty acids. The expression and activity of LPL is increased in adipose tissue in the fed state, particularly during a high-carbohydrate diet, probably through the action of insulin, whereas both expression and activity are decreased in adipose tissue during fasting and high-fat diet *(4)*. This uptake of TAG-fatty acids is probably controlled also in part through the VLDL receptor, a member of the LDL-receptor family that is expressed in adipose tissue *(5)*. It binds apoprotein E-rich lipoproteins, such as VLDL, chylomicrons, and remnants, and brings them probably in close contact with LPL, facilitating its action. Mice deficient in VLDL receptor have a decreased fat mass and are resistant to diet-induced obesity; moreover, VLDL-receptor deficiency reduces the obesity of *ob/ob* mice *(6)*. The exact role of this receptor in humans remains to be defined.

Whatever their origin, the uptake of long-chain fatty acids by adipocytes requires specific processes in order to allow them to cross the plasma membranes *(7,8)*. It is probable that both a transport by specific transporters and a passive diffusion coexist in most cells. Human white adipocytes express several fatty acid transporters that facilitate and control the transport of fatty acids: the protein CD36 (homolog to the murine fatty acid transporter [FAT]), the fatty acid transport protein (FATP), and the fatty acid binding protein plasma membrane (FABPpm), with FAT appearing as responsible for most of fatty-acid uptake *(9)*. This transport is dependent of the presence of lipid rafts in the membrane *(10)*. Insulin promotes this transport by stimulating the expression of these transporters and their trafficking to plasma membranes *(11)*.

Because fatty acids are not soluble in the cytosol and may exert toxic effects on membranes, inside the cells they are tightly bound by cytoplasmic lipid-binding proteins also called FABP. These proteins carry fatty acids from membrane to membrane or to the site of action of the enzyme Acyl-CoA synthase *(12,13)*. Human white adipocytes express two FABPs: adipocyte lipid-binding protein (ALBP or AFABP or aP2), expressed only in adipocytes, and keratinocyte lipid-binding protein that is expressed also in macrophages. aP2 is much more abundant than keratinocyte lipid-binding protein in human (and rodent) adipocytes; however, this ratio varies between different sites of adipose tissue and this may affect the metabolism of these different fat depots *(14)*. The first step in the metabolism of fatty acids after their uptake and binding by FABP is their activation in long-chain fatty acyl-CoA (LCFA-CoA) by Acyl-CoA synthase. LCFA-CoA can then be directed toward oxidation and to the synthesis of more complex lipids, such as fatty acids. As in other tissues, oxidation requires the entry of LCFA-CoA inside mitochondria through the action of the enzyme carnitine-palmitoyl transferase I. In other tissues, this step is an important site of the regulation of fatty acid metabolism through the inhibition of carnitine-palmitoyl transferase I by malonyl-CoA, the product of acetyl-CoA carboxylase (ACC) that catalyzes the first step in the lipogenic pathway *(15)*. Whether this step is also highly regulated in adipose tissue is unclear; however, the main metabolic fate of fatty acids in adipocytes appears to be re-esterification into TAG.

2.1.2. *De Novo* Lipogenesis

De novo lipogenesis (DNL) is the synthesis of new fatty acids molecules from nonlipid substrates, mainly carbohydrates in mammals. The expression and activity of the glycolytic and lipogenic pathways are therefore linked together in lipogenic tissues. The two main sites of DNL are liver and adipose tissue; the quantitative importance of this pathway, and the respective contribution of liver and adipose tissue varies between species *(16)*. Overall, DNL is less active in humans than in rodents and contributes much less than TAG dietary intake to adipose tissue lipid stores *(17)*. Indeed studies of hepatic DNL in healthy humans concluded that this pathway is a minor contributor to the fatty acids used for liver TAG synthesis and secretion and represents only about 1 to 2 g per day *(18–21)*. Liver lipogenesis is stimulated by insulin and glucose and can be largely increased (two- to fourfold) by a high-carbohydrate (CHO) diet *(18,22–24)*; it is increased in *ad libitum*-fed obese subjects *(25)*, hypertriglyceridemic type 2 diabetic subjects *(26)*, and subjects with non alcoholic fatty liver disease *(27)*, but still remains minor compared with oral TAG ingestion (usually more than 100 g/d). These new fatty acid molecules provided by hepatic DNL can be exported as TAG-VLDL for uptake and storage by adipocytes, but are a minor contributor to these stores in humans. The key enzymes for lipogenesis are also expressed in adipocytes *(28)* but the expression and activity of these enzymes are lower in human than in rat adipocytes *(24)*. From data in the literature it appears than DNL in humans is less active in adipocytes than in liver when expressed per gram of tissue but, on a whole-body basis, the contributions of liver (1.5 kg) and adipose tissue (12 to 15 kg) appear comparable (1 to 2 g/d for each tissue) *(18)*.

The regulation of DNL by hormonal (insulin, glucagon), metabolic (glucose, polyunsaturated fatty acids [PUFA]), and nutritional (total energy intake, dietary CHO-to-fat ratio) factors is less well defined in humans, either in liver or adipose tissue, than in rodents, and less well-known in adipocytes than in liver. Overall it is clear that hepatic lipogenesis is highly responsive to modifications of hormonal and nutritional conditions. Insulin and glucose stimulate it, whereas glucagon and PUFA inhibit it *(29)*. The regulation by insulin and PUFA is mediated by the transcription factor sterol response element binding protein 1c (SREBP-1c) *(30)* and also in part by liver X receptor α (LXRα; insulin and PUFA) *(31)* and carbohydrate response element binding protein (ChREBP) (PUFA) *(32)*, whereas the inhibitory action of glucagon and the stimulatory one of glucose are mediated by ChREBP *(32)*. Insulin stimulates the transcription of SREBP-1c, directly and indirectly through a stimulation of the expression of LXRα. Whether insulin stimulates also the cleavage of the precursor form of the protein SREBP-1c and the release of its mature form is debated. LXRα stimulates the expression of lipogenic genes directly and through an increase in the expression of SREBP-1c *(31)*. A full stimulation of liver lipogenesis requires the simultaneous and synergistic action of insulin and glucose *(29)*. Glucose acts by dephosphorylation of ChREBP, allowing its entry in the nucleus and its binding to specific response element in the promoter of glycolytic (L-PK) and lipogenic (FAS, ACC) genes *(33–35)*. Glucagon and PUFA, on the contrary, phosphorylate ChREBP, respectively, through protein kinase A (PKA) and AMP-dependant kinase (AMPK), inhibiting its action *(33,36)*.

The regulation of DNL in adipocytes, particularly in humans, is much less defined. It is clear that insulin increases fatty acid synthase (FAS) expression and activity in human and rodent adipocytes *(37,38)*. This action involves probably both SREBP-1c *(39)* and LXRα, although the actual role of SREBP-1c has been questioned *(40)*. Glucose also stimulates lipogenesis in adipocytes *(41)* and, as in liver, a full stimulation requires the simultaneous presence of insulin and glucose. The action of glucose could be transmitted by ChREBP, as this transcription factor is expressed in adipocytes *(24,35,42,43)*; however, a stimulatory effect of glucose on ChREBP translocation to nucleus and DNA binding activity in adipocytes has not yet been demonstrated. A stimulation of adipocyte ChREBP expression by glucose and insulin has been reported, but only in the presence of high, unphysiological, glucose levels *(42)*. In vivo, ChREBP expression in liver and adipose tissue is poorly responsive to metabolic and nutritional factors and is clearly increased only in the situation of high CHO refeeding after starvation *(42–44)*. Finally, PUFA have an inhibitory action on lipogenesis in adipose tissue, but this effect is less marked than in liver *(45)*. Overall, the expression and activity of lipogenesis appears less responsive to metabolic and nutritional factors in adipose tissue than in liver, in rodents and in humans, and is still less responsive in humans than in rodents *(18,24,43)*, although some stimulation has been observed during prolonged carbohydrate overfeeding *(46)*. It is noteworthy that the expression of ChREBP, SREBP-1c, FAS, and ACC is decreased in the adipose tissue of human obese subjects and of experimental models of obesity with long standing obesity while the expression and activity of liver lipogenesis are increased *(24,25,47)*. Whether the expression of lipogenesis in adipose tissue is increased during the initial, dynamic phase of obesity remains to be established.

A last point is the potential importance of the renin–angiotensin system in the control of lipogenesis and TAG storage. WAT expresses the components of a functional renin–angiotensin system *(48,49)*; mice with overexpression of angiotensinogen in adipose tissue have an increased fat mass with adipocyte hypertrophy *(50)*. In vitro studies have shown that angiotensin II stimulates lipogenesis in 3T3-L1 and human adipocytes *(51)*. This effect involves SREBP-1c and is mediated by the angiotensin type 2 receptor *(52)*. Deletion of this receptor results in adipocyte hypotrophy and resistance to diet-induced obesity *(53)*. These mice have a reduced expression in adipocytes of SREBP-1c, FAS but also of LPL, FAT, and aP2, suggesting that angiotensin II stimulates several pathways of TAG storage. Angiotensinogen is overexpressed in the adipose tissue of obese subjects *(54)*, particularly in visceral adipose tissue, and could therefore have a role in the development of obesity.

2.2. Sources of G3P

TAG synthesis requires G3P for the initial step of fatty acid esterification. Glycerokinase activity is very low in adipocytes and G3P is produced either from glucose, through the first steps of glycolysis, or from gluconeogenic precursors, through glyceroneogenesis *(55)*. Glucose enters adipocytes through the glucose transporters 1 and 4 (Glut-1 and Glut-4) responsible, respectively, for basal glucose and insulin-stimulated glucose uptake. Insulin acutely stimulates glucose uptake by promoting the translocation of Glut-4 from an intracellular pool to the membrane, an effect mediated through the PI-3 kinase Akt pathway *(56)*. Glucose uptake is also stimulated by acylation-stimulating protein *(57)*. The other source of G3P is glyceroneogenesis, an abbreviated version of gluconeogenesis that provides G3P from gluconeogenic substrates such as lactate and pyruvate *(55)*. The regulatory step of this way is controlled by the cytosolic form of phosphoenol pyruvate carboxy kinase (PEPCK). PEPCK-C expression and activity are increased by PUFA and by the peroxisome proliferators-activated receptor γ-agonist thiazolidinediones and inhibited by glucocorticoids *(58)*. The relative contribution of glycolysis and glyceroneogenesis to G3P production varies thus with nutritional and pharmacological factors. The overall availability of G3P controls the esterification rate of fatty acids provided by DNL and circulating lipids but also the partial re-esterification of fatty acids released by the lipolysis of stored TAG.

2.3. TAG Synthesis

This biosynthesis results from the successive esterification of the alcoholic groups of G3P by different enzymes: G3P acyltransferases (GPATs), 1-acylglycerol-3-phosphate acyltransferases (AGPATs) and diacylglycerol acyltransferases (DGATs) *(59–62)*. All these enzymes exist in different isoforms and are encoded by different genes. The isoforms GAPT1, GAPT2, AGAPT2, and DGAT1 and 2 are present in adipose tissue *(63)* but the tissue repartition and substrate specificity of these different isoforms are not yet fully clarified. The expression of DGAT1 and 2 is stimulated in adipose tissue by glucose and insulin *(64)* and both insulin and glucose increase TAG synthesis. Acylation-stimulating protein also stimulates adipocyte TAG synthesis *(57)*. The important role of these enzymes in controlling adipose TAG stores is demonstrated by studies of mice lacking

DGAT and of subjects with congenital lipodystrophy *(63,65,66)*. A point discussed is the intracellular site of TAG synthesis and how new TAG molecules are directed to lipid droplets for storage. Classically, TAG synthesis occurs in the endoplasmic reticulum. However, there is recent evidence that most of this synthesis takes place in a subclass of caveolae in the plasma membrane *(67)*. These caveolae also contain perilipin, a protein coating lipid droplets, and that could, in addition to its regulatory role of lipolysis (*see* Subheading 3.3.), be involved in the incorporation of newly synthesized TAG into lipid droplets.

3. LIPOLYSIS AND RELEASE OF FATTY ACIDS

During intracellular lipolysis, TAG is hydrolyzed successively into diacylglycerol (DAG) and monoacylglycerol (MAG) to finally release three molecules of fatty acid and one molecule of glycerol per molecule of TAG. This hydrolysis is usually complete, although some DAG and MAG can accumulate. Because adipose tissue has very low glycerol kinase activity, the end product, glycerol, is released in the circulation for use by other tissues. This release of glycerol depends in part on adipose tissue aquaporin (AQPap), a channel-forming integral protein of the cell membrane. AQPap is a member of a family of at least 11 proteins that function as water channels *(68)*. Its expression is increased during fasting and reduced by refeeding and insulin *(69)*, whereas thiazolidinediones stimulate it *(70)*. Deletion of AQPap in mice induces a lack of plasma glycerol increase in response to β-adrenergic stimulation and during fasting, with hypoglycemia during fasting *(71)*, and results in obesity *(72)*. Missense mutations resulting in the loss of transport activity have been described in humans *(73)*. One subject homozygous for such a mutation had a normal body weight and normal basal plasma glycerol concentration but a lack of increase during exercise *(73)*, suggesting that AQPap has a role in glycerol efflux in humans but is not the only mechanism. Fatty acids released by the hydrolysis of TAG, on the contrary, can be either released or re-esterified into TAG without appearing in the circulation. This intracellular recycling of fatty acids depends on the availability of G3P and of the expression and activity of esterification enzymes. In the basal, postabsorptive state, this recycling appears limited *(74)*, but high re-esterification rates can occur during exercise *(75)* or in pathological situations such as hyperthyroidism *(74)* and stress *(76)*. The mechanisms responsible for the transport of fatty acids released by lipolysis to plasma membrane are debated. aP2 is probably involved, as it forms a complex with hormone-sensitive lipase (HSL) and aP2 –/– mice have a decreased release of fatty acids from adipose tissue *(77)*. The efflux of fatty acids probably involves both diffusion and transport by specific plasma membrane proteins as their uptake.

Lipolysis is controlled mainly by the enzyme HSL, whose activity is regulated principally by catecholamines and insulin through the cAMP–PKA pathway. However, it is now clear that HSL is also controlled by other mechanisms and that other lipases are involved in adipocyte TAG hydrolysis.

3.1. Hormone-Sensitive Lipase

HSL was first characterized in rats as an 84-kDa protein with 768 amino acids. In human adipose tissue, HSL is an 88-kDa immunoreactive protein of 775 amino acids encoded by nine exons and whose gene is on chromosome 19. It is expressed also in brown adipose tissue, steroidogenic cells, skeletal muscle, heart, insulin-secreting β-cells, mammary glands, and, at least in rodents, in macrophages *(78)*. HSL is a serine protease that can hydrolyze TAG, DAG, and cholesterol esters. In adipose tissue it hydrolyzes TAG and DAG with a higher activity for DAG and, when acting on TAG, a preference for the *sn*1-ester and 3-ester bonds *(78)*. MAGs are hydrolyzed by a different enzyme, a MAG lipase that releases glycerol, and the last fatty acid that has no known regulatory role.

Analysis of the structure of HSL has shown several functional domains. The N-terminal part of approximately 300 amino acids is involved in the dimerization of HSL *(79)* and therefore in the activity of HSL, as there is evidence that its functional form is a homodimer *(80)*. Residues 192 to 200 are critical for the interaction with aP2 *(78)*, interaction that has probably a role in the efflux of fatty acids released by HSL and in preventing the inhibition of HSL activity by these fatty acids. The C-terminal part contains the catalytic and regulatory domains. The active serine of the catalytic triad is at position 423 in the rat and 424 in humans, located in a Gly-Xaa-Ser-Xaa-Gly motif found in lipases and esterases *(81)*. This serine is encoded by exon 6. A truncated, short form of HSL of 80 kDA generated by alternative splicing of exon 6 during the processing of HSL mRNA has been described in human but not rodent tissue. This short from lacks serine 424 and is devoid of activity *(82)*. The presence of this variant in some obese subjects is associated with a decreased in vitro HSL activity and a reduced maximal lipolytic response to catecholamines *(83)*. The other amino acids of the catalytic triad are Asp 703 and His 733 in rats (Asp 693 and His 723 in humans) *(84)*. The regulatory domain is encoded principally by exon 7 and most of exon 8. It runs from residue 521 to 669 in rats *(78)* and contains the serines (serine 563, 565, 659, and 660 in rats) whose phosphorylation status controls the activity of HSL.

HSL activity is stimulated by catecholamines through the classic adenylate cyclase–cAMP–PKA pathway. Actually catecholamines stimulate lipolysis through their β-receptors and inhibit it through α-receptors; the net result depends on the balance between the two actions and, in physiological situations in humans, is usually a stimulation of lipolysis *(85)*. Regional differences between different adipose tissue sites in the proportion of α- and β-receptors result in differences in the response to catecholamines *(86)* and regional differences in the regulation of adipose tissue metabolism *(87)*. Stimulation of lipolysis results from the phsophorylation of serine 563 that is the regulatory site *(88)*. Serine 565 (basal site) is phosphorylated in basal conditions. The two sites are mutually exclusive and the basal site can block the phosphorylation of serine 563 and thus exerts an antilipolytic action *(89)*. Serine 565 can be phosphorylated by several kinases, particularly the AMPK *(89)*. Compounds activating AMPK, such as metformin, may thus have an antilipolytic action *(90)*. Last, evidence has been provided that serines 659 and 660 are also phosphorylated by cAMP-dependent protein kinase in vitro in rat adipocytes and that this phosphorylation could also stimulate lipolysis *(91)*.

Other pathways of phosphorylation have been described. Increased cAMP concentration can activate the mitogen-activated protein kinase/extracellular-regulated kinase (ERK) pathway *(92,93)*. Activated ERK phosphorylates HSL at serine 600 and increases its activity *(92)*. Finally, the natriuretic peptides atrial natriuretic peptide and brain natriuretic peptide have been shown to phosphorylate HSL and stimulate lipolysis *(94)*. This effect is present only in primates. Atrial natriuretic peptide and brain natriuretic peptide activate guanylate cyclase and stimulate cGMP-dependent protein kinases. They probably play a role in the stimulation of lipolysis during exercise *(94)*.

Dephosphorylation of the regulatory site(s) inhibits HSL. Insulin, the main antilipolytic hormone, acts by phosphorylating and stimulating the activity of phosphodiesterase 3B that breaks down cAMP and reduces thus the phosphorylation of HSL *(95)*. This action is mediated by the PI3kinase–PKB pathway *(96)*. Insulin also activates ERK in adipocytes, but the relation of this action with the regulation of lipolysis is unclear. Ser-563 can also be dephosphorylated by the protein phosphatases 2A and 2C *(97)* and insulin could stimulate these phosphatases. Other mechanisms are also possible: internalization of β-adrenoreceptors *(98)* or disruption by insulin of β-adrenergic signaling *(99)*.

3.2. Other Lipases

HSL was long considered as the only, and therefore regulatory, enzyme hydrolyzing adipose tissue TAG. This view was challenged by studies of mice lacking HSL. These mice do not develop obesity and have a reduced fat mass *(100,101)*. They always have a marked basal lipolysis and a response of lipolysis to β-adrenergic stimuli *(100–102)*. These findings suggested than another lipase was present and active in the absence of HSL. The finding that DAG accumulates in adipocytes of these mice *(103)* suggested that this lipase had a preference for the hydrolysis of TAG, and was rate-limiting for this first step of lipolysis, whereas HSL was limiting for the hydrolysis of DAG. Such a lipase, named adipose tissue lipase (ATGL), has been described recently *(104)*. This enzyme is identical to the protein desnutrin *(105)* and to the calcium-independent phospholipase A2ζ *(106)*, described nearly simultaneously.

In mice and humans ATGL is expressed predominantly in white and brown adipose tissue, localized to the adipocyte lipid droplet. It is also present, to a lesser extent, in heart, skeletal muscle, and testis *(104,105)*. It hydrolyzes specifically TAG, has low activity against DAG, and has little or no activity against cholesterol esters. Its expression is increased by fasting and glucocorticoids and reduced by refeeding *(105)*. ATGL expression increases during the differentiation of human preadipocytes in adipocytes, simultaneously with HSL expression, and the expression of these two lipases appears coregulated in human adipose tissue *(107)*. Its expression is also reduced in the adipose tissue of *ob/ob* and *db/db* mice *(105)*. To our knowledge, no data on ATGL expression in human obesity are available. ATGL contains in its N-terminal part a patatin domain and belongs thus to a large family of proteins with patatin domains that have acylhydrolase activity (*see* ref. *108* for more details on this family of proteins). The N-terminal part also contains a consensus sequence Gly-Xaa-Ser-Xaa-Gly for serine lipase, with the possible active serine at position 47. Another domain, between residues 309 and 391, contains large amounts of hydrophobic residue, suggesting that it could be a lipid/membrane binding site; this domain could be responsible for the constitutive presence of ATGL on the lipid droplet. ATGL can be phosphorylated. This phosphorylation is independent of PKA *(104)* and the kinases involved and the consequences on enzyme activity remain to be established.

ATGL appears thus as the lipase that is probably responsible for the physiological hydrolysis of most of TAG and of the residual lipolytic activity in mice lacking HSL; however, much still has to be learned about the physiological regulation of its activity and about possible abnormalities of its expression and activity in pathology such as obesity and type 2 diabetes. Other potential lipases have been described in adipocytes.

Carboxyl esterase 3 (also known as hepatic triglyceride hydrolase) is present in adipocytes *(109–111)* but its quantitative contribution to lipolysis remains to be determined. Adiponutrin is expressed exclusively in adipose tissue and has high sequence homology with ATGL with a patatin domain, the consensus sequence for serine hydrolase, and possible lipid/membrane binding domains *(112)*. The regulation of its expression is quite different, however, as it is repressed during fasting and increased in *fa/fa* rats *(112,113)*. Divergent results on a possible TAG hydrolase activity of adiponutrin have been reported *(104,106)* and its role in adipose tissue lipolysis remains uncertain. Finally, two other members of the adiponutrin family (GS2, GS2-Like) that were recently described could also be involved in lipolysis *(114)*.

3.3. Perilipin and HSL Translocation

Phosphorylation of purified HSL induces only a modest two- to threefold increase in activity, whereas the stimulation of lipolysis in intact adipocytes by β-adrenergic agents induces a much larger increase of lipolytic rate. A first explanation for this discrepancy appeared when it was demonstrated that phosphorylation of HSL induced, in addition to a stimulation of its activity, its translocation from the cytosol to the surface of lipid droplets, where it can hydrolyze TAG *(115)*. This requires the phosphorylation of serines 659 and 660 *(116)*. A second explanation emerged when it appeared that PKA phosphorylated not only HSL but also perilipin, a protein surrounding lipid droplets and which acts as a gatekeeper for the access of HSL to TAG. Perilipin is one of the numerous proteins surrounding lipid droplets and belongs with adipophilin (or adipocyte-related differentiation protein) and TIP-47 to the PAT family (for a recent review of PAT proteins *see* ref. *117*). Adipocyte-related differentiation protein is expressed in all cells storing lipids *(118)*. In adipocytes, it is highly expressed during the differentiation of the cells and the constitution of lipid droplets, and its expression decreases in mature adipocytes. Its role is still unclear but it could be involved in the transport of lipids to droplets *(119)*. Perilipins are expressed in adipocytes, steroidogenic cells *(120)*, and foam cells of atheroma plaques *(121)*. Perilipin expression appears during the differentiation of adipocytes and is high in mature adipocytes. This expression requires the presence, and intracellular metabolism, of fatty acids *(122)* and is also stimulated by peroxisome proliferators-activated receptor γ-agonists *(123)*. There are at least three forms of perilipins, A, B, and C, resulting from different splicings of a common pre-messenger RNA, and sharing a common N protein part *(120)*. Perilipin A and B are expressed in adipocytes, A being the predominant form. Perilipins are phosphorylated on multiple serine sites by PKA (three serines on the N-terminal part common to perilipin A and B and three other on the C-terminal part specific to perilipin A). In the basal, unphosphorylated state, perilipin opposes the hydrolysis of TAG by HSL *(124)*. The phosphorylation of perilipin facilitates the interaction of HSL with TAG *(125)*, and their hydrolysis, probably through translocation of the phosphorylated perilipin from the surface of lipid droplets to the cytosol *(126)*. This role of perilipin is demonstrated by the studies of perilipin-null mice. These mice have a reduced fat mass and are resistant to genetic and diet-induced obesity *(127,128)*. Their lipolysis is increased in the basal state but the response to β-adrenergic stimulation is reduced *(127,128)*. Perilipins are expressed in human adipose tissue and evidence for a role in the regulation of lipolysis in humans was provided *(129–131)*. These studies showed that a low total perilipin content was associated with a high basal lipolytic rate of isolated adipocytes and high concentrations of glycerol and NEFA in vivo, thus supporting a role for perilipin in the regulation of lipolysis in humans *(130)*. The possible role of perilipin in human obesity remains unclear; both decreased *(130,131)* and increased expression *(129)* in obese subjects have been reported.

4. CHOLESTEROL METABOLISM

Adipocytes store TAG but also relatively large amounts of cholesterol (1 to 5 mg/g of total lipids) *(132)*. Contrary to what is observed in steroidogenic cells and foam cells, most (about 95%) of this cholesterol is in the free, nonesterified, form. This cholesterol is present in two major pools, the plasma membrane and the phospholipid monolayer surrounding the lipid droplets. Because the cholesterol synthetic rate is very low in adipocytes *(133)*, most of the adipocyte cholesterol comes from plasma lipoproteins. The LDL-receptor LRP and the scavenger receptor BI (SR-BI) are expressed by adipocytes but their respective quantitative importance in the uptake of cholesterol has not been defined. Interestingly, the expression of SR-BI is stimulated during the differentiation of adipocytes *(134)* and most of the cholesterol taken up through this receptor is targeted in mature adipocytes toward lipid droplets *(135)*. In addition, insulin and angiotensin induce the translocation of SR-BI from intracellular pools to the plasma membrane and stimulate the uptake of cholesterol from high-density lipoprotein; these actions are mediated by the PI3-kinase pathway *(134)*. There is a strong correlation between fat cell size and its cholesterol content. This content thus increases during replenishment of lipid droplets and increases further in hypertrophic adipocytes in obese states *(136)*. Thus adipose tissue can store large amounts of cholesterol, particularly during obesity. Adipocytes express the transporter ABCA1 *(137)* and can also release cholesterol. However, a significant increase in this efflux is observed in vitro only during prolonged stimulation of lipolysis by lipolytic agents *(137)*. Whether this influx is increased during reduction of total body fat mass and how it is regulated in this situation remains to be investigated. Overall these data suggest that adipose tissue could play a significant role in whole-body cholesterol metabolism and have a buffering role of not only plasma TAG but also plasma cholesterol.

Increased fat cell size results also in modifications of the intracellular repartition of cholesterol: more of the cholesterol is present on the surface of lipid droplets and despite the increase in cell total cholesterol content, the membrane of hypertrophied adipocytes contains less cholesterol *(132)*. This depletion in membrane cholesterol results in a stimulation of the expression of SREBP-2 and its target genes HMG-CoA reductase and synthase and LRL-r, whereas expression of ABCA1 is repressed *(136)*, modifications aimed at restoring the membrane pool of cholesterol. FAS expression is also stimulated; because SREBP-1c is not modified, this increase of FAS expression is perhaps mediated by LXRα. In addition, relative cholesterol depletion in the membrane of the adipocyte decreases the expression of Glut-4, with reduced glucose uptake and metabolism, and increases those of tumor necrosis factor-α, interleukin-6, and angiotensinogen *(136)*; all these modifications favor the development of insulin resistance. These data suggest the interesting possibility that cholesterol might be a sensor for the amount of fat stored in adipocytes and serves as a link between the increase in fat stores and some of the modifications of metabolism observed in obesity *(132,136)*.

5. CONCLUSIONS

The past years have brought important and exciting insights into adipocyte metabolism and changed our view of how processes such as lipolysis are controlled. Further developments are expected soon, particularly on the role of proteins surrounding lipid droplets and also on adipocyte cholesterol metabolism. These data could also shed a new light on the physiopathology of obesity and its complications and open the way to new therapeutic approaches.

REFERENCES

1. Elia M. In: Kinney JM, Tucker HN, eds. *Energy Metabolism: Tissue Determinants and Cellular Corollaries*. Raven Press, New York: 1992; pp. 61–79.
2. Schaffer J. Curr Opin Lipidol 2003;14:281–287.
3. Mead J, Irvine S, Ramji DJ. Mol Med 2002;80:753–769.
4. Braun JE, Severson DL. Biochem J 1992;287:337–347.
5. Tacken P, Hofker M, Havekes L, et al. Curr Opin Lipidol 2001;12:275–279.
6. Goudriaan J, Tacknen P, Dahlmans V, et al. Arterioscler Thromb Vasc Biol 2001;21:1488–1493.
7. Luiken J, Coort S, Koonen D, et al. Pflugers Arch 2004;448:1–5.
8. Bernlhor D, Ribarick-Coe N, LiCata V. Cell Dev Biol 1999;10:43–49.
9. Ibrahimi A, Abumrad N. Curr Opin Clin Nutr Metab Care 2002;5:139–145.
10. Pohl J, Ring A, Korkmaz U, et al. Mol Cell Biol 2005;16:24–31.
11. Czech M. Mol Cell 2002;9:695–696.
12. Weisiger R. Mol Cel Biochem 2002;39:35–42.
13. Storch S, Veerkamp J, Hsu K. Mol Cell Biochem 2002;239:25–33.
14. Fisher R, Thorne A, Hamsten A, et al. Mol Cel Biochem 2002;239:95–100.
15. MacGarry JD, Foster D. Ann Rev Biochem 1980;49:395–411.
16. Gondret F, Ferré P, Dugail I. J Lipid Res 2001;42:106–113.
17. Marin P, Hogh-Christiansen I, Jansson S, et al. Am J Physiol 1992;263:E473–E480.
18. Diraison F, Yankah V, Letexier D, et al. J Lipid Res 2003;44:846–853.
19. Diraison F, Beylot M. Am J Physiol 1998;274:E321–E327.
20. Hellerstein M, Christiansen M, Kaempfer S, et al. J Clin Invest 1991;87:1841–1852.
21. Faix D, Neese R, Kletke C, et al. J Lipid Res 1993;34:2063–2075.
22. Aarsland A, Chinkes D, Wolfe R. Am J Clin Nutr 1997;65:1174–1182.
23. Hudgins LC, Hellerstein MK, Seidman C, et al. J Clin Invest 1996;98:2081–2091.
24. Letexier D, Pinteur C, Large V, et al. J Lipid Res 2003;44:2127–2134.
25. Diraison F, Dusserre E, Vidal H, et al. Am J Physiol 2002;282:E46–E51.
26. Forcheron F, Cachefo A, Thevenon S, et al. Diabetes 2002;51:3486–3491.
27. Diraison F, Beylot M, Moulin P. Diabetes Metab 2003;29:478–485.
28. Shrago E, Spennetta T, Gordon E. J Biol Chem 1969;244:905–912.
29. Foufelle F, Ferré P. Biochem J 2002;366:377–391.
30. Foretz M, Guichard C, Ferre P, et al. Proc Natl Acad Sci USA 1999;96:12,737–12,742.
31. Joseph S, Laffitte B, Patel P, et al. J Biol Chem 2002;29:11,019–11,025.
32. Uyeda K, Yamashita H, Kawaguchi T. Biochem Pharmacol 2002;63:13,476–13,478.
33. Towle H. Proc Natl Acad Sci USA 2001;98:13,476–13,478.
34. Kawaguchi T, Takenoshita M, Kabashima T, et al. Proc Natl Acad Sci USA 2001;98:13,710–13,715.
35. Iizuka K, Bruick R, Liang G, et al. Proc Natl Acad Sci USA 2004;101:7281–7286.
36. Kawaguchi T, Osatomi K, Yamashita H, et al. J Biol Chem 2002;277:3829–3835.
37. Moustaid N, Jones B, Taylor J. J Nutr 1996;126:865–870.
38. Claycombe K, Jones B, Standridge M, et al. Am J Physiol 1998;274:R1253–R1259.
39. LeLay S, Lefrere I, Trautwein C, et al. J Biol Chem 2002;277:35,625–35,634.
40. Palmer D, Rutter G, Tavare J. Biochem Biophys Res Commun 2002;291:429–433.
41. Foufelle F, Gouhot B, Pegorier J, et al. J Biol Chem 1992;267:20,543–20,546.

42. He Z, Jiang T, Wang Z, et al. Am J Physiol 2004;287:E424–E430.
43. Letexier D, Peroni O, Pinteur C, et al. Diabetes Metab 2005;31:558–566.
44. Dentin R, Pegorier J, Benhamed F, et al. J Biol Chem 2004;279:20,314–20,326.
45. Fukuda H, Iritani N, Sugimoto T, et al. Eur J Biochem 1999;260:505–511.
46. Minehira K, Vega N, Vidal H, Acheson K, et al. Int J Obes Relat Metab Disord 2004;28:1291–1298.
47. Nadler S, Stoehr J, Schueler K, et al. Proc Natl Acad Sci USA 2000;97:11,371–11,375.
48. Karlsson C, Lindell K, Otosson M, et al. J Clin Endocrinol Metab 1998;83:3925–3929.
49. Engeli S, Gorzelniak K, Kreutz R, et al. J Hypertens 1999;17:555–560.
50. Massiera F, Bloch-Faure M, Celler D, et al. FASEB J 2001;115:2727–2729.
51. Kim S, Dugail I, Stanbridge M, et al. Biochem J 2001;357:899–904.
52. Jones B, Stanbridge M, Moustaid N. Endocrinology 1997;138:1512–1519.
53. Yvan-Charvet L, Even P, Bloch-Faure M, et al. Diabetes 2005;54:991–999.
54. Van Harmelen V, Ariapart P, Hoffstedt J, et al. Obes Res 2000;8:337–341.
55. Reshef L, Olswang Y, Cassuto H, et al. J Biol Chem 2003;278:30,413–30,418.
56. Tanti J, Grillo S, Gremeaux T, et al. Endocrinology 1997;138:2005–2010.
57. Sniderman A, Maslowska M, Cianflone K. Curr Opin Lipidol 2000;11:291–296.
58. Cadoudal T, Leroyer S, Reis AF, et al. Biochimie 2005;87:27–32.
59. Cases S, Stone S, Zheng Y, et al. Proc Natl Acad Sci USA 1998;95:13,018–13,023.
60. Cases S, Stone S, Zhou P, et al. J Biol Chem 2001;276:38,870–38,876.
61. Bell R, Coleman R. Annu Rev Biochem 1980;2:504–513.
62. Leung D. Front Biosci 2001;6:944–953.
63. Agarwal A, Garg A. Trends Endocrinol Metab 2003;14:214–221.
64. Meegalla R, Billheimer J, Cheng D. Biochem Biophys Res Commun 2002 298:317–323.
65. Chen H, Farese RJ. Arterioscler Thromb Vasc Biol 2004;25:482–486.
66. Smith S, Cases S, Jensen D, et al. Nat Genet 2000;25:87–90.
67. Öst A, Örtegren U, Gustavsson J, et al. J Biol Chem 2005;280:5–8.
68. Stroud R, Savage D, Miercke J, et al. FEBS Lett 2003;555:79–84.
69. Kishida K, Shimomura I, Kondo H, et al. J Biol Chem 2001;276:36,251–36,260.
70. Kishida K, Nishizawa H, Shimomura I, et al. J Biol Chem 2001;276:48,572–48,579.
71. Maeda N, Hibuse T, Funahashi T, et al. Proc Natl Acad Sci USA 2004 101:17,801–17,806.
72. Hibuse T, Funahashi T, Maeda N, et al. Proc Natl Acad Sci USA 2005 102:10,993–10,998.
73. Kondo H, Shimomura I, Kishida K, et al. Eur J Biochem 2002;269:1814–1826.
74. Beylot M, Martin C, Laville M, et al. J Clin Endocrinol Metab 1991;73:42–49.
75. Bahr R, Hansson P, Sejersted O. Metabolism 1990;39:993–999.
76. Wolfe R, Herndon D, Jahoor F, et al. N Engl J Med 1987;317:403–408.
77. Frayn K, Fielding B, Karpe F. Curr Opin Lipidol 2005;18:409–415.
78. Yeaman S. Biochem J 2004;379:11–22.
79. Osterlund T, Beussman D, Julenius K, et al. J Biol Chem 1999;274:15,382–15,388.
80. Shen W, Pate S, Hong R, et al. Biochemistry 2000;39:2392–2398.
81. Holm C, Davis R, Osterlund T, et al. FEBS Lett 1994;344:234–238.
82. Laurell H,Grober L, Vindis C, et al. Biochem J 1997;328:137–143.
83. Ray H, Arner P, Holm C, et al. Diabetes 2003;52:1417–1422.
84. Osterlund T, Contreras J, Holm C. FEBS Lett 1997;403:259–262.
85. Large V, Arner P. Diabetes Metab 1998;24:409–418.
86. Arner P, Hellström L, Warhenberg H, et al. J Clin Invest 1990;86:1595–1600.
87. Giorgino F, Laviola L, Eriksson J. Acta Physiol Scand 2005;183:13–30.
88. Garton A, Campbell D, Cohen P, et al. FEBS Lett 1988;229:68–72.
89. Garton A, Campbell D, Carling D, et al. Eur J Biochem 1989;179:249–254.
90. Daval M, Diot-Dupuy F, Bazin R, et al. J Biol Chem 2005;280:25,250–25,257.
91. Anthonsen M, Ronnstand L, Wernstedt D, et al. J Biol Chem 1998;273:215–221.
92. Greenberg A, Shen W, Mullro K, et al. J Biol Chem 2001;276:45,456–45,461.
93. Vossier S, Emmison N, Borthwick A, et al. Cell 1997;89:73–82.
94. Lafontan M, Moro C, Sengenes C, et al. Arterioscler Thromb Vasc Biol 2005;24:2032–2042.
95. Hagström-Toft E, Bolindr J, Eriksson S, et al. Diabetes 1995;44:1170–1175.

96. Kitamuta T, Kitamura Y, Kuroda S, et al. Moll Cell Biol 1999;19:6286–6296.
97. Wood S, Emmison N, Borthwick A, et al. Biochem J 1993;295:531–535.
98. Engfeldt P, Hellmer J, Wahrenberg H, et al. J Biol Chem 1988;263:15,553–15,560.
99. Zhang J, Hupfeld C, Taylor S, et al. Nature 2005;437:569–573.
100. Zimmermann R, Haemmerle G, Wagner E, et al. J Lipid Res 2003;44:2089–2099.
101. Osuga J, Ishibashi S, Oka T, et al. Proc Natl Acad Sci USA 2000;97:787–792.
102. Okazaki H, Osuga J, Tamura Y, et al. Diabetes 2002;51:3368–3375.
103. Haemmerle G, Zimmerman R, Hayn M, et al. J Biol Chem2002;277:4806–4815.
104. Zimmermann R, Strauss J, Haemmerle G, et al. Science 2004;306:1383–1386.
105. Villena J, Roy S, Sarkadi-Nagy E, et al. J Biol Chem 2004;2004:47,066–47,075.
106. Jenkins C, Mancuso D, Yan W, et al. J Biol Chem 2004;279:48,968–48,975.
107. Langin D, Dicker A, Tavernier G, et al. Diabetes 2005;54:3190–3197.
108. Zecher Z, Strauss J, Haemmerle G, et al. Curr Opin Lipidol 2005;16:333–340.
109. Soni K, Lehner R, Metalnikov P, et al. J Biol Chem 2004 279:40,683–40,689.
110. Lehner R, Vance D. Biochem J 1999;343:1–10.
111. Lehner R, Verger R. Biochemistry 1997;36:1861–1868.
112. Baulande S, Lasnier F, Lucas M, et al. J Biol Chem 2001;276:33,336–33,344.
113. Liu Y, Moldes M, Bastard J, et al. J Clin Endocrinol Metab 2004;89:2684–2689.
114. Lake A, Sun Y, Li J, et al. J Lipid Res 2005;46:2477–2487.
115. Brasaemle D, Levin D, Adler-Wailes D, et al. Biochim Biophys Acta 2000;1493:251–262.
116. Sue C, Sztalryd C, Contreras J, et al. J Biol Chem 2003;41:2408–2416.
117. Londos C, Sztalryd C, Tansey J, et al. Biochimie 2005;87:45–49.
118. Brasaemle D, Barber T, Wolins N, et al. J Lipid Res 1997;38:2249–2263.
119. Gao J, Serrero G. J Biol Chem 1999;274:16,825–16,830.
120. Londos C, Gruia-Gray J, Brasaemle D, et al. Int J Obesity 1996;20:S97–S101.
121. Forcheron F, Legedz L, Chinetti G, et al. Arterioscler Thromb Vasc Biol 2005;25:1711–1717.
122. Brasaemble D, Barber T, Kimmel A, et al. J Biol Chem 1997;272:9378–9387.
123. Dalen K, Schoonjans K, Ulven S, et al. Diabetes 2004;53:1243–1252.
124. Brasaemle D, Rubin B, Harten I, et al. J Biol Chem 2000;275:38,486–38,493.
125. Sztalryd C, Xu G, Dorward H, et al. J Cell Biol 2003;161:1093–1103.
126. Clifford G, Londos C, Kraemer F, et al. J Biol Chem 2000;275:5011–5015.
127. Martinez-Botas J, Anderson J, Tessier D, et al. Nat Genet 2000;26:474–479.
128. Tansey J, Sztalryd C, Gruia-Gray J, et al. Proc Natl Acad Sci USA 2001;98:6494–6499.
129. Kern P, Di Gregorio G, Lu T, et al. J Clin Endocrinol Metab 2004;89:1352–1358.
130. Mottagui-Tabar S, Ryden M, Lofgren P, et al. Diabetologia 2003;16:789–797.
131. Wang Y, Sullivan S, Trujillo M, et al. Obes Res 2003;11:930–936.
132. Le Lay S, Ferré P, Dugail I. Biochem Soc Trans 2004;32:103–106.
133. Kovanen P, Nikkila E, Miettenen T. J Lipid Res 1975;16:211–223.
134. Tondu A, Robichon C, Yvan-Charvet L, et al. J Biol Chem 2005;280:33,536–33,540.
135. Dagher G, Donne N, Klein C, et al. J Lipid Res 2003;44:1811–1820.
136. Le Lay S, Kreif S, Farneir C, et al. J Biol Chem 2001;276:16,904–16,910.
137. Le Lay S, Robichon C, Le Liepvre X, et al. J Lipid Res 2003;44:1499–1507.

3 Leptin

Malaka B. Jackson and Rexford S. Ahima

Abstract

Recent evidence has shown that adipose tissue is an active participant in maintaining energy and glucose homeostasis and plays crucial roles in controlling neuroendocrine, autonomic, and immune functions. Leptin is a hormone secreted by adipose tissue. Deficiency of leptin or its receptor results in hyperphagia, morbid obesity, insulin resistance, hyperlipidemia, hypothalamic hypogonadism, and immunosuppression. These abnormalities are reversed by leptin treatment in patients with congenital leptin deficiency or lipodystrophy. In contrast, diet-induced obesity is associated with elevated leptin levels and blunted response to leptin. Leptin resistance has been ascribed to the reduced entry of leptin into the brain or impairment of leptin signal transduction. This chapter focuses on the current understanding of leptin's actions, with particular emphasis on transport across the blood–brain barrier, signaling via JAK-STAT and PI3-kinase, neuropeptide targets, and electrophysiological effects.

Key Words: Adipocyte; adipokine; leptin; hypothalamus; neuropeptide; obesity.

1. ADIPOCYTE BIOLOGY AND LEPTIN

Adipose tissue is a type of loose connective tissue made up of an extensive network of blood vessels, collagen fibers, fibroblasts, and immune cells that surround lipid-laden cells, known as adipocytes. In humans, the principal form of adipose tissue is white adipose tissue, whose adipocytes have an eccentric nucleus and a single lipid droplet *(1)*. Until recently, adipose tissue was viewed primarily as a specialized tissue for storing energy mainly in the form of triglycerides. However, it is clear that adipose tissue plays an active role in energy homeostasis as well as the control of neuroendocrine, autonomic, and immune functions *(1)*. Adipose tissue synthesizes and secretes "adipokines" such as leptin, adiponectin, and resistin. Additionally, adipocytes secrete proinflammatory cytokines, such as tumor necrosis factor (TNF)-α and interleukins, and proteins involved in coagulation and vascular function *(1)*. Furthermore, adipose stromal cells mediate the metabolism of glucocorticoids and sex steroids, which exert profound local and systemic effects on adipogenesis, glucose and lipid metabolism, and cardiovascular function *(1,2)*. This chapter will focus on the biology of leptin—in particular, its role in metabolism and control of neuroendocrine function.

As early as the 1950s, it was proposed that that a factor existed whose circulating levels increased with energy stores and signaled the brain to inhibit feeding and decrease body weight and fat *(3)*. The discovery of recessive mutations of *obese (ob)* and

From: *Nutrition and Health: Adipose Tissue and Adipokines in Health and Disease*
Edited by: G. Fantuzzi and T. Mazzone © Humana Press Inc., Totowa, NJ

diabetes (db) loci in mice, coupled with hypothalamic lesion and parabiosis studies, led to the suggestion that the *ob* locus encoded a circulating "satiety factor," whereas the *db* locus was required for the response to this factor *(4,5)*. These predictions were confirmed more than four decades later by the discovery of the leptin (*lep*) and leptin receptor (*lepr*) genes *(6–9)*. Leptin is expressed mainly in adipose tissue, although low levels have been found in the placenta, skeletal muscle, gastric fundic mucosa, and mammary epithelium *(10)*. Leptin has a relative mass of approx 16 kDa and circulates as both a bound and a free hormone, the latter likely representing the bioavailable hormone. The primary and complex structures of leptin are highly conserved in mammals *(6,10)*. Leptin concentration is dependent on the quantity of stored energy in fat, as well as the status of energy balance *(10)*. As such, plasma leptin is higher in obese than in lean individuals, falls rapidly during fasting, and increases after feeding. This nutritional regulation of leptin is controlled at least partly by insulin *(10)*.

The concept that leptin acts as an "antiobesity hormone" was based on two key observations: first, rodents and humans deficient in either leptin or its receptors developed insatiable appetite and morbid obesity, associated with insulin resistance, diabetes, and hyperlipidemia *(6,11)*. Second, leptin administered peripherally and more potently via intracerebroventricular (icv) injection in rodents, decreased food intake, body weight and fat content, consistent with the idea of negative feedback action in the brain *(12–14)*. The latter was supported by the localization of leptin receptors in the hypothalamus and various CNS nuclei associated with feeding and energy balance *(15)*. However, it was soon obvious that leptin expression in adipose tissue and plasma leptin levels increased in obesity, and the rise in endogenous leptin levels did not prevent obesity *(16)*. In the absence of obvious leptin receptor abnormalities, leptin treatment in humans and rodents with "common" (diet-induced) obesity has little effect to suppress feeding or reduce weight *(16)*. This blunted response to leptin, termed "leptin resistance," may underlie the progression of obesity, impaired insulin action, diabetes, and elevated lipid levels, characteristic of "metabolic syndrome" *(16)*. Leptin resistance may result from diminution of leptin transport across the blood–brain barrier (BBB) or impairment of leptin signal transduction in the brain *(16)*.

There is strong evidence showing that leptin acts mainly as a "starvation hormone" *(17)*. Leptin falls rapidly during fasting or chronic food deprivation, leading to suppression of thermogenesis and of thyroid, sex, and growth hormones *(10,17)*. Moreover, low leptin levels during fasting mediates immunosuppression, torpor, and activation of the hypothalamic–pituitary–adrenal axis (HPA) in rodents *(18,19)*. These fasting-induced changes mediated by low leptin resemble the metabolic phenotype of congenital leptin deficiency in $Lep^{ob/ob}$ and or lack of leptin response in $Lepr^{db/db}$ mice, suggesting that leptin deficiency is perceived as a state of unmitigated starvation, leading to adaptations, such as hyperphagia, decreased metabolic rate, and changes in hormones, intended to restore energy balance *(16)*. As predicted, leptin treatment prevents the suppression of reproductive axis, thyroid and growth hormones and energy expenditure, and hyperphagia typical of food deprivation or sustained weight reduction *(17–22)*.

2. LEPTIN'S EFFECTS ON THE NEUROENDOCRINE AXIS

Congenital leptin deficiency in rodents and humans is associated with hypothalamic hypogonadism and lack of pubertal development *(10,23,24)*. Leptin treatment restores gonadotropin and sex steroid levels and puberty, confirming a primary role in reproduction *(10,23,24)*. However, studies in *Lep*$^{ob/ob}$ mice indicate that leptin is not needed for gestation, parturition, or lactation *(25)*. Leptin exerts a permissive action to restore normal hypothalamic–pituitary–gonadal axis function during caloric deprivation *(26–29)*. In rodents and nonhuman primates, these actions involve stimulation of gonadotropins as well as interaction with other metabolic signals *(28,29)*. Women with hypothalamic amenorrhea have low leptin levels associated with blunted luteinizing hormone (LH) pulses *(20,30)*. Leptin administration in these patients restores LH pulses and menstrual cycles, confirming that leptin signals the status of energy stores to the neuroendocrine axis *(20)*.

Lipodystrophy is characterized by generalized or partial loss of adipose tissue, excess lipid accumulation in nonadipose tissues (steatosis), hypoleptinemia, insulin resistance, diabetes, dyslipidemia, hyperandrogenism, and amenorrhea, the last being of a central origin *(31)*. In affected females, treatment with recombinant methionyl human leptin in physiological doses reduced steatosis, insulin resistance, glucose and plasma lipids, and free testosterone, increased sex hormone binding globulin, and produced robust LH pulses and integrated levels *(32,33)*. Importantly, most patients who had amenorrhea prior to leptin therapy developed normal menses after treatment *(33)*. Nonetheless, polycystical ovaries associated with lipodystrophy were not reversed by leptin *(33)*. Serum testosterone tended to increase and sex hormone binding globulin increased following leptin treatment therapy in male lipodystrophic patients *(33)*. In contrast to congenital leptin deficiency, hypoleptinemia did not inhibit pubertal development in lipodystrophy, pointing to a crucial role of other metabolic factors *(23,24,33)*.

The control of thyroid hormone is coupled to energy balance *(34)*. Under normal fed conditions, a fall in thyroid hormone stimulates the synthesis and secretion of thyrotropin-releasing hormone (TRH) and thyroid-stimulating hormone (TSH). Conversely, a rise in thyroid hormone suppresses TRH and TSH. This feedback response is disrupted during fasting and illness, culminating in low thyroxine (T4) and triiodotyronine (T3) levels, low or normal TSH, and suppression of TRH. The dampening of the hypothalamic–pituitary–thyroid axis response during caloric deprivation has been termed "euthyroid sick syndrome." Leptin deficiency reduces the response of pituitary thyrotropes to TRH stimulation, whereas leptin treatment reverses the suppression of T3, TSH, and TRH levels in congenital leptin deficiency and fasting *(34)*. Studies indicate that low leptin levels during fasting control the thyroid axis directly through hypophysiotrophic TRH neurons in the paraventricular nucleus (PVN) or indirectly through neurons expressing melanocortin 4 receptors (MC4R) *(35–37)*.

Leptin and growth hormone receptors belong to a family of cytokine receptors coupled to the JAK–STAT pathway *(38)*. In rodents, growth hormone is decreased in states of leptin deficiency *(10)*. Growth hormone pulses are diminished during fasting, and restored by leptin replacement *(39,40)*. In contrast, immunoneutralization of leptin decreased growth hormone secretion in fed rats *(40)*. Leptin receptors and STAT3 have been colocalized with growth hormone-releasing hormone (GHRH) and somatostatin, suggesting a direct interaction *(40)*. Leptin infused into the hypothalamus stimulated growth hormone release more robustly in fasted than fed animals; this was associated with an increase in GHRH and a decrease in somatostatin *(41)*.

Koutkia et al. *(41)* investigated the regulation of leptin and growth hormone secretion and pulsatility by sampling plasma overnight every 20 min. There was synchronicity between growth hormone and leptin with a lag time of 39 min *(41)*. Ghrelin is a gastric peptide that stimulates appetite, glucose oxidation, and lipogenesis, in contrast with growth hormone, which promotes lipolysis, glucose production, and insulin secretion *(42)*. Plasma ghrelin levels increase at night in lean individuals and in synchrony with leptin *(42)*. This nocturnal rise of ghrelin is blunted in obesity *(42)*. The HPA is activated in leptin-deficient rodents *(10,11)*. Moreover, elevated leptin levels during fasting or immobilization stress lead to corticotropin-releasing hormone (CRH), adrenocorticotropin (ACTH), and glucocorticoid inhibition *(17,43)*. Diurnal and pulsatile leptin secretions are inversely related to ACTH and cortisol in humans *(44)*, yet the HPA axis is not significantly altered in congenital leptin deficiency, lipodystrophy, or fasting *(21,24)*.

Although prolactin and leptin are both influenced by body fat content, the link between these hormones in humans is unclear *(45)*. In contrast, a constant infusion of leptin increases prolactin in fed rats, and in particular during fasting *(46)*. Because the leptin receptor is very scant in lactotropes and direct leptin infusion into the arcuate nucleus and median eminence stimulates prolactin secretion, it has been proposed that leptin controls prolactin release through a hypothalamic target *(47)*.

3. LEPTIN RECEPTORS AND SIGNAL TRANSDUCTION IN THE BRAIN

Soon after it was shown that low doses of leptin, when administered directly into the brain, decreased food intake and body weight *(14)*, leptin receptors were discovered through expression cloning in mouse brain and by positional cloning of the *db* locus *(7,8)*. The leptin receptor (LEPR) belongs to the cytokine receptor class I superfamily *(38)*. Five alternatively spliced isoforms, a, b, c, d, e, differing in the lengths of their carboxy termini, have been identified (refs. *10,38*; Fig. 1). The short leptin receptor isoform, LEPRa, is expressed in several peripheral tissues, choroid plexus, and brain microvessels, and thought to be involved in the transport of leptin across the BBB or efflux from the brain *(48–50)*. The long leptin receptor (LEPRb) that has intracellular domains necessary for signaling via the JAK–STAT pathway is highly expressed in the hypothalamus—e.g., arcuate (Arc), dorsomedial (DMN), ventromedial (VMN), and ventral premamillary nuclei (PMN). Moderate LEPRb expression is found in the periventricular and lateral hypothalamic areas, and nucleus solitarius and various brainstem nuclei *(51)*. The PVN, which integrates energy balance, neuroendocrine function, and glucose homeostasis, has a very low level of LEPRb *(51)*. LEPRb has been colocalized with neuropeptides involved in energy homeostasis *(52)*. Neuropeptide Y (NPY) and agouti-related protein (AGRP), which stimulate feeding, are present in the same neurons in the medial Arc. An increase in leptin directly suppresses NPY and AGRP. Leptin increases the levels of anorectic peptides, α-melanin-stimulating hormone (α-MSH) derived from proopiomelanocortin (POMC), and cocaine and amphetamine-regulated transcript (CART), in the lateral Arc *(53)*. Second-order neurons that synthesize CRH, TRH, and oxytocin in the PVN are controlled indirectly by leptin targets in the Arc, and mediate the inhibitory effects of leptin on food intake, stimulation of thermogenesis, and neuroendocrine secretion *(52)*. Other orexigenic peptides, such as melanin-concentrating hormone (MCH) and orexins, expressed in the lateral hypothalamus, are inhibited indirectly by leptin *(51)*. Outside the hypothalamus, LEPRb mRNA has also been found in the thalamus and cerebellum *(52)*.

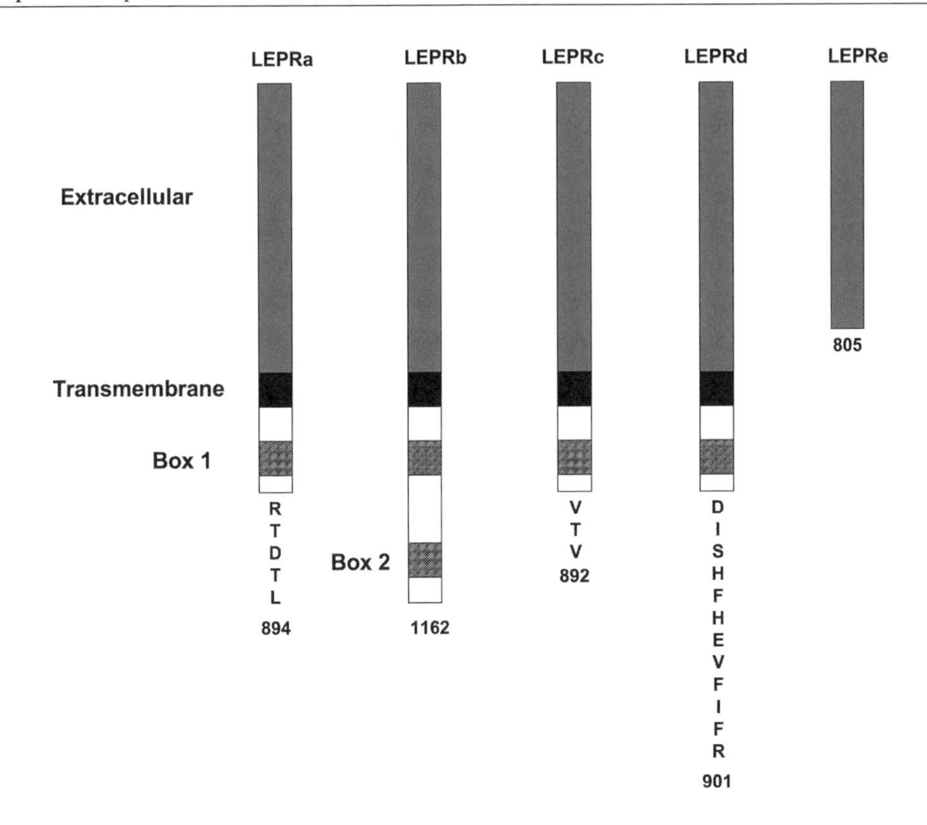

Fig. 1. Leptin receptors. Five leptin receptor isoforms result from alternate splicing of *lepr* mRNA transcript. Receptors LEPRa–LEPRd contain identical extracellular ligand binding and cytoplasmic signaling domains. Despite having the same extracellular N-terminal, each isoform has a different C-terminal. However, unlike the other leptin isoforms, LEPRe has neither a transmembrane nor intracellular domains. LEPRa is the principal "short leptin receptor" and lacks the cytoplasmic domain necessary for signaling through the JAK–STAT pathway. LEPRb, the "long leptin receptor," is primarily responsible for the leptin-mediated effects on energy homeostasis and endocrine function through activation of the JAK–STAT pathway. Terminal amino acid residues for each isoform are represented by the alphabetic code.

Binding of leptin to LEPRb in hypothalamic and brainstem neurons results in rapid activation of intracellular JAK-2, leading to tyrosine phosphorylation of LEPRb on amino acid residues 985 and 1138, which provide binding motifs for src homology 2 (SH2)-domain containing proteins—i.e., STAT-3 and SH-2-domain-phosphotyrosine phosphatase (SHP-2) (ref. *54*; Fig. 2). STAT-3 binds to Y1138, becomes tyrosine-phosphorylated by JAK-2, then dissociates and forms dimers in the cytoplasm, which are translocated to the nucleus to regulate gene transcription (Fig. 2). The importance of Y1138 has been demonstrated in mice by replacing this amino acid with serine *(55)*. The Y1138S (LeprS1138) mutation disrupted STAT-3 activation, resulting in hyperphagia, impairment of thermoregulation, and obesity *(55)*. However, in contrast to $Lepr^{db/db}$ mice, Y1138S mutation did not affect sexual maturation and growth, and glucose levels were lower, pointing to a specific physiological role of this domain *(55)*.

Leptin regulates insulin receptor substrate 1 (IRS-1) and IRS-2, mitogen-activated protein kinase, extracellular-regulated kinase, Akt, and phosphatidylinositol-3 (PI3)-kinase

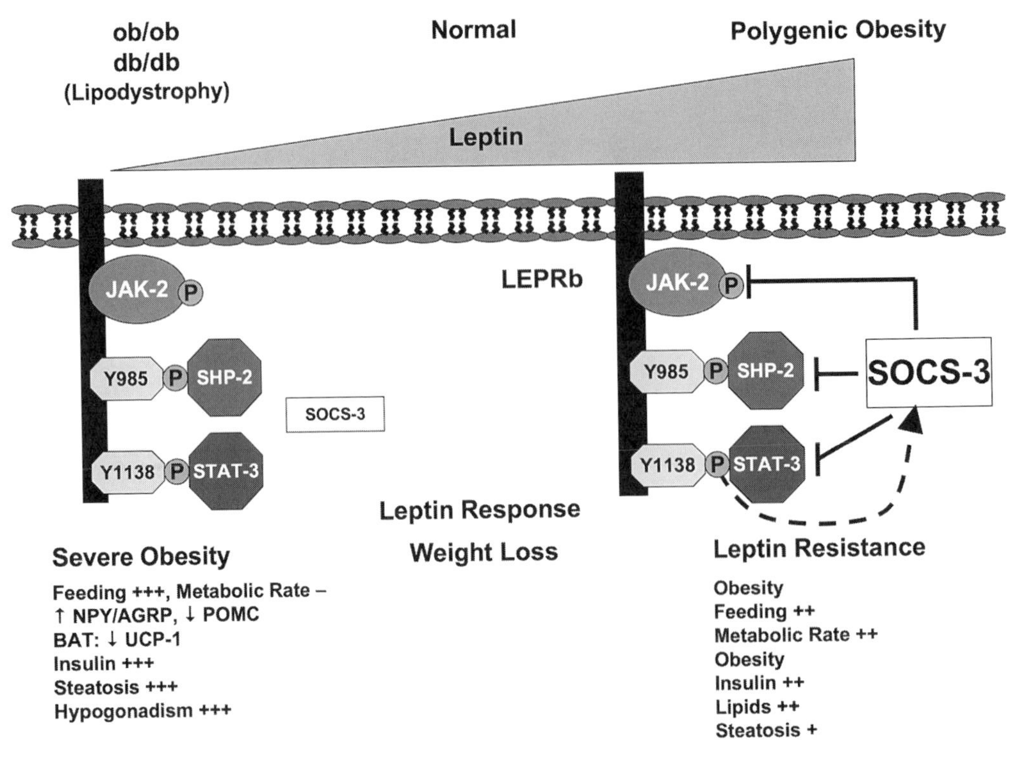

Fig. 2. Leptin signal transduction. Leptin binding to the hypothalamic LEPRb results in activation and autophosphorylation of JAK-2. Subsequently STAT-3 is phosphorylated then translocated to the nucleus, where it acts together with a variety of transcription factors to regulate the expression of neuropeptides and other genes. In congenital or acquired leptin deficiency or leptin receptor defect, failure to activate LEPRb results in increased orexigenic neuropeptide expression (NPY and AGRP), and decreased anorexigenic neuropeptide expression (POMC), and uncoupling protein (UCP)-1 expression in brown adipose tissue (BAT). The net effect is hyperphagia and a decrease in metabolic rate, resulting in severe obesity, markedly elevated insulin levels and steatosis. Another manifestation of leptin deficiency is hypothalamic hypogonadism. The fall in leptin during fasting triggers hyperphagia, metabolic and neuroendocrine responses similar to congenital leptin deficiency. An increase in leptin within the physiological range inhibits feeding and increases the metabolic rate, resulting in weight loss. In contrast, polygenic (diet-induced) obesity results in elevated leptin levels, which promote phosphorylation of Tyr1138 on the intracellular domain of LEPRb, resulting in inhibition of leptin signaling by induction of SOCS-3 and inhibition of JAK–STAT and SHP-2 pathways. These mechanisms result in "leptin resistance" that blunts the ability of leptin to suppress feeding and increase metabolic rate, resulting in obesity. The degree of obesity, steatosis, insulin resistance, and hyperlipidemia in diet-induced obesity is less severe than that seen in congenital leptin deficiency or lipodystrophy.

through LEPRb, raising the possibility of crosstalk between leptin and insulin *(56)*. Leptin enhances IRS-2-mediated activation of PI3-kinase in the hypothalamus, concomitant with its ability to inhibit food intake *(56,57)*. In contrast, blockade of PI3-kinase activity prevents the anorectic action of leptin *(56)*.

Bjørbaek et al. *(58,59)* first proposed a role of SOCS-3 in leptin signaling, based on the observation that leptin rapidly increased the levels of this cytokine mediator in hypothalami of *Lep*$^{ob/ob}$ but not *Lepr*$^{db/db}$ mice. *In situ* hybridization for SOCS-3 mRNA in the rat and mouse brains showed that leptin-induced SOCS-3 expression colocalized with LEPRb and neuropeptides in the hypothalamus *(58)*. Furthermore, expression of SOCS-3 prevented the tyrosine phosphorylation of LEPRb and downstream signaling *(59)*. STAT-3 DNA binding elements are present in the *socs-3* promoter, and because the lack of the STAT-3 binding site on LEPRb prevents induction of SOCS-3 mRNA by leptin, this indicates that leptin stimulates *socs-3* transcription via the STAT-3 pathway *(59)*. The crucial role of SOCS-3 as a negative regulator of leptin signaling was demonstrated in two studies *(60,61)*. SOCS-3 haploinsufficiency increased leptin sensitivity and prevented diet-induced obesity in mice *(60)*. More specifically, neuron-specific ablation of SOCS-3 enhanced leptin sensitivity, resulting in activation of STAT3, increase in hypothalamic POMC expression, reduction in food intake, and resistance to obesity, hyperlipidemia, and diabetes *(61)*.

Protein tyrosine phosphatase 1B (PTP1B), an insulin receptor phosphatase that inhibits insulin signaling, was implicated in leptin action, based on the finding that mice lacking PTP1B were less hyperphagic and resistant to obesity despite having low serum leptin levels *(62–64)*. In vitro studies revealed that PTP1B directly inhibited JAK2 kinase, and leptin-induced tyrosine phosphorylation of JAK-2 and STAT-3 was attenuated in cells overexpressing PTP1B *(63)*. PTP1B mRNA is colocalized with STAT-3 and neuropeptides in the Arc and various hypothalamic nuclei *(64)*. Importantly, STAT-3 phosphorylation in the hypothalamus is enhanced following leptin treatment in mice lacking PTP1B (PTP1B–/–), suggesting that PTP1B inhibits the signal transduction of leptin in the brain *(63,64)*. This idea was tested by comparing the effects of leptin treatment on energy balance in wild-type (WT) and PTP1B –/– mice. As predicted, PTP1B –/– and heterozygotes PTP1B+/– mice exhibited greater leptin sensitivity than WT *(63)*.

4. IMPLICATIONS OF LEPTIN SIGNAL TRANSDUCTION

As mentioned earlier, the fall in leptin during fasting is a potent signal to the brain to increase feeding, reduce energy expenditure, and mediate changes in hormone levels designed to conserve energy *(10,17)*. Low leptin level directly stimulates the expression of NPY and AGRP in the Arc, and indirectly increases MCH in the lateral hypothalamus, leading to hyperphagia and restoration of body weight *(10,52)*. Concurrently, low leptin disinhibits POMC neurons in the Arc, thereby decreasing the synthesis and release of the anorectic peptide α-MSH *(10,52)*. α-MSH normally acts on melanocortin 3 and 4 receptors (MC3/4-R) in the PVN and various hypothalamic nuclei to mediate the satiety effect of leptin *(52)*. α-MSH is antagonized by AGRP, resulting in hyperphagia and weight increase *(52)*. CART, another potent inhibitor of feeding, is directly suppressed by low leptin level during fasting *(52)*. NPY/AGRP and POMC/CART neurons in the Arc send extensive projections to the PVN, perifornical, and lateral hypothalamic areas *(52)*. There are compelling data in rodents showing that leptin controls energy balance and the neuroendocrine axis indirectly via CRH, TRH, vasopressin, and oxytocin producing neurons in the PVN *(52)*. The fall in leptin during fasting decreases TRH mRNA expression in the PVN, resulting in suppression of TSH and thyroid hormone *(34,36)*. Reduced leptin level also suppresses brown adipose tissue thermogenesis (ref. *19*; Fig. 2).

In contrast to the robust response to energy deprivation mediated centrally by low leptin, an increase in leptin in the overfed or obese range has little effect on neuronal or physiological functions *(65)*. This resistance to leptin in diet-induced obesity may involve inhibition of leptin signal transduction through leptin-mediated activation of SOCS-3 or PTP1B *(59,63,67)*. Expression of neuropeptide targets of leptin such NPY and POMC have been shown to be dysregulated and the transport of leptin across the BBB is impaired in obese rodents, but whether these are relevant to the onset or progression of human obesity remain to be determined *(68–70)*.

AMP-activated protein kinase (AMPK) is another CNS leptin target of interest *(71)*. AMPK is phosphorylated and activated in the hypothalamus in response to fasting, leading to increased fatty acid oxidation and inhibition of anabolic pathways in peripheral tissues *(71)*. Conversely, feeding inactivates AMPK and promotes fatty acid synthesis. In the hypothalamus, AMPK is colocalized with STAT-3, NPY, and other peptides implicated in energy balance *(71)*. The activity of AMPK appears to be linked to feeding via the MC4R.

Aside from the regulation of gene expression, leptin has rapid effects on neurotransmission and neuropeptide secretion *(72–75)*. Initial studies showed that leptin rapidly inhibited NPY secretion from hypothalamic explants *(73)*. Subsequently, leptin was shown to depolarize hypothalamic POMC neurons through a nonspecific cation channel, and to decrease the inhibitory tone of γ-amino butyric acid on POMC neurons *(74,75)*. Conversely, leptin hyperpolarized and inactivated NPY neurons in the Arc *(74,75)*. The fall in leptin during fasting increases the action potential frequency in NPY/AGRP neurons *(76)*. The latter is blunted by leptin treatment *(76)*. Similar to fasting, $Lep^{ob/ob}$ and $Lepr^{db/db}$ mice, which exhibit hyperphagia, have increased spike frequency in NPY/AGRP *(76)*. Leptin inhibits the activity of NPY/AGRP neurons in $Lep^{ob/ob}$ but not $Lepr^{db/db}$, confirming a direct action of leptin *(76)*. Significant differences in excitatory and inhibitory synapses in the Arc related to feeding have been demonstrated between $Lep^{ob/ob}$ and WT mice *(77)*. These changes are reversed by leptin treatment, suggesting an important role in synaptic plasticity *(77)*. Leptin also activates adenosine triphosphate (ATP)-sensitive potassium (K^+) channels, which may play a role in the hyperpolarization of VMN neurons through activation of IRS-associated PI3-kinase activity *(78,79)*. Electrophysiological effects of leptin have also been noted in the supraoptic nucleus as well as vagal afferents in the gastrointestinal tract.

Leptin increases insulin sensitivity in peripheral tissues via a CNS mechanism *(80–82)*. Our studies and others have demonstrated a profound trophic action of leptin in the brains of adult $Lep^{ob/ob}$ mice *(83,84)*. Leptin deficiency is associated with deficits in brain weight and neuronal and glial proteins, some of which are restored by leptin treatment *(83,84)*. Analogous to the mouse studies, treatment of leptin-deficient humans with recombinant methionyl human leptin not only decreases body weight, but also increases the volume of the anterior cingulate gyrus, inferior parietal lobule, and the cerebellum, within 6 mo *(85)*. These brain volume increases were maintained over 18 mo, demonstrating that leptin can have sustained effects on the human brain *(85)*. Bouret et al. *(86)* reported that projections from the Arc to PVN were disrupted in $Lep^{ob/ob}$ mice. Leptin treatment reversed this defect in neonatal $Lep^{ob/ob}$ but not adults, indicating that leptin plays a neurotrophic role during a critical stage of hypothalamic development *(86)*.

5. LEPTIN ACTION IN PERIPHERAL TISSUES

Ablation of the Arc, $Lepr^{db/db}$, or loss of LEPRb in neurons and specifically in POMC neurons have all confirmed that leptin acts primarily in the brain *(36,87–90)*.

Nonetheless, numerous studies have demonstrated leptin signaling in blood cells, pancreatic β-cells, pituitary, kidney, hepatocytes, muscle, and adipocytes *(18,91–100)*. Apart from immune cells, most of these tissues lack functional LEPRb or at best express very low levels, suggesting that short-form leptin receptors may mediate leptin signaling *(91–100)*. Ex vivo studies of isolated T-lymphocytes from mice and humans indicate that leptin promotes cellular survival and enhances immunity, especially during starvation *(91)*. Leptin inhibits insulin secretion from isolated pancreatic islets, although the opposite effect has been reported *(92,93)*. Furthermore, leptin induces LH and follicle-stimulating hormone (FSH) release from pituitary explants and sympathetic nerve activity to the kidneys *(94,97)*.

Kim et al. *(95)* found that intravenously injected leptin increased STAT1 and STAT3 phosphorylation and, to a lesser extent, MAP-kinase and PI3-kinase, in adipose tissue. Leptin did not affect signaling in *Lepr*$^{db/db}$ mice, supporting a role for LEPRb in adipocyte leptin action *(95)*. The latter is consistent with the presence of LEPR on human and rodent adipocytes *(96)*. Leptin has no direct effect on glucose uptake in adipocytes, but induces lipolysis in adipose explants and isolated adipocytes *(99–101)*. Furthermore, leptin antagonizes the effects of insulin to inhibit lipolysis; this response is abolished in Zucker *fa/fa* rats or *Lepr*$^{db/db}$ mice, both of which lack functional leptin receptors *(96,100)*. Direct effects of leptin on hepatic gluconeogenesis have been reported *(102)*; however, a major role of leptin in the liver is doubtful, given the apparently normal phenotype of mice with targeted ablation of LEPR in the liver *(88)*.

Minokoshi et al. *(103)* demonstrated a biphasic action of leptin on muscle after intravenous injection. The initial increase in AMPK activity in soleus muscle occurred rapidly within 15 min, and was not affected by sympathetic blockade *(103)*. In contrast, a later, more sustained, increase in AMPK activation (60 min to 6 h) was mimicked by intrahypothalamic leptin injection and abolished by sympathetic blockade *(103)*. A direct action of leptin was confirmed by incubating soleus muscle with and without leptin and demonstrating a robust leptin-dependent stimulation of AMPK activity. Direct or indirect activation of muscle LEPRb by leptin results in phosphorylation and activation of AMPK *(103)*. AMPK phosphorylates acetyl-CoA carboxylase (ACC), leading to inhibition of ACC activity and thus decreasing formation of malonyl-CoA, which in turn disinhibits carnitine palmitoyltransferase 1 (CPT-1), a critical step for translocation of fatty acids into mitochondria to undergo β-oxidation *(104)*. It has been proposed that this leptin–AMPK pathway may play a role in protecting nonadipocytes from lipid accumulation (steatosis) and lipotoxicity *(105)*. Obesity and aging are associated with leptin resistance, leading to steatosis, lipotoxicity, and pancreatic β-cell failure, diabetes, cardiomyocellular damage, and dysfunction of various organs *(105)*.

6. CONCLUSIONS

Major advances have been made over the past decade in understanding the actions of leptin on energy homeostasis, neuroendocrine function, and a variety of physiological functions. Although the signaling mechanisms of leptin have been delineated in cell lines and rodents, crucial questions remain as to how leptin acts specifically through JAK, STAT3, SCOS-3, PI3K, and AMPK to exert these diverse roles in humans; how the leptin signal integrates with hypothalamic and various CNS neurons; the mechanims underlying leptin resistance in obesity; and how leptin and insulin interact in the brain and peripheral tissues leading to metabolic abnormalities associated with obesity. Characterization of the biology of leptin in normal and disease states could benefit the diagnosis and treatment of obesity and related metabolic disorders.

REFERENCES

1. Ahima RS, Flier JS. Trends Endocrinol Metab 2000;11:327–332.
2. Kershaw EE, Flier JS. J Clin Endocrinol Metab 2004;89:2548–2556.
3. Kennedy GC. Proc R Soc Lond B Biol Sci 1953;140:578–596.
4. Hervey G. J Physiol 1959;145:336–352.
5. Coleman DL. Diabetologia 1978;14:141–148.
6. Zhang Y, Proenca R, Maffei M, et al. Nature 1994;372:425–432.
7. Chen H, Charlat O, Tartaglia LA, et al. Cell 1996;84:491–495.
8. Tartaglia LA, Dembski M, Weng X, et al. Cell 1995;83:1263–1271.
9. Isse N, Ogawa Y, Tamura N, et al. J Biol Chem 1995;270:27,728–27,733.
10. Ahima RS, Flier JS. Annu Rev Physiol 2000;62:413–437.
11. Friedman JM, Halaas JL. Nature 1998;395:763–770.
12. Halaas JL, Gajiwala KS, Maffei M, et al. Science 1995;269:543–546.
13. Pelleymounter MA, Cullen MJ, Baker MB, et al. Science 1995;269:540–543.
14. Campfield LA, Smith FJ, Guisez Y, et al. Science 1995;269:546–549.
15. Ahima RS, Saper CB, Flier JS, et al. Front Neuroendocrinol 2000;21:263–307.
16. Flier JS. Obes Cell 2004;116:337–350.
17. Ahima RS, Prabakaran D, Mantzoros C, et al. Nature 1996;382:250–252.
18. Lord GM, Matarese G, Howard JK, et al. Nature 1998;394:897–901.
19. Gavrilova O, Leon LR, Marcus-Samuels B, et al. Proc Natl Acad Sci USA 1999;96:14,623–14,628.
20. Welt CK, Chan JL, Bullen J, et al. N Engl J Med 2004;351:987–997.
21. Chan JL, Heist K, DePaoli AM, et al. J Clin Invest 2003;111:1409–1421.
22. Rosenbaum M, Goldsmith R, Bloomfield D, et al. J Clin Invest 2005;115:3579–3586.
23. Farooqi IS, Jebb SA, Langmack G, et al. N Engl J Med 1999;341:879–884.
24. Farooqi IS, Matarese G, Lord GM, et al. J Clin Invest 2002;110:1093–1103.
25. Mounzih K, Qiu J, Ewart-Toland A, et al. Endocrinology 1998;139:5259–5262.
26. Ahima RS, Dushay J, Flier SN, et al. J Clin Invest 1997;99:391–395.
27. Chehab FF, Mounzih K, Lu R, et al. Science 1997;275:88–90.
28. Barash IA, Cheung CC, Weigle DS, et al. Endocrinology 1996;137:3144–3147.
29. Finn PD, Cunningham MJ, et al. Endocrinology 1998;139:4652–4662.
30. Laughlin GA, Yen SS. J Clin Endocrinol Metab 1997;82:318–321.
31. Garg A. N Engl J Med 2004;350:1220–1234.
32. Oral EA, Simha V, Ruiz E, et al. N Engl J Med 2002;346:570–578.
33. Musso C, Cochran E, Javor E, et al. Metabolism 2005;54:255–263.
34. Legradi G, Emerson CH, Ahima RS, et al. Endocrinology 1997;138:2569–2576.
35. Guo F, Bakal K, Minokoshi Y, et al. Endocrinology 2004;145:2221–2227.
36. Legradi G, Emerson CH, Ahima RS, et al. Neuroendocrinology 1998;68:89–97.
37. Fekete C, Marks DL, Sarkar S, et al. Endocrinology 2004;145:4816–4821.
38. Tartaglia LA. J Biol Chem 1997;272:6093–6096.
39. Carro E, Senaris R, Considine RV, et al. Endocrinology 1997;138:2203–2206.
40. Carro E, Senaris RM, Seoane LM, et al. Neuroendocrinology1999;69:3–10.
41. Koutkia P, Canavan B, Johnson ML, et al. Am J Physiol Endocrinol Metab 2003;285:E372–E379.
42. Yildiz BO, Suchard MA, Wong ML, et al. Proc Natl Acad Sci USA 2004;101:10,434–10,439.
43. Heiman ML, Ahima RS, Craft LS, et al. . Endocrinology 1997;138:3859–3863.
44. Licinio J, Mantzoros C, Negrao AB, et al. Nat Med 1997;3:575–579.
45. Kopelman PG. Int J Obes Relat Metab Disord 2000;24(Suppl2):S104–S108.
46. Watanobe H, Schioth HB, Suda T. Brain Res 2000;887:426–431.

47. Watanobe H, Suda T, Wikberg JE, et al. Biochem Biophys Res Commun 1999;263:162–165.
48. Bjørbæk C, Elmquist JK, Michl P, et al. Endocrinology 1998;139:3485–3491.
49. Hileman SM, Pierroz DD, Masuzaki H, et al. Endocrinology 2002;143:775–783.
50. Hileman SM, Tornoe J, Flier JS, et al. Endocrinology 2000;141:1955–1961.
51. Elias CF, Kelly JF, Lee CE, et al. J Comp Neurol 2000;423:261–281.
52. Elmquist JK, Maratos-Flier E, Saper CB, et al. Nature Neurosci 1998;1:445–450.
53. Elias CF, Aschkenasi C, Lee C, et al. Neuron 1999;23:775–786.
54. Myers MG, Jr. Recent Prog Horm Res 2004;59:287–304.
55. Bates SH, Stearns WH, Dundon TA, et al. Nature 2003;421:856–859.
56. Niswender KD, Baskin DG, Schwartz MW. Trends Endocrinol Metab 2004;15:362–369.
57. Xu AW, Kaelin CB, Takeda K, et al. J Clin Invest 2005;115:951–958.
58. Bjørbæk C, Elmquist JK, Frantz, JD, et al. Mol Cell 1998;1:619–625.
59. Bjørbæk C, El-Haschimi K, Frantz JD, et al. J Biol Chem 1999;274:30,059–30,065.
60. Howard JK, Cave BJ, Oksanen LJ, et al. Nat Med 2004;10:734–738.
61. Mori H, Hanada R, Hanada T, et al. Nat Med 2004;10:739–743.
62. Elchebly M, Payette P, Michaliszyn E, et al. Science1999;283:1544–1548.
63. Cheng A, Uetani N, Simoncic PD, et al. Dev Cell 2002;2:497–503.
64. Zabolotny JM, Bence-Hanulec KK, Stricker-Krongrad A, et al. Dev Cell 2002;2:489–495.
65. Ahima RS, Kelly J, Elmquist JK, et al. Endocrinology 1999;140:4923–4931.
66. El-Haschimi K, Pierroz DD, Hileman SM, et al J Clin Invest 2000;105:1827–1832.
67. Van Heek M, Compton DS, France CF, et al. J Clin Invest 1997;99:385–390.
68. Bergen HT, Mizuno T, Taylor J, et al. Brain Res 1999;851:198–203.
69. Takahashi N, Patel HR, Qi Y, et al. Horm Metab Res 2002;34:691–697.
70. Banks WA, Farrell CL. Am J Physiol Endocrinol Metab 2003;285:E10–E15.
71. Minokoshi Y, Alquier T, Furukawa N, et al. Nature 2004;428:569–574.
72. Glaum SR, Hara M, Bindokas VP, et al. Mol Pharmacol 1996;50:230–235.
73. Stephens TW, Basinski M, Bristow PK, et al. Nature 1995;377:530–532.
74. Cowley MA, Smart JL, Rubinstein M, et al. Nature 2001;411:480–484.
75. Cowley MA, Cone RD, Enriori P, et al. Ann NY Acad Sci 2003;994:175–186.
76. Takahashi KA, Cone RD. Endocrinology 2005;146:1043–1047.
77. Pinto S, Roseberry AG, Liu H, et al. Science 2004;304:110–115.
78. Spanswick D, Smith MA, Groppi VE, et al. Nature 1997;390:521–525.
79. Spanswick D, Smith MA, Mirshamsi S, et al. Nat Neurosci 2000;3:757–758.
80. Kamohara S, Burcelin R, Halaas JL, et al. Nature 1997;389:374–377.
81. Muzumdar R, Ma X, Yang X, et al. FASEB J 2003;17:1130–1132.
82. Pocai A, Morgan K, Buettner C, et al. Diabetes 2005;54:3182–3189.
83. Ahima RS, Bjørbæk C, Osei S, et al. Endocrinology 1999;140:2755–2762.
84. Steppan CM, Swick AG. Biochem Biophys Res Commun 1999;256:600–602.
85. Matochik JA, London ED, Yildiz BO, et al. J Clin Endocrinol Metab 2005;90:2851–2854.
86. Bouret SG, Draper SJ, Simerly RB. Science 2004;304:108–110.
87. Clement K, Vaisse C, Lahlou N, et al. Nature 1998;392:398–401.
88. Cohen P, Zhao C, Cai X, et al. J Clin Invest 2001;108:1113–1121.
89. Balthasar N, Coppari R, McMinn J, et al. Neuron 2004;42:983–991.
90. Ghilardi N, Ziegler S, Wiestner A, et al. Proc Natl Acad Sci USA 1996;93:6231–6235.
91. Howard JK, Lord GM, Matarese G, et al. J Clin Invest 1999;104:1051–1059.
92. Ishida K, Murakami T, Mizuno A, et al. Regul Pept 1997;70:179–182.
93. Kieffer TJ, Heller RS, Leech CA, et al. Diabetes 1997;46:1087–1093.
94. Yu WH, Kimura M, Walczewska A, et al. Proc Natl Acad Sci USA 1997;94:1023–1028.
95. Kim YB, Uotani S, Pierroz DD, et al Endocrinology 2000;141:2328–2339.
96. Kawaji N, Yoshida A, Motoyashiki T, et al. J Lipid Res 2001;42:1671–1677.
97. Haynes WG, Morgan DA, Djalali A, et al. Hypertension 1999;33:542–547.
98. Berti L, Kellerer M, Capp E, et al Diabetologia 1997;40:606–609.
99. Elimam A, Kamel A, Marcus C. Horm Res 2002;58:88–93.
100. Fruhbeck G, Aguado M, Martinez JA. Biochem Biophys Res Commun 1997;240:590–594.

101. Zierath JR, Frevert EU, Ryder JW, et al. Diabetes 1998;47:1–4.
102. Cohen B, Novick D, Rubinstein M. Science 1996;274:1185–1188.
103. Minokoshi Y, Kim YB, Peroni OD, et al. Nature 2002;415:339–343.
104. Hardie DG. Endocrinology 2003;144:5179–5183.
105. Unger RH. Biochimie 2005;87:57–64.

4 Adiponectin

Aimin Xu, Yu Wang, and Karen S. L. Lam

Abstract

Adiponectin is a glycosylated adipokine selectively secreted from adipocytes. The native adiponectin forms several oligomeric complexes, including trimer, hexamer, and high-molecular-weight (HMW) 12–18 multimers. In the past 5 yr, numerous clinical studies have demonstrated a close association between low plasma levels of adiponectin (hypoadiponectinemia) with obesity-related diseases, including dyslipidemia, atherosclerotic cardiovascular diseases, type II diabetes, hypertension, fatty liver, and certain types of cancers. Mounting experimental evidence shows that adiponectin is an endogenous insulin sensitizer with potent antidiabetic, anti-atherogenic, and anti-inflammatory properties. It also has profound protective actions against hepatic and cardiac injury. Different oligomeric forms of adiponectin might act on different tissue targets and stimulate distinct cellular pathways. Two putative adiponectin receptors (adipoR1 and adipoR2) have recently been cloned. In vitro studies indicate that adipoR1 and adipoR2 mediate the metabolic actions of adiponectin via activation of AMP-activated protein kinase. Although the pathophysiological relevance of adiponectin and its receptors need to be further investigated, this adipokine may represent a novel target for the prevention and treatment of obesity-related pathologies.

Key Words: Adipokine; adiponectin; obesity; metabolic syndrome; diabetes.

1. INTRODUCTION

Adiponectin—also termed ACRP30, AdipoQ, Apm1, and GBP28—is one of the most abundant adipokines produced by adipocytes. This adipokine was first characterized in mice as a transcript selectively expressed during the differentiation of preadipocytes into mature adipocytes *(1)*. The human homolog was subsequently identified as the most abundant transcript in adipose tissue during large-scale random sequencing of a human adipose tissue cDNA library *(2)*. The human adiponectin gene was localized to chromosome 3q27, a region highlighted as a genetic susceptibility locus for type 2 diabetes and metabolic syndrome.

Since 2001, adiponectin has attracted much attention because of its potential antidiabetic, antiatherogenic, and anti-inflammatory activities. Numerous animal experiments and clinical studies have demonstrated the utility of adiponectin as a plasma biomarker of metabolic syndrome and a possible therapeutic target for the treatment of type 2 diabetes and cardiovascular disease. Furthermore, research on adiponectin has recently been expanded to explore its potential roles in chronic liver diseases and cancers. Here,

From: *Nutrition and Health: Adipose Tissue and Adipokines in Health and Disease*
Edited by: G. Fantuzzi and T. Mazzone © Humana Press Inc., Totowa, NJ

we will review the current knowledge on structure and biological functions of adiponectin and discuss its relevance with respect to human health and disease.

2. STRUCTURAL FEATURES OF ADIPONECTIN

2.1. Primary Sequence, Domain Organization, and Post-Translational Modifications

Adiponectin structurally belongs to the soluble defense collagen superfamily sharing significant homology with collagen X, collagen VIII, and complement factor C1q *(3)*. The primary sequence of adiponectin is composed of a signal sequence, a hypervariable NH_2-terminal domain, followed by a collageneous domain comprising of 22 Gly-X-Y repeats and a COOH-terminal C1q-like globular domain (Fig. 1). The detailed structural information for full-length adiponectin is not available at this stage. Nevertheless, the crystal structure for the globular domain of murine adiponectin has been resolved at a resolution of 2.1 Å *(4)*. Notably, this structure reveals an unexpected homology to the tumor necrosis factor (TNF) family of cytokines. Despite the lack of homology at the primary amino acid sequence level, the structural features between TNF-α and globular adiponectin are highly conserved. Both TNF-α and globular adiponectin have a 10-β strand jellyroll folding topology and form bell-shaped homotrimeric oligomers.

Adiponectin is modified at the post-translational level during its secretion from adipocytes *(5)*. Several lysine and proline residues within the collagenous domain are hydroxylated (Fig. 1). Hydroxylysine residues at position 68, 71, 80, and 104 are further modified by a glucosyl α(1–2)galactosyl group, as determined by nuclear magnetic resonance analysis *(6)*. Notably, each of these four lysines and their surrounding consensus motif (GXKGE[D]) are highly conserved across all species of adiponectin. Post-translational modifications on these four lysines have been shown to play important roles in enhancing adiponectin's ability to inhibit gluconeogenesis in hepatocytes *(5)*. In addition, adiponectin was reported to be an α2,8-linked disialic acid-containing glycoprotein *(7)*, although the functional relevance of this post-translational modification remains to be determined.

2.2. Oligomerization of Adiponectin and Its Regulation

In the circulation, adiponectin is present predominantly as three different oligomeric complexes *(8,9)*. The monomeric form of adiponectin has never been detected in native conditions. The basic building block of adiponectin is a tightly associated homotrimer, which is formed via hydrophobic interactions within its globular domains (Fig. 1). Freeze–etch electron microscopy showed the trimer to exhibit a ball-and-stick-like structure containing a large globular sphere, an extended collagen stalk, and a smaller sphere on the opposite end of the stalk *(10)*. Two trimers self-associate to form a disulfide-linked hexamer, which further assembles into a bouquet-like higher molecular-weight (HMW) multimeric complex that consists of 12 to 18 protomers.

The assembly of hexameric and HMW forms of adiponectin depends on the formation of a disulfide bond mediated by an NH_2-terminal conserved cysteine residue within the hypervariable region. Substitution of this cysteine with either alanine or serine leads exclusively to trimer formation *(10–12)*. Notably, adiponectin produced from different sources have different compositions of the oligomeric complexes. The full-length

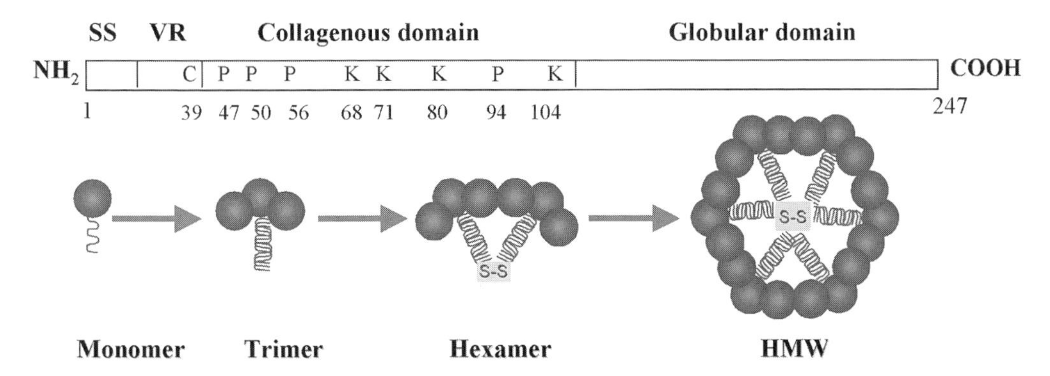

Fig. 1. Structural features of adiponectin protein. **Upper panel:** the schematic diagram of the primary amino acid sequence of murine adiponectin. The numbers below correspond to the sites of hydroxylated prolines or lysines. Hydroxylated lysines are further glycosylated. SS, signal sequence; VR, hypervariable region. **Lower panel:** oligomeric complex formation of adiponectin. The disulfide bridge "S–S" is mediated by cysteine residue 39 at the hypervariable region.

adiponectin expressed in mammalian cells can form all three oligomeric forms, a pattern reminiscent of those observed in the circulation *(8,12)*. On the other hand, full-length adiponectin produced from *Escherichia coli* can form the trimeric and hexameric forms, but not the HMW species. These findings suggest that post-translational modifications at the collagenous domain might contribute to the formation and/or stabilization of the HMW oligomeric complex. An adiponectin mutant lacking hydroxylation and glycosylation on the conserved lysine residues cannot form the HMW oligomers even when expressed in mammalian cells (Wang and Xu, unpublished data).

A growing body of evidence suggests that different oligomers of adiponectin possess distinct biological activities. Earlier studies from Tsao's group showed that the trimeric adiponectin, but not the hexameric and HMW forms, could activate AMP-activated protein kinase (AMPK) in skeletal muscle *(10)*. On the other hand, the HMW oligomeric complex of adiponectin has been the major bioactive form responsible for inhibition of hepatic glucose production *(13)*, and for protection of endothelial cells from apoptosis *(14)*. It has recently been proposed that the oligomeric complex distribution, but not the absolute amount of total adiponectin, determines insulin sensitivity *(15)*.

In both human and rodents, the ratio of HMW to total adiponectin in females is much higher than in males *(9,11)*. This gender difference is primarily attributed to the selective inhibition of testosterone on secretion of the HMW species from adipocytes *(9)*. In addition, acute treatments with insulin and glucose selectively decrease plasma levels of the HMW form and the ratio of HMW/total adiponectin in mice *(11)*. Neverthelss, the mechanism that regulates the formation of adiponectin oligomeric complexes remain largely obscure at this stage.

3. PLEIOTROPIC BIOLOGICAL FUNCTIONS OF ADIPONECTIN

Over the past several years, the functions of adiponectin have been extensively studied in numerous animal models and in vitro systems. It is now appreciated that adiponectin is a multifunctional protein that regulates insulin sensitivity, energy homeostasis, vascular reactivity, inflammation, cell proliferation, and tissue remodeling. Thus

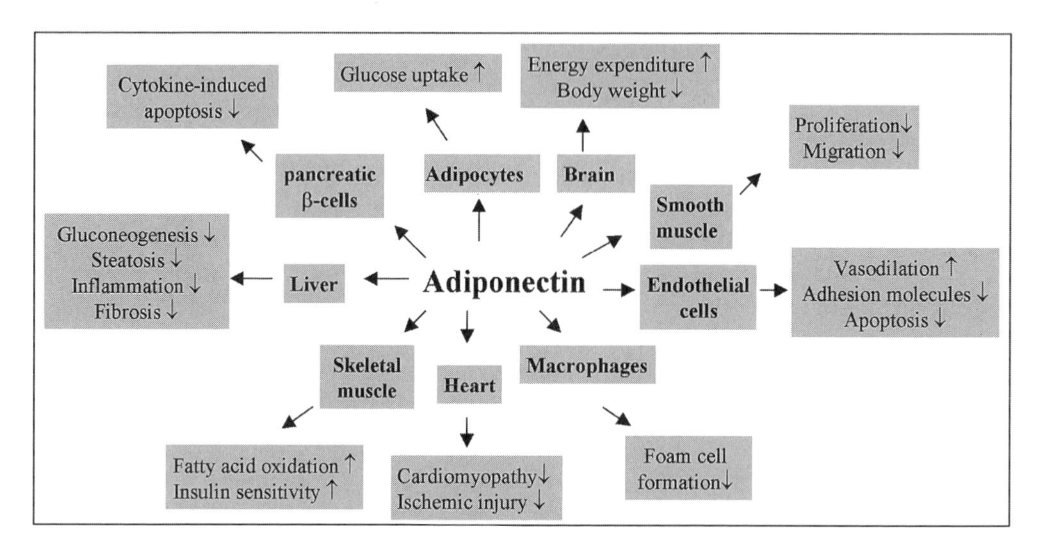

Fig. 2. Major target tissues and biological actions of adiponectin.

far, the identified targets of adiponectin include liver, skeletal muscle, adipose tissue, heart, brain, pancreas, macrophages, and blood vessels (Fig. 2).

3.1. Adiponectin as Insulin-Sensitizing Hormone

The role of adiponectin as an important regulator of insulin sensitivity was first reported by Fruebis and colleagues in 2001 *(16)*. The authors found that injection of a COOH-terminal globular adiponectin into mice acutely decreased postprandial blood glucose levels and enhance lipid clearance via increasing fatty acid β-oxidation in skeletal muscles. This observation was subsequently confirmed and extended by several pharmacological studies using different forms of recombinant adiponectin. Yamauchi's group demonstrated that chronic infusion of full-length or globular adiponectin produced from *E. coli* significantly ameliorated insulin resistance and improved lipid profiles in both lipoatrophic diabetic mice and diet-induced obese mice *(17)*. On the other hand, Berg's group showed that intraperitoneal injection of full-length adiponectin expressed in mammalian cells triggered a significant and transient decrease in basal blood glucose levels by inhibiting the rates of endogenous glucose production in both wild type mice and several diabetic mouse models *(18)*.

The chronic effects of adiponectin on insulin sensitivity and energy metabolism were also investigated in adiponectin transgenic mice or adiponectin knockout (KO) mice. Scherer's group generated a transgenic mouse model with approximately threefold elevation of native adiponectin oligomers *(19)*. The authors demonstrated that hyperadiponectinemia significantly increased lipid clearance and lipoprotein lipase activity, and enhanced insulin-mediated suppression of hepatic glucose production, thereby improving insulin sensitivity. Kadowaki's group showed that transgenic overexpression of globular adiponectin in the genetic background of *ob/ob* obese mice led to partial amelioration of insulin resistance, hyperinsulinemia, and hyperglycemia *(20)*. Conflicting results have been obtained from adiponectin KO mice studies. Yamauchi et al. found no impact of adiponectin depletion on insulin sensitivity under either normal chow or after 7 mo of feeding with a high-fat diet *(21)*. In contrast, adiponectin KO mice reported by Maeda et al. exhibited more severe high-fat diet induced insulin resistance and dyslipidemia, despite having normal glucose tolerance when fed with regular chow *(22)*. Kubota et al. observed mild insulin resistance in the heterozygous adiponectin KO mice and moderate insulin resistance in the homozygous adiponectin KO mice even when fed with a regular chow *(23)*. The latter two studies support the role of adiponectin as an endogenous insulin sensitizer in mice.

The insulin-sensitizing effect of adiponectin appears to be primarily attributed to its direct actions in skeletal muscle and liver, through the activation of AMPK and peroxisome proliferator-activated receptor (PPAR)α *(24,25)*. In liver, stimulation of AMPK by full-length adiponectin leads to decreased expression of gluconeogenic enzymes, such as phosphoenolpyruvate carboxykinase and glucose-6-phosphatase, which may account for its glucose-lowering effect in vivo *(19,24)*. In skeletal muscle, activation of AMPK by globular or full-length adiponectin causes increased expression of proteins involved in fatty acid transport (such as CD36), fatty acid oxidation (such as acyl-coenzyme A oxidase) and energy dissipation (such as uncoupling protein-2), resulting in enhanced fatty acid oxidation and energy dissipation, and decreased tissue triglyceride (TG) accumulation. Excessive tissue TG accumulation has been proposed to be a major causative factor of insulin resistance in skeletal muscle *(26)*. Therefore, reduction of tissue TG's contents by adiponectin might be the major contributor to the insulin-sensitizing activity of this adipokine.

In addition to liver and muscle, adiponectin can also act in an autocrine manner on adipocytes. It can antagonize the inhibitory effect of TNF-α on insulin-stimulated glucose uptake *(27)*, and blocks the release of insulin resistance-inducing factors from adipocytes *(28)*. Furthermore, it has recently been suggested that adiponectin also acts in the brain to increase energy expenditure and cause weight loss *(29)*.

3.2. Antiatherogenic Actions

In addition to its insulin-sensitizing effect, adiponectin possesses direct antiatherogenic properties *(30,31)*. Both adenovirus-mediated overexpression of full-length adiponectin *(32)* and transgenic overexpression of globular adiponectin *(20)* have been shown to inhibit atherosclerotic lesion formation in the aortic sinus of *apoE*-deficient mice. On the other hand, disruption of the adiponectin gene results in impaired vasoreactivity *(33)* and increased neointimal thickening in response to external vascular cuff injury *(23,34)*.

Adiponectin can act directly on the vascular system to exert its vasoprotective functions. In endothelial cells, full-length adiponectin enhances eNOS activity and increases nitric oxide (NO) production, which in turn improves endothelium-dependent vasodilation *(35)*. This protein also suppresses TNF-α-induced production of proinflammatory chemokines and adhesion molecules, including interleukin-8 *(36)*, intracellular adhesion molecule-1, vascular cellular adhesion molecule-1, and E-selectin *(37)*. Globular adiponectin has been shown to inhibit cell proliferation and suppress superoxide release induced by oxidized LDL in bovine aortic endothelial cells *(38)*. In addition, the HMW form of adiponectin can protect endothelial cells from apoptosis by activation of AMPK *(14)*.

In cultured aortic smooth muscle cells, adiponectin inhibits cell proliferation and migration induced by several atherogenic growth factors, including heparin-binding epidermal growth factor-like growth factor, platelet-derived growth factor BB, and basic fibroblast growth factor *(12,39)*. Adiponectin oligomers interact with these growth factors and subsequently block their binding to the respective cell membrane receptors. In macrophages, adiponectin prevents lipid accumulation and suppresses foam cell transformation by inhibiting expression of the class A scavenger receptors and uptake of acetylated low density lipoprotein particles *(40)*. It also blocks the attachment of monocytes to endothelial cells, which is the first and crucial step of atherosclerosis *(41)*. A more recent study has shown that adiponectin increases tissue inhibitor of metalloproteinase-1 through inducing interleukin-10 expression in primary human macrophages *(42)*.

3.3. Hepatoprotective Effects

Several recent studies suggest that adiponectin is protective against various types of liver injuries, steatosis, and fibrosis *(43–46)*. Our group has examined the potential roles of adiponectin in alcoholic and nonalcoholic fatty liver diseases in mice *(43)*. In both *ob/ob* obese mice and mice fed with a high fat–ethanol diet, chronic treatment with recombinant adiponectin dramatically alleviated hepatomegaly and steatosis (fatty liver), and also significantly attenuated inflammation and the elevated levels of serum alanine aminotransferase (ALT), a marker of liver injury. These protective effects were partly attributed to adiponectin's ability to increase carnitine palmitoyltransferase I activity and enhance hepatic fatty acid oxidation, and to decrease hepatic lipogenesis and TNF-α production *(43)*.

Masaki et al. investigated the effect of adiponectin on liver injury induced by D-galactosamine/LPS (GalN/LPS) in KK-Ay obese mice *(44)*. The authors found that pretreatment with adiponectin ameliorated the GalN/LPS-induced elevation of serum AST and ALT levels, and also decreased the apoptotic and necrotic changes in hepatocytes, resulting in a marked reduction in lethality. Kamada et al. demonstrated that adiponectin KO mice were more susceptible to liver fibrosis induced by carbon tetrachloride (CCl_4), whereas adenovirus-mediated overexpression of adiponectin prevented the development of this disease *(45)*. In cultured rat hepatic stellate cells, adiponectin suppressed cell proliferation and migration and attenuated transforming growth factor (TGF)-β1-induced nuclear translocation of Smad2 *(45,46)*. In addition, adiponectin has been shown to accelerate the apoptosis of activated hepatic stellate cells *(46)*, and protect hepatocytes from TNF-α-induced death *(47)*.

3.4. Protection Against Myocardial Injury

The beneficial effects of adiponectin on heart diseases have recently been demonstrated in several different animal models. Shibata et al. reported that pressure overload in adiponectin KO mice resulted in enhanced concentric cardiac hypertrophy and increased mortality that was associated with increased extracellular signal-regulated kinase (ERK) and diminished AMPK signaling in the myocardium *(48)*. Consistent with this finding, a recent study from Liao et al. also showed the exacerbation of heart failure in adiponectin-KO mice *(49)*. On the other hand, adenovirus-mediated supplementation of adiponectin attenuated cardiac hypertrophy in response to pressure overloads in adiponectin KO, wild-type and *db/db* diabetic mice.

Shibata et al. also examined the role of adiponectin in myocardial remodeling in response to acute injury. They demonstrated that ischemia-reperfusion in adiponectin KO mice resulted in enlarged myocardial infarction size and apoptosis, and enhanced

TNF-α expression compared with wild-type mice *(50)*. Adiponectin treatment diminished infarction size, apoptosis, and TNF-α production in both adiponectin KO mice and wild-type mice, through the activation of AMPK and induction of cyclooxygenase (COX)2-dependent synthesis of prostaglandin E_2.

More recently, Takahashi et al. studied the effects of adiponectin replacement therapy on myocardial damage in *ob/ob* obese mice with acute viral myocarditis *(51)*. The results from this study demonstrated that intraperitoneal injection of encephalomyocarditis virus into *ob/ob* obese mice led to elevated cardiac weights and severe inflammatory myocardial damage; these abnormalities were reversed following treatment with adiponectin.

4. PUTATIVE ADIPONECTIN RECEPTORS

4.1. Adiponectin Receptor 1 and Adiponectin Receptor 2

These two putative receptors were identified from human skeletal muscle cDNA library by screening for adiponectin binding *(52)*. Both adiponectin receptor 1 (adipoR1) and adiponectin receptor 2 (adipoR2) are predicted to contain seven transmembrane domains that are structurally and functionally distinct from classical G protein-coupled receptors (GPCRs). Unlike all the other GPCRs reported, adipoR1 and adipoR2 have an inverted membrane topology with a cytoplasmic NH_2-terminus and a short extracellular COOH-terminal domain of approx 25 amino acids *(53)*. AdipoR1 is a high-affinity receptor for globular adiponectin and also a low-affinity receptor for full-length adiponectin. On the other hand, adipoR2 is an intermediate-affinity receptor for full-length and globular adiponectin. In C2C12 myotubes, siRNA-mediated suppression of both adipoR1 and adipoR2 expression abolishes adiponectin-mediated increase in AMPK activation, fatty-acid oxidation and glucose uptake, suggesting that at least some of its metabolic functions are mediated by these two receptors *(52)*.

In mice, adipoR1 is ubiquitously expressed with the highest expression levels in skeletal muscle, whereas adipoR2 is most abundantly expressed in the liver. The mRNA expression levels of adipoR1 and adipoR2 in the liver and skeletal muscles increases after fasting, and decreases after refeeding *(54)*. The expressions of adipoR1/R2 in *ob/ob* mice are significantly decreased in skeletal muscle and adipose tissue, which is correlated with decreased adiponectin binding to membrane fractions of skeletal muscle and reduced AMPK activation by adiponectin. These data suggest that decreased expression of adipoR1/adipoR2 might contribute to adiponectin resistance observed in *ob/ob* mice.

Both hyperglycemia and hyperinsulinemia decrease adipoR1 expression in skeletal muscle *(54–56)*. On the other hand, hyperinsulinemia increases adipoR2 expression in L6 rat myotubes *(56)*. Interestingly, hyperinsulinemia-induced suppression of adipoR1 and upregulation of adipoR2 in L6 myotubes is associated with diminished sensitivty of the cells to globular adiponectin but increased sensitivity toward full-length adiponectin.

In humans, both adipoR1 and adipoR2 are highly expressed in skeletal muscle, with the expression ratio of adipoR1 to adipoR2 approx 6:1 *(57,58)*. A significant correlation between the expression levels of these two receptors and insulin sensitivity was reported in nondiabetic Mexican Americans with or without a family history of type 2 diabetes *(57)*.

In addition, the expression of adiponectin receptors were found to be decreased in obese subjects and type 2 diabetic patients *(59,60)*. Notably, two more recent studies have found that the effect of globular adiponectin to stimulate fatty acid oxidation and glucose uptake in skeletal muscles was blunted in obese subjects *(58,61)*. However, whether adiponectin resistance observed in obese subjects is caused by the defects of adipoR1 and adipoR2 remains to be clarified.

4.2. T-Cadherin

By using an expression cloning strategy, Hug's group has identified T-cadherin, a glycosylphosphatidylinol-linked cell surface molecule, as a potential receptor for adiponectin *(62)*. In contrast to adipoR1 and adipoR2, only eukaryotically expressed adiponectin binds to T-cadherin, implying that post-translational modifications of adiponectin are critical for binding. In addition, T-cadherin binds only to hexameric and HMW oligomers of adiponectin, but not to its trimeric and globular forms. T-cadherin is highly expressed in the heart, smooth muscle, and vascular endothelium, which are also the targets of adiponectin. However, the functional relevance of adiponectin binding to T-cadherin remains to be demonstrated.

5. CLINICAL STUDIES ON ADIPONECTIN

5.1. Plasma Adiponectin Levels and Adiposity

Unlike most adipokines, adiponectin levels in the circulation are paradoxically decreased in obese subjects *(63)*. On the other hand, weight reduction by gastric partition surgery or calorie restriction leads to an increase in plasma levels of total adiponectin *(64)*. There is a strong inverse correlation between plasma levels of adiponectin and measures of adiposity, including body mass index (BMI) and total fat mass *(15)*.

Besides total fat mass, body fat distribution appears to be another important determinant of adiponectin production. Intra-abdominal fat is an independent negative predictor of plasma adiponectin *(65)*. In both lean and obese individuals, adiponectin mRNA abundance and protein levels in intra-abdominal fat are much lower than in subcutaneous fat. Furthermore, hypoadiponectinemia has been found to be closely associated with both congenital and HIV-related lipodystrophy, a disease characterized by body fat redistribution *(15)*.

5.2. Correlation of Hypoadiponectinemia With Insulin Resistance

Low plasma adiponectin concentrations are observed in several forms of diabetes with insulin resistance, including type 2 diabetes *(66)*, gestational diabetes *(67)*, and diabetes associated with lipodystrophy *(68)*. Hyperinsulinemic–euglycemic studies show that plasma levels of adiponectin are positively associated with insulin-stimulated glucose disposal *(69)*, but inversely related with basal and insulin-stimulated hepatic glucose production *(70)*, suggesting a potential role of adiponectin as an endogenous insulin sensitizer in humans. Multivariate analysis demonstrates that hypoadiponectinemia is more closely associated with the degree of insulin resistance and hyperinsulinemia than with the degree of glucose intolerance and adiposity *(71)*. Case-controlled studies show that subjects with low concentrations of adiponectin are more likely to develop type 2 diabetes than those with high concentrations *(72,73)*.

The causative role of hypoadiponectinemia in the development of insulin resistance and type 2 diabetes is further supported by the data from human genetic studies on the adiponectin gene *(74–77)*. Single nucleotide polymorphisms (SNPs) at positions 45, 276, and in the proximal promoter region and exon 3 of the adiponectin gene have been found to be closely associated with insulin resistance and type 2 diabetes in several ethnic groups. Notably, subjects with the G/G phenotype at position 276 in intron 2 have lower adiponectin levels and higher insulin resistance index, and are more susceptible to type 2 diabetes than those with the T/T phenotypes *(74–76,78)*. In addition, eight rare missense mutations in the collagenous domain (G84R, G90S, R92X, and Y111H) and globular domain (R112C, I164T, R221S, and H241P) of the adiponectin gene have been detected, some of which are closely related to hypoadiponectinemia and insulin resistance *(75,77)*. Among these mutations, R112C and I164T mutants are associated with impaired trimeric formation and secretion of adiponectin from the cells *(79)*. G84R and G90S mutations can form trimers and hexamers, but lack capacity to form the HMW oligomers, suggesting that impaired oligomerization might also be an important causative factor for type 2 diabetes.

5.3. Adiponectin Deficiency and Cardiovascular Diseases

Adiponectin is inversely correlated with a panel of traditional cardiovascular risk factors, including blood pressure, heart rate, and total and low-density lipoprotein (LDL) cholesterol and triglyceride levels, and is positively related to high-density lipoprotein (HDL) cholesterol levels *(80,81)*. Hypoadiponectinemia has been shown to be an independent risk factor for endothelial dysfunction and hypertension, regardless of insulin resistance *(82,83)*. In addition, the association of hypoadiponectinemia with coronary heart disease *(84)*, ischemic cerebrovascular disease *(85)*, and coronary artery calcification *(86)* was also reported to be independent of classical cardiovascular risk factors, such as diabetes, dyslipidemia, and hypertension.

A recent report by Kumada et al. showed that the prevalence of coronary artery disease in male subjects with hypoadiponectinemia (<4 μg/mL) was 2.05-fold higher than those with adiponectin concentrations of more than 7.0 μg/mL, after adjustment for classical cardiovascular risk factors *(87)*. In a large nested case–control study, Pischon et al. showed that high plasma levels of adiponectin are associated with a significantly decreased risk of myocardiac infarction over a follow-up period of 6 yr among 18,225 male participants without previous history of cardiovascular disease *(88)*. This association was independent of hypertension, diabetes, or inflammation, and was only partly explained by changes in lipid profiles. Taken together, these data suggest a protective effect of adiponectin against cardiovascular disease in human subjects.

5.4. Plasma Adiponectin Levels and Chronic Liver Diseases

Many recent studies have demonstrated a close association of hypoadiponectinemia with nonalcoholic fatty liver disease (NAFLD) and nonalcoholic steatohepatitis (NASH) *(89–92)*. Plasma levels of adiponectin are inversely correlated with hepatic fat contents, the grade of hepatic necroinflammation, and measures of liver injury, such as serum alanine aminotransferase and γ-glutamyltranspeptodase *(43,93–95)*. Importantly, this association remains significant even after adjustment for sex, age, BMI, and insulin resistance. In a multiple logistic regression analysis model, low adiponectin level was found to be the only independent predictor of NAFLD in men *(96)*.

In contrast to NAFLD and NASH, plasma levels of adiponectin are significantly elevated in patients with liver fibrosis and cirrhosis *(97,98)*. This increase is thought to reflect one of the body's compensatory responses against these diseases, although it is unclear whether the mechanism by which this occurs involves enhanced adiponectin expression or decreased protein clearance.

5.5. Hypoadiponectinemia and Cancers

Low levels of plasma adiponectin are closely correlated with several obesity- and insulin resistance-related cancers. A strong inverse association between plasma adiponectin levels and the risk of both breast cancer and endometrial cancer has recently been reported in two case–control studies *(99,100)*. Importantly, these associations are independent of adiposity, insulin resistance, and other classical risk factors. Another study by Miyoshi et al. has shown that breast cancers arising in women with low adiponectin levels are more likely to show a biologically aggressive phenotype *(101)*. Plasma adiponectin levels in patients with gastric cancer, especially those with upper gastric cancer, were also reported to be much lower than in control subjects *(102)*. Interestingly, in patients with undifferentiated cancer, serum adiponectin showed a negative correlation with pathological findings such as tumor size, depth of invasion, as well as tumor stage. In a large prospective study comprising 18,225 subjects, hypoadiponectinemia was found to be an independent predictor for the future development of colorectal cancer *(103)*. Together these data suggest that adiponectin deficiency might partly account for the increased risk of the aforementioned cancers in obese and insulin resistant subjects.

6. ADIPONECTIN AS A POTENTIAL THERAPEUTIC TARGET

As discussed above, promising results obtained from numerous animal experiments and human epidemiological studies support the role of adiponectin as a potential drug target for developing novel therapeutics against a panel of obesity-related chronic diseases. However, adiponectin is an abundant plasma protein (5–30 μg/mL). The production of recombinant adiponectin is also challenging because of the complex tertiary and quaternary structure of the protein and the distinct activities of the different isoforms. Direct supplementation of recombinant adiponectin in human subjects would be extremely expensive. An alternative approach is to use pharmacological or dietary intervention to increase the suppressed endogenous adiponectin production in obesity, or to enhance adiponectin actions in its target tissues. In this respect, it is interesting to note that the PPARγ agonists thiazolidinediones (TZDs), such as rosiglitazone and pioglitazone, which increase adiponectin production in both humans and rodents, demonstrate many of the therapeutic effects of adiponectin, such as insulin-sensitizing, vasoprotective, and anti-inflammatory properties *(13,17)*. Whether the therapeutic effects of the PPARγ agonists are mediated via induction of adiponectin remain to be investigated. In addition, metformin, another commonly used antidiabetic drug, has been shown to mimic the action of adiponectin in stimulating AMPK in liver *(104)*.

7. CONCLUSIONS

Adiponectin is a multifunctional adipokine that forms several oligomeric complexes in the circulation. This adipokine possesses antidiabetic, antiatherogenic, anti-inflammatory, and antitumor properties, and also protects against various cardiac and liver injuries. The pleiotropic effects of adiponectin are dependent on its oligomerization status. Data from clinical studies support a causative role of hypoadiponectinemia in the development of a wide spectrum of obesity-related metabolic and cardiovascular disorders, liver diseases, and certain types of cancers. Therefore, adiponectin holds great promise to serve as an attractive therapeutic target as well as a diagnostic parameter for these diseases. However, many fundamental questions remain to be answered on the relationship between the structure and function of adiponectin. In particular, further investigation is needed to address how oligomeric complex formation of adiponectin is regulated under various pathophysiological conditions and why different adiponectin oligomers possess distinct functions. The physiological relevance of the newly identified putative adiponectin receptors remains to be demonstrated. The mechanism of adiponectin resistance, which has recently been reported in obese subjects *(58,61)*, needs to be elucidated. Further studies on these research areas will help transform our current knowledge on adiponectin into clinical practice.

REFERENCES

1. Scherer PE, Williams S, Fogliano M, et al. J Biol Chem 1995;270:26,746–26,749.
2. Maeda K, Okubo K, Shimomura I, et al. Biochem Biophys Res Commun 1996;221:286–289.
3. Hu E, Liang P, Spiegelman BM. J Biol Chem 1996;271:10,697–10,703.
4. Shapiro L, Scherer PE Curr Biol 1998;8:335–338.
5. Wang Y, Xu A, Knight C, et al. J Biol Chem 2002;277:19,521–19,529.
6. Wang Y, Lu G, Wong WP, et al. Proteomics 2004;12:3933–3942.
7. Sato C, Yasukawa Z, Honda N, et al. J Biol Chem 2001;276:28,849–28,856.
8. Tsao TS, Murrey HE, Hug C, et al. J Biol Chem 2002;277:29,359–29,362.
9. Xu A, Chan KW, Hoo RL, et al. J Biol Chem 2005;280:18,073–18,080.
10. Tsao TS, Tomas E, Murrey HE, et al. J Biol Chem 2003;278:50,810–50,817.
11. Pajvani UB, Du X, Combs TP, et al. J Biol Chem 2003;278:9073–9085.
12. Wang Y, Lam KS, Xu JY, et al. J Biol Chem 2005;280:18,341–18,347.
13. Pajvani UB, Hawkins M, Combs TP, et al. J Biol Chem 2004;279:12,152–12,162.
14. Kobayashi H, Ouchi N, Kihara S, et al. Circ Res 2004;94:e27–e31.
15. Trujillo ME, Scherer PE. J Intern Med 2005;257:167–175.
16. Fruebis J, Tsao TS, Javorschi S, et al. Proc Natl Acad Sci USA 2001;98:2005–2010.
17. Yamauchi T, Kamon J, Waki H, et al. Nat Med 2001;7:941–946.
18. Berg AH, Combs TP, Du X, et al. Nat Med 2001;7:947–953.
19. Combs TP, Pajvani UB, Berg AH, et al. Endocrinology 2003;23:23.
20. Yamauchi T, Kamon J, Waki H, et al. J Biol Chem 2003;278:2461–2468.
21. Ma K, Cabrero A, Saha PK, et al. J Biol Chem 2002;277:34,658–34,661.
22. Maeda N, Shimomura I, Kishida K, et al. Nat Med 2002;8:731–737.
23. Kubota N, Terauchi Y, Yamauchi T, et al. J Biol Chem 2002;277:25,863–25,866.
24. Yamauchi T, Kamon J, Minokoshi Y, et al. Nat Med 2002;8:1288–1295.
25. Tomas E, Tsao TS, Saha AK, et al. Proc Natl Acad Sci USA 2002;99:16,309–16,313.
26. Hegarty BD, Furler SM, Ye J, et al. Acta Physiol Scand 2003;178:373–383.
27. Wu X, Motoshima H, Mahadev K, et al. Diabetes 2003;52:1355–1363.
28. Dietze-Schroeder D, Sell H, Uhlig M, et al. Diabetes 2005;54:2003–2011.
29. Qi Y, Takahashi N, Hileman SM, et al. Nat Med 2004;11:11.
30. Fasshauer M, Paschke R, Stumvoll M. Biochimie 2004;86:779–784.
31. Lam KS, Xu A. Curr Diab Rep 2005;5:254–259.
32. Okamoto Y, Kihara S, Ouchi N, et al. Circulation 2002;106:2767–2770.
33. Ouchi N, Ohishi M, Kihara S, et al. Hypertension 2003;42:231–234.

34. Matsuda M, Shimomura I, Sata M, et al. J Biol Chem 2002;277:37,487–37,491.
35. Chen H, Montagnani M, Funahashi T, et al. J Biol Chem 2003;278:45,021–45,026.
36. Kobashi C, Urakaze M, Kishida M, et al. Circ Res, 2005;Epub Nov 3.
37. Ouchi N, Kihara S, Arita Y, et al. Circulation 2000;102:1296–1301.
38. Motoshima H, Wu X, Mahadev K, et al. Biochem Biophys Res Commun 2004;315:264–271.
39. Arita Y, Kihara S, Ouchi N, et al. Circulation 2002;105:2893–2898.
40. Ouchi N, Kihara S, Arita Y, et al. Circulation 2001;103:1057–1063.
41. Ouchi N, Kihara S, Arita Y, et al. Circulation 1999;100:2473–2476.
42. Kumada M, Kihara S, Ouchi N, et al. Circulation 2004;109:2046–2049.
43. Xu A, Wang Y, Keshaw H, et al. J Clin Invest 2003;112:91–100.
44. Masaki T, Chiba S, Tatsukawa H, et al. Hepatology 2004;40:177–184.
45. Kamada Y, Tamura S, Kiso S, et al. Gastroenterology 2003;125:1796–1807.
46. Ding X, Saxena NK, Lin S, et al. Am J Pathol 2005;166:1655–1669.
47. Sennello JA, Fayad R, Morris AM, et al. Endocrinology 2005:146:2157–2164. Epub 2005 Jan 27.
48. Shibata R, Ouchi N, Ito M, et al. Nat Med 2004;10:1384–1389. Epub 2004 Nov 21.
49. Liao Y, Takashima S, Maeda N, et al. Cardiovasc Res 2005;67:705–713.
50. Shibata R, Sato K, Pimentel DR, et al. Nat Med 2005;11:1096–1103. Epub 2005 Sep 11.
51. Takahashi T, Saegusa S, Sumino H, et al. J Int Med Res 2005;33:207–214.
52. Yamauchi T, Kamon J, Ito Y, et al. Nature 2003;423:762–769.
53. Kadowaki T, Yamauchi T Endocr Rev 2005;26:439–451.
54. Tsuchida A, Yamauchi T, Ito Y, et al. J Biol Chem 2004;279:30,817–30,822.
55. Inukai K, Nakashima Y, Watanabe M, et al. Am J Physiol Endocrinol Metab 2005;288:E876–E882. Epub 2004 Dec 21.
56. Fang X, Palanivel R, Zhou X, et al. J Mol Endocrinol 2005;35:465–476.
57. Civitarese AE, Jenkinson CP, Richardson D, et al. Diabetologia 2004;47:816–820.
58. Chen MB, McAinch AJ, Macaulay SL, et al. J Clin Endocrinol Metab 2005;90:3665–3672.
59. Wang H, Zhang H, Jia Y, et al. Diabetes 2004;53:2132–2136.
60. Kaser S, Moschen A, Cayon A, et al. Gut 2005;54:117–121.
61. Bruce CR, Mertz VA, Heigenhauser GJ, et al. Diabetes 2005;54:3154–3160.
62. Hug C, Wang J, Ahmad NS, et al. Proc Natl Acad Sci USA 2004;101:10,308–10,313. Epub 2004 Jun 21.
63. Arita Y, Kihara S, Ouchi N, et al. Biochem Biophys Res Commun 1999;257:79–83.
64. Yang WS, Lee WJ, Funahashi T, et al. J Clin Endocrinol Metab 2001;86:3815–3819.
65. Gavrila A, Chan JL, Yiannakouris N, et al. J Clin Endocrinol Metab 2003;88:4823–4831.
66. Hotta K, Funahashi T, Arita Y, et al. Arterioscler Thromb Vasc Biol 2000;20:1595–1599.
67. Worda C, Leipold H, Gruber C, et al. Am J Obstet Gynecol 2004;191:2120–2124.
68. Kosmiski L, Kuritzkes D, Lichtenstein K, et al. Antivir Ther 2003;8:9–15.
69. Stefan N, Vozarova B, Funahashi T, et al. Diabetes 2002;51:1884–1888.
70. Stefan N, Stumvoll M, Vozarova B, et al. Diabetes Care 2003;26:3315–3319.
71. Weyer C, Funahashi T, Tanaka S, et al. J Clin Endocrinol Metab 2001;86:1930–1935.
72. Lindsay RS, Funahashi T, Hanson RL, et al. Lancet 2002;360:57–58.
73. Spranger J, Kroke A, Mohlig M, et al. Lancet 2003;361:226–228.
74. Hara K, Boutin P, Mori Y, et al. Diabetes 2002;51:536–540.
75. Vasseur F, Helbecque N, Dina C, et al. Hum Mol Genet 2002;11:2607–2614.
76. Stumvoll M, Tschritter O, Fritsche A, et al. Diabetes 2002;51:37–41.
77. Kondo H, Shimomura I, Matsukawa,Y, et al. Diabetes 2002;51:2325–2328.
78. Menzaghi C, Ercolino T, Di Paola R, et al. Diabetes 2002;51:2306–2312.
79. Waki H, Yamauchi T, Kamon J, et al. J Biol Chem 2003;278:40,352–40,363.
80. Kazumi T, Kawaguchi A, Sakai K, et al. Diabetes Care 2002;25:971–976.
81. Hotta K, Funahashi T, Bodkin NL, et al. Diabetes 2001;50:1126–1133.
82. Iwashima Y, Katsuya T, Ishikawa K, et al. Hypertension 2004;43:1318–1323.
83. Tan KC, Xu A, Chow,WS, et al. J Clin Endocrinol Metab 2004;89:765–769.
84. Schulze MB, Shai I, Rimm EB, et al. Diabetes 2005;54:534–539.
85. Chen MP, Tsai JC, Chung FM, et al. Arterioscler Thromb Vasc Biol 2005;25:821–826. Epub 2005 Feb 3.
86. Maahs DM, Ogden LG, Kinney GL, et al. Circulation 2005;111:747–753.

87. Kumada M, Kihara S, Sumitsuji S, et al. Arterioscler Thromb Vasc Biol 2003;23:85–89.
88. Pischon T, Girman CJ, Hotamisligil GS, et al. JAMA 2004;291:1730–1737.
89. Pagano C, Soardo G, Esposito W, et al. Eur J Endocrinol 2005;152:113–118.
90. Yoon D, Lee SH, Park HS, et al. J Korean Med Sci 2005;20:421–426.
91. Targher G, Bertolini L, Zenari L. Diabetes Care 2004;27:2085–2086.
92. Matsubara M. Endocr J 2004;51:587–593.
93. Hui JM, Hodge A, Farrell GC, et al. Hepatology 2004;40:46–54.
94. Yokoyama H, Hirose H, Ohgo H, et al. J Hepatol 2004;41:19–24.
95. Lopez-Bermejo A, Botas P, Funahashi T, et al. Clin Endocrinol (Oxf) 2004;60:256–263.
96. Sargin H, Sargin M, Gozu H, et al. World J Gastroenterol 2005;11:5874–5877.
97. Kaser S, Moschen A, Kaser A, et al. J Intern Med 2005;258:274–280.
98. Tacke F, Wustefeld T, Horn R, et al. J Hepatol 2005;42:666–673.
99. Mantzoros C, Petridou E, Dessypris N, et al. J Clin Endocrinol Metab 2004;89:1102–1107.
100. Petridou E, Mantzoros C, Dessypris N, et al. J Clin Endocrinol Metab 2003;88:993–997.
101. Miyoshi Y, Funahashi T, Kihara S, et al. Clin Cancer Res 2003;9:5699–5704.
102. Ishikawa M, Kitayama J, Kazama S, et al. Clin Cancer Res 2005;11:466–472.
103. Wei EK, Giovannucci E, Fuchs CS, et al. J Natl Cancer Inst 2005;97:1688–1694.
104. Zhou G, Myers R, Li Y, et al. J Clin Invest 2001;108:1167–1174.

II

Adipokines as Regulators of Immunity and Inflammation

5 Vasoactive Factors and Inflammatory Mediators Produced in Adipose Tissue

Gema Frühbeck

Abstract

In the search for mechanisms of obesity-mediated vascular pathology, attention has been focused on the role played by adipose tissue, a multifunctional organ involved not only in fat storage but also in the production of numerous hormones, growth factors, and cytokines with pleiotropic features. In the last decade the list of adipose-derived factors shown to be implicated either directly or indirectly in the regulation of vascular homeostasis through effects on blood pressure, inflammation, atherogenesis, coagulation, fibrinolysis, angiogenesis, proliferation, apoptosis, and immunity has increased at a phenomenal pace. By definition, adipocytokines are cytokines produced by adipocytes. Although adipose tissue secretes a wide variety of factors, strictly speaking, not all of them can be contemplated as cytokines. Interleukin-6, tumor necrosis factor-α, leptin, adipsin, resistin, adiponectin, and visfatin fall within the category that satisfies the stricter requirements to be properly classified as adipocytokines. However, the less strict term of adipokines has been coined to include a wider range of factors such as PAI-1, C-reactive protein, monocyte chemoattractant protein-1, serum amyloid A, and vascular endothelial growth factor, among others. Adipokines are known to contribute to the low-grade inflammation state observed in obese patients at the same time as participating in the development of obesity-related comorbidities, such as insulin resistance, metabolic syndrome, and atherogenesis. The molecular mechanisms linking the adiposity–inflammation–immunity cluster are complex. The triad of obesity–insulin resistance–cardiovascular disease is interwoven in a setting of inflammation, endothelial dysfunction and atherosclerosis in which adipokines act as markers of the acute phase reaction at the same time as being directly involved as causative factors in an extensive crosstalk between adipocytes and elements of the stroma vascular fraction. The current knowledge in this field is reviewed with a broad perspective approach.

Key Words: Adipokines; proinflammatory cytokines; adipocytes; vascular reactivity; obesity; cardiovascular disease.

1. INTRODUCTION

Already in 1998, in response to epidemiological data linking excess body weight to coronary heart disease (CHD), the American Heart Association reclassified obesity as a major, modifiable risk factor for CHD *(1)*. In fact, the incidence of obesity and its main associated comorbidities, such as type 2 diabetes mellitus (T2DM), hypertension (HTA), and cardiovascular disease (CVD), has increased in the last decade, reaching epidemic proportions *(2,3)*. The multifaceted influences involved in CVD appearance include ordinary lifestyle habits that share common ground with obesity and metabolic

From: *Nutrition and Health: Adipose Tissue and Adipokines in Health and Disease*
Edited by: G. Fantuzzi and T. Mazzone © Humana Press Inc., Totowa, NJ

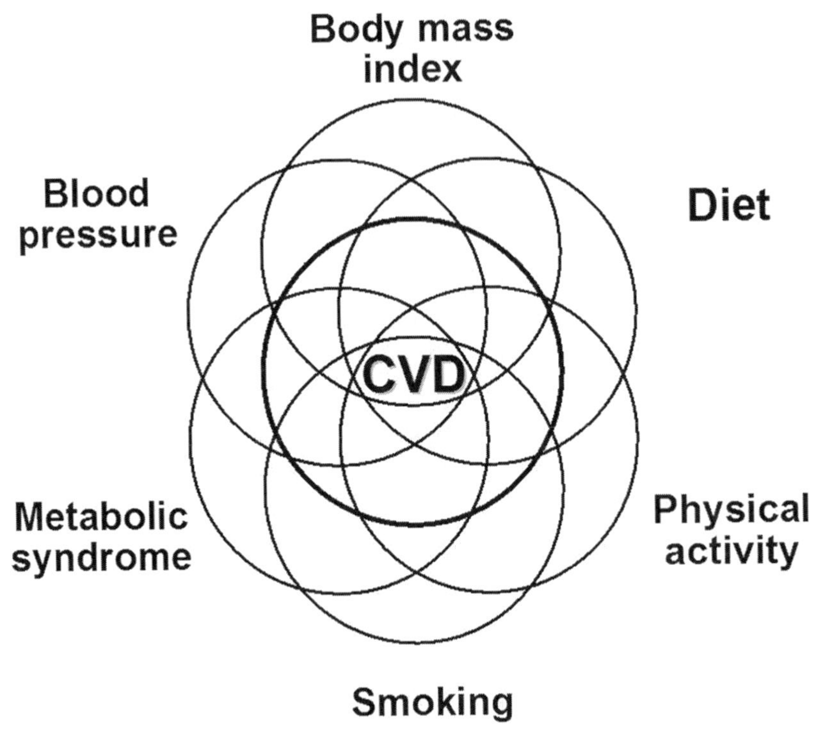

Fig. 1. Multifactorial influences involved in the development of cardiovascular disease (CVD).

syndrome (Fig. 1). Interestingly, an increased adiposity has been shown to lie at the center of the clustering of insulin resistance, dyslipidemia, HTA, and atherothrombosis, among others. In recent years the notion that white adipose tissue (WAT) represents an important determinant of a chronic low-grade, inflammatory state has gained more preponderance at the same time as providing a potential link between obesity and CVD (Fig. 2). A mounting body of knowledge has drawn attention to the pathophysiological relation of adipose tissue-derived factors in the development of a low-grade systemic inflammation through the induction of insulin resistance and endothelial dysfunction that contribute to obesity-associated vasculopathy and cardiovascular risk *(4–24)*. Because blood vessels and the heart express receptors for most of adipose tissue-derived factors (Fig. 3), this extremely active endocrine organ seems to play a key role in cardiovascular physiology and pathophysiology through the existence of a network of local and systemic signals. This chapter reviews the current knowledge in this field in the broader perspective of vasoactivity control and inflammation.

2. ADIPOSE TISSUE AS A DYNAMIC ENDOCRINE AND PARACRINE ORGAN

The primary role of adipose tissue has been related to its ability to store triglycerides during periods of positive energy balance and to mobilize this reserve when expenditure exceeds intake. Adipose tissue is a special loose connective tissue composed not only of adipocytes, but also of other cell types, termed the stromavascular fraction, comprising

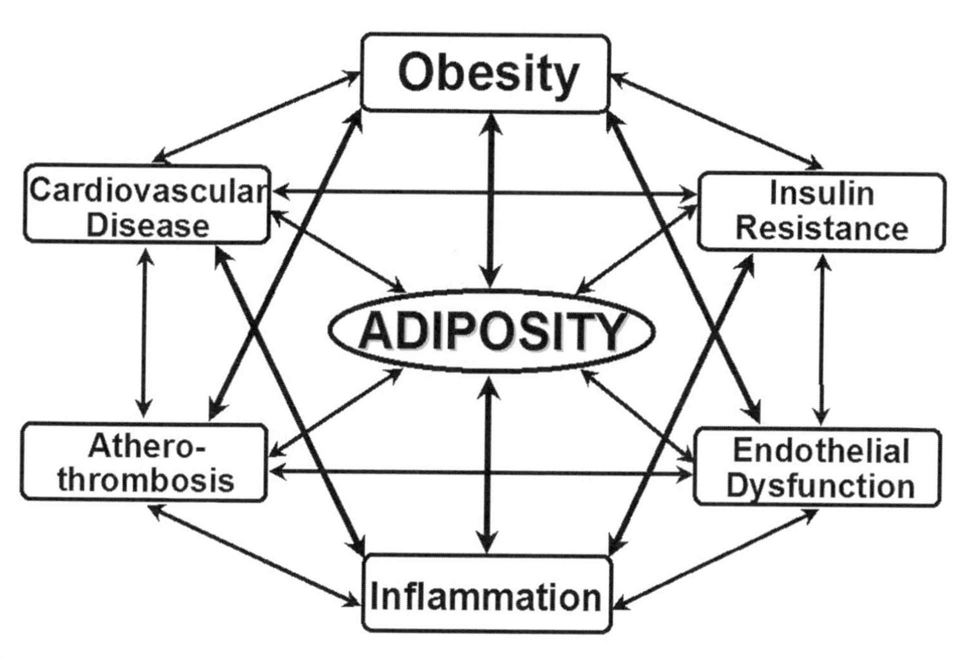

Fig. 2. Diagram of how increased adiposity lies at the center of metabolic and cardiovascular derangements.

blood cells, endothelial cells, pericytes, and adipose precursor cells, among others, which warrant an extensive crosstalk at a local and systemic level in response to specific external stimuli or metabolic changes. It is now widely recognized that adipose tissue lies at the heart of a complex autocrine, paracrine, and endocrine network participating in the regulation of a variety of quite diverse biological functions. Furthermore, the multicellular nature of adipose tissue provides the basis for the secretion of an extreme diversity of molecules including a wide variety of group types, which ranges from hormones to cytokines and chemokines, at the same time as encompassing transcription factors, proinflammatory peptides, participants of the coagulation cascade, growth factors, enzymes involved in glucose and lipid metabolism, complement factors, as well as elements of the renin–angiotensin–aldosterone system. A key group is represented by proinflammatory–proliferative–atherosclerotic–vascular factors such as tumor necrosis factor (TNF)-α, plasminogen activator inhibitor (PAI)-1, tissue factor, nitric oxide (NO), angiotensinogen, metallothionein, C-reactive protein (CRP), and interleukins (IL)—in particular, IL-6, IL-1, IL-10 and IL-8. Growth factors include insulin-like growth factor (IGF)-1, macrophage colony-stimulating factor, transforming growth factor (TGF)-β, vascular endothelial growth factor (VEGF), heparin-binding epidermal growth factor, leukemia inhibitory factor, nerve growth factor, and bone morphogenetic protein. Glucocorticoids, sex steroids, prostaglandins, adipsin, leptin, resistin, and adiponectin/Acrp30/adipoQ are now among the better known secretions of adipose tissue. In the category of nonsecreted factors, perilipin, adiponutrin, adipophilin, uncoupling proteins, and the membrane channel proteins Glut-4 and aquaporin-7 stand out.

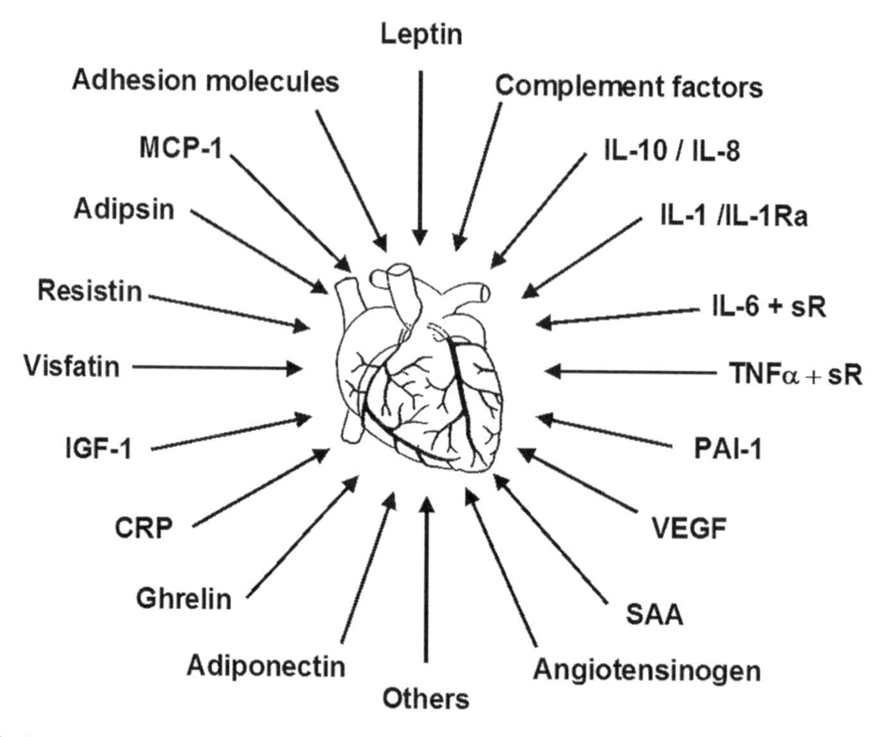

Fig. 3. Schematic representation of the multiple adipose-derived factors impinging on the cardiovascular sphere.

The observation that regional distribution of body fat is an important determinant in the development of metabolic and cardiovascular alterations was formulated more than a century ago *(25)*. Since then, numerous prospective studies have revealed that android or male-type obesity, characterized by increased visceral fat accumulation, correlates more often with an elevated mortality and risk for the development of T2DM, dyslipidemia, hypertension, and atherosclerosis than gynecoid or female-type obesity with fat depots predominantly located in the subcutaneous femorogluteal region *(26)*.

3. ADIPOKINES WITH VASOACTIVE OR INFLAMMATORY EFFECTS

IL-6 and TNF-α are among the well-known cytokines consistently found to be increased in obesity, comprising elevations both at the adipose tissue expression level and the bloodstream *(11,27)*. The list of factors shown to be implicated either directly or indirectly in the regulation of vascular homeostasis through effects on blood pressure, inflammation, atherogenesis, coagulation, fibrinolysis, angiogenesis, proliferation, apoptosis, and immunity has increased at a phenomenal pace *(5–12,14–18,21,24)*. By definition, adipocytokines are cytokines produced by adipocytes. Although adipose tissue secretes a variety of factors, strictly speaking, not all of them can be contemplated as cytokines. Therefore, the less strict term of "adipokines" has been coined to include a wider range of factors. Leptin, adipsin, resistin, adiponectin, and visfatin fall within the category that satisfy the more strict requirements to be properly classified *(11)*.

3.1. Interleukin-6

IL-6 exhibits pleiotropic effects on a variety of tissues, including stimulation of acute phase protein synthesis, activation of the hypothalamic–pituitary axis, and thermogenesis *(4)*. The participation of IL-6 as an inflammatory mediator as well as a stress-induced cytokine merits particular interest *(10,11)*. During acute inflammation IL-6 might favor the resolution of the neutrophilic infiltrate and the initiation of the immune response, whereas in chronic inflammation it might increase the mononuclear-cell infiltrate, thus facilitating leukocyte recruitment transition and a shift in chemokine production, which contribute to the progress and pathogenetic manifestations of inflammatory diseases *(28)*.

The proinflammatory role of IL-6 and its involvement in CHD development are based on an increased production of IL-6 at both the mRNA level and the circulating concentrations *(8)*. A rapid and sustained production of IL-6, with concomitant expression of IL-6R and gp130, supports their participation in a local inflammatory cascade after myocardial ischemia and reperfusion, revealing the involvement of these cytokines in the signaling cascade of postischemic myocardium *(29)*. IL-6 has a proved pro-inflammatory effect by itself as well as through increasing IL-1 and TNF-α, with all three cytokines being implicated in atherogenesis *(30)*. In addition, the effects of IL-6 on vascular smooth muscle cells (VSMC), platelets as well as macrophages, and their production of chemokines and adhesion molecules further impair endothelial dysfunction at the same time as inducing a procoagulant state *(8)*. Therefore, IL-6 represents an important mechanistic link relating obesity, inflammation, atherothrombosis, and CHD.

Adipocytes themselves secrete IL-6, with approx 30% of total circulating IL-6 concentrations originating from adipose tissue in obese states *(8,31)*. However, other cell types within the adipose tissue mass also contribute to the high release of IL-6. Interestingly, adipocytes isolated from the omental depot secrete more IL-6 than fat cells from the subcutaneous site *(31)*. Modulators of IL-6 expression in fat depots include TNF-α, glucocorticoids, and catecholamines *(4,31)*. Although TNF-α, noradrenaline, isoprenaline, and β-adrenergic receptor activation stimulate IL-6 gene expression and protein secretion, dexamethasone markedly suppresses its production *(4,8)*.

3.2. Tumor Necrosis Factor-α

TNF-α was originally identified as a macrophage product implicated in the metabolic disturbances of chronic inflammation and malignancy. Later on, its biological actions were shown to further extend to anorexia, weight loss, and insulin resistance *(7)*. Elevated adipose tissue expression of TNF-α mRNA has been reported in different rodent models of obesity as well as in clinical studies involving obese patients *(23)*. TNF-α mRNA expression is positively correlated with body adiposity as well as with hyperinsulinemia, showing positive associations with fasting insulin and triglyceride concentrations. TNF-α inhibits the expression of the transcription factor CCAAT/enhancer binding protein-α (CEBPα) and the nuclear receptor peroxisome proliferator-activated receptor (PPAR)γ2 *(8,12,14)*. Furthermore, TNF-α stimulates the nuclear factor-κB transcription factor (NFκB), which orchestrates a series of inflammatory events, including expression of adhesion molecules on the surface of both endothelial cells and VSMC *(8,11,24)*. Regarding the impact on the cardiac sphere, the deleterous effects of TNF-α on left ventricular dysfunction reportedly take place in the short term through stimulation of the neutral sphingomyelinase pathway, whereas the more delayed response is mediated by a NO-induced blunting of the β-adrenergic signaling, resulting in a negative inotropism *(17)*. Further inflammatory cytokines, such as IL-1, IL-18, and IL-6, are also known to be released and activated in heart failure.

3.3. Leptin

The involvement of leptin in the pathogenesis of atherosclerotic disease and other cardiovascular complications started to unfold shortly after its identification as a cytokine *(7,8,32)*. In this respect, the functional relationship between leptin and NO emerges as a relevant signaling pathway *(2,33)*. Leptin administration has been observed to increase circulating concentrations of NO *(33,34)*. The leptin–NO relation has an impact on blood pressure control, with leptin simultaneously producing a neurogenic pressor action and an opposing NO-mediated depressor effect *(33)*. Furthermore, in VSMC and aortic rings leptin reportedly counteracts the angiotensin II-induced intracellular calcium increase and vasoconstriction via an NO-dependent mechanism *(35)*.

The effects of leptin on inflammation and immunity are complex. The consensus attributes to leptin a proinflammatory role, although the increase of leptin mRNA WAT expression and circulating concentrations triggered by inflammatory stimuli in experimental animals have not consistently been observed in humans *(36)*. The immune-modulating properties of leptin are plentiful. This adipokine protects T-lymphocytes from apoptosis and modulates T-cell proliferation by elevating the proliferation of naive T-cells at the same time as it reduces the proliferation of memory T-cells *(11,37,38)*. Furthermore, leptin has been shown to modulate cell-derived cytokine production and to upregulate the expression of CD25 and CD71, activation markers of $CD4^+$ and $CD8^+$ cells. The effects of leptin on monocytes also include upregulation of cytokine production, activation markers, and phagocytosis *(11)*. On endothelial cells leptin has been shown to upregulate endothelin-1 and NO synthase, as well as to stimulate the expression of adhesion molecules and monocyte chemoattractant protein (MCP)-1 at the same time as inducing oxidative stress *(16)*. In addition, leptin has been observed to stimulate angiogenesis, platelet aggregation, and atherothrombosis. Furthermore, leptin exerts a direct effect on macrophages, promoting cholesterol accumulation in them at the same time as leading to an increased release of monocyte colony-stimulating factor.

3.4. Adipsin

Adipocytes and monocytes–macrophages express several components of both the classical and alternative complement cascade. Adipsin, which corresponds to complement factor D in humans, represents the rate-limiting enzyme in the alternative pathway of complement activation *(11)*. Adipsin was originally identified as a highly differentiation-dependent gene in 3T3-L1 adipocytes, with its expression being markedly downregulated in rodent models of obesity, probably owing to increased concentrations of insulin and glucocorticoids *(7)*. Whereas in murine models of obesity adipsin expression has been shown to be decreased, in obese patients either unchanged or increased levels have been reported *(39)*. In cultured fat cell lines and other tissues the activity of the alternative complement pathway requires stimulation by cytokines. However, the proximal pathway is fully functional in adipose tissue fragments even without stimulation by cytokines, probably as a result of endogenous cytokine production. Interestingly, the complement system has been evoked as a key effector in the cascade mediating post-myocardial ischemia reperfusion injury.

3.5. Resistin

The name of this adipokine was derived from the seminal observation that it induced insulin resistance in mice *(40)*. Circulating resistin concentrations were shown to be increased in genetically obese rodents (*ob/ob* and *db/db* mice) as well as in high-fat-diet-induced obesity. Immunoneutralization of resistin was shown to improve hyperglycemia and insulin resistance in high-fat-induced obese mice, whereas recombinant resistin administration impaired glucose tolerance and insulin action in normal mice. Insulin, TNF-α, epinephrine, β-adrenoreceptor stimulation, and thiazolidinediones reportedly decrease resistin gene expression. However, insulin, β_3-adrenoreceptor stimulation, and thiazolidinediones have been also observed to increase the expression of resistin, together with other factors such as glucose, growth hormone, and glucocorticoids *(41)*. The real contribution of resistin in human pathophysiology remains controversial. Although resistin transcripts have been found in WAT of obese patients, no correlation between resistin mRNA levels with body weight, adiposity, and insulin resistance was obtained *(8)*. Resistin has been shown to be expressed in the stromovascular fraction of WAT and in peripheral blood monocytes, but its mRNA is undetectable in human adipocytes of lean, insulin resistant, obese, and diabetic patients *(41)*. Given the expression of FIZZ1/RELMα in inflammatory regions as well as in inflammatory cells, resistin emerges as a critical mediator of the insulin resistance associated with sepsis and possibly other inflammatory settings *(42,43)*. Resistin has been shown to increase the expression of MCP-1 and cell adhesion molecules (VCAM-1 and ICAM-1) in endothelial cells *(44)*, key processes in early atherosclerotic lesion formation. Moreover, resistin-treated cells have been reported to express lower TNF-α receptor-associated factor (TRAF-3), a potent inhibitor of CD40 ligand-mediated endothelial cell activation *(16)*. Interestingly, the resistin-induced upregulation of adhesion molecules is antagonized by adiponectin *(45)*. Furthermore, the participation of resistin in endothelial dysfunction of insulin resistance patients has been related to its direct effect on endothelial cells promoting the release of endothelin-1 *(16)*. The proliferative effect of resistin on VSMC has been suggested to underlie the increased incidence of restenosis common among diabetic patients.

3.6. Adiponectin

Adiponectin (also identified as Acrp30, AdipoQ, apM1, or GBP28) is completely different from other known adipokines in that it is the only one known so far to improve insulin sensitivity, inhibit vascular inflammation, and exhibit a cardioprotective effect *(7,11,12,21,24,30,41)*. An evident hypoadiponectinemia has been observed in pathological conditions such as obesity and the insulin resistance accompanying a prediabetic state and manifest T2DM *(7)*. Adiponectin paralleled the decreased insulin sensitivity and remained suppressed when frank diabetes had developed. Surprisingly, no independent association between adiponectin and insulin concentrations has been observed. In lean subjects and women, WAT adiponectin expression has been found to be higher and to be associated with higher degrees of insulin sensitivity and lower expression of TNF-α *(46)*. Adiponectin gene expression in human visceral adipose tissue is inhibited by glucocorticoids, TNF-α, and IL-6, whereas it is increased by insulin, IGF-1, and PPARγ agonists. Three putative adiponectin receptors have been cloned so far and shown to be abundantly expressed in skeletal muscle in the case of AdipoR1, predominantly present in liver for AdipoR2, whereas a third one was observed in endothelial cells and smooth muscle *(47)*.

During the last years adiponectin has been shown to exhibit particular cardioprotective properties *(16)*. Adiponectin has been proved to participate in processes related to atherosclerotic plaque formation prevention such as inhibition of monocyte adhesion to endothelial cells by decreasing NFκB signaling through a cAMP-dependent pathway *(45,48)*. In addition, adiponectin has been reported to protect atherosclerotic rodent models such as *ob/ob* and apoE-deficient mice from both atherosclerosis and T2DM *(8,16)*. The findings relating hypoadiponectinemia with obesity-associated metabolic syndrome and atherosclerosis in animals have a clinical parallel in humans as evidenced by studies showing a negative correlation of adiponectin with markers of inflammation *(49,50)*. The hypoadiponectinemia characteristic of obesity is inversely correlated to insulin resistance and CRP levels *(16)*. In addition, patients with CHD exhibit lower adiponectin concentrations compared with age- and body mass index (BMI)-adjusted controls. Moreover, adiponectin has also been shown to control vascular inflammation via a direct effect on endothelial cells and by decreasing VSMC proliferation and migration by reducing the effects of growth factors such as platelet-derived growth factor and heparin-binding epidermal growth factor *(7)*. Even more, high plasma adiponectin concentrations have been reported to be associated with lower risk of myocardial infarction in men *(51)*.

An anti-inflammatory activity of adiponectin on macrophages has been observed, thus strengthening the detrimental effects of hypoadiponectinemia in obesity, T2DM, and CVD. Adiponectin exerts its antiatherogenic characteristics through suppression of the endothelial inflammatory response, inhibiting VSMC proliferation and decreasing VCAM-1 expression *(16)*. Adiponectin further supresses transformation of macrophages into foam cells. Furthermore an increased neointimal proliferation in response to injury has been observed in adiponectin-deficient mice *(16)*.

3.7. Visfatin

Visfatin is a recently discovered adipokine, which is apparently produced and secreted mainly by visceral WAT, with putative antidiabetogenic properties by binding to the insulin receptor and exerting an insulinomimetic effect both in vitro and in vivo *(52,53)*. Visfatin was originally identified as pre-B-cell colony-enhancing factor (PBEF), a cytokine with increased presence in the bronchoalveolar lavage fluid of animal models of acute lung injury as well as in the neutrophils of septic patients *(11)*. Despite its name, plasma concentrations of visfatin and visceral visfatin mRNA expression have been reported to correlate with measures of obesity but not with visceral fat mass or waist-to-hip ratio. Moreover, no differences in visfatin mRNA expression between the visceral and subcutaneous fat depots have been observed *(54)*. Regarding its regulation, it has been reported that IL-6 exerts an inhibitory effect on visfatin expression, which is in part mediated by the p44/42 mitogen-activated protein kinase *(55)*. In T2DM patients a twofold increase in circulating concentrations of visfatin has been recently reported *(56)*. However, the independent association between visfatin and

T2DM disappeared after adjustment for body mass index and waist-to-hip ratio. Currently, the pathophysiological relevance of visfatin remains unclear, with the paradoxical effects simultaneously favoring fat accretion and promoting insulin sensitivity, thus deserving further analysis *(57)*. Visfatin may participate in the mechanisms facilitating fat accumulation in the intra-abdominal depot or as a feedback control preventing the detrimental effects of increased visceral fat on insulin sensitivity, or may simply represent an epiphenomenon that might be useful as a surrogate marker of increased omental adipose tissue.

4. OTHER ADIPOSE-DERIVED FACTORS

In response to various infectious and inflammatory signals, WAT has been shown to induce the expression and secretion of several acute-phase reactants and mediators of inflammation different from those mentioned previously. These include diverse interleukins to IL-6, such as IL-1β, IL-8, IL-10, IL-15, IL-17, and IL-18. Leukemia inhibitory factor, hepatocyte growth factor, haptoglobin, complement factors, and prostaglandin E2 also share the inflammatory modulator properties. Although for many of them their area of influence is restricted to autocrine and paracrine effects, part of the cytokines or chemokines secreted from adipocytes and adipose-resident macrophages are known to make a significant contribution to systemic inflammation *(12)*. Some of the main factors are briefly discussed in this section.

4.1. Plasminogen Activator Inhibitor-1

Although PAI-1 is primarily derived from platelets and the endothelium, it has been demonstrated that most of the elevated concentrations of this regulatory protein of the coagulation cascade in inflammatory and obese states is attributable to an upregulated expression by adipose tissue itself *(12,58)*. Therefore, WAT represents a quantitatively relevant source of PAI-1 production, with consequently increased circulating concentrations present in obesity. Stromal cells have been shown to be the main PAI-1 producing cells in human fat, with a fivefold higher expression in the visceral than in the subcutaneous depots, which is in agreement with the strong relationship observed between circulating PAI-1 concentrations and visceral fat accumulation *(59)*. However, whether adipose tissue itself directly contributes to circulating PAI-1, or whether it exerts an indirect effect via adipokines, such as TNF-α , IL-1β, and TGF-β, to stimulate PAI-1 production by other cells has not been clearly established. Moreover, PAI-1 gene expression is increased by hyperglycemia, insulin, glucocorticoids, angiotensin II, thrombin, and LDL cholesterol, whereas IL-6, estrogens, and β-adrenoceptor agonists exert an inhibitory effect *(8)*. PAI-1 activity has been shown to play a key role in thrombus formation upon unstable atherosclerotic plaque rupture through the inhibition of fibrin clot breakdown. In addition, by altering the fibrinolytic balance, PAI-1 also contributes to vascular structure remodeling. A close correlation between plasma PAI-1 with visceral adiposity and increased myocardial infarct was established a decade ago *(7)*. In fact, in humans plasma PAI-1 concentrations correlate with atherosclerotic events and mortality, with some studies suggesting that PAI-1 may be an independent risk factor for coronary artery disease. In this respect, increased plasma PAI-1 concentrations have been observed in obese patients and a close correlation with an abdominal pattern of adipose tissue distribution in both men and women, as well as a positive association with other components of the insulin resistance syndrome, have been reported *(7,59)*.

4.2. C-Reactive Protein

The liver represents the main source of production of CRP, with its synthesis being largely regulated by IL-6 *(60)*. Hepatic CRP production is also responsive to TNF-α, which simultaneously induces IL-6 expression. However, it has been shown that human adipose tissue itself expresses CRP mRNA, which is negatively correlated with adiponectin expression in the same tissue *(61)*. Therefore, hypoadiponectinemia can be responsible for a low-level systemic chronic inflammation state, which is closely related to increased high-sensitivity CRP concentrations. In both cross-sectional and longitudinal studies, CRP concentrations have been observed to be associated with CHD, whereby plasma CRP predicted CV events or CHD mortality during variable follow-up periods encompassing between 2 and 17 yr *(8)*. These observations support the assumption that increased CRP concentrations may predict atheroma progression and plaque instability. Interestingly, whether the relationship between CVD and CRP reflects vascular wall inflammation or an inflammatory response originating in more remote sites, with secondary effects on the vascular system through cytokines and other mediators, has remained debatable.

Even before myocardial infarction, circulating CRP has proven atherogenic effects on vascular endothelium and smooth muscle *(12)*. Furthermore, CRP reportedly induces the endothelial adhesion molecules VCAM-1 and ICAM-1, as well as E- and P-selectin. CRP appears to operate as an amplifier of vascular inflammation even in the absence of tissue injury through activation of endothelial NFκB, induction of endothelial IL-6, TNF-α, IL-1β, PAI-1, MCP-1, endothelin-1, and tissue factor, together with inhibition of endothelial NO synthase and NO signaling *(62–64)*. Moreover, CRP accumulation within damaged myocardium has been shown to participate in postinfarction pathology through its complement-activating and opsonizing activities *(12)*. All together, these findings support the notion that CRP operates as an active mediator of inflammatory vasculopathy.

4.3. Monocyte Chemoattractant Protein-1

The mounting evidence of the relevance of inflammation in vascular disease has focused attention on the molecules that modulate leukocyte migration from the bloodstream to the vessel wall. Macrophages play a wide array of roles in atherogenesis, operating as scavengers, immune mediators, or as a source of chemokines and cytokines *(65)*. MCP-1 attracts monocytes bearing the chemokine receptor CCR-2. In fact, macrophage expression of cyclooxygenase-2, a key enzyme in inflammation, has been shown to stimulate atherosclerotic lesion formation, whereas chemokines reportedly promote migration of monocytes into the arterial intima *(65,66)*. Monocytes differentiate into macrophages in the arterial intima, where they accumulate cholesterol esters to form lipid-laden foam cells. Visceral adipose tissue has the ability to produce large amounts of MCP-1, which has been shown to relate to morphological and functional echocardiographic abnormalities *(67)*. Myocardial infarction is known to be associated with an inflammatory response leading to leukocyte recruitment. MCP-1 is directly involved in ventricular remodeling with elevated circulating concentrations being observed in patients at risk for coronary artery disease *(68)*. Recent evidence further supports the contribution of MCP-1 to thrombin generation and thrombus formation via tissue factor production *(66)*. Given their pivotal roles in monocyte recruitment in vascular diseases, MCP-1 and CCR-2 may turn out as attractive therapeutic targets to counteract vascular disease pathogenesis.

4.4. Serum Amyloid A

Serum amyloid A (SAA) corresponds to an acute-phase reactant protein secreted by diverse cell types, including adipocytes, which has been associated with systemic inflammation at the same time as being linked to atherosclerosis and to serve as a predictor for coronary disease and cardiovascular outcome *(16)*. Circulating SAA concentrations are increased in obese and diabetic patients *(12)*. Under normal conditions WAT is known to express low levels of SAA, which are extraordinarily upregulated in obesity *(69)*. Mechanistically the detrimental effects of elevated SAA seem to be related to the displacement of apolipoprotein A_1 from HDL cholesterol, increasing its binding to macrophages, thus decreasing the availability of cardioprotective HDL cholesterol *(16)*. Furthermore, SAA operates as a chemoattractant, an inducer of remodeling metalloproteinases, and a stimulator of T-cell cytokine production *(12)*.

4.5. Vascular Endothelial Growth Factor

VEGF is a well-recognized angiogenic factor that induces migration and proliferation of vascular endothelial cells *(12)*. VEGF is encoded by a single gene; however, four isoforms are produced by alternative splicing, which have been implicated in both normal blood vessel development and in pathogenic neovascularization and atherosclerosis. In obese patients serum concentrations of the 164-amino-acid-long isoform has been observed to be dependent on intra-abdominal fat accumulation *(70)*. VEGF mRNA expression has been identified in various cell types, including endothelial, epithelial, and mesenchymal cells. Recently, attention has been focused on the altered expression profile of VEGF in omental WAT obtained from obese individuals *(71)*. Given the growth potential of adipose tissue, it is not surprising that adipocytes express an angiogenic factor like VEGF implicated in vascular bed expansion to support fat mass accretion. The participation of VEGF in vascular inflammation and remodeling through increased subendothelial macrophage accumulation and intima media thickening merits further consideration in the context of atheroma initiation and restenosis events.

5. MOLECULAR LINKS UNDERLYING THE ADIPOSITY–INFLAMMATION–IMMUNITY CLUSTER

The body of knowledge of the pathophysiological effects of adipokines on vasoactivity and inflammation has been gathered from different types of studies, including epidemiological observations, animal model experiments, in vitro approaches, and microarray application. The consistent interrelationship between several adipokines with known effects on inflammation, atherosclerosis, and insulin resistance provides support for the crucial role of adipose tissue in the regulation of obesity-linked CV derangements. Although the existence of inflammatory events is well-known to exert a relevant role in CHD etiology, the exact mechanisms underlying this relationship have not been completely disentangled. For instance, it has not been established whether adipokines simply act as markers of the acute phase reaction or are directly involved as causative factors. Most probably both statements are applicable, with adipokines operating as markers and simultaneously exhibiting a causal implication.

The molecular mechanisms linking the adiposity–inflammation–immunity cluster are complex and their understanding is in continous evolution. The triad of obesity–insulin resistance–CVD is interwoven in a setting of inflammation, endothelial dysfunction, and atherosclerosis. Obesity is characterized by a low-grade systemic inflammation and a hypercoagulable state contributing to atherosclerosis. Endothelial dysfunction represents an early pivotal event in the pathogenesis of atherosclerosis, with a manifest imbalance between endothelium-dependent vasodilation and vasoconstriction taking place in a prothrombotic environment. The endothelial vasodilatory capacity is maintained by NO, which opposes the vasoconstrictor influences of angiotensin II and endothelin-1 *(72)*. The maintenance of the endothelium as a smooth nonthrombotic barrier is further acomplished by prostacyclins and the inhibitory effect on leukocyte and platelet activation and aggregation *(16)*. Adhesion molecules such as the vascular (VCAM-1) and intercellular (ICAM-1) ones are known to be increased in response to inflammatory insults. The elevated expression of P- and E-selectins contributes—together with MCP-1 and integrins—to increase leukocyte recruitment, transmigration, and subsequent adherence to the intima media *(73)*. Phagocytosis of oxidized LDL particles by monocytes leads to foam cell formation and the development of fatty streaks and plaques together with VSMC proliferation *(16)*. In addition, the inflammatory IκB kinase-β/NFκB pathway has been demostrated to also mediate insulin resistance *(74)*. Adipokines are known to impinge on all these pathways, which provide the complex molecular links underlying obesity and the development of CV alterations (Fig. 4).

6. CONCLUSIONS

In conclusion, the notion of adipose tissue as a passive bystander in energy homeostasis has been surpassed. This almost ubiquitously distributed, extraordinarily active endocrine organ has emerged as a dynamic and pleiotropic tissue. The multifunctional nature of adipose tissue is based on the ability of its cellular constituents to secrete a large number of hormones, growth factors, enzymes, cytokines, complement factors, and matrix proteins, collectively termed adipokines or adipocytokines, at the same time as they express receptors for most of these factors. Only recently has attention been devoted to the vasoactive factors produced by adipose tissue. The extensive crosstalk at both a local and systemic level in response to neuroendocrine stimuli as well as metabolic changes warrants a key role in processes well beyond body weight control and insulin resistance, such as inflammation, coagulation, fibrinolysis, and atherosclerosis.

The exact contribution of the complex network of bioactive mediators on vasoactivity and inflammation remains to be fully determined. Furthermore, it will be interesting to gain more insight into the mechanisms involved in the activation and integration of the diverse signaling pathways. It will be also worthwhile to focus on how the known vasoactive factors are related to the more recently discovered hormones,

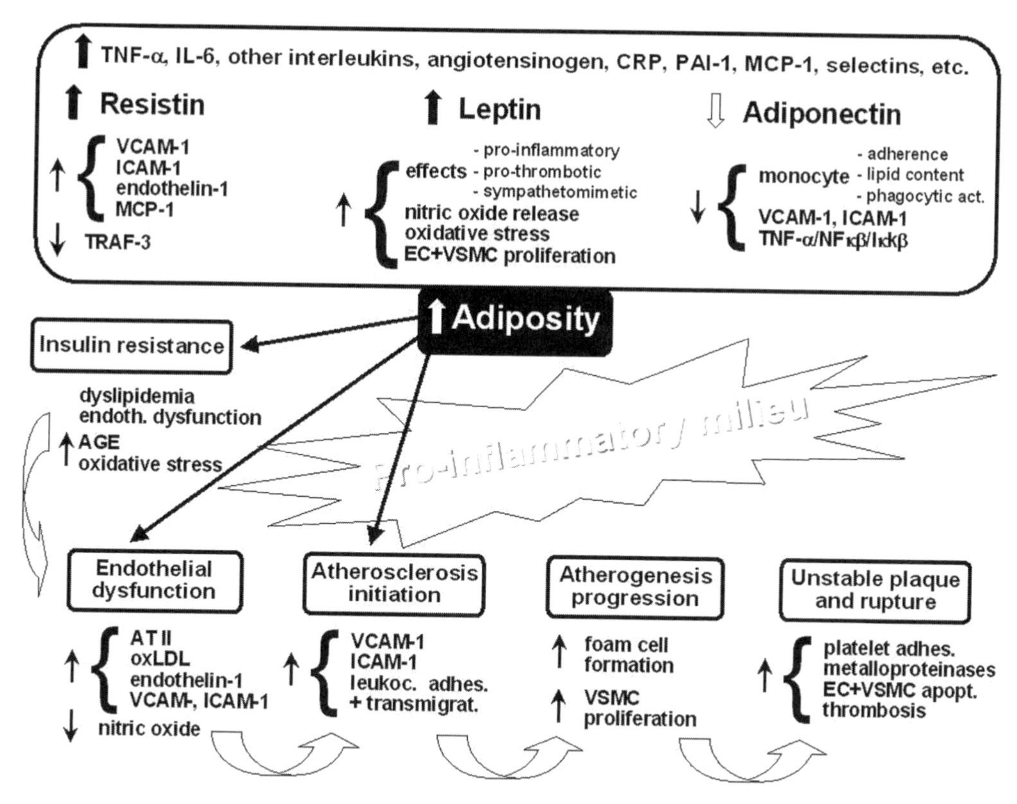

Fig. 4. Molecular links underlying the complex adiposity–inflammation–atherosclerosis relation.

adipokines, receptors, channels, and peptides such as obestatin, adrenomedullin, hypoxia-sensitive molecules, aquaporins, and caspases. In addition, major advances in disentangling the molecular mechanisms underlying inflammation and atherogenesis are to be expected. Undoubtedly, given adipose tissue's versatile and ever-expanding list of activities, additional and unexpected vasoactive peptides are sure to emerge. The intense studies under way on many different frontiers of cardiovascular investigation will add more information to the already large body of knowledge representing a fertile and exciting research field.

ACKNOWLEDGMENTS

This work was supported by grant 20/2005 from the Department of Health of the Gobierno de Navarra, and the Instituto de Salud Carlos III-FIS of Spain.

REFERENCES

1. Eckel RH, Krauss RM. For the AHA Nutrition Committee. Circulation 1998;97:2099–2100.
2. Mokdad AH, Ford ES, Bowman BA, et al. JAMA 2003;289:76–79.
3. Gregg EW, Cheng YJ, Cadwell BL, et al. JAMA 2005;293:1868–1874.
4. Frühbeck G, Gómez-Ambrosi J, Muruzábal FJ, et al. Am J Physiol Endocrinol Metab 2001;280: E827–E847.
5. Marette A. Curr Opin Clin Nutr Metab Care 2002;5:377–383.
6. Blake GJ, Ridker PM. J Intern Med 2002;252:283–294.

7. Frühbeck G, Salvador J. Nutr Res 2004;24:803–826.
8. Frühbeck G. Curr Med Chem—Cardiovasc Hematol Agents 2004;2:197–208.
9. Hauner H. Physiol Behav 2004;83:653–658.
10. Granger JP. Am J Physiol Reg Integr Comp Physiol 2004;286:R989–R990.
11. Fantuzzi G. J Allergy Clin Immunol 2005;115:911–919.
12. Berg AH, Scherer PE. Circ Res 2005;96:939–949.
13. Gimeno RE, Klaman LD. Curr Opin Pharmacol 2005;5:122–128.
14. Juge-Aubry CE, Henrichot E, Meier CA. Best Pract Res Clin Endocrinol Metab 2005;19:547–566.
15. Kougias P, Chai H, Lin PH, et al. J Surg Res 2005;126:121–129.
16. Lau DCW, Dhillon B, Yan H, et al. Am J Physiol Heart Circ Physiol 2005;288:H2031–H2041.
17. Matsuzawa Y. Best Pract Res Clin Endocrinol Metab 2005;19:637–647.
18. Nawrocki AR, Scherer PE. Drug Discovery Today 2005;10:1219–1230.
19. Hutley L, Prins JB. Am J Med Sci 2005;330:280–289.
20. Trayhurn P. Acta Physiol Scand 2005;184:285–293.
21. Schäffler A, Schölmerich J, Büchler C. Nat Clin Pract Gastroenterol Hepatol 2005;2:103–111.
22. Yu Y-H, Ginsberg HN. Circ Res 2005;96:1042–1052.
23. Rudin E, Barzilai N. J Immunity Ageing 2005;2:1–3.
24. Matsuzawa Y. Nat Clin Pract Cardiovasc Med 2006;3:35–42.
25. Vague J. Am J Clin Nutr 1956;4:20–34.
26. Arner P. Biochem Soc Trans 2001;29 Part 2:72–75.
27. Fernández-Real JM, Ricart W. Endocrine Rev 2003;24:278–301.
28. Kaplanski G, Marin V, Montero-Julian F, et al. Trends Immunol 2003;24:25–29.
29. Chandrasekar B, Mitchell DH, Colston JT, et al. Circulation 1999;99:427–433.
30. Lyon CJ, Law RE, Hsueh WA. Endocrinology 2003;144:2195–2200.
31. Fried SK, Bunkin DA, Greenberg AS. J Clin Endocrinol Metab 1998;83:847–850.
32. Correia ML, Haynes WG. Curr Opin Nephrol Hypertens 2004;13:215–223.
33. Frühbeck G. Diabetes 1999;48:903–908.
34. Frühbeck G. Biochem J 2006;393:7–20.
35. Fortuño A, Rodríguez A, Gómez-Ambrosi J, et al. Endocrinology 2002;143:3555–3560.
36. Fantuzzi G, Faggioni R. J Leukoc Biol 2000;68:437–446.
37. La Cava A, Matarese G. Nat Rev Immunol 2004;4:371–379.
38. La Cava A, Alviggi C, Matarese G. J Mol Med 2004;82:4–11.
39. Maslowska M, Vu H, Phelis S, et al. Eur J Clin Invest 1999:29:679–686.
40. Steppan CM, Lazar MA. J Intern Med 2004;255:439–447.
41. Koerner A, Kratzsch J, Kiess W. Best Pract Res Clin Endocrinol Metab 2005;19:525–546.
42. Gómez-Ambrosi J, Frühbeck G. Ann Int Med 2001;135:306–307.
43. Lehrke M, Reilly MP, Millington SC, et al. PloS Medicine 2004;1:161–168.
44. Verma S, Li SH, Wang CH, et al. Circulation 2003;108:736–740.
45. Kawanami D, Maemura K, Takeda N, et al. Biochem Biophys Res Commun 2004;314:415–419.
46. Yamauchi T, Kamon J, Ito Y, et al. Nature 2003;423:762–769.
47. Kadowaki T, Yamauchi T. Endocrine Rev 2005;26:439–451.
48. Ouchi N, Kihara S, Arita Y, et al. Circulation 2000;102:1296–1301.
49. Xydakis AM, Case CC, Jones PH, et al. J Clin Endocrinol Metab 2004;89:2697–2703.
50. Pischon T, Girman CJ, Hotamisligil GS, et al. JAMA 2004;291:1730–1737.
51. Wellen KE, Hotamisligil GS. J Clin Invest 2003;112:1785–1788.
52. Fukuhara A, Matsuda M, Nishizawa M, et al. Science 2005;307:426–430.
53. Sethi JK, Vidal-Puig A. Trends Mol Med 2005;11:344–347.
54. Berndt J, Kloting N, Kralisch S, et al. Diabetes 2005;54:2911–2916.
55. Kralisch S, Klein J, Lossner U, et al. Am J Physiol Endocrinol Metab 2005;289:E586–E590.
56. Chen M-P, Chung F-M, Chang D-M, et al. J Clin Endocrinol Metab 2006;91:295–299.
57. Arner P. J Clin Endocrinol Metab 2006;91:28–30.
58. Skurk T, Havner H. Int J Obes Relat Metab Disord 2004;28:1357–1364.
59. Alessi MC, Peiretti F, Morange P, et al. Diabetes 1997;46:860–867.
60. Yudkin JS, Stehouwer CD, Emeis JJ, et al. Arterioscler Thromb Vasc Biol 1999;19:972–978.

61. Ouchi N, Kihara S, Nagai R, et al. Circulation 2003;107:671–674.
62. Verma S, Badiwala MV, Weisel RD, et al. J Thorac Cardiovasc Surg 2003;126:1886–1891.
63. Verma S, Kuliszewski MA, Li SH, et al. Circulation 2004;109:2058–2067.
64. Labarrere CA, Zaloga GP. Am J Med 2004;117:499–507.
65. Linton MF, Fazio S. Int J Obes Relat Metab Disord 2003;27(Suppl 3):S35–S40.
66. Charo IF, Taubman MB. Circ Res 2004;95:858–866.
67. Malavazos AE, Cereda E, Morricone L, et al. Eur J Endocrinol 2005;153:871–877.
68. Martinovic I, Abegunewardene N, Seul M, et al. Circ J 2005;69:1484–1489.
69. Clément K, Viguerie N, Poitou C, et al. FASEB J 2004;18:1657–1669.
70. Miyazawa-Hoshimoto S, Takahashi K, Bujo H, et al. Diabetologia 2003;46:1483–1488.
71. Gómez-Ambrosi J, Catalán V, Diez-Caballero A, et al. FASEB J 2004;18:215–218.
72. Verma S, Anderson TJ. Circulation 2002;105:546–549.
73. Libby P, Ridker PM, Maseri A. Circulation 2002;105:1135–1143.
74. Shoelson SE, Lee J, Yuan M. Int J Obes Relat Metab Disord 2003;27(Suppl 3):S49–S52.

6 Regulation of the Immune Response by Leptin

Víctor Sánchez-Margalet, Patricia Fernández-Riejos, Carmen González-Yanes, Souad Najib, Consuelo Martín-Romero, and José Santos-Alvarez

Abstract

Adipose tissue is no longer considered as a mere energy store, but an important endocrine organ that produces many signals in a tightly regulated manner. Leptin is one of the most important hormones secreted by the adipocyte, with a variety of physiological roles related with the control of metabolism and energy homeostasis. One of these functions is the connection between nutritional status and immune competence. Leptin's modulation of the immune system is exerted at the development, proliferation, anti-apoptotic, maturation, and activation levels. The role of leptin in regulating the immune response has been assessed in vitro as well as in clinical studies. Both the innate and adaptive immune responses are regulated by leptin. Every cell type involved in immunity can be modulated by leptin. In fact, leptin receptors have been found in neutrophils, monocytes, and lymphocytes, and the leptin receptor belongs to the family of class I cytokine receptors. Moreover, leptin activates similar signaling pathways to those engaged by other members of the family. The overall leptin action in the immune system is a proinflammatory effect, activating proinflammatory cells, promoting T-helper 1 responses, and mediating the production of other proinflammatory cytokines, such as tumor necrosis factor-α, interleukin (IL)-2, or IL-6. Leptin receptor is also upregulated by proinflammatory signals. Thus, leptin is a mediator of the inflammatory response, and could have also a permissive role in the development of autoimmune diseases.

Key Words: Leptin; cellular immnunology; immnune response; lymphocytes; monocytes; neutrophils; inflammation; leptin receptor; leptin signaling.

1. INTRODUCTION

During the last 10 yr, evidence has been accumulating from data obtained from basic work in animal models, ex vivo cell experiments, and clinical studies, demonstrating that the control of orexigenic–anorexigenic circuits regulates not only body weight but also other important physiological functions, mainly acting at the central level in the hypothalamus. One of the functions that is regulated by the control of energy balance is the immune system.

From: *Nutrition and Health: Adipose Tissue and Adipokines in Health and Disease*
Edited by: G. Fantuzzi and T. Mazzone

Different cytokines, hormones, and neuropeptides play important roles in both metabolic and immune processes. However, even though the correlation of nutritional status and immune competence has been known for many decades, the molecular and physiological mechanisms underlying this connection have not been unraveled until very recently.

Cytokines are circulating signaling molecules that mediate intercellular communication contributing to the regulation of the innate and adaptative immunity, inflammation, and metabolic function. These cytokines exert their function within a complex net of interactions in which a cytokine may influence both the production and the response of other cytokines.

White adipose tissue (WAT) plays a very important role in the energetic balance of mammals. This tissue is specialized to store lipids and supply fuels to the whole body whenever it is necessary. In order to face energetic requirements, adipocytes regulate fatty acid mobilization in response to catabolic and anabolic stimuli. However, adipose tissue is not only a reserve organ; it is also an endocrine organ that is able to release hormones, peptides, and cytokines that affect both the energetic status and the immune system. That is why these adipocyte-derived molecules are called adipokines. Members of this family of adipokines include interleukin (IL)-1, IL-6, IL-8, interferon (IFN)-γ, tumor necrosis factor (TNF)-α, transforming growth factor (TGF)-β, leukemia inhibitory factor (LIF), monocyte chemoattractant protein (MCP)-1, macrophage inflammation protein-1, and leptin. Leptin is one of the most important hormones secreted by adipose tissue *(1)* and its implication in energetic homeostasis has been largely described *(2)*.

Leptin is a signal of starvation, in that a falling serum leptin concentration leads to neurohumoral and behavioral changes, trying to preserve energy reserves for vital functions. Thus, during a fasting period and after reduction of body fat mass, there is a decrease in leptin levels that leads to a reduction in total energy expenditure to provide enough energy for the function of vital organs—i.e., the brain, the heart, and the liver *(3)*. Even though these effects of leptin decrease are aimed at improving the survival chances under starving conditions, the fall in leptin levels may lead to immune suppression *(4)*, in addition to other neuroendocrine alterations affecting adrenal, thyroid, and sexual/reproductive function *(5)*. At least, these alterations observed during fasting parallel the decrease in circulating leptin levels. In fact, both *ob/ob* mice (lacking leptin secretion) and *db/db* mice (lacking leptin receptor) are not only obese but they also show the immune/endocrine deficiencies observed during starvation *(4,6)*. Moreover, it has been recently shown that leptin withdrawal during 8 d in experimental animals leads to the same effects regarding central control of endocrine systems, including sexual function *(7)*. Even in humans, it has been found that leptin levels are associated with immune response in malnourished infants *(8)*.

The leptin modulation of the immune system is also mediated by the regulation of hematopoiesis and lymphopoiesis *(6,9)*, which are the topics of chapters 11 and 13.

In this chapter we will summarize data from literature that demonstrate the regulation of the immune response by leptin, as well as the known signaling mechanisms whereby leptin activates immune cells *(10)*.

2. LEPTIN MODULATION OF INNATE IMMUNITY

The primary amino acid sequence of leptin indicated that it could belong to the long-chain helical cytokine family *(11)*, such as IL-2, IL-12, and growth hormone (GH). In fact, leptin receptor (Ob-R) shows sequence homology to members of the class I cytokine receptor (gp130) superfamily *(12)*, which includes the receptor for IL-6, LIF, and granulocyte colony-stimulating factor (G-CSF). Moreover, Ob-R has been shown to have the signalling capabilities of IL-6-type cytokine receptors *(13)*; this is detailed later in this chapter. In this context, a role for leptin in the regulation of innate immunity has been proposed *(10,14)*. Figure 1 summarizes the role of leptin modulating the function of cells involved in both innate and adaptive immunity.

Consistent with this role of leptin in the mechanisms of immune response and host defense, circulating leptin levels are increased upon infectious and inflammatory stimuli such as lipopolysaccharide (LPS), turpentine, and cytokines *(15,16)*. On the other hand, unlike other members of the IL-6 family, it is not clear that leptin may induce the expression of acute phase proteins, and contradictory data have been provided *(16,17)*. The role of leptin regulating innate immunity has been previously reviewed *(5)* and it has been depicted in Fig. 1.

2.1. Leptin Regulation of Monocytes/Macrophages

Studies of rodents with genetic abnormalities in leptin or leptin receptors revealed obesity-related deficits in macrophage phagocytosis and the expression of proinflammatory cytokines both in vivo and in vitro, whereas exogenous leptin upregulated both phagocytosis and the production of cytokines *(18)*. Furthermore, phenotypic abnormalities in macrophages from leptin-deficient, obese mice have beeen found *(19)*. More important, leptin deficiency increases susceptibility to infectious and inflammatory stimuli and is associated with dysregulation of cytokine production *(16)*. More specifically, murine leptin deficiency alters Kupffer cell production of cytokines that regulate the innate immune system. In this context, leptin levels increase acutely during infection and inflammation, and may represent a protective component of the host response to inflammation *(20)*.

Human leptin was found to stimulate proliferation and activation of human circulating monocytes in vitro, promoting the expression of activation markers CD69, CD25, CD38, CD71, in addition to increasing the expression of monocytes surface markers, such as HLA-DR, CD11b and CD11c *(21)*. In addition, leptin potentiates the stimulatory effect of LPS or phorbol myristate acetate on the proliferation and activation of human monocytes. Moreover, leptin dose-dependently stimulates the production of proinflammatory cytokines by monocytes, i.e., TNF-α and IL-6 *(21)*. The presence of both isoforms of the leptin receptor was also assessed. Later, it was found that leptin directly induces the secretion of interleukin 1 receptor antagonist in human monocytes *(22)* and upregulates IP-10 (interferon-γ-inducible protein) in monocytc cells *(23)*. In alveolar macrophages, leptin augments leukotriene synthesis *(24)*.

A possible role of leptin as a trophic factor to prevent apoptosis has also been found in serum-depleted human monocytes *(25)*, further supporting the role of leptin as a growth factor for the monocyte. Moreover, leptin regulates monocyte function, as assessed by in vitro experiments measuring free-radical production. Thus, leptin was shown to stimulate the oxidative burst in control monocytes *(26)*, and binding of leptin at the macrophage cell surface increases lipoprotein lipase expression, through oxidative stress- and protein kinase C (PKC)-dependent pathways. In this line, leptin has been found to increase oxidative stress in macrophages *(27)*. Finally, leptin can also increase

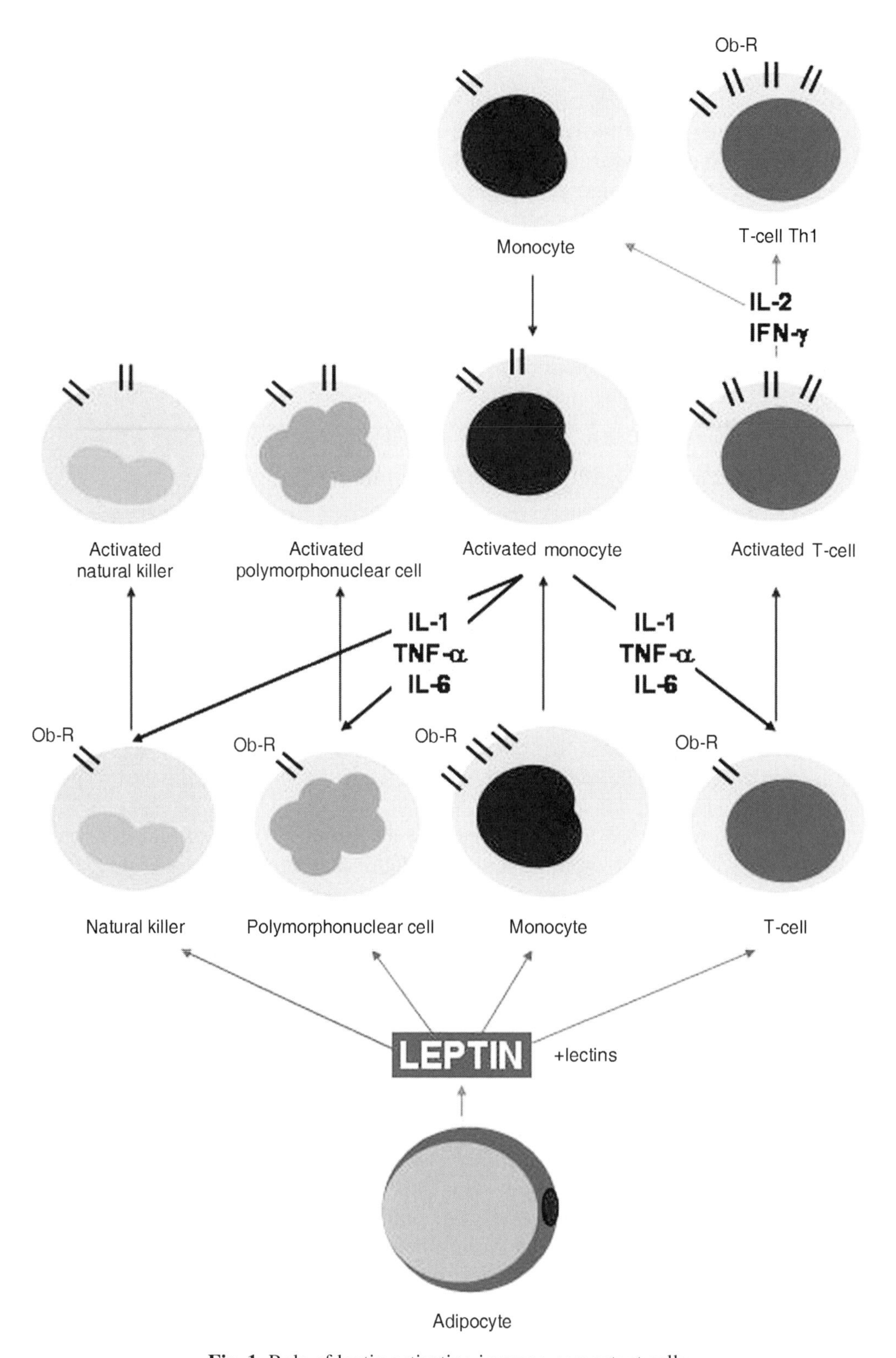

Fig. 1. Role of leptin activating immune-competent cells.

chemotaxis of blood monocytes and may mediate the inflammatory infiltrate *(28)*. On the other hand, human leptin seems to downregulate oxidative burst in previously activated monocytes *(26)*.

Dendritic cells belong to the same cell lineage as monocytes/machrophages and also present leptin receptor (OBRb) in the cell surface *(29)*. Thus, leptin has been also found to increase the production of IL-8, IL-12, IL-6, and TNF-α, whereas it decreases MIP-1-α production by dendritic cells. Similarly to leptin's effect on monocytes, it may also increase the survival of dendritic cells, and increase the expression of surface molecules, such as CD1a, CD80, CD83, or CD86.

2.2. *Neutrophils*

Human polymorphonuclear neutrophils (PMN) have been found to express leptin receptor in vitro and in vivo *(30,31)*. However, Zarkesh-Esfahani et al. *(32)* demonstrated that neutrophils express only the short form of the leptin receptor, which is enough to signal inside the cell, enhancing the expression of CD11b, and preventing apoptosis *(31,32)*. Therefore, leptin seems to behave as a survival cytokine for PMN, similar to G-CSF.

Leptin promotes neutrophil chemotaxis *(16,33)*. In fact, the chemoattractant effect is comparable to that of well-known formyl–methionyl–leucyl–phenylalanine (FMLP). Moreover, leptin also has a stimulating effect on intracellular hydrogen peroxide production in PMN, although this effect seems to be mediated by the activation of monocytes *(32)*. More specifically, leptin modulates neutrophil phagocytosis of *Klebsiella pneumoniae (34)*.

2.3. *Natural-Killer Cells*

Human natural-killer (NK) cells constitutively express both long and short forms of Ob receptor. Moreover, the leptin receptors can signal in NK cells, as leptin activates STAT3 phosphorylation in NK cells. Moreover, leptin increases IL-2 and perforin gene expression at the transcription levels in NK cells. Consistent with this role of leptin regulating NK cells, *db/db* mice have been found to have impaired NK cell function *(35,36)*.

Leptin action in NK cells include cell maturation, differentiation, activation, and cytotoxicity *(17)*.

Leptin enhances both the development and the activation of NK cells *(37)*, increasing IL-12 and reducing the expression of IL-15 *(38)*. In addition, leptin mediates the activation of NK cells indirectly by modulation of IL-1β, IL-6, and TNF-α by monocytes and macrophages *(29)*.

3. LEPTIN MODULATION OF ADAPTIVE IMMUNE RESPONSE

Faggioni et al. *(16)* have reviewed the role of leptin in cell-mediated immunity, from data obtained working with *ob/ob* mice. These mice have a decreased sensitivity of T-cells to activating stimuli. Furthermore, these animals show atrophy of lymphoid organs *(4–6)*, with a decrease in the number of circulating T-cells and an increase in the number of monocytes. In addition, *ob/ob* mice have a decrease in TNKCD4$^+$ in the liver *(39)*. The ability of leptin to prevent thymic atrophia is due to a direct antiapoptotic effect on T-cells *(6)*.

Acute deficiency of leptin has a potent effect in the immune system, which is even higher than that observed in *ob/ob* mice (genetic defect). Acute hypoleptinemic mice show a higher decrease in the total number of thymocytes, and double the number of apoptotic cells, than *ob/ob* mice. Moreover, the acute deficiency of leptin also causes a decrease in splenic cellularity, which does not occur in *ob/ob* mice, even though they have a smaller spleen than control mice *(7)*.

Both *ob/ob* and *db/db* mice show defects in cell-mediated immune response that lead to impaired reaction of delayed hypersensibility, suppression of skin allograft rejection, and inhibition of footpad swelling by recall antigens *(5,40,41)*.

Lord et al. *(4)* demonstrated that mouse lymphocytes express the long form of leptin receptor, and that leptin modulates cytokine production in these cells. Leptin also regulates the number and activation of T-lymphocytes. The proliferative response to leptin seems to be produced in naïve T-cells ($CD4^{+}CD45RA^{+}$), whereas it has been shown that leptin inhibits proliferation of memory T-cells ($CD4^{+}CD45RO^{+}$) *(4)*. Leptin provides a survival signal in double-positive T-cells ($CD4^{+}CD8^{+}$) and simple positive $CD4^{+}CD8^{-}$ thymocytes during thymic maturation *(6)*. In addition, leptin promotes the expression of adhesion molecules in $CD4^{+}$ T-cells, such as VLA-2 (CD49b) or ICAM-1 (CD54) *(4,15)*.

More recently, we have reviewed the role of human leptin on T-cell response *(10)*. Human leptin alone is not able to activate human peripheral blood lymphocytes in vitro *(21)*. However, when T-lymphocytes are co-stimulated with PHA or concanavalin A (Con A), leptin dose-dependently enhances the proliferation and activation of cultured T-lymphocytes, achieving maximal effect at 10 n*M* concentration *(42)*. Thus, leptin increases the expression of early activation markers such as CD69, as well as the expression of late activation markers, such as CD25, or CD71 in both $CD4^{+}$ and $CD8^{+}$ T-lymphocytes in the presence of suboptimal concentrations of activators such as PHA (2 μg/mL). However, when maximal concentrations of PHA or Con A are employed, leptin has no further effect. These effects of leptin on T-lymphocytes are observed even in the absence of monocytes, suggesting a direct effect of human leptin on circulating T-lymphocytes when they are co-stimulated *(42)*. The activation of T-cells induces the expression of the long isoform of the Ob receptor *(43)*. The need for co-stimulation with mitogens to get the effect of leptin in lymphocytes may be partly explained by this effect of activation, increasing leptin receptor expression in T-lymphocytes. Besides, these data suggest that the leptin receptor may be regulated in a similar way to other cytokine receptors, such as the IL-2 receptor (CD25).

Human leptin not only modulates the activation and proliferation of human T-lymphocytes, but it also enhances cytokine production induced by submaximal concentrations of PHA *(42)*. Thus, human leptin enhances the production of IL-2 and IFN-γ in stimulated T-lymphocytes. It had been previously shown in mice that leptin can enhance cognate T-cell response, skewing cytokine responses toward a Th1 phenotype in mice *(4)*. These data are in agreement with the observation of the leptin effect on anti-CD3 stimulation of T-cells, which increases the production of the proinflammatory cytokine IFN-γ *(44)*. The effect of leptin polarizing T-cells toward a Th1 response seems to be mediated by the stimulation of IL-12 production and the inhibition of IL-10 production *(29)*.

These data regarding leptin modulation of Th1 type cytokine production are in line with the observed effects of leptin stimulating TNF-α and IL-6 production by monocytes

(21), further suggesting the possible role of human leptin in the regulation of the immune system inducing a proinflammatory response.

On the other hand, leptin promotes T-cell survival by modulating the expression of anti-apoptotic proteins, such as Bcl-xL, in stress-induced apoptosis *(45)*. This trophic effect of leptin on T-cell is consistent with the reduction in lymphocyte numbers observed in fasted mice, that might be explain by the acute decrease in leptin levels *(6,15)*.

The role of leptin modulating T-cell function in human being has been finally defined by clinical studies in specific and rare cases of monogenic obesity. In human obesity, due to congenital leptin deficiency, there is a T-cell hyporesponsiveness (in addition to neuroendocrine/metabolic dysfunction); leptin treatment in these patients not only is effective to lower body weight, but also can revert T-cell response to mitogen activation in vitro *(46)*.

4. MECHANISMS OF LEPTIN ACTION IN IMMUNE CELLS

Leptin receptor (Ob-R) belongs to the family of class I cytokine receptors, which include receptors for IL-2, IL-3, IL-4, IL-6, IL-7, LIF, G-CSF, growth hormone-releasing hormone, prolactin, and erythropoietin *(12)*. As mentioned in the previous section, Ob-R expression is present in hematopoietic cells as well as the cells that participate in the innate and adaptive immune response. Leptin signaling has recently been reviewed *(47,48)*, and we have previously reviewed leptin signaling in mononuclear cells *(10)*. Most members of the cytokine family of receptors stimulate tyrosine phosphorylation of signal transducers and activators of transcription (STAT) proteins by activating JAK kinases, which are associated with the intracellular part of the transmembrane receptor *(49,50)*.

Tyrosine phosphorylation of the activated leptin receptor has already been reported in other systems *(51,52)*. In human blood mononuclear cells, we have also found that human leptin stimulates tyrosine phosphorylation of the long form of the leptin receptor *(53)*, as assessed by immunoprecipitation with an antibody against the C-terminal of the protein and immunoblotting with antibodies against phosphotyrosine. This effect is dependent on the dose at 5 min incubation. Maximal phosphorylation can be observed with 10 n*M* leptin.

The leptin receptor lacks intrinsic tyrosine kinase activity, but requires the activation of receptor-associated kinases of the Janus family (JAKs) *(54)*, which initiate downstream signalling including members of the STAT family of transcription factors *(13,51)*. After ligand binding, JAKs autophosphorylate and tyrosine phosphorylates various STATs. Activated STATs by leptin stimulation in the hypothalamus dimerize and translocate to the nucleus, where specific gene responses are elicited *(55,56)*. In this context, we have studied the JAK–STAT signaling pathway triggered by leptin stimulation in human peripheral blood mononuclear cells *(10,57)*. To investigate the activation of JAK kinases by leptin receptor in human peripheral blood mononuclear cells, we studied the effect of leptin on tyrosine phosphorylation of immunprecipitated JAK proteins by immunoblot. We found that both JAK-2 and JAK-3 are transiently activated, with maximal response at 5 min after leptin stimulation. Moreover, both isoforms have been found physically associated with the leptin receptor, as assessed by coimmunoprecipitation studies. This association turned out to be constitutive, and it occurs both in the absence and presence of leptin in peripheral blood mononuclear cells. Preassociation of

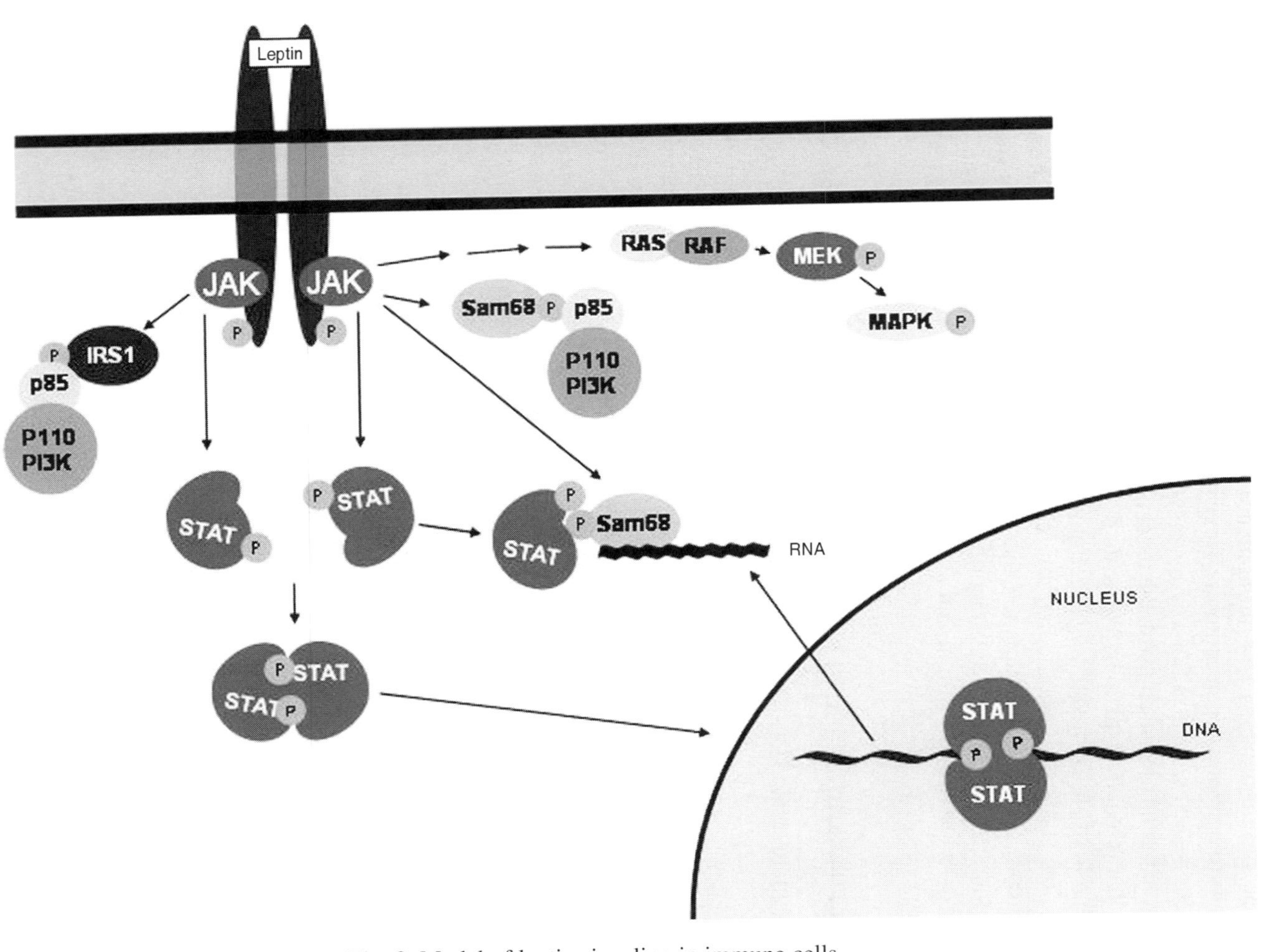

Fig. 2. Model of leptin signaling in immune cells.

JAK proteins with cytokine receptors has been described for other members of the family *(58)*, and for the leptin receptor itself with JAK-2 *(51)*. The relative contribution of each JAK isoform in leptin receptor signaling in human peripheral blood mononuclear cells, however, remains to be assessed.

The possible activation of STAT-3 by human leptin in mononuclear cells has also been studied at different time-points in peripheral blood mononuclear cells *(57)* by antiphosphotyrosine immunoblotting of anti-STAT-3 immunoprecipitates. The effect of leptin promoting STAT-3 tyrosine phosphorylation is maximal at 10 min, and the effect is dose-dependent, reaching maximal effect at 10 n*M* leptin. Leptin stimulation of mononuclear cells not only activates and phosphorylates STAT-3 but also promotes the tyrosine phosphorylation of the RNA binding protein Sam68, which has been previously found to be recruited upon TCR, and insulin receptor activation *(59,60)* associates with STAT-3 in the same signaling complex *(57)*. Tyrosine phosphorylation of Sam68 by the Src family kinase p59fyn has been shown previously to regulate negatively its association with RNA *(61)*. In mononuclear cells we have found that leptin not only promotes the tyrosine phosphorylation of Sam68 *(57)*, but also inhibits the RNA binding efficiency of this protein *(53)*. This effect of leptin was also found to be dependent on the dose. Even though we do not know the role of Sam68 in leptin signaling and leptin action, the effect of leptin regulating the RNA binding capacity of Sam68 may be involved in the post-transcriptional modulation of RNA, regulating either the metabolism, the splicing, or the localization of RNA. In this way, Sam68 has been proposed to provide the means for a rapid pathway to regulate protein expression by modifying the mRNA stability and/or mRNA translation.

Leptin has been shown to promote the translocation of STAT-3 to the nucleus in the rat hypothalamus *(55,56)*. Some STAT-3 can be detected in nuclear lysates in basal conditions, but a significant increase in the amount of STAT-3 is found in the nuclear extract from cells stimulated with human leptin (10 n*M*) for 15 min, suggesting the translocation of STAT-3 to the nucleus upon leptin stimulation. Other STAT isoforms (STAT-1, STAT-5, STAT-6) have also been shown to be activated by the leptin receptor, although only in transfected systems *(62,63)*. However, other STAT forms have not been assayed in lymphocytes and therefore we cannot rule out the possible implication of STATs other than STAT-3. Nevertheless, STAT-3 is the only STAT that has been shown to be activated by leptin in the hypothalamus *(64)*.

In this context, the tyrosine phosphorylation of leptin receptor and the activation of JAK-2 and STAT-3 by leptin stimulation has been confirmed in a murine macrophage cell line *(65)*.

Different pathways in addition to STATs are known to be involved in leptin receptor signaling in a similar way to other members of the cytokine family. Thus, leptin has been shown to activate mitogen-activated protein kinase (MAPK) *(52,66,67)* and phosphatidylinositol 3-kinase (PI3K) *(68,69)*. More precisely, leptin receptor signaling in mononuclear cells has been shown to activate MAPK and p70 S6K pathways *(70)*. We have also found that leptin stimulates tyrosine/threonine phosphorylation of MAPK in blood mononuclear cells, as assessed by immunoblot *(53)*. Both Erk-1 and Erk-2 were phosphorylated after 10 min incubation with leptin. This effect was dose-dependent and maximal response was obtained at 10 n*M* leptin. The activation of MAPK was transient and decreased after 15 to 30 min incubation *(53)*. More important, we have found that

MAPK activation by leptin is necessary for the antiapoptotic effect in human mononuclear cells *(25)*.

Evidence that leptin initiates a signaling cascade involving MAPK-dependent pathway has been also found in neutrophils *(31)*, whereby leptin also inhibits apoptotic pathways. Neutrophils seem to express only the short isoform of leptin receptor, Ob-Ra *(32)*, which does not signal via the JAK–STAT pathway, but may be sufficient to stimulate the MAPK pathway.

We have also explored the PI3K pathway in peripheral blood mononuclear cells (PBMC) in response to human leptin. Thus, PI3K activity associated with tyrosine phosphorylated proteins is found to be increased more than threefold after 10 n*M* leptin stimulation *(53)*. PI3K activation is regulated by the association of tyrosine phosphorylated proteins with the SH2 domains of p85. Thus, in response to leptin, a band corresponding to the molecular mass of the insulin receptor substrate (IRS)-1 and several bands of 60 to 70 kDa (including Sam68) are phosphorylated and associated with p85 in a dose–response manner, similar to our previous data in response to insulin *(53,60,71)*. Maximal response is observed at 10 n*M* leptin and 5 min incubation time. The activation of PI3K in a macrophage cell line has also been demonstrated *(65)*. Leptin also initiates the activation of PI3K pathway in neutrophils, even though they express only the short isoform of leptin receptor *(31)*. This signaling pathway is involved in transducing leptin-mediated antiapoptotic signals into neutrophils.

Finally, new mechanisms by which leptin can promote inflammatory responses have been recently provided, at least in alveolar macrophages—i.e., the upregulation of phospholipase A2 activity and phospholipase A2γ protein levels *(24)*.

5. LEPTIN AND PATHOPHYSIOLOGY OF THE IMMUNE SYSTEM

It seems to be generally accepted that leptin may be an important signal that connects energy stores with the immune system *(3,5,10)*. In fact, leptin has been considered an indicator of nutritional status. Thus, falling leptin levels in starvation may alert the organism to avoid energy wasting and to seek energy storing. Therefore, the lack or decrease of leptin levels may play a role in the immunosuppression of starvation. In this context, leptin may serve to signal the organism to conserve energy by shutting down nonessential systems. Moreover, leptin might be not only a signal for the adaptation of starvation, but in addition, leptin could have played a role in selection, helping the better nourished to survive under starving conditions, allowing the defense against infection to those better prepared.

The same immune deficiency observed in starvation has been found in human obesity syndromes caused by a deficiency of leptin production or leptin action *(46)*. In addition, leptin administration in obesity caused by lack of leptin can restore immune function as observed in a similar way as that seen in starving conditions *(6–8)*.

On the other hand, an excess of leptin in the circulation that occurs in obesity and overweight could have a role in pathological conditions mediated by an excess of immune response *(73)*. This "proinflammatory state" may be relevant for cardiovascular disease and the risk for myocardial infarction that is increased in obese people. Moreover, leptin, together with other adipokines, could be a common link between obesity and cardiovascular risk in metabolic syndrome. Because obese people do not respond to leptin properly, central versus peripheral leptin resistance may underlie the pathophysiology of obesity, and therefore, the study of leptin signaling at central versus peripheral levels may improve our understanding of the mechanisms involved in the metabolic and immune alterations in the metabolic syndrome that lead to increased cardiovascular risk. These molecular defects may connect the thrifty phenotype with the proinflammatory phenotype in a common trait that turns out to be lethal in the Westernized way of life.

Recent data also suggest the possible role of leptin in the pathophysiology of some autoimmune diseases. This has already been demonstrated in some animal models, such as the experimental autoimmune encephalomyelitis, antigen-induced arthritis, models of type 1 diabetes, autoimmune colitis, and experimental hepatitis, as well as some clinical studies *(15,17,72–74)*. However, this is the subject of other chapters in this book, and we are not reviewing the possible role of leptin in the pathophysiology of immune diseases.

Finally, leptin administration has been proposed as a possible treatment, not only for the immunodeficiency of obesity syndromes caused by a deficit of leptin *(46)*, but also in some immunodeficiency syndromes, such as the common variable immunodeficiency, to improve both function of T-cells and the synthesis of immunoglobulins. In this line, recent data obtained using cells from patients in vitro support this hypothesis *(75)*.

6. CONCLUSIONS

In conclusion, leptin may be considered as a therapeutic target in some clinical situations, such as proinflammatory states or autoimmune diseases, to control an excess of immune response, as well as in other clinical situations, such as starving, excess of exercise, or immune deficiencies, to improve the impaired immune response. That is why the investigation of the role of leptin in the regulation of the immune response is still a challenge for the future.

REFERENCES

1. Zhang Y, Proenca R, Maffei M, et al. Nature 1994;372:425–432.
2. Flier JS. Cell 1995;80:15–18.
3. Ahima RS, Prabakaran D, Mantzoros C, et al. Nature 1996;382:250–252.
4. Lord GM, Matarese G, Howard JK, et al. Nature 1998;394:897–901.
5. Fantuzzi G, Faggioni R. J Leukoc Biol 2000;68:437–446.
6. Howard JK, Lord GM, Matarese GJ. Clin Invest 1999;104:1041–1059.
7. Montez JM, Soukas A, Asilmaz E, et al. Proc Natl Acad Sci USA 2005;102:2537–2542.
8. Palacio A, Lopez M, Perez-Bravo F, et al. J Clin Endocrinol Metab 2002;87:3040–3046.
9. Bennet BD, Solar GP, Yuan JQ, et al. Curr Biol 1996;6:1170–1180.
10. Sánchez-Margalet V, Martín-Romero C, Santos-Alvarez J, et al. Clin Exp Immunol 2003;133:11–19.
11. Madej T, Boguski MS, Bryant SH. FEBS Lett 1995;373:13–18.
12. Tartaglia LA, Dembski M, Weng X, et al. Cell 1995;83:1263–1271.
13. Baumann H, Morella KK, White DW, et al. Proc Natl Acad Sci USA 1996;93:8374–8378.
14. Lord GM, Matarese G, Howard JK. Science 2001;292:855–856.
15. Matarese G. European Cytokine Network 2000;11:7–13.
16. Faggioni R, Feingold KR, Grunfeld C. FASEB J 2001;15:2565–2571.
17. Matarese G, Moschos S, Mantzoros CS. J Immunol 2005;173:3137–3142.
18. Loffreda S, Yang SQ, Lin HZ, et al. FASEB J 1998;12:57–65.
19. Lee FY, Li Y, Yang EK, et al. Am J Physiol 1999 ;276:C386–C394.
20. Sarraf P, Frederich RC, Turner EM, et al. J Exp Med 1997;185:171–175.
21. Santos-Alvarez J, Goberna R, Sánchez-Margalet V. Cell Immunol 1999;194:6–11.
22. Gabay C, Dreyer M, Pellegrinelli N, et al. J Clin Endocrinol Metab 2001;86:783–791.

23. Meier CA, Chicheportiche R, Dreyer M, et al. Cytokine 2003;21:43–47.
24. Mancuso P, Canetti C, Gottschalk A, et al. Am J Physiol Lung Cell Mol Physiol 2004;L287:497–502.
25. Najib S, Sanchez-Margalet V. Cell Immunol 2002;220:143–149.
26. Sánchez-Pozo C, Rodriguez-Bano J, Dominguez-Castellano A, et al. Clin Exp Immunol 2003;134: 464–469.
27. Maingrette F, Renier G. Diabetes 2003;52:2121–2128.
28. Curat CA, Miranville A, Sengenes C, et al. Diabetes 2004;53:1285–1292.
29. Mattioli B, Straface E, Quaranta MG, et al. J Immunol 2005;174:6820–6828.
30. Caldefie-Chezet F, Poulin A, Tridon A, et al. J Leukoc Biol 2001;69:414–418.
31. Bruno A, Conus S, Schmid I, et al. J Immunol 2005;174:8090–8096.
32. Zarkesh-Esfahani H, Pockley AG, Wu Z, et al. J Immunol 2004;172:1809–1814.
33. Caldefie-Chezet F, Poulin A, Vasson MP. Free Radic Res 2003;37:809–814.
34. Moore SI, Huffnagle GB, Chen GH, et al. Infect Immun 2003;71:4182–4185.
35. Tian Z, Sun R, Wei H, et al. Biochem Biophys Res Commun 2002;298(3):297–302.
36. Zhao Y, Sun R, You L, et al. Biochem Biophys Res Commun 2003;300:247–252.
37. La Cava A, Matarese G. Nat Rev Immunol 2004;4:371–379.
38. Li Z, Lin H, Yang S, et al. Gastroenterology 2002;123:1304–1310.
39. Guebre-Xabier M, Yang S, Lin HZ, et al. Hepatology 2000;31:633–640.
40. Mandel MA, Mahmoud AAF. J Immunol 1978;120:1375–1377.
41. Chandra RK, Au B. Int Arch Allergy Appl Immunol 1980;62:94–98.
42. Martín-Romero C, Santos-Alvarez J, Goberna R, et al. Cell Immunol 2000;199:15–24.
43. Sánchez-Margalet V, Martín-Romero C, González-Yanes C, et al. Clin Exp Immunol 2002;129:119–124.
44. Lord GM, Matarese G, Howard JK, et al. J Leukoc Biol 2002;72:330–338.
45. Fujita Y, Murakami M, Ogawa Y, et al. Clin Exp Immunol 2002;128:21–26.
46. Farooqi IS, Matarese G, Lord GM, et al. J Clin Invest 2002;110:1093–1103.
47. Myers MG Jr. Recent Prog Horm Res 2004;59:287–304.
48. Ahima RS, Osei SY. Physiol Behav 2004;81:223–241.
49. Kishimoto T, Taga T, Akira S. Cell 1994;76:253–262.
50. Darnell JE Jr. Science 1997;277:1630–1635.
51. Ghilardi N, Skoda RC. Mol Endocrinol 1997;11:393–399.
52. Bjørbaek C, Uotani S, da Silva B, et al. J Biol Chem 1997;272:32,686–32,695.
53. Martín-Romero C, Sánchez-Margalet V. Cell Immunol 2001;212:83–91.
54. Tartaglia LA. J Biol Chem 1997;272:6093–6096.
55. Schwartz MW, Seeley RJ, Campfield LA, et al. J Clin Invest 1996;98:1101–1106.
56. Vaisse C, Halaas JL, Horvath CM, et al. Nat Genet 1996;14:95–97.
57. Sánchez-Margalet V, Martin-Romero C. Cell Immunol 2001;211:30–36.
58. Ihle JN. Nature 1995;377:591–594.
59. Fusaki N, Iwamatsu A, Iwashima M, et al. J Biol Chem 1995;272:6214–6219.
60. Najib S, Martin-Romero C, Gonzalez-Yanes C, et al. Cell Mol Life Sci 2005;62:36–43.
61. Hartmann AM, Nayler O, Schwaiger FW, et al. Mol Biol Cell 1999;10:3909–3926.
62. Rosenblum CI, Tota M, Cully D, et al. 1996;137:5178–5181.
63. Wang Y, Kuropatwinski KK, White DW, et al. J Biol Chem 1997;272:16,216–16,223.
64. McCowen KC, Chow JC, Smith RJ. Endocrinology 1998;139:4442–4447.
65. O'Rourke L, Yeaman SJ, Shepherd PR. Diabetes 2001;50:955–961.
66. Bjørbaek C, Buchholz RM, Davis SM, et al. J Biol Chem 2001;276:4747–4755.
67. Takahashi Y, Okimura Y, Mizuno I, et al. J Biol Chem 1997;272:12,897–12,900.
68. Kellerer M, Koch M, Metzinger E, et al. Diabetologia 1997;40:1358–1362.
69. Harvey J, McKay NG, Walker KS, et al. J Biol Chem 2000;275:4660–4669.
70. Van den Brink GR, O'Toole T, Hardwick JC, et al. Mol Cell Biol Res Commun 2000;4:144–150.
71. Sánchez-Margalet V, Najib S. FEBS Lett 1999;455:307–310.
72. Matarese G, Alviggi C, Sanna V, et al. J Clin Endocrinol Metab 2000;85:2483–2487.
73. Matarese G, La Cava A, Sanna V, et al. Trends Immunol 2002;23:182–187.
74. Fantuzzi G, Sennello JA, Batra A, et al. Clin Exp Immunol 2005;142:31–38.
75. Goldberg AC, Eliaschewitz FG, Montor WR, et al. Clin Immunol 2005;114:147–153.

7 Leptin in Autoimmune Diseases

Giuseppe Matarese

Abstract

Over the last few years, a series of molecules known to play a function in metabolism have also been shown to play an important role in the regulation of the immune response. In this context, the adipocyte-derived hormone leptin has been shown to regulate the immune response both in normal as well as in pathological conditions. More specifically, it has been shown that conditions of reduced leptin production are associated with increased infection susceptibility. Conversely, immune-mediated disorders such as autoimmune diseases are associated with increased secretion of leptin and production of pro-inflammatory pathogenic cytokines. In this context, leptin could represent the "missing link" between immune response, metabolic function, and nutritional status. Strategies aimed at interfering with the leptin axis could represent innovative therapeutic tools for infections and autoimmune disorders. This chapter reviews the most recent advances in the role of leptin in autoimmune responses.

Key Words: Leptin; metabolism; autoimmunity; inflammation; immune tolerance.

1. INTRODUCTION

Over the past century, improved hygienic and nutritional conditions have significantly reduced the incidence of infectious diseases, at least in the most developed countries *(1)*. In parallel with the improvement in nutritional status, however, an increase in susceptibility to autoimmune disorders has emerged *(2,3)*. Recently, it has been proposed that the lifestyle in developed countries, with reduced exposure to environmental pathogens, could be relevant to the increase in the prevalence of autoimmune disorders *(3)*. Conversely, in less-affluent societies, exposure to microorganisms, pathogens, and other environmental influences might promote the development of T-regulatory responses that protect against autoimmune responses *(3,4)*. Leptin, an adipocyte-derived hormone of the long-chain helical cytokine family, has recently been proposed to act as a link between nutritional status and immune function *(5,6)*. Leptin has multiple biological effects on nutritional status, metabolism, and the neuro–immuno–endocrine axis. The circulating concentration of leptin is proportional to fat mass *(6)*, and reduced body fat or nutritional deprivation—typically associated with hypoleptinemia—is a direct cause of secondary immunodeficiency and increased susceptibility to infections *(5,7,8)*. The reason for this association was not apparent until recently. Now, it can be hypothesized that a low concentration of serum leptin increases susceptibility to infectious diseases by reducing T-helper (Th)-cell priming and direct effects on thymic function

From: *Nutrition and Health: Adipose Tissue and Adipokines in Health and Disease*
Edited by: G. Fantuzzi and T. Mazzone © Humana Press Inc., Totowa, NJ

(8,9). Furthermore, congenital deficiency of leptin has been found to be associated with increased frequency of infection and related mortality *(9,10)*. By contrast, the Th1-promoting effects of leptin have been linked recently to enhanced susceptibility to experimentally induced autoimmune diseases, such as experimental autoimmune encephalomyelitis (EAE), type 1 diabetes (T1D), antigen-induced arthritis (AIA), and others. These latter observations suggested a novel role for leptin in determining the gender bias of susceptibility to autoimmunity, because female mice and humans, which are relatively hyperleptinemic, have an increased frequency of autoimmune diseases compared with males, which are relatively hypoleptinemic *(5,6)*. In view of these findings, we suggest leptin as novel candidate able to explain at least in part the increased frequency of autoimmune disorders in the more affluent countries and in females. Further experimental evidence is needed to address the precise role of leptin in autoimmune disease susceptibility.

2. LEPTIN IN MULTIPLE SCLEROSIS

Multiple sclerosis (MS) is a chronic, immune-mediated, inflammatory disorder of the central nervous system (CNS) myelin. The most studied model of multiple sclerosis in animals is the EAE that can be induced in susceptible strains of mice by immunization with self-antigens derived from myelin. The disease is characterized by autoreactive T-cells that traffic to the brain and the spinal cord and injury myelin, with the result of chronic or relapsing-remitting paralysis, depending on the antigen and the strain of mice used. It is known that myelin-reactive Th1 $CD4^+$ cells induce and/or transfer disease and that Th1 cytokines are present in inflammatory EAE lesions in the central nervous system. In contrast, Th2 cytokines are associated with recovery from EAE or protection from the disease.

Leptin is involved in both the induction and progression of EAE in mice *(11–13)*. Analysis of the disease susceptibility in naturally leptin-deficient *ob/ob* mice before leptin replacement revealed resistance to both active and adoptive EAE that was reversed by leptin administration. Leptin replacement converted Th2- to Th1-type response and shifted IgG antibodies from IgG1 to IgG2a. In addition, leptin administration to susceptible wild-type C57BL/6J mice worsened the disease by increasing proinflammatory cytokine release and IgG2a production *(11)*. In addition, it has also been recently observed that a serum leptin surge precedes the onset of EAE in susceptible strains of mice *(12)*. This peak in serum leptin is correlated with inflammatory anorexia, weight loss, and development of a pathogenic T-cell response against myelin *(12)*. In animals with EAE, inflammatory brain infiltrates have also been shown to be a source of leptin, attesting to an *in situ* leptin production in active lesions *(12)*. Systemic and/or *in situ* leptin secretion was not observed in EAE-resistant mice. Taken together, these data show an involvement of leptin in the pathogenesis of central nervous system autoimmunity in the EAE model.

In human MS, it has been reported that secretion of leptin is increased in serum and cerebrospinal fluid (CSF) of naïve-to-treatment MS patients and positively correlated with the secretion of interferon (IFN)-γ in CSF and inversely with the percentage of circulating regulatory T-cells (T_{Regs}), a key cellular subset in the suppression of immune and autoimmune responses, involved in the maintenance of T-cell tolerance *(14)*. In addition, T_{Regs} in patients with MS were not only inversely related to the leptin levels but also were reduced in percentage and absolute numbers when compared with healthy controls *(14)*. This suggests that the number of T_{Regs} can be affected by leptin secretion. Because of thymic generation of T_{Regs}, it is possible to speculate that leptin produced in the mediastinic perithymic adipose tissue could be able to affect T_{Regs} generation/function in autoimmunity-prone subjects. This hypothesis is object of extensive investigation.

The evidence that a significant increase of leptin secretion occurs in the acute phase of MS and that this event positively correlates with CSF production of IFN-γ is of particular interest for the pathogenesis and clinical follow-up of patients with MS. Increased secretion was present in both the serum and CSF of MS patients and determined loss of correlation between leptin and body mass index (BMI) *(14)*. Moreover, the increase of leptin in the CSF was higher than in serum, possibly secondary to *in situ* synthesis of leptin in the central nervous system (CNS) and/or an increased transport across the blood–brain barrier (BBB), upon enhanced systemic production.

Recently, gene-microarray analysis of Th1 lymphocytes and active MS lesions in humans revealed elevated transcripts of many genes of the neuro–immuno–endocrine axis, including leptin *(15)*. Its transcript was also abundant in the gene-expression profile of human Th1 clones, demonstrating that the leptin gene is induced in and associated with polarization toward Th1 responses, commonly involved in T-cell-mediated autoimmune diseases such as MS *(15)*. Recently, Sanna et al. reported *in situ* leptin secretion by inflammatory T-cells and macrophages in active EAE lesions *(12)*. More recently, Matarese et al. reported that autoreactive human myelin basic protein (hMBP)-specific T-cells from patients with MS can produce immunoreactive leptin and upregulate the leptin receptor after activation, possibly explaining in part the increased *in situ* CSF leptin levels in patients with MS *(14)*. Interestingly, both anti-leptin as well as anti-leptin receptor-blocking antibodies reduced the proliferative responses of hMBP-specific T-cell lines, underlying possibilities of leptin-based intervention on this autocrine loop *(14)*. In addition, recent reports *(16)* have shown increased secretion of serum leptin before relapses in patients with MS during treatment with IFN-β and the capacity of leptin to enhance in vitro secretion of tumor necrosis factor (TNF)-α, IL-6, and IL-10 by peripheral blood mononuclear cells (PBMCs) of patients with MS in the acute phase of the disease but not in patients in stable phase *(16)*. In view of the above considerations, we suggest that in MS leptin may be part of a wider scenario in which several proinflammatory soluble factors may act in concert in driving the pathogenic (autoreactive) Th-1 responses targeting neuroantigens.

3. LEPTIN IN TYPE 1 AUTOIMMUNE DIABETES

Leptin is involved in other autoimmune conditions *(17)*. Leptin accelerates autoimmune diabetes in female NOD/LtJ mice *(18)*. Fluctuations in serum leptin levels have also been observed in a study performed by our group in an animal model of $CD4^+$ T-cell-mediated autoimmune disease, such as T1D. Nonobese diabetic (NOD/LtJ) female mice, spontaneously prone to the development of β-cell autoimmunity, have higher serum leptin levels, as compared with NOD/LtJ males and nonsusceptible strains of mice, and show a serum leptin surge preceding the appearance of hyperglycemia *(19)*.

Furthermore, leptin administration early in life significantly anticipated the onset of diabetes and increased mortality and inflammatory infiltrates in beta-islets; this phenomenon correlated with increased secretion of IFN-γ in leptin-treated NOD mice *(20)*. More recently, it has been found that a natural leptin receptor mutant of the NOD/LtJ strain of mice (named NOD/LtJ-db5J) displays reduced susceptibility to T1D *(21)*. These data further support the role of leptin in the pathogenesis of T1D. These NOD-db5J mice are obese, hyperphagic, and show hyperglycemia associated with hyperinsulinemia. The leptin receptor mutation affects the extracellular domain of the leptin receptor, probably impairing leptin binding and/or receptor dimerization. This effect is likely able to alter the intracellular signaling machinery, thus impairing the pathogenicity of anti-islets autoreactive T-cells. Indeed, these mice show low-grade infiltration of the islets. This model nicely complements the previously published data from our group, hypothesizing a key role for leptin in the development of T1D. Further studies are needed to address the molecular machinery determining the phenotype of resistance observed in these mice, as well as the possibility to interfere with T1D pathogenesis by blocking the leptin axis.

4. LEPTIN IN RHEUMATOID ARTHRITIS

AIA is a model of immune-mediated joint inflammation induced by administration of methylated bovine serum albumin (mBSA) into the knees of immunized mice *(21)*. The severity of arthritis in leptin- and leptin receptor-deficient mice was reduced in this model *(22)*. The milder form of AIA seen in *ob/ob* and *db/db* mice, as compared with their controls, was accompanied by decreased synovial levels of interleukin (IL)-1β and TNF-α, decreased proliferative response to antigen of lymph node cells in vitro, and a switch toward production of Th2 cytokines *(22)*. Serum levels of all isotypes of anti-mBSA antibodies were significantly decreased in arthritic *ob/ob* mice as compared with controls. Thus, in AIA leptin probably contributes to joint inflammation by regulating both humoral and cell-mediated immune responses. However, joint inflammation in AIA depends on the adaptive immune response, which is known to be impaired in *ob/ob* and *db/db* mice, so recent studies investigated the effect of leptin and leptin receptor deficiency on inflammatory events of zymosan-induced arthritis (ZIA), a model of proliferative arthritis, which is restricted to the joint injected with zymosan A and is not dependent on the adaptive immune response *(23)*. The results of these experiments showed that, in contrast to AIA, ZIA was not impaired in *ob/ob* and *db/db* mice. On the contrary, the resolution of acute inflammation appeared to be delayed in the absence of leptin or leptin signaling, suggesting that leptin may exert anti-inflammatory properties and chronic leptin deficiency may interfere with adequate control of the inflammatory response in ZIA *(23)*.

In patients with rheumatoid arthritis (RA) it was reported that fasting leads to an improvement of different clinical and biological measures of disease activity, which were associated with a marked decrease in serum leptin, a decreased $CD4^+$ lymphocyte activation, and a shift toward Th2 cytokine production, such as IL-4 *(24)*. These features, resembling those seen during AIA in *ob/ob* mice, suggest that leptin may also influence the inflammatory mechanisms of arthritis in humans through the induction of Th1 responses. However, the same investigators showed that a 7-d ketogenic diet reduced serum leptin concentrations in patients with RA without significant changes in any clinical or biological measurements of disease activity. In this context, Bokarewa et al. *(25)* reported increased leptin plasma levels in 76 patients with RA as compared with healthy controls. The authors also observed that leptin levels in synovial fluid were reduced as compared with matched plasma samples and that the difference between plasma and synovial fluid was particularly pronounced in nonerosive arthritis. In addition, the authors suggested a local consumption of leptin in the joint, which may exert a protective effect against the destructive course of RA. Consistent with this hypothesis, these investigators previously observed that treatment with recombinant leptin reduced both the severity of joint manifestations in *Staphylococcus aureus*-induced arthritis and the inflammatory response, as measured by serum IL-6 levels, without affecting the survival of bacteria in vivo *(26)*. Contrasting results were obtained by Popa et al. *(27)*. The authors found a significant inverse correlation between inflammation and leptin concentrations in patients with active RA, although plasma leptin concentrations did not significantly differ from those observed in healthy controls. They suggest that active chronic inflammation may lower plasma leptin concentrations *(27)*. Other groups showed that serum levels of leptin were not increased in patients with RA as compared with controls. In addition, they did not find any correlation between leptin levels and either clinical or biological signs of disease activity, whereas a positive correlation was seen between leptin and BMI or the percentage of body fat *(28)*. Another group reported lower plasma leptin levels in patients with RA than in controls *(29)*. In patients with RA from this cohort, leptin did not correlate with BMI, C-reactive protein (CRP), total fat mass, or disease activity score *(29)*. Despite the presence of methodological problems and contrasting results, all these data suggest that leptin may influence the disease process in two opposing ways, either by enhancing the expression of Th1 cytokines or by limiting the inflammatory responses. Thus, further longitudinal studies including patients with early RA are still needed to clarify the potential influence of leptin on disease outcome, and particularly the progression of joint damage.

5. LEPTIN IN LIVER AND KIDNEY IMMUNE-MEDIATED DISORDERS

Protection of *ob/ob* mice from autoimmune damage is also observed in experimentally induced hepatitis (EIH). Activation of T-cells and macrophages is one of the initial events during viral or autoimmune hepatitis. Activated T-cells are directly cytotoxic for hepatocytes and release proinflammatory cytokines, which mediate hepatocyte damage in several models. A well-described mouse model of T-cell-dependent liver injury is the iv injection of the T-cell mitogen concanavalin A (Con A), which results in fulminant hepatitis. During Con A-induced hepatitis, TNF-α is a crucial cytokine, as specific neutralization of this cytokine reduces liver damage. The injection of TNF-α causes acute inflammatory hepatocellular apoptosis followed by organ failure, and TNF-α appears to cause hepatoxicity in several experimental models. Siegmund et al. *(30)* showed that leptin-deficient *ob/ob* mice were protected from Con A-induced hepatitis. Moreover, TNF-α and IFN-γ levels, as well as expression of the activation marker CD69, were not elevated in *ob/ob* mice following administration of Con A, suggesting that resistance is associated with reduced levels of those selected proinflammatory cytokines and a lower percentage of intrahepatic natural-killer (NK) T-cells, which also contribute to disease progression. Similar results were obtained in EIH induced by *Pseudomonas aeruginosa* exotoxin A administration *(31)*. Also in this case, leptin administration restores responsiveness of *ob/ob* mice to EIH, and T-lymphocytes and TNF-α were required for induction of liver injury. The authors also demonstrated that leptin plays an important role in the production of two proinflammatory cytokines, such as TNF-α and IL-18, in the liver.

In human inflammatory liver disorders, such as steatohepatitis, conflicting results have been reported, in that leptin levels were increased or decreased according to the clinical stage, sex, and age of patients. The reasons accounting for these differences may be ascribed to different patients enrolling and different study settings. Further investigation is needed to address these points.

Most recently, protection from autoimmunity in *ob/ob* mice has been observed in experimentally induced glomerulonephritis (EIG) *(32)*. In this immune-complex-mediated inflammatory disease induced by the injection of sheep antibodies specific for mouse glomerular basement membrane into mice preimmunized against sheep IgG, the authors observed renal protection of *ob/ob* mice associated with reduced glomerular-crescent formation, reduced macrophage infiltration, and glomerular thrombosis. These protective effects were associated with concomitant defects of both adaptive and innate immune response (testified by reduced in vitro proliferation of splenic T-cells and reduced humoral responses to sheep IgG, respectively). In spite of this trend, in one experiment, *ob/ob* mice developed histological injury, but they were still protected from disease, indicating that defects in effector responses were present in *ob/ob* mice, in line with in vitro experiments that have indicated defective phagocytosis and cytokine production.

It has been also demonstrated that leptin is a renal growth and profibrogenic factor and therefore can contribute to renal damage, characterized by endocapillary proliferation and subsequent development of glomerulosclerosis *(33)*. This is more likely to happen in pathophysiological situations characterized by high circulating leptin levels, such as diabetes and obesity *(33)*. This evidence suggests that leptin may also exert propathogenic effects in immune-mediated disorders of the kidney. Strategies aimed at reducing leptin may envisage novel therapeutic approaches able to delay immune-mediated kidney disorders.

6. LEPTIN IN ENDOMETRIOSIS

Pelvic endometriosis (PE) is one of the most common benign gynecological diseases, characterized by implantation and growth of endometrial tissue in the pelvic peritoneum and associated with infertility and pain *(34,35)*. The most accredited pathogenetic theory of endometriosis considers the transport of endometrial tissue to the peritoneal cavity throughout the tubes (retrograde menstruation), adherence to the peritoneal wall, followed by proliferation and formation of endometriotic lesions *(34)*. However, an abnormal peritoneal microenvironment is thought to be a "permissive" condition for implantation and growth of refluxed endometrium. In this context, local and systemic abnormalities in immune response have been described, such as increased frequency of autoimmune disorders and the presence of autoantibodies *(34,35)*. Endometriosis is, in fact, an immune-related chronic inflammatory disease, characterized by production of many proinflammatory cytokines, such as interleukin (IL)-1, IL-6, TNF-α, and vascular endothelial growth factor (VEGF) *(34,36,37)*. These molecules, together with other angiogenic factors, are thought to be of fundamental importance in the pathogenesis of the disease. Leptin may be one of these factors, as it promotes angiogenesis and proinflammatory cytokine production and induces the expression of intercellular adhesion molecule 1 (ICAM-1) and matrix metalloproteinases *(34,36,37)*. Evidence from the literature demonstrates that serum and peritoneal fluid leptin is significantly higher in endometriosis patients and correlates with the stage of disease *(38)*. Interestingly, endometrial cells of women with endometriosis have altered expression of ICAM-1, matrix metalloproteinases, and increased secretion of TNF-α, IL-1, and IL-6 *(34)*. Eutopic endometrium from patients with endometriosis might be more invasive and prone to peritoneal implantation, as a consequence of altered production of proteolytic enzymes such as matrix metalloproteinases *(39)*. In addition, recent reports suggest that endometrial cells from women with PE show a decrease in apoptosis and in sensitivity to macrophage-mediated cytolysis *(40)*. Indeed, an increase in endometrial expression of the anti-apoptotic gene *Bcl-2* has been demonstrated in such patients *(40)*. It is possible to speculate that, in physiological conditions, leptin is actively involved in the regulation of cytokine production and in the control of the apoptotic rate through its receptor expressed in normal endometric tissue and in the abundant adipose tissue hypersecreting leptin present in the peritoneal cavity *(41)*.

Finally, the possibility to interfere with the peritoneal microenvironment makes endometriosis a particularly interesting model of human inflammatory disorder to be utilized for testing innovative therapeutic approaches. Indeed, intraperitoneal injection of leptin antagonists could be more approachable than systemic injection. Studies on experimental models of endometriosis in mice and rats are under investigation to explore this possibility.

7. LEPTIN AND THE HYGIENE/AFFLUENCE HYPOTHESIS IN AUTOIMMUNITY

Fewer infections and more autoimmunity, observed in affluent countries, lead us to postulate the so-called "leptin hypothesis" to explain this phenomenon *(42)*. During the past century, in the industrialized world, the incidence of infections has diminished greatly because of improved hygienic conditions, better nutrition, vaccination, and the use of antibiotics *(42)*. Interestingly, in affluent and more-developed societies, epidemiological studies have revealed a parallel increase in the incidence of autoimmune diseases, whereas these diseases have become less common in less-developed nations. Thus, susceptibility to infection and autoimmunity appear to be inversely related *(42)*. Several factors other than nutrition might contribute to this relationship, such as the environment, genetic background, other hormones, stress, and exposure to specific pathogens (Fig. 1). Nevertheless, changes in diet and calorie intake and, subsequently, serum leptin concentrations should be taken into account to explain the complex network connecting nutritional status and susceptibility to autoimmune and infectious diseases *(42)* (Fig. 1). Animal studies provide support for this concept. In some murine models of systemic lupus erythematosus (SLE), T1D, and EAE, the induction and progression of disease can be prevented by starvation and/or reduced calorie intake, or by

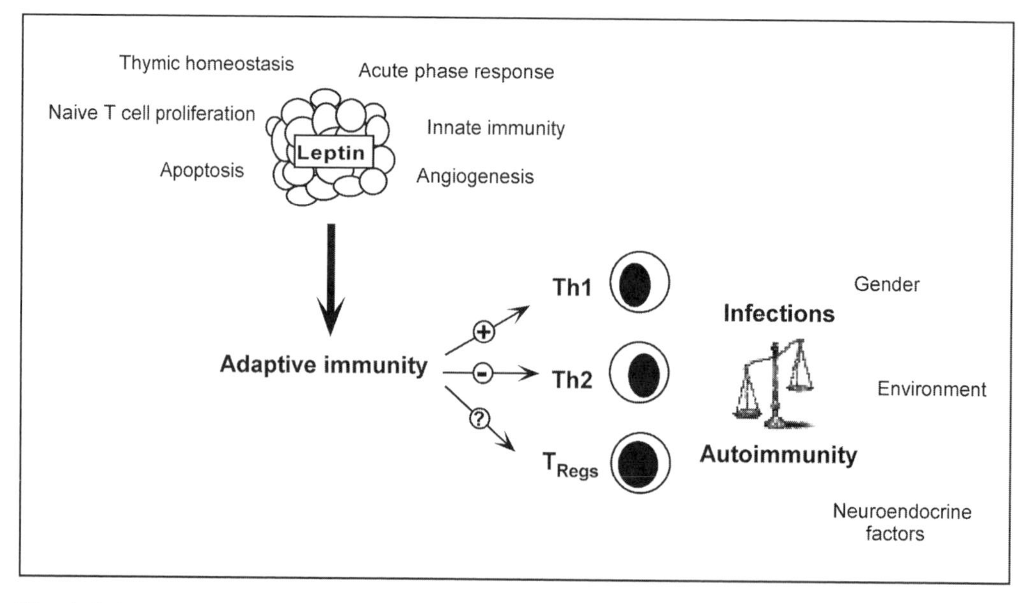

Fig. 1. Schematic representation of the pleiotropic effects of leptin in immunity and pathogenesis of autoimmune responses. Leptin secreted by adipose tissue is involved in thymic homeostasis and prevention of apoptosis, naïve T-cell proliferation/IL-2 secretion, acute phase response, innate immunity, and angiogenesis. More specifically, on adaptive immunity, leptin favors the switch toward Th1 immunity by inhibiting Th2 cytokine release. The effect of leptin on T_{Regs} is still under investigation. These pleiotropic actions of leptin may contribute to the different autoimmune vs infectious disease susceptibility observed in affluent countries together with the contribution of environmental, genetic, and neuroendocrine factors. Th, T helper; T_{Regs}, regulatory T-cells.

administering nutrients such as polyunsaturated fatty acids, which are able to reduce the inflammatory response and leptin secretion *(42)*. In humans, a similar observation has been reported by Bruining, who described an increased incidence of T1D at younger ages in affluent countries, where affluence is associated with increased postnatal growth and abundant nutrition *(43)*. More specifically, children that developed diabetes had a greater gain in BMI in the first year of life compared with healthy siblings and the early presence of autoantibodies specific for pancreatic islet tyrosine phosphatase (IA-2). Leptin, with its pleiotropic functions, including the promotion of Th1 responses, reduction of the apoptotic rate of thymocytes, reversal of acute-starvation-induced immunosuppression, and induction of expression of adhesion molecules (e.g., ICAM-1 and CD49b [integrin 2]) (Fig. 1), is a good candidate for contributing to the pathogenesis and maintenance of autoimmunity in genetically predisposed individuals. Conversely, malnutrition and nutritional deficiency might protect individuals from autoimmunity by lowering circulating leptin concentrations, but predispose to infections, such as candidiasis, tuberculosis, pneumonia, and bacterial and viral diarrhea. Last but not least, the most common form of human obesity, characterized by hyperleptinemia causing central and peripheral leptin resistance, is associated with an increased frequency of infections (Fig. 1) *(42)*. In this context, leptin receptor desensitization is perceived by T-cells as a condition of leptin deficiency, leading to immune dysfunction in a similar manner to malnutrition and genetic leptin deficiency.

8. CONCLUSIONS

Because of the influence of leptin on food intake and metabolism, the findings reported here underscore the role of molecules at the interface between metabolism and immunity in the control not only of inflammation but also of autoimmune reactivity. Recently, molecules with orexigenic activity such as ghrelin *(44)* and neuropeptide Y (NPY) *(45)* have been shown to mediate effects opposite to leptin not only in the hypothalamic control of food intake, but also on peripheral immune responses. Indeed, ghrelin blocks leptin-induced secretion of proinflammatory cytokines by human T-cells *(44)* and NPY ameliorates clinical score and progression of EAE *(45)*. Given these considerations, we may envisage a situation in which several metabolic regulators, including leptin, might broadly influence vital functions not limited to caloric tuning but rather affecting immune responses and the interaction of the individual with the environment. Since its discovery in 1994, leptin has attracted increasing interest in the scientific community because of its remarkably pleiotropic functions. Many related aspects remain to be investigated. However, current information on leptin function and regulation already suggests novel possibilities, using as yet unexploited leptin-based therapeutic tools for the treatment of both infection and autoimmune diseases.

ACKNOWLEDGMENTS

This work was supported by grants from the Juvenile Diabetes Research Foundation (JDRF)-Telethon-Italy (n. GJT04008) and from Fondazione Italiana Sclerosi Multipla (FISM) (n. 2002/R/55). The authors wish to thank Claudio Procaccini and Veronica De Rosa for helpful discussion and revising the manuscript.

REFERENCES

1. Blackburn GL. Nature 2001;409:379–401.
2. Harbige LS. Nutr Health 1996;10:285–312.
3. Black P. Trends Immunol 2001;22:354–355.
4. Yazdanbakhsh M, van den Biggelaar A, Maizels RM. Trends Immunol 2001;22:372–377.
5. Matarese G. Eur Cytokine Netw 2000;11:7–13.
6. Friedman JM, Halaas JL. Nature 1998;395:763–770.
7. Chan J, Tian Y, Tanaka KE, et al. Proc Natl Acad Sci USA 1996;93:14,857–14,861.
8. Lord GM, Matarese G, Howard JK, et al. Nature 1998;394:897–901.
9. Howard JK, Lord GM, Matarese G, et al. J Clin Invest 1999;104:1051–1059.
10. Ozata M, Ozdemir IC, Licinio J. J Clin Endocrinol Metab 1999;84:3686–3695.
11. Matarese G, Di Giacomo A, Sanna V, et al. J Immunol 2001;166:5909–5916.
12. Sanna V, Di Giacomo A, La Cava A, et al. J Clin Invest 2003;111:241–250.
13. Matarese G, Sanna V, Di Giacomo A, et al. Eur J Immunol 2001;31:1324–1332.
14. Matarese G, Carrieri PB, La Cava A, et al. Proc Natl Acad Sci USA 2005;102:5150–5155.
15. Lock C, Hermans G, Pedotti R, et al. Nat Med 2002;8:500–508.
16. Frisullo G, Angelucci F, Mirabella M, et al. J Clin Immunol 2004;24:287–293.
17. La Cava A, Matarese G. Nat Rev Immunol 2004;4:371–379.
18. Leiter EH, Lee CH. Diabetes 2005;54(Suppl 2):S151–S158.
19. Matarese G, Sanna V, Lechler RI, et al. Diabetes 2002;51:1356–1361.
20. Lee CH, Reifsnyder PC, Naggert JK, et al. Diabetes 2005;54:2525–2532.
21. Palmer G, Gabay C. Ann Rheum Dis 2003;62:913–915.
22. Busso N, So A, Chobaz-Peclat V, et al. J Immunol 2002;168:875–882.
23. Bernotiene E, Palmer G, Talabot-Ayer D, et al. Arthritis Res Ther 2004;6:R256–R263.

24. Fraser DA, Thoen J, Reseland J, et al. Clin Rheumatol 1999;18:394–401.
25. Bokarewa M, Bokarew D, Hultgren O, et al. Ann Rheum Dis 2003;62:952–956.
26. Hultgren OH, Tarkowski A. Arthritis Res 2001;3:389–394.
27. Popa C, Netea MG, Radstake TR, et al. Ann Rheum Dis 2005;64:1195–1198.
28. Anders HJ, Rihl M, Heufelder A, et al. Metabolism 1999;48:745–748.
29. Tokarczyk-Knapik A, Nowicki M, Wyroslak J. Pol Arch Med Wewn 2002;108:761–767.
30. Siegmund B, Lear-Kaul KC, Faggioni R, et al. Eur J Immunol 2002;32:552–560.
31. Faggioni R, Jones-Carson J, Reed DA, et al. Proc Natl Acad Sci USA 2000;97:2367–2372.
32. Tarzi RM, Cook HT, Jackson I, et al. Am J Pathol 2004;164:385–390.
33. Wolf G, Hamann A, Han DC, et al. Kidney Int 1999;56:860–872.
34. Matarese G, De Placido G, Nikas Y, et al. Trends Mol Med 2003;9:223–228.
35. De Placido G, Alviggi C, Carravetta C, et al. Hum Reprod 2001;16:1251–1254.
36. Akoum A, Lemay A, Maheux R. J Clin Endocrinol Metab 2002;87:5785–5792.
37. Piva M, Horowitz GM, Sharpe-Timms KL. J Clin Endocrinol Metab 2001;86:2553–2561.
38. Matarese G, Alviggi C, Sanna V, et al. J Clin Endocrinol Metab 2000;85:2483–2487.
39. Chung HW, Lee JY, Moon HS, et al. Fertil Steril 2002;78:787–795.
40. Meresman GF, Vighi S, Buquet RA, et al. Fertil Steril 2000;74:760–766.
41. Kitawaki J, Koshiba H, Ishihara H, et al. J Clin Endocrinol Metab 2000;85:1946–1950.
42. Matarese G, La Cava A, Sanna V, et al. Trends Immunol 2002;23:182–187.
43. Bruining GJ. Lancet 2000;356:655–656.
44. Dixit VD, Schaffer EM, Pyle RS, et al. J Clin Invest 2004;114:57–66.
45. Bedoui S, Miyake S, Lin Y, et al. J Immunol 2003;171:3451–3458.

8 Leptin and Gastrointestinal Inflammation

Arvind Batra and Britta Siegmund

Abstract

Leptin is a 16-kDa protein, predominantly produced by adipose tissue with serum leptin concentrations correlating to body fat mass. Initially described as a regulator of appetite, it is now well established that this mediator exerts pro-inflammatory effects on various immune as well as non-immune cells. In vitro as well as in vivo studies suggest an involvement of leptin in the regulation of intestinal inflammation. The data leading to this finding are the subject of discussion in this chapter. In particular, the various effects of leptin on different cell populations that are involved in the induction and persistence of intestinal inflammation, i.e., T-cells, antigen presenting cells, and epithelial cells, are outlined in detail. In addition, we propose a significant role for the potential interaction of the adipose tissue and the mucosal immune system in the pathophysiology of inflammatory bowel disease.

Key Words: Leptin; adipokines; cytokines; inflammatory bowel disease; experimental colitis; adipocytes; mucosal immunity.

1. INTRODUCTION

Inflammatory bowel disease (IBD) represents a chronic disease accompanying patients throughout their life. The etiology and pathogenesis have not been finally clarified. Various studies have led to the consensus hypothesis that genetic predisposition and exogenous factors together result in a chronic state of dysregulated mucosal immune function that is further modified by specific environmental factors *(1)*. Ulcerative colitis (UC) and Crohn's disease (CD) are the two major types of inflammatory bowel disease. Both diseases are characterized by a particular cytokine profile, which led in the past to the conclusion that CD represents a T-helper cell type 1 (Th1)- and UC a T-helper cell type 2 (Th2)-mediated disease *(2)*. Considering the breaking paradigm of Th1 and Th2, this general classification has to be discarded. Nevertheless, in CD, diseases severity can be ameliorated by neutralizing antibodies to either tumor necrosis factor (TNF)-α or interleukin (IL)-12 *(3,4)*. For ulcerative colitis, IL-13 neutralizing compounds have been proven efficacious in animal models of experimental colitis and are currently under clinical investigation *(5)*.

Leptin, the product of the *ob* gene, is a 16-kDa mediator predominantly produced by adipocytes, with leptin levels in the circulation correlated to body mass index (BMI). The primary role of leptin is considered to be its inhibitory effect on appetite, as absence

From: *Nutrition and Health: Adipose Tissue and Adipokines in Health and Disease*
Edited by: G. Fantuzzi and T. Mazzone © Humana Press Inc., Totowa, NJ

of leptin signaling results in obesity *(6)*. However, leptin-deficient (*ob/ob*) mice as well as mice deficient in the long isoform of the leptin receptor (*db/db*) are not merely obese. These mice also develop a complex syndrome characterized by an abnormal reproductive function, hormonal imbalances, alterations in the hematopoietic system, and in particular, a dysfunctional immune system *(7,8)*. Interestingly, *ob/ob* as well as *db/db* mice exhibit immune defects similar to those observed in starved animals, malnourished humans, as well as leptin-deficient humans, all conditions associated with low or absent circulating leptin concentrations *(9–12)*. Of particular interest with regard to IBD is the evidence that *ob/ob* as well as *db/db* mice have reduced T-cell function (for a summary *see* Table 1; *[13]*). This observation can be explained by the enhancing effect of leptin on T-cell stimulation: leptin promotes the responsiveness of naïve T-cells, whereas it has no activating effect on memory T-cells *(14)*. Furthermore, leptin favors a shift to a predominantly Th1-type response. Because this phenomenon is not observed in cells from *db/db* mice, this effect seems to be mediated by the long isoform of the leptin receptor (OB-Rb), which has been proven to be involved in signal transduction, as further discussed in the following paragraph *(14,15)*. In agreement with these findings *ob/ob* mice have been shown to be protected in several models of inflammation such as concanavalin A (Con A)-induced hepatitis, experimental autoimmune encephalomyelitis, and collagen-induced arthritis, suggesting a decreased responsiveness of the immune system in the absence of this mediator *(16–18)*.

Based on its immune modulatory effects, recently the concept emerged that leptin should be classified as a cytokine/adipokine *(19,20)*. Although leptin shares structural similarities with long-chain helical cytokines such as IL-6, IL-12, or IL-15, the Ob-R, of which six splice variants (Ob-Ra to Ob-Rf) exist, belongs—based on its sequence homology—to the class I cytokine receptor family *(21,22)*. Leptin is ubiquitously present throughout the body, as adipocyte-derived leptin, the main source of leptin, is secreted into the plasma. The different isoforms of the leptin receptor show tissue-specific distribution. Most common is the Ob-Ra isoform, which is expressed on various tissues such as lung, spleen, and kidney, as well as macrophages. As for the other short isoforms of the OB-R, its biological significance is not completely understood, but it seems to be involved in the transport and degradation of leptin. Although the Ob-Ra does contain only a short intracellular signaling domain, signal transduction through this receptor has been reported for Ob-Ra transfected cells *(23,24)*. However, the pathways involved in signal transduction have been studied in detail solely for the longest splice variant Ob-Rb, which contains the longest (302 amino acids) cytoplasmatic domain. Initially the Ob-Rb had been described in the hypothalamus, where it is expressed in areas that regulate appetite and body weight, but it is actually expressed as well by endothelial cells, ovary cells, pancreatic β-cells, $CD34^+$ hematopoietic bone-marrow precursors, and immune cells like monocytes/macrophages, $CD4^+$ and $CD8^+$ lymphocytes, as well as colonic epithelial cells *(7,25)*. The Ob-Rb cytosolic domain includes binding motifs associated with the activation of the Janus-activated kinase–signal transducer and activator of transcription (STAT) pathway *(7)*. Leptin activates STAT-1, STAT-3, and STAT-5 *(7)*. Of particular interest for IBD is STAT-3 activation, as hyperactivation of STAT-3—but not STAT-1 or STAT-5—results in severe colitis *(26)*. In addition, increased levels of phosphorylated STAT-3 were detected in colon samples from patients with CD and UC, as well as in mice after experimental dextran sulfate sodium (DSS)-induced colitis *(26)*.

Table 1
Overview of Effects of Leptin on Immune and Nonimmune Cells That Might Contribute to Induction and Persistence of Intestinal Inflammation

Systemic leptin Adipocyte-derived		*Colonic intraluminal leptin Epithelial cell-derived*
Adaptive immune system	*Antigen-presenting cells*	*Non-immune cells*
T-cells	***Dendritic cells***	***Epithelial cells***
Proliferation of naïve T-cells ↑ *(14)*	↑ Production of IL-10 resulting in promotion of Th1 polarization *(46)*	Induction of proinflammatory cytokines and colonic epithelial cell damage ↑ *(34)*
Stimulation of LPMC and intraepithelial lymphocytes ↑ *(32)*		
Shift to Th1 ↑ *(14)*	***Macrophages/Monocytes***	
	Survival ↑ *(45)*	
	Phagocytosis ↑ *(43)*	

Thus, the stimulatory effect on T-cells, the OB-Rb expression profile, the intracellular signaling pathways, and the results from the other in vivo models suggest a role of leptin in intestinal inflammation.

2. LEPTIN AS PIVOTAL MEDIATOR OF INTESTINAL INFLAMMATION

In several experimental colitis models, such as trinitrobenzene sulfonic acid (TNBS) as well as indomethacin-induced colitis and in spontaneously developing colitis in IL-$2^{-/-}$ mice, serum leptin concentrations are increased *(27,28)*. However, the elevated serum concentration of a mediator during active colitis does not necessarily imply its key function in the inflammatory process. To prove a critical role, neutralization experiments or the use of knockout animals is required. Thus, we compared disease severity in wild-type (WT) and leptin-deficient *ob/ob* mice using the model of DSS-induced colitis, which represents a model of chemically induced intestinal inflammation. The exposure of the colon to DSS results in disruption of the epithelial barrier, followed by phagocytosis of otherwise unavailable antigen from the normal mucosal microflora by antigen-presenting cells located in the lamina propria. Presentation of these antigens subsequently results in activation of T-cells *(29–31)*. These experiments indicated that *ob/ob* mice were protected from DSS-induced colitis *(32)*. This unresponsiveness toward DSS was associated with a decrease in local production of proinflammatory cytokines and an increase in apoptosis of lamina propria mononuclear cells (LPMC), both mechanisms known to be associated with protection in intestinal inflammation. To confirm that the prevention of colitis induction observed was in fact caused by leptin deficiency rather than induced by the obese phenotype, *ob/ob* mice were reconstituted with leptin and compared to accordingly obese *ob/ob* as well as WT mice in the same model of DSS-induced colitis. Although *ob/ob* mice were protected, leptin-reconstituted *ob/ob* mice exhibited equal disease severity when compared with the WT group, thus indicating that leptin deficiency and not the obesity *per se* is responsible for the protection observed *(32)*.

To further evaluate the role of leptin on intestinal T-cell populations, experiments with TNBS-induced colitis were conducted. It is hypothesized that the ethanol used as vehicle in the rectal administration of TNBS disrupts the epithelial barrier, enabling this hapten to bind covalently to proteins of colonic epithelial cells and modify cell surface proteins. Fragments of these haptenized cells can be taken up by macrophages and dendritic cells for presentation to T-cells as antigens, resulting in a Th1-dominated colitis *(33)*. Again, *ob/ob* mice were protected in this model when compared with WT mice.

Besides T-cell activation, leptin can induce the activation of the NFκB pathway in intestinal epithelial cells, a key signaling system to proinflammatory stimuli *(34)*. In a recent study by Sitaraman and colleagues, rectally administered leptin induced epithelial cell wall damage and neutrophil infiltration that represent characteristic histological findings of acute intestinal inflammation. These observations provide evidence for an intraluminal biological function of leptin and a new pathophysiological role for leptin during states of intestinal inflammation such as IBD *(34)*.

3. DIRECT ROLE OF LEPTIN ON T-CELL ACTIVATION IN INTESTINAL INFLAMMATION

Leptin has primarily been described as a regulator of appetite that exerts its regulatory function via a negative feedback mechanism in the hypothalamus *(31)*. Because of this systemic effect of leptin it was necessary to evaluate whether the alterations in the immune system in *ob/ob* mice are induced by a central effect of leptin or whether leptin can directly act on cells in the periphery. The in vitro data described in the introduction strongly suggested that leptin in fact can directly stimulate T-cells as well as epithelial cells *(14,25)*. In line with these results, data from our own group indicated that leptin exerts direct stimulatory effects on intraepithelial lymphocytes as well as LPMC in vitro *(32)*. However, in vivo data were lacking. In order to approach this question in vivo, the $CD4^+CD45Rb^{high}$ transfer model of colitis was chosen. In this model, naïve T-cells, characterized by the $CD4^+CD45Rb^{high}$ phenotype, are transferred into mice with *severe combined immunodeficiency* (*scid*), lacking functional T-cells. After transfer, the cells proliferate and home to the gut, leading to induction of intestinal inflammation. Th1 cytokines, such as interferon (IFN)-γ are the major players in the induction of this disease, whereas IL-10 and transforming growth factor (TGF)-β play a protective role *(35,36)*.

In order to characterize the effect of leptin on naïve T-cells in intestinal inflammation, either naïve WT or leptin receptor-deficient *db/db* cells were transferred into *scid* mice. Whereas recipients of WT cells developed as expected a transmural colitis 4 wk post-cell transfer, the recipients of *db/db* cells were still healthy and did not show any signs of inflammation at this time point *(37)*. However, 12 wk after cell transfer, recipients of both WT and *db/db* cells had developed colitis with comparable disease severity. Thus, leptin does not seem to be an essential mediator in the inflammatory process but is rather an important costimulatory factor for intestinal immune cells *(37)*. In addition, these data provide the first proof in vivo for a direct stimulatory capacity of leptin on peripheral T-cells.

4. BIOLOGICAL SIGNIFICANCE OF LEPTIN PRODUCED AT THE SITE OF INFLAMMATION

Adipocytes represent the primary source of leptin *(38)*. However, several studies have reported local leptin production by lymphocytes and other cells at inflamed areas. The first of these studies demonstrated local leptin production in inflammatory infiltrates around neurons in the brain and spinal cord of mice with experimental autoimmune encephalomyelitis *(39)*. Another study showed leptin production of colon epithelial cells at the sites of inflammation *(34)*. Our own group demonstrated leptin production by LPMC as well as T-cells from mesenteric lymph nodes in the $CD4^{+}CD45Rb^{high}$ transfer model of colitis *(37)*. However, it remained unclear whether or not leptin released by these activated lymphocytes is of significance for the inflammatory process itself. In order to address this question, the $CD4^{+}CD45Rb^{high}$ transfer model of colitis was used. Cells isolated from either WT or leptin-deficient *ob/ob* mice were transferred into *scid* mice and were compared for their ability to induce inflammation. With regard to macroscopic parameters, weight loss was more pronounced in recipients of *ob/ob* cells, whereas there was no difference in stool consistency and rectal bleeding between recipients of WT and *ob/ob* cells. Furthermore, the histological score indicated equal inflammation in both experimental groups *(40)*. These macroscopic and histological findings allow for the conclusion that lack of T-cell-derived leptin does not protect from intestinal inflammation. In summary, the data obtained suggest that T-cell-derived leptin does not play a major role in the mucosal inflammatory process, thus implying that other cellular sources of leptin—most likely, the adipose tissue—are critical in leptin-dependent modulation of the immune system.

However, as mentioned above, local leptin production by colonic epithelial cells occurs within the inflamed colon *(34)*. Interestingly, those cells secrete leptin directly into the lumen, resulting in a more than 15-fold elevation of luminal concentrations of leptin in patients with mild to severe UC or CD. Whereas intraluminal leptin in the small intestine is expected to be involved in absorption, it is suggested that colonic leptin acts as a proinflammatory cytokine on colonic epithelial cells by induction of NFκB *(41,42)*. Furthermore, at least in mice, luminal leptin induces colonic epithelial wall damage and formation of crypt abscesses. Taken together, besides the stimulating effect systemic leptin exerts on immune cells, local leptin production at the sites of inflammation promotes inflammation by directly affecting epithelial cells *(34)*.

5. INFLUENCE OF LEPTIN ON ANTIGEN PROCESSING AND PRESENTING CELLS

Although T-cells are key players in the development and persistence of intestinal inflammation, these cells from the adaptive immune system rely on support by antigen processing and presenting cells for activation. As mentioned previously, OB-R isoforms are expressed on a variety of cell types, including macrophages/monocytes. A study using *ob/ob* mice in the experimental infection model of Gram-negative pneumonia has shown that in the absence of leptin signaling the phagocytic capacity, as well as the production of leukotrienes, is reduced in alveolar macrophages *(43)*. Furthermore, in vitro studies have demonstrated that leptin enhances survival of monocytes in a concentration-dependent manner *(44,45)*. Macrophages/monocytes are sensitive to leptin stimulation, and because their activity is increased by this mediator, their efficacy as antigen-presenting cells might be affected by leptin levels in vivo as well. In addition, a stimulating effect of leptin on dendritic cells was recently demonstrated. Leptin promotes dendritic cell survival and affects cytokine levels induced in stimulated dendritic cells, including downregulation of IL-10, ultimately driving polarization of T-cells stimulated by these antigen-presenting cells toward a Th1 phenotype *(46)*.

In summary, these studies suggest that leptin not only affects effector cells from the adaptive immunes system directly, but also positively influences the activity and survival of a variety of antigen-presenting cells (Table 1). However, although the findings described above suggest a regulatory role for leptin on antigen-presenting cells in the mucosal immune system, the relevance of leptin-dependent changes on antigen-presenting cells and the subsequent impact on intestinal inflammation have not been studied in detail so far.

6. SPONTANEOUS INTESTINAL INFLAMMATION

Despite the important observations described above, the role of leptin in regulating inflammatory or autoimmune responses developing spontaneously as a result of alterations in the immune system—a situation that more closely reflects human autoimmune disease—has not been studied in great detail. The only published study about this topic demonstrates that in the nonobese diabetic model of type 1 diabetes, administration of exogenous leptin accelerates destruction of insulin-producing β-cells and promotes an increased production of IFN-γ *(47)*. Concerning intestinal inflammation, it has been demonstrated that IL-$2^{-/-}$ mice that develop colitis spontaneously have increased leptin serum concentrations *(28)*.

IL-10-deficient mice spontaneously develop chronic intestinal inflammation, which is mediated by $CD4^+$ T-cells and is associated with enhanced Th1 responses in the early course of disease, with Th2 cytokines progressively increasing later on *(48)*. As colitis in IL-10-deficient mice is dependent on the very cell types and cytokines that are thought to be regulated by leptin, leptin-deficient IL-10 knockout mice were generated to evaluate the role of leptin in a model of spontaneously developing intestinal inflammation. However the results of this study indicate that leptin deficiency does not alter development of intestinal inflammation in IL-10-deficient mice *(49)*. The data obtained indicate that in the absence of leptin and IL-10, the net effect of the lack of the anti-inflammatory mechanisms and subsequent lack of tolerance to colonic bacteria usually mediated by IL-10 is predominant and cannot effectively be counter-regulated by the absence of leptin. It should be noted that the data presented herein on the IL-10-deficient mouse model of spontaneous colitis may not necessarily be transferred to other models of spontaneous colitis in which the specific balance of inflammatory mediators may well leave a critical role for leptin, as suggested in previous studies in the nonobese type I diabetes model *(47)*.

7. INFECTIOUS COLITIS MODELS

The role of leptin in infectious colitis models is not very well defined. There is only one study that examined the role of leptin in *Clostridium* toxin A-induced intestinal inflammation, the causative agent of antibiotic-related colitis *(50)*. It could be demonstrated that *Clostridium* toxin A-induced colitis in WT mice resulted in an increase of circulating leptin concentrations. Furthermore, OB-Rb expression in the mucosa could be induced by *Clostridium* toxin A at the mRNA as well as at the protein level. Thus, to further explore the function of leptin in this model of intestinal inflammation, WT mice were compared to leptin-deficient *ob/ob* as well as leptin receptor-deficient *db/db* mice. Remarkably, in the absence of leptin or leptin signaling mice were partially protected against *Clostridium* toxin A-induced colitis. By including studies with adrenalectomized *db/db* as well as WT mice it could be demonstrated that corticosteroid-dependent as well as -independent mechanisms are involved *(50)*.

8. HUMAN INFLAMMATORY BOWEL DISEASE

Abnormalities of adipose tissue in the mesentery, including adipose tissue hypertrophy and fat wrapping, have been long recognized on surgical specimens as characteristic features of Crohn's disease. However, the importance, origin, and significance of the mesenteric fat hypertrophy in this chronic inflammatory disease are unknown. Desreumaux and colleagues evaluated this phenomenon and quantified intra-abdominal fat in patients with CD vs UC by using magnetic resonance imaging *(51)*. By applying this technique they were able to demonstrate a significant accumulation of intra-abdominal fat in patients with CD. This mesenteric obesity, present from the onset of disease, is associated with overexpression of PPARγ as well as TNF-α mRNA, as evaluated by RT-PCR studies *(51)*. In a subsequent study, the same group could demonstrate an overexpression of leptin mRNA in the mesenteric adipose tissue in inflammatory bowel disease, whereas no difference could be detected between UC and CD. The increase in leptin went in parallel with an increase in TNF-α mRNA expression *(27)*.

Comparable with other cytokines known to play a local role, reports about alterations in leptin plasma levels in IBD are inconsistent so far. Whereas some authors could detect no increase in circulating leptin levels in IBD *(52,53)*, one group noted a significant increase in patients with acute UC, an observation supported by a further study where, in patients with active CD, a nonsignificant increase of serum leptin was detected *(54,55)*. However, as leptin exerts stimulatory effects on a variety of cells involved in the induction and persistence of IBD (summarized in Fig. 1) a closer characterization of leptin levels during stages of IBD might be crucial to understanding its contribution to these inflammatory diseases.

9. FUTURE PERSPECTIVE

Although the question of whether the increase of mesenteric fat as seen as in CD supports inflammation through potentially increased leptin production is not solved yet owing to the inconsistency of reports, one has to consider that adipose tissue can play a role in the immune system besides its capability to produce leptin. On the one hand, adipocytes are the main energy storage of the body and the enlargement of lymphatic tissue during immune stimulation is supported by supply with fatty acids and other nutrients from nearby adipose tissue tissue *(56)*. Furthermore adipocytes are potent producers of various cytokines besides leptin, including IL-10, TNF-α, macrophage migration inhibitory factor, IL-18, and plasminogen activator inhibitor-1, as well as IL-6 with, for example adipocyte-derived IL-6 even accounting for 15 to 35% of circulating levels

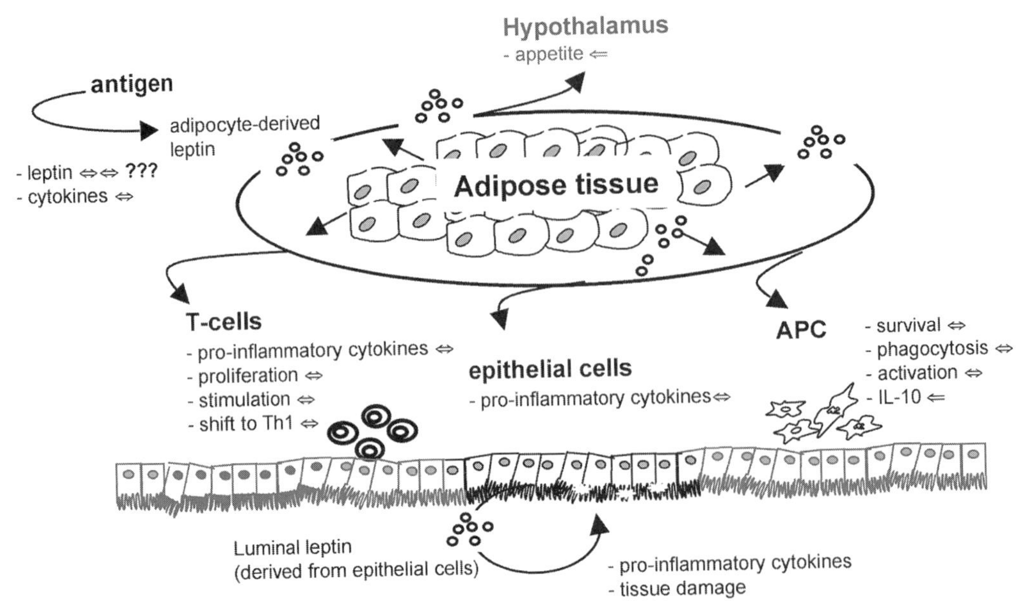

Fig. 1. Summary of the biological effects of leptin with regard to the relevance in intestinal inflammation. Levels of circulating adipocyte-derived leptin influence the activity not only of T-cells but also of antigen-presenting epithelial cells with additional action of intraluminal leptin on colonic epithelial cells, as discussed further in the text.

of this cytokine *(57–59)*. Based on this potent involvement of adipose tissue in immunity and the observation that preadipocytes can convert into macrophage-like cells that are capable of phagocytosing antigens *(60)*, our current studies focus on whether adipocytes and their precursors are capable of directly interacting with pathogens. Preliminary data from these studies indicate that adipocytes as well as preadipocytes can be considered as new members of the innate immune system, as they not only express a variety of pattern recognition receptors but, in addition, can also functionally respond to receptor specific stimulation (Batra et al., unpublished data). This observation might be of particular interest because a recent study has confirmed a long-known concept indicating that the mesenteric adipose tissue of patients with Crohn's disease is colonized by bacterial flora, previously described as bacterial translocation *(61)*. Thus the further characterization of the biological significance of fat hypertrophy in inflammation might be of particular relevance for the understanding of mechanisms underlying IBD.

10. CONCLUSIONS

The data described in this chapter point to a strong regulatory role of leptin in intestinal inflammation. More importantly, based on these results, a critical interaction between adipose tissue and immune cells can be suggested and will open new avenues in understanding the pathophysiology of intestinal inflammation.

ACKNOWLEDGMENTS

The work was supported by a grant from the Emmy Noether Program of the German Research Foundation (SI 749/3-3).

REFERENCES

1. Fiocchi C. Gastroenterology 1998;115:182–205.
2. Neurath MF, Finotto S, Fuss I, et al. Trends Immunol 2001;22:21–26.
3. Mannon PJ, Fuss IJ, Mayer L, et al. N Engl J Med 2004;351:2069–2079.
4. Targan SR, Hanauer SB, van Deventer SJ, et al. N Engl J Med 1997;337:1029–1035.
5. Heller F, Fuss IJ, Nieuwenhuis EE, et al. Immunity 2002;17:629–638.
6. Pelleymounter MA, Cullen MJ, Baker MB, et al. Science 1995;269:540–543.
7. Fantuzzi G, Faggioni R. J Leukoc Biol 2000;68:437–446.
8. Matarese G, La Cava A. Trends Immunol 2004;25:193–200.
9. Chandra RK. Am J Clin Nutr 1980;33:13–16.
10. Farooqi IS, Matarese G, Lord GM, et al. J Clin Invest 2002;110:1093–1103.
11. Palacio A, Lopez M, Perez-Bravo F, et al. Am J Physiol Regul Integr Comp Physiol 2001;281:R753–R759.
12. Ozata M, Ozdemir IC, Licinio J. J Clin Endocrinol Metab 1999;84:3686–3695.
13. Mandel MA, Mahmoud AA. J Immunol 1978;120:1375–1377.
14. Lord GM, Matarese G, Howard JK, et al. Nature 1998;394:897–901.
15. Martin-Romero C, Santos-Alvarez J, Goberna R, et al. Cell Immunol 2000;199:15–24.
16. Busso N, So A, Chobaz-Peclat V, et al. J Immunol 2002;168:875–882.
17. Matarese G, Sanna V, Di Giacomo A, et al. Eur J Immunol 2001;31:1324–1332.
18. Siegmund B, Lear-Kaul KC, Faggioni R, et al. Eur J Immunol 2002;32:552–560.
19. Baumann H, Morella KK, White DW, et al. Proc Natl Acad Sci USA 1996;93:8374–8378.
20. Zhang F, Basinski MB, Beals JM, et al. Nature 1997;387:206–209.
21. Tartaglia LA, Dembski M, Weng X, et al. Cell 1995;83:1263–1271.
22. Zabeau L, Lavens D, Peelman F, Eyckerman S, et al. FEBS Lett 2003;546:45–50.
23. Murakami T, Yamashita T, Iida M, et al. Biochem Biophys Res Commun 1997;231:26–29.
24. Yamashita T, Murakami T, Otani S, et al. Biochem Biophys Res Commun 1998;246:752–759.
25. Hardwick JCH, Van den Brink GR, Offerhaus GJ, et al. Gastroenterology 2001;121:79–90.
26. Suzuki A, Hanada T, Misuyama K, et al. J Exp Med 2001;193:471–481.
27. Barbier M, Vidal H, Desreumaux P, et al. Gastroenterol Clin Biol 2003;27:987–991.
28. Gaetke LM, Oz HS, de Villiers WJ, et al. J Nutr 2002;132:893–896.
29. Murthy SN, Cooper HS, Shim H, et al. Dig Dis Sci 1993;38:1722–1734.
30. Saubermann LJ, Beck P, De Jong YP, et al. Gastroenterology 2000;119:119–128.
31. Takizawa H, Shintani N, Natsui M, et al. Digestion 1995;56:259–264.
32. Siegmund B, Lehr HA, Fantuzzi G. Gastroenterology 2002;122:2011–2025.
33. Neurath M, Fuss I, Strober W. Int Rev Immunol 2000;19:51–62.
34. Sitaraman S, Liu X, Charrier L, et al. FASEB J 2004;18:696–698. Epub 2004 Feb 20.
35. Groux H, Powrie F. Immunol Today 1999;20:442–445.
36. Powrie F, Leach MW, Mauze S, et al. Immunity 1994;1:553–562.
37. Siegmund B, Sennello JA, Jones-Carson J, et al. Gut 2004;53:965–972.
38. Gregoire FM, Smas CM, Sul HS. Physiol Rev 1998;78:783–809.
39. Sanna V, Di Giacomo A, La Cava A, et al. J Clin Invest 2003;111:241–250.
40. Fantuzzi G, Sennello JA, Batra A, et al. Clin Exp Immunol 2005;142:31–38.
41. Rouet-Benzineb P, Aparicio T, Guilmeau S, et al. J Biol Chem 2004;279:16,495–16,502.
42. Doi T, Liu M, Seeley RJ, et al. Am J Physiol Regul Integr Comp Physiol 2001;281:R753–R759.
43. Mancuso P, Gottschalk A, Phare SM, et al. J Immunol 2002;168:4018–4024.
44. Lee FY, Li Y, Yang EK, et al. Am J Physiol 1999;276:C386–C394.
45. Sanchez-Margalet V, Martin-Romero C, Santos-Alvarez J, et al. Clin Exp Immunol 2003;133:11–19.
46. Mattioli B, Straface E, Quaranta MG, et al. J Immunol 2005;174:6820–6828.
47. Matarese G, Sanna V, Lechler RI, et al. Diabetes 2002;51:1356–1361.
48. Spencer DM, Veldman GM, Banerjee S, et al. Gastroenterology 2002;122:94–105.
49. Siegmund B, Sennello JA, Lehr HA, et al. J Leukoc Biol 2004;76:782–786.
50. Mykoniatis A, Anton PM, Wlk M, et al. Gastroenterology 2003;124:683–691.
51. Desreumaux P, Ernst O, Geboes K, et al. Gastroenterology 1999;117:73–81.
52. Ballinger A. Gut 1999;44:588–589.

53. Hoppin AG, Kaplan LM, Zurakowski D, et al. J Pediatr Gastroenterol Nutr 1998;26:500–505.
54. Bannerman E, Davidson I, Conway C, et al. Clin Nutr 2001;20:399–405.
55. Tuzun A, Uygun A, Yesilova Z, et al. J Gastroenterol Hepatol 2004;19:429–432.
56. Pond CM, Mattacks CA. Br J Nutr 2003;89:375–383.
57. Juge-Aubry CE, Somm E, Pernin A, et al. Cytokine 2005;29:270–274.
58. Mohamed-Ali V, Goodrick S, Rawesh A, et al. J Clin Endocrinol Metab 1997;82:4196–4200.
59. Rudin E, Barzilai N. Immun Ageing 2005;2:1.
60. Charriere G, Cousin B, Arnaud E, et al. J Biol Chem 2003;278:9850–9855.
61. Gay J, Tachon M, Neut C, et al. Gastroenterology 2005;128.

9 Adiponectin and Inflammation

Yuji Matsuzawa

Abstract

Adipose tissue has been recognized as the organ not only storing excess energy but also secreting a variety of bioactive substances named adipocytokines. Adipocytokines include tumor necrosis factor-α and interleukin-6, which are secreted from adipose tissue and may induce inflammatory response in various tissues including vascular walls and muscles. We found a novel adipocytokine named adiponectin, which has a strong ability of anti-inflammation. Adiponectin inhibits several inflammatory responses in macrophages and endothelial cells. This chapter presents a variety of anti-inflammatory functions and also discusses the mechanism of antidiabetic, anti-atherogenic, and anti-oncogenic properties of adiponectin.

Key Words: Visceral fat; metabolic syndrome; hypoadiponectinemia; C-reactive protein; acute coronary syndrome; insulin resistance.

1. INTRODUCTION

Excess body fat, especially intra-abdominal visceral fat accumulation, is associated with a number of disease conditions, including dyslipidemia, type 2 diabetes, hypertension, and cardiovascular disease *(1)*. Therefore, visceral fat accumulation—estimated by waist circumference—is adopted as an essential component of metabolic syndrome, which has been recently classified as a highly atherogenic state. Recent research, including ours, has shown that adipose tissue secretes various bioactive substances, collectively referred to as adipocytokines, that may directly contribute to the pathogenesis of conditions associated with obesity *(2)*. Thus, adipose tissue seems to be an endocrine organ that can affect the function of other organs, including the vascular walls, through secretion of various adipocytokines. These adipocytokines include heparin-binding epidermal growth factor-like growth factor (HB-EGF), leptin, tumor necrosis factor (TNF)-α, plasminogen activator inhibitor type 1 (PAI-1), and angiotensinogen. The expression and plasma levels of these adipocytokines increase with visceral fat accumulation; these molecules are implicated in insulin resistance and atherosclerosis *(3,4)*. In addition to the analysis of known genes, we identified adipose tissue's most abundant gene transcript, which we named apM-1 *(5)* and named the protein encoded by apM-1 adiponectin *(6)*. In contrast with other adipocytokines, adiponectin levels are inversely correlated to visceral adiposity and low levels have been associated with visceral obesity, type 2 diabetes, hypertension, and cardiovascular disease *(7)*.

From: *Nutrition and Health: Adipose Tissue and Adipokines in Health and Disease*
Edited by: G. Fantuzzi and T. Mazzone © Humana Press Inc., Totowa, NJ

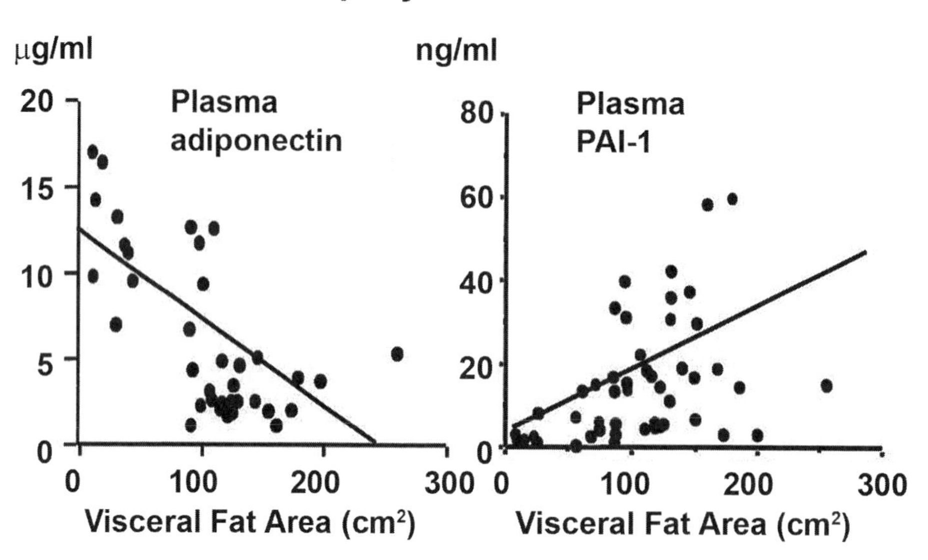

Fig. 1. Correlation between visceral adiposity and plasma adiponectin or PAI-1 *(3,7)*.

This article discusses the physiological roles of adiponectin in the prevention of obesity-related diseases, especially focusing on its anti-inflammatory function.

2. DISCOVERY OF ADIPONECTIN AND ITS CLINICAL SIGNIFICANCE

When we started the comprehensive analysis of expressed genes in human adipose tissue, 40% of expressed genes were unknown—in other words, they were novel genes. The molecule encoded by apM-1 possesses a signal peptide, collagen-like motif, and a globular domain, and has notable homology with collagen X and VIII, and complement factor C1q *(5)*. We termed the collagen-like protein adiponectin. The mouse homolog of adiponectin has been cloned as AdipoQ or ACRP30 *(8,9)*.

We established a method for measurement of plasma adiponectin levels using enzyme-linked immunosorbent assay *(10)*. The average levels of adiponectin in human plasma are extremely high—up to 5 to 10 μg/mL *(10)*. Plasma concentrations are negatively correlated with body mass index (BMI), whereas leptin increases with BMI. The negative correlation of adiponectin levels and visceral adiposity is stronger than between adiponectin levels and subcutaneous adiposity (Fig. 1) *(7)*.

The mechanism by which plasma adiponectin levels are reduced in individuals with visceral fat accumulation is not yet clarified. Coculture with visceral fat inhibits adiponectin secretion from subcutaneous adipocytes *(11)*. This finding suggests that some inhibiting factors for adiponectin synthesis or secretion are released from visceral adipose tissue. TNF-α was reported to be a strong inhibitor of adiponectin promoter activity *(12)*. The negative correlation between visceral adiposity and adiponectin levels might be explained at least in part by the increased secretion of this cytokine from accumulated visceral fat.

Plasma adiponectin concentrations are lower in people who have type 2 diabetes mellitus than in BMI-matched controls *(13)*. Plasma adiponectin concentrations have been shown to correlate strongly with insulin sensitivity, which suggests that low plasma concentrations are related to insulin resistance *(14)*. In a study of Pima Indians, individuals with high levels of adiponectin were less likely than those with low concentrations to develop type 2 diabetes *(15)*. High adiponectin concentration was, therefore, a notable protective factor against development of type 2 diabetes.

Studies on adiponectin knockout (KO) mice support observations in humans. The KO mice showed no specific phenotype when they were fed a normal diet, but a high-sucrose and high-fat diet induced a marked elevation of plasma glucose and insulin levels *(16)*, although other groups have reported that the KO mouse develops insulin resistance on a regular diet *(17)*. Insulin resistance, estimated by the insulin tolerance test during a high-sucrose with high-fat diet, also developed in adiponectin KO mice. Supplementation of adiponectin by adenovirus transfection clearly improved insulin resistance *(16)*. Although I will not mention the details of molecular mechanisms, adiponectin has been shown to exert its actions on muscle fatty acid oxidation and insulin sensitivity by activation of AMP-activated protein kinase *(18)*. Plasma levels of adiponectin are also decreased in hypertensive humans, irrespective of the presence of insulin resistance *(19)*. Endothelium-dependent vasoreactivity is impaired in people with hypoadiponectinemia, which might be at least one mechanism of hypertension in visceral obesity *(20)*.

Most important, plasma concentrations of adiponectin are lower in people with coronary heart disease than in controls, even when BMI and age are matched. Kaplan-Meier analysis in Italian individuals with renal insufficiency demonstrated that those with high adiponectin concentrations were free from cardiovascular death for longer than other groups *(21)*. A case–control study performed in Japan demonstrated that the group with plasma adiponectin lower than 4 μg/mL has increased risk of coronary artery disease (CAD) and multiple metabolic risk factors, which indicates that hypoadiponectinemia is a key factor in metabolic syndrome *(22)*. A prospective study by Pischon et al. *(23)* confirmed that high adiponectin concentrations are associated with reduced risk of acute myocardial infarction in men. In addition to hypoadiponectinemia accompanied with visceral fat accumulation, genetic hypoadiponectinemia caused by a missense mutation has been reported to be associated with the clinical phenotype of metabolic syndrome *(24)*.

This clinical evidence shows that hypoadiponectinemia is a strong risk factor for cardiovascular disease.

3. ADIPOSE TISSUE AND INFLAMMATION

Recent studies have revealed that inflammatory responses contribute to the development of a variety of common diseases, including atherosclerosis and metabolic diseases, including diabetes mellitus. On the other hand, adipose tissue has been recognized to secrete bioactive substances that relate to inflammation. TNF-α is a typical cytokine that plays a major role in inflammatory cellular phenomena. Since Uysal et al. first reported that adipose tissue secretes this cytokine and proposed it as one of the candidates for molecules inducing insulin resistance *(25)*, TNF-α has come to be recognized as an important adipocytokine. It has been shown that adipose TNF-α mRNA and plasma TNF-α protein are increased in most animal models and in humans with obesity and insulin resistance *(25,26)*. Neutralizing TNF-α in obese rats with a soluble TNF-α receptor-immunoglobulin G fusion protein markedly improves insulin resistance *(27)*.

These results indicate that higher production of TNF-α in accumulated adipose tissue may be causative for obesity-associated insulin resistance. In addition to TNF-α, interleukin (IL)-1β, IL-6, macrophage migration factor, nerve growth factor, and haptoglobin have been shown to be secreted from adipose tissue and linked to inflammation and the inflammatory response. More recently one typical marker of inflammation, C-reactive protein (CRP), was found to be produced in adipose tissue and CRP mRNA expression was found to be enhanced in adipose tissue in adiponectin KO mice *(28)*.

The elevated production of these inflammation-related adipocytokines is increasingly considered to be important in the development of diseases linked to obesity, particularly type 2 diabetes and cardiovascular disease. Namely, adipose tissue is involved in extensive crosstalk with other organs and multiple metabolic systems through the various adipocytokines.

4. ADIPONECTIN AS A POTENT ANTI-INFLAMMATION ADIPOCYTOKINE

As already mentioned, adiponectin has multiple functions for prevention of metabolic diseases and cardiovascular diseases. More recently, adiponectin was shown to prevent liver fibrosis and some kinds of cancer such as endometrial cancer, breast cancer, and colon cancer, in animal models *(29–32)*. Besides these well-characterized biological functions, recent evidence supports a strong anti-inflammatory function. We first reported that adiponectin strongly suppresses production of the potent proinflammatory cytokine TNF-α in macrophages *(33)*. Treatment of cultured macrophages with adiponectin significantly inhibits their phagocytic activity and their lipopolysaccharide-induced production of TNF-α. Suppression of phagocytosis by adiponectin is mediated by one of the complement C1q receptors, C1qRp, because this function was completely abrogated by the addition of an anti-C1qRp monoclonal antibody *(33)*. These observations suggest that adiponectin is an important negative regulator in immune and inflammatory systems, indicating that it may be involved in terminating inflammatory responses through its inhibitory functions.

In the process of development of atherosclerosis, macrophages play crucial roles in plaque formation. Adiponectin attenuates cholesterol ester accumulation in macrophages. Adiponectin-treated macrophages contain fewer lipid droplet stained by oil red O *(34)*. Adiponectin suppresses expression of class A macrophage scavenger receptor (MSR) at both mRNA and protein levels, without affecting the expression of CD36, as demonstrated by flow cytometry *(35)*. Adiponectin was also shown to attenuate expression of MSR and TNF-α in atherosclerotic lesions *(34)*. Adiponectin inhibits TNF-α-induced mRNA expression of monocyte adhesion molecules without affecting the interaction between TNF-α and its receptors in human aortic endothelial cells. Adiponectin suppresses TNF-α-induced IκB-α phosphorylation and subsequent NFκB activation without affecting other TNF-α-mediated signals, including Jun N-terminal kinase, p38 kinase, and Akt kinase *(36)*. This inhibitory effect of adiponectin is accompanied by cAMP accumulation and is blocked by either an adenylate cyclase inhibitor or a protein kinase A (PKA) inhibitor. These observations suggest that adiponectin, which is naturally present at very high amounts in the bloodstream, modulates the inflammatory response of both macrophages and endothelial cells through crosstalk between cAMP–PKA and

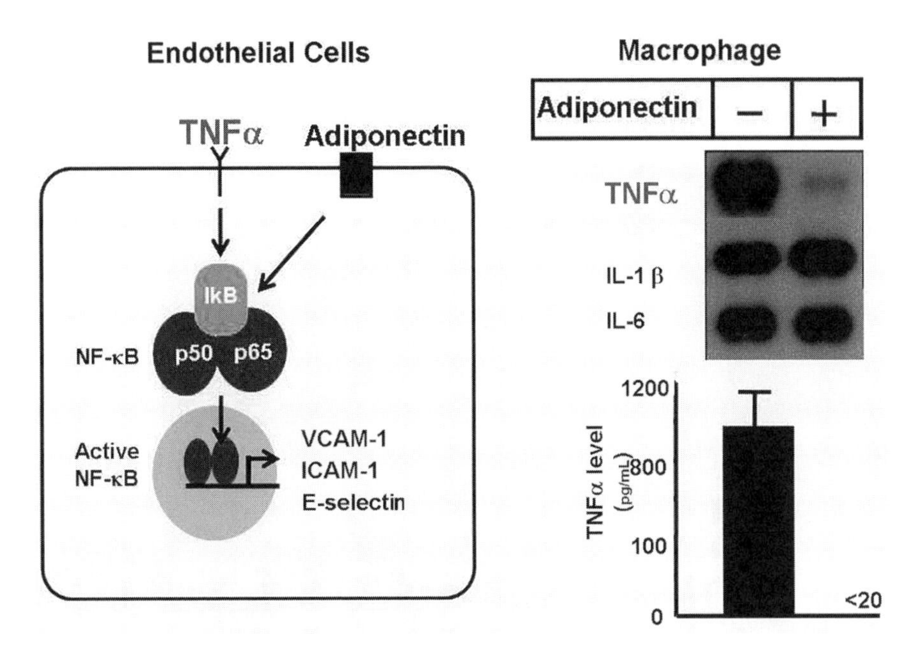

Fig. 2. Anti-inflammatory functions of adiponectin *(33,36)*.

NFκB signaling pathways (Fig. 2). The anti-inflammatory function of adiponectin may result in the prevention of atherogenic cell phenomena, such as monocyte adhesion to endothelial cells, differentiation of monocytes to macrophages, and foam cell formation.

Recent studies demonstrate that adiponectin also induces various anti-inflammatory cytokines, such as IL-10 and IL-1 receptor antagonist *(37)*. Acute coronary syndrome is considered to determine the prognosis of cardiovascular disease, in which vulnerability of the plaque is an important determinant of plaque rupture. In this process, matrix metalloproteinases (MMP) secreted by macrophages are thought to play an important role in plaque vulnerability. Tissue inhibitor of metalloproteinase (TIMP) is thought to act as protector of plaque rupture by inhibition of MMP activity. Adiponectin increases the expression of mRNA and protein production of TIMP in macrophages *(38)*. Prior to the induction of TIMP formation and secretion, adiponectin induces IL-10 synthesis in macrophages, suggesting that adiponectin induces TIMP formation and secretion via induction of IL-10 synthesis in an autocrine manner in macrophages and may also inhibit MMP activity (Fig. 3) *(38)*. These functions may result in the prevention of acute coronary syndrome.

As mentioned before, CRP is a typical marker of inflammation; increased levels of CRP are now considered to be a risk factor for atherosclerosis *(39)*. A number of studies have shown that there is a reciprocal association of adiponectin and CRP in plasma in healthy subjects and in subjects with a variety of diseases, including type 2 diabetes, metabolic syndrome, and end-stage kidney disease *(40,41)*. The mechanism of this reciprocal association of CRP and adiponectin remains to be clarified. Our group detected the expression of adiponectin mRNA in human adipose tissue and demonstrated a significant inverse correlation between the CRP and adiponectin mRNA *(27)*. In addition, CRP mRNA levels in white adipose tissue in adiponectin KO mice are

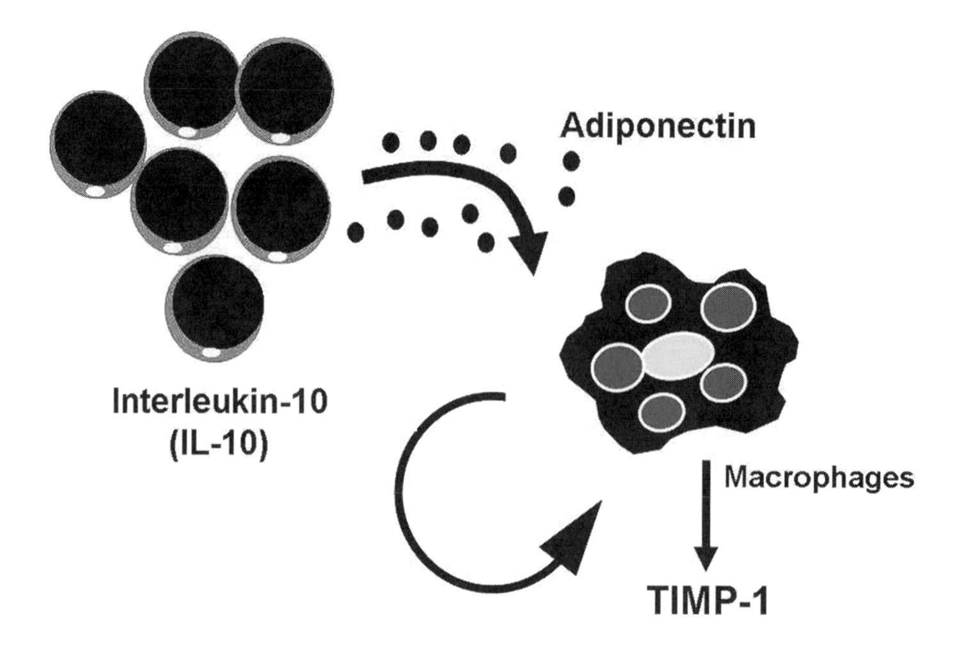

Fig. 3. Adiponectin induces TIMP-1 secretion via induction of IL-10 secretion in macrophages *(38)*.

higher than those of wild-type mice *(27)*. The reciprocal association of adiponectin and CRP levels in both human plasma and adipose tissue might participate in the development of atherosclerosis via inflammatory responses.

5. CONCLUSIONS

Adipose tissue has long been considered to be an organ that stores excess energy and supplies stored energy when food is lacking, in order to favor survival. However, recent studies have revealed that adipose tissue secretes a variety of bioactive substances—adipocytokines—controlling other organs and cells. A substantial proportion of adipocytokines are involved in inflammatory stimulation and response, as either proinflammatory adipocytokines or anti-inflammatory adipocytokines. As mentioned above, adiponectin is the most potent anti-inflammatory adipocytokine. Visceral fat accumulation causes a decrease in adiponectin plasma levels together with an increase of proinflammatory adipocytokines, such as TNF-α and PAI-1. An insufficient amount of adiponectin may be the background of enhanced inflammatory response and may result in the development of metabolic and cardiovascular diseases.

REFERENCES

1. Matsuzawa Y. Diabetes/Metab Rev 1997;13:3–13.
2. Maeda K, Okubo K, Shimomura I, et al. Gene 1997;190:227–235.
3. Shimomura I, Funahashi T, Takahashi M, et al. Nature Med 1997;2:1–5.
4. Masumoto S, Kishida K, Shimomura I, et al. Biochem Biophys Res Commun 2002;292:781–786.
5. Maeda K, Okubo K, Shimomura I, et al. Biochem Biophys Res Commun 1996;221:286–289.
6. Ouchi N, Kihara S, Arita Y, et al. Circulation 1999;100:2473–2476.
7. Matsuzawa Y. Atheroscler Thromb Vas Biol 2004;24:29–33.

8. Scherer EP, Williams S, Fogliano M. J Biol Chem 1995;270:26,746–26,749.
9. Hu E, Liang P, Spiegelman BM. J Biol Chem 1996;271:10,697–10,703.
10. Arita Y, Kihara S, Ouchi N, et al. Biochem Biophys Res Commun 1999;257:79–83.
11. Halleux CNM, Lee WJ, Delporte ML, et al. Biochem Biophys Res Commun 2001;288:1102–1107.
12. Maeda N, Takahashi M, Funahashi T, et al. Diabetes 2001;50:2094–2099.
13. Hotta K, Funahashi T, Arita Y, et al. Arterioscl Thromb Vas Biol 2000;20:1595–1599.
14. Hotta K, Funahashi T, Bodkin NL, et al. Diabetes 2001;50:1126–1133.
15. Lindsay RS, Funahashi T, Hanson RL, et al. Lancet 2002;360:57–58.
16. Maeda N, Shimomura I, Kishida K, et al. Nat Med 2002;8:731–737.
17. Kubota N, Terauchi Y, Yamauchi T, et al. J Biol Chem 2002;277:25,863–25,866.
18. Tomas E, Tsao TS, Saha AK, et al. Proc Natl Acad Sci USA 2002;99:16,309–16,313.
19. Iwashima Y, Katsuya T, Ishikawa K, et al. Hypertension 2004;43:1318–1323.
20. Ouchi N, Ohishi M, Kihara S, et al. Hypertension 2003;42:231–234.
21. Zoccali C, Mallamaci F, Tripepi G, et al. J Am Soc Nephrol 2002;13:134–141.
22. Kumada M, Kihara S, Sumitsuji S, et al. Arterioscler Thromb Vasc Biol 2003;23:85–89.
23. Pischon T, Girman CJ, Hotamisligil GS, et al. JAMA 2004;291:1730–1737.
24. Kondo H, Shimomura I, Matsukawa Y, et al. Diabetes 2002;51:2325–2328.
25. Ho RC, Davy KP, Hickery MS, et al. Cytokine 2005;30:14–21.
26. Moon YS, Kin DH, Song DK. Metabolism 2004;53:863–867.
27. Uysal KT, Wiesblock SM, Mario MW, et al. Nature 1997;389:610–614.
28. Ouchi N, Kihara S, Funahashi T, et al. Circulation 2003;107:671–674.
29. Kamada Y, Tamura S, Kiso S, et al. Gastroenterology 2003;125:1796–1807.
30. Petridou E, Mantzoros C, Dessypris N, et al. J Clin Endcrinol Metab 2003;88:993–997.
31. Miyoshi Y, Funahashi T, Kihara S, et al. Clin Cancer Res 2003;9:5699–5704.
32. Otake S, Takeda H, Suzuki Y, et al. Clin Cancer Res 2005;11:3642–3646.
33. Yokota T, Oritani K, Takahashi I, et al. Blood 1999;96:1727–1732.
34. Okamoto Y, Kihara S, Ouchi N, et al. Circulation 2002;26:2767–2770.
35. Ouchi N, Kihara S, Arita Y, et al. Circulation 2001;103:1057–1063.
36. Ouchi N, Kihara S, Arita Y, et al. Circulation 2000;102:1296–1301.
37. Tilg H, Wolf AM. Expert Opin Ther Targets 2005;9:245–251.
38. Kumada M, Kihara S, Ouchi N, et al. Circulation 2004;109:2046–2049.
39. Ridker PM, Burning JE, Shih J, Matias M, Hennekens CH. Circulation 1998;98:731–733.
40. Matsushita K, Yatsuya H, Tamakoshi K, et al. Arterioscler Thromb Vas Biol 2006;26:871–876.
41. Ignacy WW, Chudek J, Adamczak M, et al. Nephron Clin Prac 2005;101:c18–c24.

III

Interactions Between Adipocytes and Immune Cells

10 Macrophages, Adipocytes, and Obesity

Anthony W. Ferrante, Jr.

Abstract

Obesity induces an inflammatory response, which is implicated in the pathogenesis of obesity-associated complications. Adipose tissue is an important, perhaps an initiating, site in the development of obesity-induced inflammation. Macrophages are present in adipose tissue and their number and activation status correlate with measures of adiposity. Production of pro-inflammatory molecules and activation of intracellular pathways that regulate inflammatory responses in adipose tissue macrophages implicate these cells in development of obesity-induced complications, including insulin resistance.

More clearly elucidating the physiology of macrophages associated with adipose tissue and their contribution to obesity-induced inflammation will provide important insights into the physiology of adipose tissue and the pathophysiology of obesity and its complications. A detailed characterization of obesity-induced alterations in monocytes and macropahges will likely identify new therapeutic strategies to combat obesity-induced complications, such as atherosclerosis, non-alcoholic fatty liver disease and diabetes.

Key Words: Macrophages, inflammation, cytokines.

1. INTRODUCTION

Evidence for the association of obesity with inflammation dates back at least to the 1960s, when population studies found that obesity increases the circulating concentration of fibrinogen and other acute phase factors *(1–4)*. Studies of patients with type 2 diabetes mellitus (T2DM) similarly found that acute-phase proteins, including fibrinogen, are elevated in the circulation of subjects with T2DM *(5–7)*. The circulating concentration of more than a dozen proinflammatory cytokines (tumor necrosis factor [TNF]-α, monocyte chemoattractant protein [MCP]-1, interleukin [IL]-6), acute-phase reactants (C-reactive protein [CRP], serum amyloid A [SAA]-3, lipocalin) and procoagulant proteins (plasminogen activator inhibitor [PAI]-1, Factor VII) are now known to be increased by obesity *(8)*. In addition, the circulating concentration of at least one protein with strong anti-inflammatory properties, adiponectin, is inversely related to adiposity and insulin sensitivity *(9–12)*. A mechanistic role for inflammation in the development of obesity-induced complications in humans is suggested by the ability of circulating concentration of some factors (e.g., IL-6, adiponectin, CRP) to predict the development of obesity-associated complications in prospective human studies *(13–15)*, and by the protection from obesity-associated complications of mice genetically modified to reduce proinflammatory signaling molecules (e.g. TNF-α, inducible nitric oxide synthase [iNOS], inhibitor of κB kinase β [IκBKB]) *(16–20)*.

From: *Nutrition and Health: Adipose Tissue and Adipokines in Health and Disease*
Edited by: G. Fantuzzi and T. Mazzone © Humana Press Inc., Totowa, NJ

In the early 1990s Hotamisligil and colleagues reported that mouse adipose tissue expresses TNF-α and that obesity markedly increases the expression of this prototypical inflammatory cytokine *(21)*. Treatment of insulin-responsive cells, adipocytes and myocytes, with TNF-α impairs insulin signaling by increasing inhibitory phosphorylation of insulin receptor substrate (IRS) proteins *(22,23)*. These findings led Hotamisligil and his colleagues to propose a model in which obesity induces insulin resistance through adipose tissue overexpression of TNF-α *(24)*. They tested this hypothesis by demonstrating that genetic deficiency or pharmacological antagonism of TNF-α signaling improved insulin sensitivity in obese, insulin-resistant mice *(16)*. In humans, obesity also increases expression of TNF-α by adipose tissue *(25)*. However, in contrast to the mouse studies, TNF-α neutralizing therapies have not proved consistently effective at improving insulin sensitivity *(26–28)*. Nonetheless, recognition that adipose tissue is a source of a potent proinflammatory molecule had two important effects. It suggested a novel paradigm in which obesity-induced complications, and specifically insulin resistance, were due at least in part to inflammatory processes in adipose tissue. The studies of TNF-α also motivated searches for other proinflammatory molecules produced by adipose tissue and positively regulated by adiposity. More than a dozen have been identified, including IL-6, MCP-1, IL-1β, PAI-1, and CRP *(8)*.

In rodents, the increased expression by adipose tissue of inflammatory proteins has been mechanistically implicated in the development of important obesity-induced complications. Similar to the initial findings for TNF-α, genetic deficiency of iNOS or PAI-1 ameliorates insulin resistance in dietary and genetic models of murine obesity *(17,29)*. Our recent findings that deficiency of the MCP receptor, CCR2, also ameliorates insulin resistance and nonalcoholic fatty liver disease (*see* Headings 3 and 5) further support a model in which inflammation plays a causative role in obesity-induced complications. Conversely, obesity-induced decreases in adiponectin, an adipocyte-derived protein with anti-inflammatory properties, contribute to insulin resistance, nonalcoholic fatty liver disease, and atherosclerosis *(30–33)*. Compared with wild-type mice fed a high-fat diet, two strains of mice deficient in adiponectin developed greater insulin resistance *(30,31)*. Although the mechanistic role of adipose tissue-derived secretory inflammatory factors in human disease has not been as well defined, prospective clinical studies have shown that for any given body mass index or measure of adiposity, circulating concentrations of adiponectin, IL-6, and IL-1β predict future development of diabetes *(13,34)*. These findings are consistent with inflammation playing a causative role in obesity-induced complications in humans.

2. OBESITY ACTIVATES INTRACELLULAR PATHWAYS THAT REGULATE INFLAMMATORY RESPONSES

NFκB is a transcriptional complex that regulates the expression of many proinflammatory proteins and also regulates the transcriptional response to proinflammatory stimuli. In the cytosol the NFκB complex includes either NFκB1 or NFκB2, a REL protein (REL, RELA or RELB), and an inhibitor protein, either NFκBIA or NFκBIB (also known as IκBa and IκBb). IκBa and IκBb prevent translocation of the NFκB complex into the nucleus, and thereby negatively regulate NFκB-dependent transcription *(35–37)*. Many inflammatory stimuli—including activation of Toll-like receptors, reactive oxygen species, ultraviolet radiation, and proinflammatory cytokines—cause the translocation of NFκB into the nucleus *(38–41)*. These stimuli activate the IκB kinase complex (IκBKA, IκBKB, and IκBKG), which phosphorylates the IκBa or b component of NFκB. Phosphorylation targets IκBa or b for ubiquitin-dependent degradation, and thus induces translocation of the active NFκB complex into the nucleus. Once in the nucleus, the NFκB complex coordinates the transcription of a large number of genes in a cell-type specific manner *(35)*.

Obesity increases the nuclear localization of active NFκB subunits in liver and skeletal muscle *(18,20)*. Targeted genetic inactivation of IκB kinase, through deletion of the *Ikbkb* gene in hepatocytes, reduces hepatic expression of proinflammatory molecules (e.g., IL-1β, IL-6, Suppressor of cytokine signaling [SOCS]-1, SOCS-3) and reduces circulating concentrations of inflammatory cytokines IL-1β and IL-6 in dietary and genetic models of obesity *(19,20)*. The reduction in hepatic and systemic inflammation in obese mice with target mutations in *Ikbkb* also reduces insulin resistance *(19,20)*. Salicylates, including aspirin at high doses, impair NFκB nuclear translocation by inhibiting IκB degradation. Consistent with genetic studies, inhibition of NFκB activity by treatment with high-dose salicylates improves insulin sensitivity in both obese rodents and humans *(18,42)*. Arkan and colleagues also provided direct evidence that immune cells also contribute to obesity-induced inflammation and insulin resistance. They demonstrated that targeted deletion of *Ikbkb* in myeloid cells of obese mice ameliorates insulin resistance *(19)*.

JNK kinases are members of the mitogen activated protein (MAP) superfamily of kinases and key regulators of inflammation that are expressed by most cell types, including macrophages. In mammals the JNK family of kinases consists of three structurally related serine/threonine kinases (JNK1, JNK2, JNK3) that phosphorylate the activation domain of the c-JUN transcription factor *(43,44)*. JNKs were originally identified in studies of ultraviolet radiation-induced transcription, but are now known to be activated in response to many cellular stresses, including fatty acid treatment, reactive oxidant species, and endoplasmic reticulum stress, and in response to treatment with proinflammatory cytokines, including TNF-α *(45–47)*. Obesity increases JNK activity in liver, muscle, and adipose tissue in a manner similar to that seen for NFκB *(48)*. JNK kinases have also been demonstrated to phosphorylate the insulin-signaling protein IRS-1, thereby attenuating insulin signaling *(48,49)*.

Genetic studies have identified JNK1 as an important contributor to obesity-related phenotypes. Genetic deletion of JNK1 in mice attenuates the development of obesity in dietary and genetic models of obesity, and deficiency of JNK1 also ameliorates insulin resistance in these rodent obesity models *(48)*. Systemic treatment of obese, insulin-resistant mice with a peptide that inhibits JNK kinase activity also reduces insulin resistance in murine models of obesity *(50)*. Impairment of hepatic JNK activity is sufficient to improve insulin sensitivity in obese mice in the absence of effects on weight *(51)*. Although JNK kinase activity and NFκB activity are increased in liver and skeletal muscle by obesity, there is no evidence that obesity-induced NFκB activation is JNK-dependent. Like NFκB, JNK kinases are active in hematopoietic cells and macrophages specifically, However, unlike NFκB-dependent signaling, there are no published studies implicating hematopoietic JNK kinases in obesity-induced inflammation and insulin resistance.

3. MACROPHAGE PHYSIOLOGY

Macrophages are mononuclear phagocytes found in almost all tissues. They are derived from a common pool of circulating monocytes. Although different macrophage populations possess distinct tissue-specific morphologies and phenotypes, they also share common functions that are required to maintain tissue homeostasis. As "professional phagocytes," macrophages clear apoptotic and necrotic cells, cellular debris, and foreign pathogens *(52)*. They are the primary coordinators of the innate immune response and act as immune sentries that often are the first to recognize exogenous pathogens. Through a series of pattern recognition receptors, including Toll-like and complement receptors and nucleotide-binding oligomerization domain and mannose-binding proteins, macrophages are able to identify almost all common microbial pathogens and orchestrate the initial inflammatory response. Macrophages are also responsible for scavenging endogenous debris, including cellular byproducts of abnormal metabolism (e.g., ferrous sulfate in iron-overloaded states), and apoptotic and necrotic cells. Through the use of scavenger receptors, both cellular debris and dead cells are recognized and rapidly cleared *(53)*.

Two populations of macrophages exist within tissues—resident and elicited macrophages. Resident macrophages are cells found in tissues in the absence of pathological stimuli. They represent a stable population of macrophages that turnover at a basal rate. Elicited macrophages arrive in tissues in response to stimuli that have classically been defined as immunological *(54)*. However, the accumulation and maturation of macrophages in atherosclerotic lesions is an example of a population of elicited macrophages that are recruited in response to a stimulus that can more accurately be described as metabolic *(55–57)*. It seems reasonable to hypothesize that with the onset of obesity the accumulation and maturation of macrophages occurs through a process that entails the same basic steps of recruitment, differentiation, and activation that are characteristic of macrophage populations elicited by classical immunological provocations.

The initial step in the accumulation of elicited macrophages is the recruitment of monocytes. Monocytes and other circulating leukocytes possess on their cell surface two classes of molecules, β-integrins and mucins, that are critical for their attachment to endothelial cells and transmigration through vessel walls. Endothelial cells express the counter-receptors—selectins that bind mucins and receptors for β-integrins. In response to underlying activation signals, such as injury, microbial infection, or lipid accumulation, endothelial cells become "activated" *(58,59)*. Upon activation, endothelial cells release the contents of Weibel-Palade granules that contain P-selectin and upregulate the expression of E-selectin, and they also upregulate the expression of the integrin receptors, including intercellular adhesion molecule (ICAM)-1 and vascular cell adhesion molecule (VCAM)-1. The selectins are deposited on the surface of endothelial cells and allow initial attachment and rolling of monocytes along the endothelium. Firm adherence and efficient extravasation require the binding of β-integrins on monocytes by ICAM-1 or VCAM-1. For example, during the early evolution of an atherosclerotic plaque, endothelial cells overlying fatty streaks become activated and increase P- and E-selectin and ICAM-1 expression *(60–62)*. Genetic deletion of either selectins or ICAM-1 markedly attenuates the development of atherosclerotic lesions in mouse models of atherosclerosis *(63,64)*.

Following attachment and extravasation, the migration of monocytes into and through tissues is dependent on MCPs *(65,66)*. These chemokines are released from activated endothelia, macrophages, and mesenchymal cells in response to pathological stimuli. MCP-1 (also known as chemokine C-C motif ligand 2 [CCL2]) is the best characterized of the MCPs and binds to chemokine C-C motif receptor 2 (CCR2) *(67)*. In mice, genetic deletion of CCL2 or CCR2 impairs accumulation of elicited macrophage populations, decreases the clearance of microbial pathogens, and markedly attenuates disease in mouse models of atherosclerosis *(68,69)* and multiple sclerosis *(70,71)*.

Once monocytes have been recruited and enter a tissue, their differentiation into macrophages depends upon local signals. Colony-stimulating factors (CSFs) regulate the proliferation, differentiation, and survival of hematopoietically derived cells. M-CSF, also known as CSF-1, is the primary regulator of monocyte and macrophage differentiation *(72)*. In the absence of M-CSF, other CSFs, including GM-CSF and IL-3 (Multi-CSF), can partially compensate and support monocyte differentiation *(73,74)*. However, the compensation is typically incomplete, as demonstrated by the phenotype of a spontaneous, inactivating mutation in the M-CSF gene (*Csf1 op*) *(75,76)*. Resident macrophage populations in *Csf1 op/op* mice or in mice that carry a targeted mutation in the M-CSF receptor (*Csf1r –/–*) are reduced, some by as much as 80% *(73,77)*. In the absence of M-CSF, elicited macrophage populations are markedly reduced, so that *Csf1 op/op* mice that are also deficient for *Apoe* are completely resistant to developing atherosclerosis, despite having circulating cholesterol levels 10 times higher than *Csf1 +/+ Apoe –/–* atherogenic mice *(78)*.

Macrophages can exist in quiescent or activated states. As sentries monitoring the presence of exogenous or endogenous pathogens, quiescent macrophages produce few inflammatory signals. When their pattern recognition receptors, scavenger receptors and cytokine receptors are engaged macrophages become activated *(54,79)*. The varied activation phenotypes of macrophages are signal-dependent but are broadly characterized by the expression and secretion of proinflammatory molecules, increased motility, and phagocytosis. For example, interferon-γ and lipopolysaccharide (LPS) from Gram-negative bacteria together stimulate the production of TNF-α, IL-6, IL-1, and nitric oxide, the so-called "classical activation response." Stimulation of macrophages by IL-4 or IL-13 leads to upregulation of MHC and production of MIP1-γ and transforming growth factor (TGF)-β, in a response known as the "alternative activation response." As the effects of various stimuli have been characterized, it has become apparent that there are not a few discrete activation states but rather a broad and varied spectrum of responses that are stereotypical for individual stimuli and contexts. Although there are clearly many different activation states, NFκB activation is a common feature of many. For example, TNF-α, ER stress, and Toll-like receptor activation each activate NFκB *(41,80,81)*.

4. MACROPHAGES CONTRIBUTE TO OBESITY-INDUCED ADIPOSE TISSUE INFLAMMATION

Studies of obesity-induced inflammation have until recently focused on the effects of specific inflammatory molecules on metabolically important cell types, including hepatocytes, adipocytes, myocytes, and hypothalamic neurons *(23,82–84)*. Several recent analyses of adipocyte and nonadipocyte cell populations within adipose tissue demonstrate that much of adipose tissue inflammatory gene expression is derived from the nonadipocyte, stromal vascular cells *(85–87)*. In the original model proposed by Hotamisligil and Speigelman, obesity-induced increases in lipid content of adipocytes activate an inflammatory pathway in adipocytes. In this model, resulting proinflammatory cytokines act in autocrine fashion to impair adipocyte metabolic function, in a paracrine fashion to alter preadipocyte differentiation, and systemically to reduce insulin sensitivity. The role of classic inflammatory and immune cells has not been clearly delineated in adipose tissue physiology or obesity-associated phenotypes. However, sporadic reports have noted a correlation between adiposity and circulating leukocyte numbers, and decreases in leukocyte counts with weight loss *(88–90)*. Although the focus of these leukocytes studies was primarily on the effect of adiposity on classical immune functions—i.e., responses to infections—they suggested that obesity-induced inflammation may be, in part, attributable to immune cells.

Studies from our laboratory, and studies performed independently by Xu and colleagues, revealed that macrophages accumulate in adipose tissue of obese mice *(91,92)* and humans *(91)*, and contribute significantly to obesity-induced inflammation. In mice the percentage of macrophages contained within each depot correlates strongly with adipocyte size *(91)*. In lean animals, approx 5% of the cells in adipose tissue were macrophages, but in severely obese, leptin-deficient mice, 50% or more of the cells in mesenteric, perigonadal, and perirenal adipose tissue depots are macrophages. In these morbidly obese mice the macrophage content of subcutaneous adipose tissue is markedly increased, by about 30%, but slightly lower than that observed for visceral depots. However, the average adipocyte size is also lower in the subcutaneous compared with the visceral depots of leptin-deficient mice *(91)*. Multinucleated giant cells formed from macrophages are characteristic of chronic inflammatory conditions, such as Wegener's granulomatosis and mycobacterial infections. In adipose tissue of obese rodents and humans multinucleated giant cells are common *(91–94)*.

Preadipocytes have significant phagocytic capacity *(95)*, and previous reports had suggested that preadipocytes may transdifferentiate into authentic macrophages *(96)*. However, rescue of lethally irradiated mice with donor-tagged bone marrow demonstrated that in vivo in obese mice at least 85% of adipose tissue macrophages are bone marrow-derived *(91)*, suggesting that vast majority of cells identified as adipose tissue macrophages (ATMs) are of bone marrow origin and do not arise from resident preadipocytes.

Consistent with studies of isolated stromal vascular cells populations *(86)*, expression studies of macrophages isolated from the stromal vascular fraction (SVC) of adipose tissue reveal that ATMs are responsible for a large portion of the proinflammatory gene expression that increases with the onset of obesity *(91,92,97)*. Indeed, secreted factors produced by adipocytes can induce the expression of TNF-α and other proinflammatory proteins by macrophages, consistent with the effects of obesity on adipose tissue gene expression *(98,99)*. An important caveat to these ex vivo and in vitro studies is that separation of adipose tissue cell populations or manipulation of cell lines may induce inflammatory changes that are not physiological. Indeed, standard techniques to isolate adipocytes and stromal vascular cells do induce, at least transiently, the expression of proinflammatory genes *(100)*. Nonetheless, data support a model in which adipocytes can activate ATMs, thereby contributing to obesity-induced adipose tissue inflammation. Further experiments are required that will determine the contribution of macrophages to the inflammatory profile of adipose tissue.

Adiposity also regulates macrophage content of human adipose tissue depots. Body mass index (BMI) and adipocyte size are strong predictors of macrophage content of both subcutaneous and visceral adipose tissue depots *(91,97,101)*. Several studies have now shown that, similar to finding in rodents, the ATM content is a dynamically regulated process. Weight loss and thiazolidinedione treatment, but not metformin therapy, reduce the macrophage content of adipose tissue in humans *(93,102)*. However, Di Gregorio and colleagues suggest that insulin sensitivity may be a better predictor of ATM content than adiposity. In human subcutaneous adipose tissue they did not detect a significant ($p = 0.18$) correlation of BMI with the expression of the macrophage-expressed gene *Cd68*, but they did observe a significant negative correlation of *Cd68* expression with insulin sensitivity ($p = 0.02$). Larger studies will be required to determine which metabolic characteristics independently predict ATM content.

5. RECRUITMENT OF MONOCYTES TO ADIPOSE TISSUE IN OBESITY

Although recent studies have established that obesity is associated with the accumulation of macrophages in adipose tissue of rodents and humans, the mechanisms that regulate this process are just being studied. The first step in the recruitment of monocytes to a tissue is the adhesion to endothelial cells. In human studies, obesity and impaired insulin sensitivity are associated with elevated circulating concentrations of cellular adhesion molecules, including ICAM-1, VCAM-1, and E-selectin *(103–107)*. In vitro adipocyte-conditioned medium can directly upregulate the expression of ICAM-1 and platelet-endothelial cell adhesion molecule (PECAM), and increase adhesion, migration, and chemotaxis of monocytes *(97)*. Recently leptin and adiponectin have been shown to have opposing effects on endothelial cells: leptin increases monocyte adhesion to endothelial cells, and adiponectin reduces expression of adhesion molecules and other proinflammatory molecules by endothelium *(97,108)*.

Several lines of evidence have also implicated MCPs—in particular, CCL2 (MCP-1) and CCL7 (MCP-3)—as participating in the recruitment and chemotaxis of monocytes in adipose tissue. In both rodents and humans, adipose tissue expression of CCL2 and CCL7 are increased by obesity and reduced following weight loss or thiazolidinedione treatment *(97,102,109–112)*. CCR2 is a high-affinity receptor for CCL2 and CCL7, and is expressed on hematopoietic cells, including circulating monocytes. Genetic deficiency of CCR2 in a C57BL/6J background partially protects mice from obesity when fed a high-fat diet *(112)*. However, when obese mice are matched by body weight and adiposity, *Ccr2*-deficiency reduces the macrophage content of adipose tissue. This reduction of macrophages is associated with a reduction in proinflammatory gene expression and a coordinate upregulation of metabolically important genes in adipose tissue. Importantly, adiponectin adipose tissue expression and circulating adiponectin concentrations were increased in obese *Ccr2–/–* mice compared with wild-type obese animals. Glucose tolerance and insulin sensitivity were improved, and hepatosteatosis was reduced in obese mice deficient for *Ccr2*. Short-term pharmacological antagonism of CCR2 also reduced macrophage content of adipose tissue and improved glucose tolerance and insulin sensitivity in obese mice *(112)*. In contrast, deficiency of *Ccr2* in mice on the DBA/J background had no discernable effect on the inflammatory character of adipose tissue or systemic metabolic parameters when mice were fed a high-fat diet *(113)*. The genotype-dependent differences in these two strains of *Ccr2–/–* mice may derive from inherent differences in these strains in developing obesity-induced inflammation or from differences in the manner in which the experiments were carried out. These differences may, therefore, provide important clues as to genetic and environmental modifiers of obesity-induced inflammation and complications.

An important unanswered question is what are the primary events that induce monocyte recruitment and macrophage differentiation in adipose tissue of obese individuals. One hypothesis is that obesity accelerates adipocyte death and turnover, and that consistent with the role of macrophages in other tissue, ATMs serve to clear dead cells. Genetically engineered mice in which apoptosis is induced specifically in adipocytes by treatment with an exogenous drug demonstrate that massive adipocyte apoptosis does indeed induce macrophage recruitment to adipose tissue *(114)*. Cinti and colleagues have suggested that obesity induces adipocyte necrosis and that electron micrographic data suggest that macrophages form multinucleated giant cells specifically in response to adipocyte necrosis. They argue that ultrastructure analysis is not consistent with apoptosis but is distinctively characteristic of necrosis. They convincingly demonstrate that multinucleated giant cells contain lipid droplets that are not perilipin-coated *(115)*. Further studies are needed to clarify whether the apoptosis or necrosis are in fact mechanistically involved in monocyte/macrophage recruitment to adipose tissue in the setting of obesity. Another compelling hypothesis suggests that obesity induces metabolic derangements in adipocytes that lead to the production of factors that activate endothelial cells, direct chemotaxis and induce differentiation. At this time little direct evidence exists to support this proposition *(116)*.

6. CONCLUSIONS

Obesity induces a complex inflammatory response implicated in the pathogenesis of obesity-associated complications. Adipose tissue is an important, perhaps an initiating, site in the development of obesity-induced inflammation. The strong correlation of ATM content with measures of adiposity, and the production of proinflammatory molecules implicated in insulin obesity-induced complications by ATMs, suggest that ATMs may play a role in regulating adiposity-dependent phenotypes. More clearly elucidating the physiology of ATMs and their contribution to obesity-induced inflammation will provide important insights into the physiology of adipose tissue and the pathophysiology of obesity and its complications. A detailed characterization of obesity-induced alterations in monocytes and macrophages will likely identify new therapeutic strategies to combat obesity-induced complications, such as atherosclerosis, nonalcoholic fatty liver disease, and diabetes.

REFERENCES

1. Bennett NB, Ogston CM, McAndrew GM, et al. J Clin Pathol 1966;19:241–243.
2. Grace CS, Goldrick RB. J Atheroscler Res 1968;8:705–719.

3. Grace CS, Goldrick RB. Aust J Exp Biol Med Sci 1969;47:397–400.
4. Grace CS, Goldrick RB. Australas Ann Med 1969;18:26–31.
5. Fearnley GR, Vincent CT, Chakrabarti R. Lancet 1959;2:1067.
6. Fearnley GR, Chakrabarti R, Avis PR. Br Med J 1963;5335:921–923.
7. Abe T, Endo K, Imagawa U, et al. Tohoku J Exp Med 1967;91:129–142.
8. Wellen KE, Hotamisligil GS. J Clin Invest 2005;115:1111–1119.
9. Arita Y, Kihara S, Ouchi N, et al. Biochem Biophys Res Commun 1999;257:79–83.
10. Hotta K, Funahashi T, AritaY, et al. Arterioscler Thromb Vasc Biol 2000;20:1595–1599.
11. Berg AH, Combs TP, Scherer PE. Trends Endocrinol Metab 2002;13:84–89.
12. Kern PA, Di Gregorio GB, Lu T, et al. Diabetes 2003;52:1779–1785.
13. Lindsay RS, Funahashi T, Hanson RL, et al. Lancet 2002;360:57–58.
14. Freeman DJ, Norrie J, Caslake MJ, et al. Diabetes 2002;51:1596–1600.
15. Festa A, D'Agostino R Jr, Tracy RP, et al. Diabetes 2002;51:1131–1137.
16. Uysal KT, Wiesbrock SM, Marino MW, et al. Nature 1997;389:610–614.
17. Perreault M, Marette A. Nat Med 2001;7:1138–1143.
18. Yuan M, Konstantopoulos N, Lee J, et al. Science 2001;293:1673–1677.
19. Arkan MC, Hevener AL, Greten FR, et al. Nat Med 2005;11:191–198.
20. Cai D, Yuan M, Frantz DF, et al. Nat Med 2005;11:183–190.
21. Hotamisligil GS, Shargill NS, Spiegelman BM. Science 1993;259:87–91.
22. Hotamisligil GS, Peraldi P, Budavari A, et al. Science 1996;271:665–668.
23. Hotamisligil GS, Murray DL, Choy LN, et al. Proc Natl Acad Sci USA 1994;91:4854–4858.
24. Hotamisligil GS. Int J Obes Relat Metab Disord 2000;24(Suppl 4):S23–S27.
25. Hotamisligil GS, Arner P, Caro JF, et al. J Clin Invest 1995;95:2409–2415.
26. Ofei F, Hurel S, Newkirk J, et al. Diabetes 1996;45:881–885.
27. Paquot N, Castillo MJ, Lefebvre PJ, et al. J Clin Endocrinol Metab 2000;85:1316–1319.
28. Yazdani-Biuki B, Stelzl H, Brezinschek HP, et al. Eur J Clin Invest 2004;34:641–642.
29. Ma LJ, Mao SL, Taylor KL, et al. Diabetes 2004;53:336–346.
30. Maeda N, Shimomura I, Kishida K, et al. Nat Med 2002;8:731–737.
31. Kubota N, Terauchi Y, Yamauchi T, et al. J Biol Chem 2002;277:25,863–25,866.
32. Xu A, Wang Y, Keshaw H, et al. J Clin Invest 2003;112:91–100.
33. Okamoto Y, Kihara S, Ouchi N, et al. Circulation 2002;106:2767–2770.
34. Spranger J, Kroke A, Mohlig M, et al. Diabetes 2003;52:812–817.
35. Baldwin AS Jr. J Clin Invest 2001;107:3–6.
36. Ghosh S, Karin M. Cell 2002;109Suppl:S81–S96.
37. Bonizzi G, Karin M. Trends Immunol 2004;25:280–288.
38. Medzhitov R, Preston-Hurlburt P, Janeway CA Jr. Nature 1997;388:394–397.
39. Schneider A, Martin-Villalba A, Weih F, et al. Nat Med 1999;5:554–559.
40. Devary Y, Rosette C, DiDonato JA, et al. Science 1993;261:1442–1445.
41. Israel A, Le Bail O, Hatat D, et al. EMBO J 1989;8:3793–3800.
42. Hundal RS, Petersen KF, Mayerson AB, et al. J Clin Invest 2002;109:1321–1326.
43. Davis RJ. Cell 2000;103:239–252.
44. Dong C, Davis RJ, Flavell RA. Annu Rev Immunol 2002;20:55–72.
45. Rosette C, Karin M. Science 1996;274:1194–1197.
46. Urano F, Wang X, Bertolotti A, et al. Science 2000;287:664–666.
47. Liu ZG, Hsu H, Goeddel DV, et al. Cell 1996;87:565–576.
48. Hirosumi J, Tuncman G, Chang L, et al.Nature 2002;420:333–336.
49. Lee YH, Giraud J, Davis RJ, et al. J Biol Chem 2003;278:2896–2902.
50. Kaneto H, Nakatani Y, Miyatsuka T, et al. Nat Med 2004;10:1128–1132.
51. Nakatani Y, Kaneto H, Kawamori D, et al. J Biol Chem 2004;279:45,803–45,809.
52. Gordon S. Bioessays 1995;17:977–986.
53. Gordon S. Cell 2002;111:927–930.
54. Gordon S. Nat Rev Immunol 2003;3:23–35.
55. Hansson GK, Libby P, Schonbeck U, et al. Circ Res 2002;91:281–291.
56. Hansson GK. N Engl J Med 2005;352:1685–1695.

57. Bouloumie A, Curat CA, Sengenes C, et al. Curr Opin Clin Nutr Metab Care 2005;8:347–354.
58. Frenette PS, Wagner DD. N Engl J Med 1996;334:1526–1529.
59. Frenette PS, Wagner DD. N Engl J Med 1996;335:43–45.
60. Jonasson L, Holm J, Skalli O, et al. Arteriosclerosis 1986;6:131–138.
61. Davies MJ, Gordon JL, Gearing AJ, et al. J Pathol 1993;171:223–229.
62. Poston RN, Haskard DO, Coucher JR, et al. Am J Pathol 1992;140:665–673.
63. Bourdillon MC, Poston RN, Covacho C, et al. Arterioscler Thromb Vasc Biol 2000;20:2630–2635.
64. Collins RG, Velji R, Guevara NV, et al. J Exp Med 2000;191:189–194.
65. Gerard C, Rollins BJ. Nat Immunol 2001;2:108–115.
66. Daly C, Rollins BJ. Microcirculation 2003;10:247–257.
67. Charo IF, Myers SJ, Herman A, et al. Proc Natl Acad Sci USA 1994;91:2752–2756.
68. Boring L, Gosling J, Cleary M, et al. Nature 1998;394:894–897.
69. Gosling J, Slaymaker S, Gu L, et al. J Clin Invest 1999;103:773–778.
70. Izikson L, Klein RS, Charo IF, et al. J Exp Med 2000;192:1075–1080.
71. Huang DR, Wang J, Kivisakk P, et al. J Exp Med 2001;193:713–726.
72. Pixley FJ, Stanley ER. Trends Cell Biol 2004;14:628–638.
73. Cecchini MG, Dominguez MG, Mocci S, et al. Development 1994;120:1357–1372.
74. Wiktor-Jedrzejczak W. Leukemia 1993;7 Suppl 2:S117–S121.
75. Yoshida H, Hayashi S, Kunisada T, et al. Nature 1990;345:442–444.
76. Wiktor-Jedrzejczak W, Bartocci A, Ferrante AW Jr, et al. Proc Natl Acad Sci USA 1990;87:4828–4832.
77. Dai XM, Ryan GR, Hapel AJ, et al. Blood 2002;99:111–120.
78. Smith JD, Trogan E, Ginsberg M, et al. Proc Natl Acad Sci USA 1995;92:8264–8268.
79. Gordon S. Res Immunol 1998;149:685–688.
80. Pahl HL, Baeuerle PA. EMBO J 1995;14:2580–2588.
81. Beutler B. Nature 2004;430:257–263.
82. Senn JJ, Klover PJ, Nowak IA, et al. Diabetes 2001;50:1102–1109.
84. Luheshi GN, Gardner JD, Rushforth DA, et al. Proc Natl Acad Sci USA 1999;96:7047–7052.
85. Fried SK, Bunkin DA, Greenberg AS. J Clin Endocrinol Metab 1998;83:847–850.
86. Fain JN, Cheema PS, Bahouth SW, et al. Biochem Biophys Res Commun 2003;300:674–678.
87. Fain JN, Madan AK, Hiler M, et al. Endocrinology 2004;145:2273–2282.
88. Nanji AA, Freeman JB. Am J Clin Pathol 1985;84:346–347.
89. Field CJ, Gougeon R, Marliss EB. Am J Clin Nutr 1991;54:123–129.
90. Kullo IJ, Hensrud DD, Allison TG. Am J Cardiol 2002;89:1441–1443.
91. Weisberg SP, McCann D, Desai M, Rosenbaum M, Leibel RL, Ferrante AW Jr. J Clin Invest 2003;112:1796–1808.
92. Xu H, Barnes GT, Yang Q, et al. J Clin Invest 2003;112:1821–1830.
93. Cancello R, Henegar C, Viguerie N, et al. Diabetes 2005;54:2277–2286.
94. Cinti S, Mitchell G, Barbatelli G, et al. J Lipid Res 2005;46:2347–2355.
95. Cousin B, Munoz O, Andre M, et al. FASEB J 1999;13:305–312.
96. Charriere G, Cousin B, Arnaud E, et al. J Biol Chem 2003;278:9850–9855.
97. Curat CA, Miranville A, Sengenes C, et al. Diabetes 2004;53:1285–1292.
98. Berg AH, Lin Y, Lisanti MP, et al. Am J Physiol Endocrinol Metab 2004;287:E1178–E1188.
99. Suganami T, Nishida J, Ogawa Y. Arterioscler Thromb Vasc Biol 2005;25:2062–2068.
100. Ruan H, Zarnowski MJ, Cushman SW, et al. J Biol Chem 2003;278:47,585–47,593.
101. Bruun JM, Lihn AS, Pedersen S, Richelsen B. J Clin Endocrinol Metab 2005;90:2282–2289.
102. Di Gregorio GB, Yao-Borengasser A, Rasouli N, et al. Diabetes 2005;54:2305–2313.
103. Bagg W, Ferri C, Desideri G, et al. J Clin Endocrinol Metab 2001;86:5491–5497.
104. Ferri C, Desideri G, Valenti M, et al. Hypertension 1999;34:568–573.
105. Leinonen E, Hurt-Camejo E, Wiklund O, et al. Atherosclerosis 2003;166:387–394.
106. Pontiroli AE, Pizzocri P, Koprivec D, et al. Eur J Endocrinol 2004;150:195–200.
107. Straczkowski M, Lewczuk P, Dzienis-Straczkowska S, et al. Metabolism 2002;51:75–78.
108. Ouchi N, Kihara S, Arita Y, et al. Circulation 1999;100:2473–2476.
109. Sartipy P, Loskutoff DJ. Proc Natl Acad Sci USA 2003;100:7265–7270.
110. Takahashi K, Mizuarai S, Araki H, et al. J Biol Chem 2003;278:46,654–46,660.

111. Christiansen T, Richelsen B, Bruun JM. Int J Obes Relat Metab Disord 2005;29:146–150.
112. Weisberg SP, Hunter D, Huber R, et al. J Clin Invest 2006;116:115–124.
113. Chen A, Mumick S, Zhang C, et al. Obes Res 2005;13:1311–1320.
114. Pajvani UB, Trujillo ME, Combs TP, et al. Nat Med 2005;11:797–803.
115. Cinti S, Mitchell G, Barbatelli G, et al. J Lipid Res 2005;46:2347–2355.
116. Wellen KE, Hotamisligil GS. J Clin Invest 2003;112:1785–1788.

11 Interactions of Adipose and Lymphoid Tissues

Caroline M. Pond

Abstract

Interactions between adipose and lymphoid tissues at the molecular, cellular, and tissue levels are summarized, with emphasis on the special composition and metabolic properties of perinodal adipose tissue that is anatomically associated with lymph nodes. Perinodal adipose tissue intervenes between the diet and nutrition of the immune system, modulating the action of dietary lipids on immune function. The roles of peptide- and lipid-derived messenger molecules are complementary; precursors of prostaglandins and leukotrienes are specific fatty acids that are often essential constituents of the diet and may be supplied to lymphoid cells by paracrine interactions with adjacent adipocytes. Prolonged stimulation of paracrine interactions may induce local hypertrophy of adipose depots associated with lymphoid structures. Specialization of adipose tissue for paracrine interactions may be a unique, advanced feature of mammals that supports faster, more efficient immune processes and permits fever and other energetically expensive defences against pathogens to take place simultaneously with immune responses and unrelated functions such as lactation and exercise. The possible roles of local interactions between adipose and other tissues, and defects thereof, in human diseases, including HIV-associated lipodystrophy, Crohn's disease, lymphedema, atherosclerosis and in obesity and starvation, are briefly reviewed.

Key Words: Mesenteric; omental; popliteal; fatty acids; cytokines; paracrine interactions; perinodal; mammals; anorexia.

1. INTRODUCTION

1.1. Signal Molecules

Adipose tissue has become of interest to immunologists during the past 15 yr as a source of cytokines and other messenger molecules that modulate the immune system. The first such secretion to be identified was adipsin, a protein that closely resembles complement *(1–3)*. Since then, many more protein secretions and/or cytokine receptors have been described *(4)*. In most cases, cytokines such as tumor necrosis factor (TNF)-α and many interleukins were isolated first from the immune system and later found to be secreted by and/or taken up by adipose tissue, but others were identified first in adipose tissue and later shown to modulate immune function.

Identification and characterization of lipid-derived messenger molecules has been delayed because they are technically more difficult to characterize than proteins. Prostaglandins, leukotrienes, and thromboxanes are now regarded as comparable to

From: *Nutrition and Health: Adipose Tissue and Adipokines in Health and Disease*
Edited by: G. Fantuzzi and T. Mazzone © Humana Press Inc., Totowa, NJ

proteins in chemical diversity and physiological importance in the immune system and elsewhere *(5,6)*. They are synthesized from long-chain polyunsaturated fatty acids (PUFA) that are dietary essentials (i.e., they cannot be synthesized *de novo*) and are "stored" as components of membranes and triacylglycerols.

In addition to acting as precursors for lipid-derived messenger molecules, lipids also modulate the synthesis, secretion, and reception of peptide messengers. Nanotubes derived from the cell membrane have been suggested as a means by which immune cells exchange surface proteins and perhaps larger particles *(7)*. Nonesterified fatty acids can act as regulators of surface receptors *(8)* and gene transcription *(9)*. The incorporation of fatty acids into complex lipids does not have the equivalents of mRNA, tRNA, and ribosomes that ensure that the correct precursors are assembled into proteins, so many, including phospholipids, are structurally varied. Recent experiments on *Caenorhabiditis elegans* have demonstrated that fatty acid insufficiency alters neurotransmission and behavior *(10)*. Lipids rafts require appropriate fatty acids that are believed to have a major role in cell signaling in lymphoid cells *(11,12)*. Insufficiencies would impair cell signaling and so disrupt immune function. Caveolae are also numerous in adipocyte membranes and are essential for their responses to insulin *(13)*, and probably other signal molecules.

Adipocytes can selectively sequester and release fatty acids that differ in molecular weight and degree of unsaturation *(13)* and may supply specific fatty acid precursors to contiguous lymphoid cells *(14)*. By ensuring that appropriate precursors are available where and when they are needed for synthesis of complex molecules, adipocytes can fulfill for lipids the role of some of the protein synthesis machinery for proteins. But the anatomical relations between adipocytes and lymphoid cells are important for the efficiency of such interactions.

1.2. Cellular Interactions

The next step was the demonstration of immune cells, especially macrophages *(15,16)* and dendritic cells *(17,18)*, in adipose tissue. Although representing only a tiny fraction of the total mass of a depot, these intercalated cells are the source of large proportion of the TNF-α (and probably of many other signal molecules) synthesized in whole adipose tissue *(19)*. The fatty acid compositions of lipids in intercalated dendritic cells closely resembles those of contiguous adipocytes (Fig. 1). Changing the dietary lipids alters the fatty acid composition of both types of cells, but the correlation between values for cells that were adjacent in life remains. The simplest explanation for this similarity is that the dendritic cells take up their lipids from the contiguous adipocytes, rather than from the blood or lymph, as was previously assumed *(14)*. The site-specific differences in adipocyte composition *(20)* are thus conferred on intercalated dendritic cells, adding another source of diversity to these cells that hitherto have been classified by genes activated and proteins synthesized. It also may be an essential and hitherto unrecognized link in the mechanisms by which dietary lipids modulate immune function. Some properties of preadipocytes in vitro are strikingly similar to those of macrophages, suggesting common pathways of early development of macrophages and adipocytes *(21,22)*. Preadipocytes injected into mice acquire surface antigens characteristic of macrophages *(23)*. Receptors for bacterial products such as lipopolysaccharide have been demonstrated in 3T3-L1 adipocytes in vitro *(24,25)*.

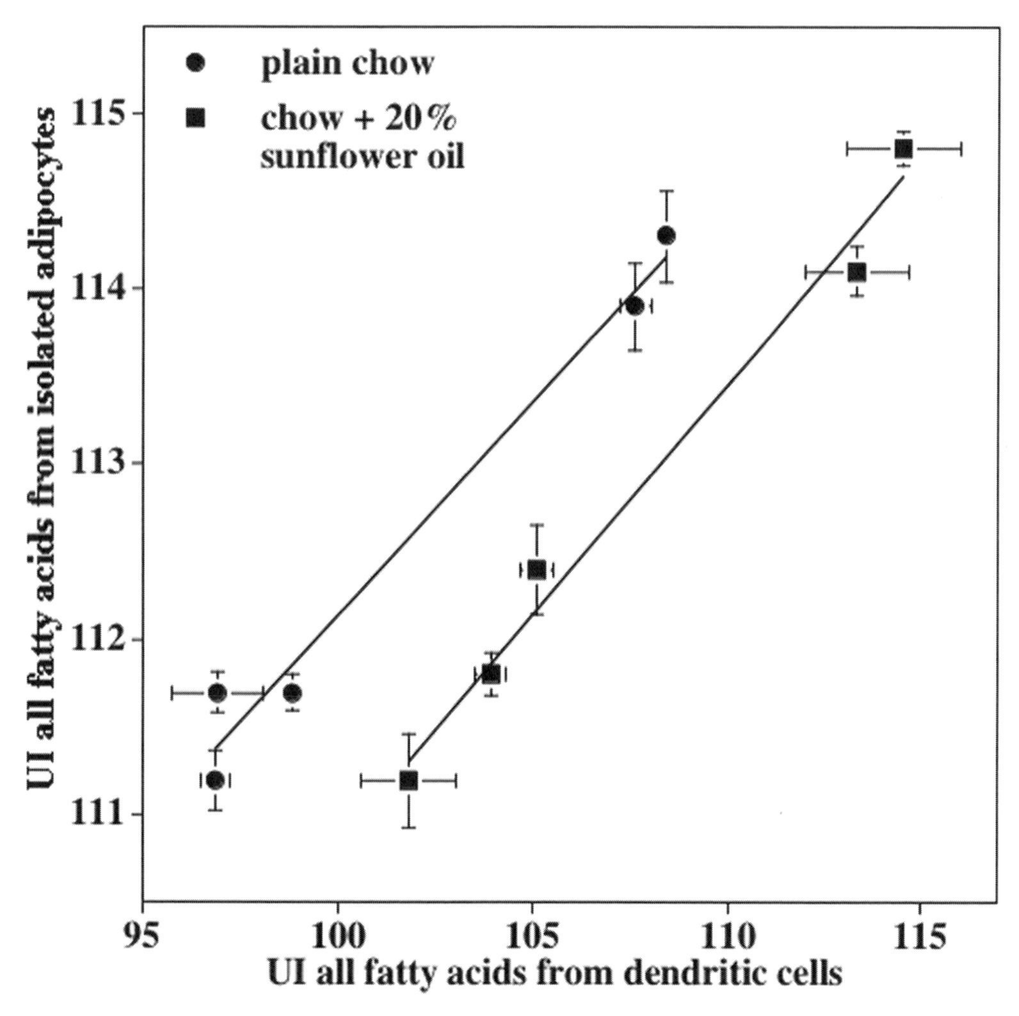

Fig. 1. The correlations between mean unsaturation indices (UI = [% monoenoic fatty acids] + 2[% dienoic] + 3[% trienoic]...etc.) of fatty acids extracted from complex lipids (mostly membrane phospholipids) in dendritic cells (DCs) and those of adipocytes (mostly triacylglycerols) isolated from the corresponding sample of adipose tissue (mesenteric perinodal and remote, omental with many or few milky spots, and popliteal remote from node) from adult male rats fed on unmodified chow (circles); rats fed on chow + 20% sunflower oil for 6 wk (squares). $N = 6$ sets of homologous samples from 3 similarly treated cage-mate rats for each dietary group. The standard errors of each mean are shown as bars. Data simplified from ref. *14*.

These properties are found only in preadipocytes extracted from white adipose tissue, not those of brown adipose tissue *(26)*, and disappear as soon as mitosis ends and the cells differentiate.

Dendritic cells interact with adjacent adipocytes *(27)*. Those extracted from the adipose tissue stimulate lipolysis, whereas those from an adjacent lymph node inhibit the process, although the effects are strong only in perinodal and milky spot-rich samples and minimal in the adipocytes extracted from sites more than 1 cm from lymph nodes. Switching from antilipolytic to lipolytic secretions seems to be among the transformations that dendritic cells undergo as they migrate between the lymph nodes and the adjacent adipose tissue, and thus should be considered as part of the maturation process *(27)*. Inducing mild inflammation by injection of lipopolysaccharide amplifies both effects, suggesting that they are integral to immune responses. Paracrine interactions between adipocytes and macrophages have also been demonstrated *(28)*. The significance of these lymphoid cells in intact animals remains unclear; a role in defense against pathogens has been suggested *(29)* and they may contribute to insulin resistance and other chronic metabolic disorders *(30)*.

1.3. Paracrine Interactions Between Adipose Tissue and Other Tissues

1.3.1. Lymph Nodes and Omental Milky Spots

In most animals, including murid rodents, adipose depots (such as perirenal, epididymal, and parametrial) that consist of "pure" adipose tissue are larger—in obese specimens, much larger—than those with embedded lymphoid structures. These depots are often the only ones providing enough tissue in mice and other small species, and are preferred for study, especially for experiments in which the tissue undergoes large changes in relative mass, as in obesity research or for the study of gene action in transgenic mice *(31)*. To avoid collateral damage, biopsy sites from humans and larger animals are always chosen for their remoteness from lymph nodes and vessels. Consequently, most of the data come from samples of large adipose depots chosen for their surgical accessibility *(32–36)* rather than for known site-specific properties. Much of the confusion and paradoxical findings about the relationship between adipose stores and immune function can be clarified by taking account of site-specific properties of adipose tissue and its interactions with adjacent lymphoid cells.

Recent reviews *(37,38)* summarize the evidence demonstrating that perinodal adipocytes selectively take up and store dietary lipids (and perhaps other precursors) that are released in response to local, paracrine signals from adjacent lymphoid cells that take them up. The depots that incorporate lymphoid tissue, especially the perinodal regions, are specialized to support the growth and metabolism of adjacent leukocytes.

Many hitherto unexplained site-specific differences in cytokine metabolism in adipose tissue are consistent with the hypothesis of paracrine interactions between lymphoid and adipose TNF-α than those from the nodeless epididymal or perirenal depots, and their secretions are less readily modulated by dietary lipids *(39)*. The role of embedded lymphoid tissues in this finding was never investigated. Genes specific to adipocytes are upregulated in the stomach wall of mice successfully immunized against *Helicobacter*, probably arising from paracrine interactions between adipose and lymphoid tissues *(40)*.

In vitro studies (Figs. 2,3) show that perinodal adipocytes respond much more strongly than others to TNF-α, interleukin (IL)-4, IL-6, and probably other cytokines *(41)*. Figure 2 shows that the response of adipocytes isolated from perinodal adipose tissue contiguous to nodes differ from those from a few millmeters distant, and the site-specific differences are amplified by the neurotransmitter norepinephrine *(41)*. Furthermore, the action of combinations of cytokines is not predictable from that of single agonists: although IL-6 is prolipolytic, especially for perinodal adipocytes (Fig. 2), when combined with IL-4 it suppressed lipolysis (Fig. 3). The same is true of interactions between IL-4 and TNF-α *(41)*. The data in Figs. 2 and 3 report only lipolysis, but similar site-specific differences in the responses to signal molecules probably occur for other

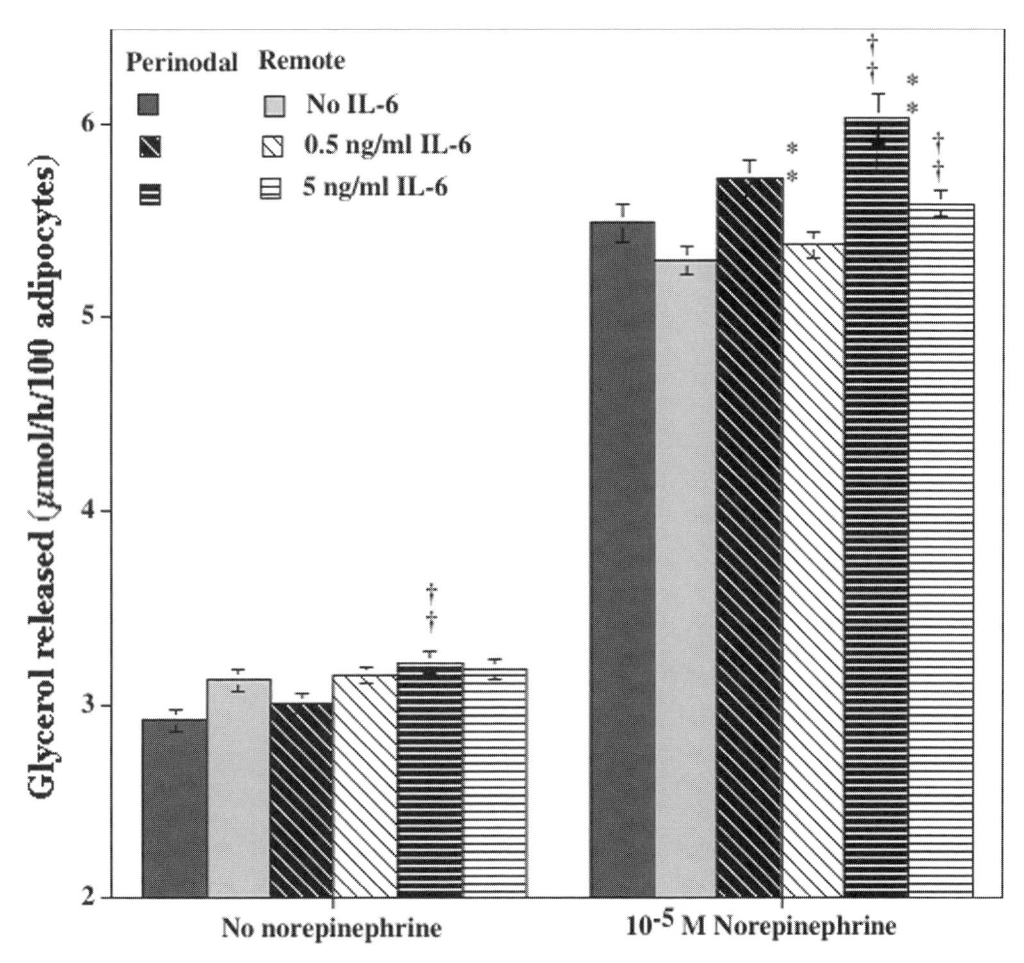

Fig. 2. Rates of glycerol release over 1 h from mesenteric perinodal (darker bars) and far from (paler bars) lymph node(s) adipocytes from adult guinea pigs ($N = 12$) with and without supramaximal norepinephrine, after preincubation for 24 h with and without interleukin-6 at two different concentrations. Asterisks denote comparison of pairs of samples from the same depot under the same conditions: ** significantly different at $p < 0.01$; * significantly different at $p < 0.05$. Daggers denote comparison of corresponding sample incubated with and without cytokines: ††† significantly different at $p < 0.001$; †† significantly different at $p < 0.01$; † significantly different at $p < 0.05$. Data simplified from ref. *41*.

metabolic pathways. Greater sensitivity to cytokines is also implicated in the larger responses of perinodal adipocytes to the presence of dendritic cells *(29)*. The interactions between the effects of cytokines and neurotransmitters on perinodal adipocytes (Figs. 2,3) suggest mechanisms by which the nervous system may modulate immune processes.

IL-6 was found to be 100 times more concentrated in the interstitial fluid of human superficial adipose tissue than in the blood plasma of the same subjects, strongly suggesting a paracrine role for this cytokine in adipose tissue *(42)*. Nonetheless, the nodeless epididymal or perirenal depots remain the favorite choice of adipose tissue for cytokine studies *(34,43,44)*. Even where depots are compared, the largest rather than those most strongly involved in cytokine secretion are chosen for study *(45)*. Matters are

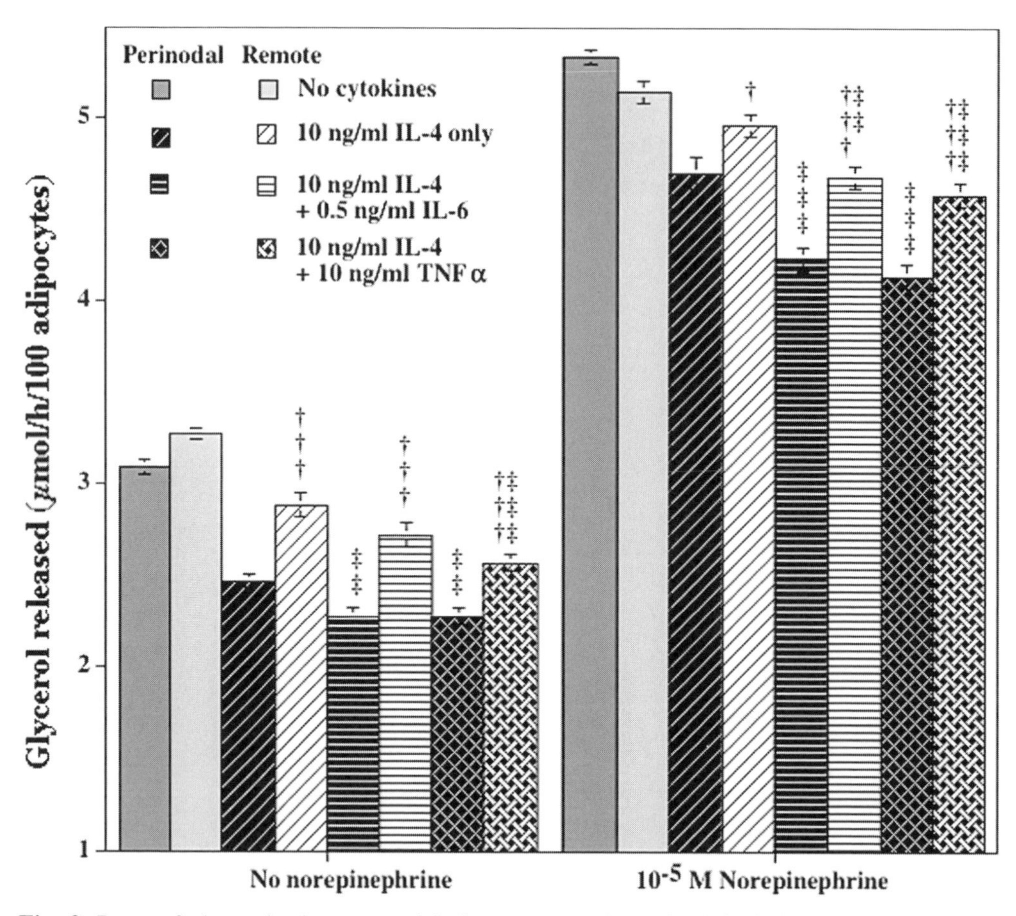

Fig. 3. Rates of glycerol release over 1 h from mesenteric perinodal (darker bars) and far from (paler bars) lymph node(s) adipocytes from adult guinea pigs ($N = 12$) with and without supramaximal norepinephrine, after preincubation for 24 h with and without interleukin-4 (10 ng/mL) alone or with interleukin-6 (0.5 ng/mL) or TNF-α (10 ng/mL). Daggers compare corresponding samples incubated with or without cytokines: ††† significantly different at $p < 0.001$; † significantly different at $p < 0.05$. For clarity, symbols indicating within-depot differences, and those indicating that all the values from "near node" adipocytes, are significantly different at $p < 0.001$ from those from the corresponding control samples incubated without cytokines are *not* shown. Double daggers compare samples incubated with two cytokines with the corresponding sample incubated with IL-4 alone: ‡‡‡ significantly different at $p < 0.001$; ‡‡ significantly different at $p < 0.01$. Data simplified from ref. *41*.

further complicated by erroneously classifying the intra-abdominal perirenal/retroperitoneal and gonadal (i.e., epididymal or parametrial) depots in the same category ("visceral") as the omentum and mesentery *(46–50)*, although the latter have many distinctive properties arising from the embedded lymphoid structures not found in the former. Data from the larger, nodeless depots may elucidate the metabolic changes associated with obesity *(34,51)*, but could be misleading as well as irrelevant to understanding the source and action of adipocytes and the nutrition of the immune system.

By releasing fatty acids only to lymphoid cells and only when and where they are required, the perinodal adipose tissue partially emancipates immune function from dependence on the quantity and lipid composition of food *(38)*. Paracrine interactions may also account for some features of the anatomy of lymph vessels and nodes *(52–54)*. The branching of lymph vessels near nodes would slow the passage of lymph and bring a greater surface area of vessels into contact with adipocytes, thus facilitating the exchange of signals and metabolites.

1.3.2. Bone Marrow

Red bone marrow has much in common with lymph nodes in mammals, as it is the origin of many of the lineages of lymphoid cells found in nodes *(55)*. Adipocytes in "yellow" marrow are smaller than those of other depots, but the roles of cell proliferation and cell expansion in tissue growth follow a similar ontogenetic time course *(56)*. Independent studies of the contribution of bone marrow adipocytes to the control of lymphohematopoiesis in vitro *(57,58)* reveal paracrine interactions similar in principle to those we have described between perinodal adipose tissue and lymph-node lymphoid cells *(37,38)*. Suggested mediators for paracrine interactions between bone marrow adipocytes and lymphoid cells include both cytokines *(59)* and polyunsaturated fatty acids *(60)*. Bone marrow adipocytes may also interact locally with osteoblasts in bone formation *(154)*. However, most of the data come from in vitro experiments, and exactly how such processes determine the course of osteoporosis in elderly people remains to be established *(61)*. Transgenic mice without detectable adipocytes in the bone marrow develop normal skeletons *(62)*.

1.3.3. Other Paracrine Interactions

Local paracrine interactions between minor depots of physiologically specialized adipocytes and anatomically contiguous tissues are increasingly recognized. Such effects have been described for major blood vessels *(63,64)*, skeletal muscle *(65)*, cardiac muscle *(66,67)*, bone *(57)*, and lymph nodes *(37,68)*. All these concepts are based on experiments using laboratory animals. Site-specific differences and paracrine interactions are difficult to detect in humans, as they have little or no blood manifestation, but they should be considered in interpreting data from the emerging field of lipidomics *(69,70)* as well as the actions of dietary and blood-borne lipids on the composition of lymphoid cells *(71,72)*.

The concept of local, paracrine exchange of signals and nutrients between functionally specialized contiguous adipose tissue and other tissues provides an explanation for why adipose tissue in mammals is partitioned into a few large and many small depots, most of the latter associated with lymphoid structures *(38)*. Lipid reserves that are dedicated to supplying the immune system may be essential to combining fever with immune responses in defense against pathogens and to enabling both processes to be combined with anorexia. Many of the polyunsaturated fatty acids selectively accumulated in triacylglycerols of perinodal adipocytes *(20)* and in phospholipids in lymph node lymphoid cells *(68)* and dendritic cells *(14)* are dietary essentials, so they can be scarce, especially during periods of high metabolic demand (e.g., pregnancy, lactation) and reduced food intake (e.g., while disabled by injury or disease). Local control of lipolysis by the immune system manages supplies of these essential fatty acids efficiently, and minimizes competition with other tissues (e.g., pyrogenic tissues such as muscles and liver) for circulating lipids.

1.4. Growth of Adipose Tissue Associated With Lymph Nodes

Control of proliferation of preadipocytes and of their maturation into adipocytes has long been studied, because of its implications for obesity in humans and domestic livestock, but emphasis has been on diet and energy balance *(73)* and angiogenic factors *(74,75)*. Local growth of adipose tissue associated with long-standing lymphatic disorders has long been known, though its cause is poorly understood *(76)*. More recently, the increasing prevalence of certain human disorders has directed attention to the possibility that inflammatory cytokines and other immune-derived factors also have a role in regulating adipogenesis *(51,77)*.

The actions of dietary lipids and hormones can be simulated in vitro, but demonstrating a role for the immune system requires in vivo experiments. Unfortunately, adipocytes form and die very slowly, and preadipocytes cannot be recognized unambiguously, so measuring adipocyte turnover in vivo is difficult *(78,79)*. The total adipocyte complement seems to differ substantially between individuals of similar body composition, even among inbred mammals *(80)*, making the detection of experimentally induced increases in numbers of adipocytes difficult unless they are very large. Major systemic immune responses induce anorexia, and eventually cachexia, which deplete the adipose tissue anyway, especially in small laboratory animals. However, by comparing the numbers of adipocytes in the popliteal depots, and locally stimulating the lymph nodes in one of them with small doses of lipopolysaccharide, the formation of more adipocytes in response to chronic inflammation could be demonstrated, corresponding to the decrease in average volume of adipocytes throughout the depot *(81)*. After experimental lipectomy of the epididymal fat pads of adult rats, compensatory regrowth of adipose tissue is significant 16 wk later in the node-containing mesenteric and inguinal depots but not in perirenal depots *(82)*.

Local hypertrophy of adipose tissue associated with lymph nodes is not reversed within 3 mo of the end of experimental inflammation *(83)*. Increases in infiltrating dendritic cells and in adipocyte apoptosis were found, as expected, in the adipose tissue around the inflamed lymph node, and also in the perinodal mesenteric and milky spot-rich parts of the omentum. This finding is another case in which adipocytes in these intra-abdominal depots have well-developed properties associated with paracrine interactions with lymphoid cells and actively participate in immune responses that in other respects are confined to remote parts of the body.

The quantities of perivascular and intermuscular adipose tissue are so small and so variable in laboratory rodents that systematic study of their growth in vivo has not yet been possible. Human postmortem data suggest the presence of cardiac depot hypertrophy in chronic heart disease *(66,67)*.

2. PARACRINE INTERACTIONS OF ADIPOSE TISSUE IN HUMAN DISEASES

2.1. Adipose Tissue Hypertrophy and Inflammation

One of the most radical developments in obesity research in the past few years is recognition that low-level chronic inflammation and other forms of immune dysfunction are associated with excess adipose tissue, at least in humans and laboratory rodents *(84,85)*. Chronic inflammation is believed to be one of several mechanisms linking obesity to increased risk of adult-onset cancer *(86)*, type 2 diabetes *(85)*, and atherosclerosis *(87,88)*, although the causal relationships are far from clear *(89)*. In this context, it may be appropriate to point out that many and varied disorders linked to pathological and artificial obesity have not been reported in natural obesity. The evidence from animals that are naturally obese for all or part of the year points to the opposite association: fatter specimens are usually more fecund and live as long or longer than average *(90)*.

The formation of additional adipocytes following prolonged, chronic inflammation *(81,83)* may be behind reports of an association between markers of viral infection and enlargement of certain adipose depots in domesticated birds, captive primates, and humans *(91)*. The fact that several different pathogens have been identified as causes of obesity *(92)* points to the immune response, rather than direct action of the organisms themselves, as the main pathological mechanism. Such "obesity" may not involve the usual metabolic correlates and is in fact the slow, selective hypertrophy of adipose depots that enclose lymphoid tissue *(18)*. The effects are most obvious in (though not necessarily confined to) the mesentery and omentum, large adipose depots that incorporate much lymphoid tissue and respond to remote as well as to local inflammation.

Hypertrophy of the mesentery and omentum eventually leads to swelling of the abdomen and to high waist/hip ratios in humans *(93)*. Thick waists are common among people of average body mass who smoke heavily, which continually exposes them to toxins and irritants in tobacco *(94)*, or who are frequently exposed to a wide variety of parasites and pathogens *(95)*. The link between hypertrophy of intra-abdominal adipose tissue and susceptibility to type 2 diabetes is long established *(96)*. More recently, chronic inflammation has been implicated in the progress of long-standing obesity to insulin resistance *(97)* and diabetes *(98)*. Obesity is accompanied by changes in blood cytokines, at least some of which may come from "non-fat cells" found in human adipose tissue *(99)*.

2.1.1. Lymphedema

Physicians have long suspected that long-standing lymphedema, such as that arising from mastectomy *(100)* and other forms of lymph blockage *(101)*, leads to local hypertrophy of adipose tissue. Studies of transgenic mice in which the lymphatic system develops abnormally indicate that leaky and mispatterned lymph vessels induce hypertrophy in anatomically associated adipose tissue, such as the mesentery *(53)*. In genetically unaltered adult rats, chronic but very mild inflammation induces the formation of additional adipocytes over several weeks, especially near a site of immune stimulation *(81)*. This hypertrophy reverses only very slowly, with local enlargement of adipose depots containing stimulated lymph nodes still detectable 12 wk after experimental inflammation ended *(83)*. The structural adipose tissue in the orbit of the eye is normally without visible lymph vessels, but these can develop where chronic inflammation is present *(102)*. The wide variation among animals in the cellularity of the intraorbital adipose tissue *(103)* may arise from local inflammation early in life, leading to lymphangiogenesis and stimulation of adipocyte proliferation in affected individuals.

2.1.2. Cardiovascular Disease

Macrophages and T-cells become more numerous at the interface between perivascular adipose tissue and the adventitia of human atherosclerotic aortas *(87)*. Secretions from perivascular adipose tissue modulates contraction of the smooth muscle of rat aortas in vitro *(63)*. Such observations have led to the concept of "vasocrine" interactions between perivascular adipose tissue and vascular tissues *(88)*.

2.1.3. HIV Lipodystrophy

Reports of bizarre site-specific enlargement of certain normally minor adipose depots following drug treatment for HIV infection began appearing in the late 1990s *(104)* and was named HIV-associated adipose redistribution syndrome (HARS), also known as lipodystrophy/lipoatrophy. The syndrome is characterized by selective atrophy of certain adipose depots and/or hypertrophy of others and is accompanied by various changes in blood chemistry, some of them characteristic of type 2 diabetes and obesity. The manifestations are variable, and patterns of adipose tissue redistribution cannot easily be correlated with particular immunological symptoms or therapeutic regimes beyond the generalization that, as HIV takes hold and immune function deteriorates, HARS gets worse. The debate over how HARS should be characterized, which began as soon as it was first noted *(105–107)*, is still not resolved *(108,109)*.

HARS develops gradually after several months, or sometimes years, of infection with the human immunodeficiency virus *(110)*. It has been described in patients in whom the progression from HIV infection to AIDS is delayed naturally *(111)*, although it is more common and worse in those treated with either or both major classes of antiviral drugs, nucleoside reverse transcriptase inhibitors (NRTIs) and protease inhibitors (PIs). HARS has been described in both sexes and in patients of all ages, including children infected perinatally *(112)*, and can accompany constant, increasing, or decreasing body mass, with or without changes in average energy intake. Typical sites of hypertrophy are the intra-abdominal depots, probably mostly or entirely the omentum and mesentery, the breasts, and the "buffalo hump" around the neck, all sites in which adipose and lymphoid tissues are in intimate contact *(113)*. These enlargements are usually accompanied by depletion of superficial adipose tissue (especially buttocks, thighs, arms, and legs), sometimes to the extent that cutaneous veins become clearly visible, and of the structural adipose tissue of the face, producing the hollow cheeks and sunken eyes characteristic of famine-starved, elderly, and terminally ill people *(110)*.

The role of antiretroviral drugs in inducing lipoatrophy has been extensively investigated and cannot be reviewed here. Adipose hypertrophy has received less attention *(109,114)*. We have proposed that the presence of antiretroviral drugs may alter the paracrine signaling between adipocytes and lymphoid cells, leading to higher rates of lipolysis in adipocytes associated with lymphoid tissues *(115)*, of which the mesentery and omentum are the largest and most active depots. The higher concentration of extracellular fatty acids in lymphoid tissue-containing depots slowly leads to their hypertrophy by cell enlargement and/or the formation and maturation of more adipocytes. The latter process has been demonstrated experimentally in rats *(81,83)*. The overspill into the blood contributes to hyperlipidemia, another common side effect of HIV infection treated with antiretroviral drugs *(110)*. This shift to local control by the immune system diminishes the sensitivity of adipocytes in depots that contain lymphoid tissue to the endocrine conditions of fasting *(41)*, so the adipose tissue thus altered becomes less available to supply other tissues' energy needs. These processes gradually withdraw lipid from the nodeless depots that normally respond strongly to fasting.

This hypothesis has not been explored systematically in animals or human patients, but certain observations are consistent with it. In monkeys experimentally infected with simian immunodeficiency virus, the postmortem viral antigen load was higher in alimentary-associated lymphoid tissues (tonsils, mesenteric and retropharyngeal lymph nodes, and the gut-associated lymphoid tissue) than in nonalimentary-associated lymphoid tissues (spleen, thymus, inguinal and axillary lymph nodes) *(116)*. The mesenteric lymph nodes and mucosal lymphoid tissue are also altered in the mouse model of AIDS *(117)*. For these reasons, and because they interact most strongly with activated lymphoid cells *(118)*, our hypothesis predicts that the mesenteric and omental perinodal adipocytes should be most altered by HIV infection and its treatment. Such effects have been shown in adults *(110,119)* and in children *(120,121)*, who normally have relatively little intra-abdominal adipose tissue. HARS is less likely to occur in those who acquired the infection through injecting drugs than in those infected sexually (in which case HIV reaches abdominal lymph nodes first) *(122)*. But women taking antiretroviral drugs whose breasts enlarged and lower limb fat wasted had the best blood indicators of effective suppression of HIV *(123)*, suggesting that strong interactions between adipose and lymphoid tissues, at the expense of "general purpose" depots, is associated with effective immune function.

Also consistent with this hypothesis is the observation that the breast tissue (in which adipocytes and lymphoid tissues come into intimate contact) enlarge in HARS. Mammary adipocytes resemble those associated with lymph nodes in expressing and responding to IL-6 and probably other cytokines *(124)*, and in selective accumulation *(125)* and release *(126)* of long-chain PUFAs. In view of these similarities, it is not surprising that breasts enlarge in HIV-infected women *(123,127,128)*, unilateral gynecomastia is much more common among HIV-infected men *(129)*, and the hypertrophied tissues do not regress after patients switch to alternative drugs *(130)*.

Detailed studies of the adipose tissue of the human face *(131)* clarify the important distinction between the corpus adiposum buccae and the orbital depots, in which lipolysis and lipogenesis are too slow to be measured, and the subcutaneous layer over the cheeks that are easily depleted during illness or fasting. The structural adipose depots in the orbit of the eye are normally metabolically inert and without lymph vessels, but after chronic, long-standing inflammation, lymphatics may form *(102)*, providing access for lymphocytes. Paracrine interactions between adipocytes of the metabolically inert structural depots and resident or infiltrating T-lymphocytes in the orbit of the eye are believed to be important to the pathogenesis of Graves' ophthalmopathy *(132)*. Similar interactions with HIV-infected lymphocytes could lead to the characteristic atrophy of the adipose tissue of the face.

2.2. Crohn's Disease

Crohn's disease (CD) is chronic inflammation of the alimentary canal, mostly the bowel, that usually first appears in children or young adults. It is relevant to this review because of a distinctive feature, the selective hypertrophy of mesenteric adipose tissue around the inflamed area of the intestine, known as "fat wrapping," first identified by Crohn himself *(133)*. It appears early in the course of the disease and persists, even though nearly all patients become thin following prolonged disruption of appetite, digestion, and absorption *(134)*. CD thus shares with HARS the peculiarity of selective expansion of certain adipose depots while others are depleted. There may also be some similarities between the inflammation of the intestinal mucosa in CD and that of enteropathies associated with long-standing HIV infection *(135)*. The symptoms can occur even in patients whose immune responses are severely compromised by long-standing, aggressive HIV infection *(136)*. Although the gross anatomy is obviously abnormal in CD, there is nothing unusual in the microscopical appearance of the adipocytes themselves *(137)*, and there is still no comprehensive explanation for its etiology.

Most investigations focus on peptide synthesis and secretion. The mRNAs for TNF-α and possibly other inflammatory cytokines are increased in the mesenteric adipocytes themselves, but not in other intra-abdominal adipocytes or those from superficial depots *(137)*. Production of TNF-α may be enhanced by increased local secretion of leptin *(138)*. Abnormal adipokines have been implicated in a wider range of intestinal diseases *(139)*.

Recent studies *(140)* show that the site-specific differences in fatty acid composition of adipose tissue lipids in the mesentery expected from animal studies *(141)* are absent from patients with CD, although they were found in similar samples from the controls. The composition of lymphoid cells in mesenteric lymph nodes resembles that of the adjacent mesenteric adipose tissue in the controls, but not in diseased patients, which suggests that their adipocytes are not supplying fatty acids to cells in the adjacent lymph nodes. Particularly striking was the finding that lymph node lymphoid cells from patients with CD contained only 23% as much arachidonic acid (C20:4*n*-6) as the controls, in spite of the slightly higher ratio of *n*-6/*n*-3 fatty acids in the adipose tissue *(140)*. This eicosanoid precursor is relatively abundant in most lymphoid cells and insufficiency may lead to abnormal immune function, especially of dendritic cells *(142)*. The discrepancy between the composition of perinodal adipocytes and that of adjacent lymphoid cells contrasts with Fig. 1 and the concept of paracrine nutrition of lymphoid cells *(37,68)*, but is consistent with reports that blood-borne mononuclear cells from patients with CD contain more, not less, *n*-3 PUFAs than those of the controls and are deficient in arachidonic acid (20:4*n*-6) *(143)*.

Clinical studies suggest that supplying more *n*-6 PUFAs by enteral diets relieves some of the inflammatory symptoms *(144)*. The fatty acid compositions of phospholipids in blood cells and storage lipids in adipose tissue of patients with CD are anomalous, especially in patients in whom active symptoms are long established *(145)*. Fish oil rich in *n*-3 polyunsaturated fatty acids has been recommended to relieve CD *(146)* and other chronic inflammatory conditions *(147,148)*. Animal experiments show that dietary lipids modulate the spread of local immune stimulation within and between node-containing depots *(149)*. But clinical evidence is paradoxical, with some studies apparently indicating that supplying more *n*-6 polyunsaturated fatty acids by enteral diets relieves inflammatory symptoms *(144)*. The data on site-specific fatty acid composition help to explain this finding: although *n*-6 polyunsaturated fatty acids are indeed more abundant in the adipose tissue lipids from patients with CD, they are actually less abundant in these patients' lymph node lymphoid cells than in those of the controls.

The abnormalities of the composition of adipose tissue and lymph node lymphoid cells in the mesentery of patients with CD probably indicate similar deficiencies in the corresponding tissues elsewhere in the body. General defects in perinodal adipose tissue leading to impaired immune function could explain the association between the bowel disorder and other symptoms *(150,151)*, such as arthritis, eczema, and rhinitis *(152)* in CD. Bone marrow adipocytes may interact locally with osteoblasts *(153)* and hemopoeitic tissues *(57)*. Defective paracrine interactions may contribute to the bone normalities that accompany CD *(154,155)*.

2.3. Energy Balance and Immune Function

It is not appropriate here to review all the voluminous literature on human and animal obesity. However, certain topics are relevant to adipokines and immunological processes, including the action of endocrine controls of appetite and energy expenditure on immune function and the possible role of infection.

Leptin was first identified as an adipocyte secretion that acts on the brain to regulate energy balance and lipid storage and was later shown to act on lymphoid cells in vitro *(156)*. More recently, other appetite regulators, including adipophilin *(157)* and ghrelin *(158)*, have been shown to act on various kinds of leukocytes, at least in vitro. The possibility of using leptin or other appetite regulators as antiobesity drugs has prompted intensive study of the interactions between mechanisms of energy storage and immune function *(159,160)*. Unfortunately much of the data come only from depots of "pure" adipose tissue that do not incorporate lymphoid structures.

In describing observations on the effects of deficiencies in the secretion and reception of leptin on immune function, Matarese et al. mention the presence of adipocytes associated with lymph nodes, thymus, and bone marrow, and implicate them in these effects, although no evidence is presented that these adipocytes secrete leptin at all *(161)*. The limited data available indicate that their contribution would be small: adipocytes in the omental, mesenteric, and other depots that incorporate lymphoid structures synthesize much less leptin per unit mass than depots with little or no lymphoid tissue *(162–165)*. Bone marrow adipocytes produce a fair amount of leptin, at least in vitro *(166)*, but so do the osteoblasts and chondrocytes themselves *(167)*. Leptin seems to have a major role in osteogenesis *(168,169)*. No net uptake of leptin was measured in the human splanchnic or pulmonary regions, in spite of the presence of many lymph nodes and other lymphoid tissues in the spleen, omentum, mesentery, gut, and lungs, although the legs, with relatively fewer lymph nodes, are net leptin exporters *(170)*. In addition to adipocytes, many other cell types are now known to secrete and/or respond to leptin. Those of the reproductive system were the first to be described *(171)* and more recently, the protein has been found in skin *(172)*, vascular endothelium *(173)*, the stomach of suckling rats *(174)*, and bone and cartilage *(167)*.

Other endocrine regulators of appetite, such as ghrelin, also act on human blood and mouse spleen leukocytes *(158)*. Similarly, endocrine regulators of adipocyte metabolism and energy utilization, such as adenosine, also regulate the immune system *(175)*. These findings support the concept of "coupling the metabolic axis to the immune system." The problem with this approach is that it creates the impression that competition between the immune system and other tissues for energy and other metabolic resources is limiting, which is incompatible with some universal and very familiar aspects of normal responses to infection.

For example, the "competition for resources" model fails to explain why fever generated by endogenous thermogenesis universally accompanies major immune responses to bacteria in mammals, and many cytokines associated with reaction to infection promote anorexia *(176)*: nutrient intake stops just as energy expenditure increases abruptly, hardly an efficient arrangement if the immune system and pyrogenic tissues were normally in direct competition for fuels and other resources. Fever and systemic immune responses combined with anorexia cause small mammals (including human infants) to lose weight and are thus regarded as deleterious. However, when adult mice were experimentally infected with *Listeria* and fed forcibly or *ad libitum* over the following days, weight loss (because of the combination of anorexia and high energy expenditure) correlated positively with survival *(177)*. Overriding anorexia by force-feeding seems to accelerate the progress of pathogens, and hasten death, the opposite of what would be expected if depletion of lipid and protein reserves suppressed immune function.

Dhurandhar, Vangipuram, and colleagues have championed the view that obesity can be promoted by long-term viral infection *(92,178)*. Various viruses in domestic poultry, certain rodents, and monkeys have been implicated, but it is too early to speculate on the mechanism by which the metabolic changes associated with viral infection lead to obesity.

2.3.1. Starvation and Cachexia

Famines, wars, and other tragedies have many times demonstrated in humans the association between undernutrition and increased susceptibility to pathogens and parasites, especially those that invade through the gut and skin, presumably as a result of immune inadequacies *(179)*. However, there are some paradoxes. Many animals manage to remain healthy and breed normally while very lean, and "stress" rather than weight loss *per se* seems more important for impaired immune function in wild animals. Although maintaining the immune system is energetically expensive *(180)*, many wild animals manage to remain healthy and breed normally while very lean, as well as while very obese *(90)*. Studies of human athletes *(181)* suggest that endocrine and paracrine changes caused by "stress," rather than weight loss *per se*, impair immune function. Under some circumstances, notably prolonged anorexia nervosa, immune function remains surprisingly efficient in spite of massive reduction in adipose tissue mass *(182)*, less fever in response to infection *(183)*, and altered plasma cytokines *(184)*.

A notable feature of naturally lean mammals is the retention of a small amount of perinodal adipose tissue around major lymph nodes. Prolonged fasting in laboratory rodents does not raise lipolysis in perinodal adipocytes as much as in adipocytes not anatomically associated with lymphoid tissue *(41)*. As long as local interactions between adipose and lymphoid tissues are unimpaired, the immune system can probably function over a wide range of body compositions. Obvious cachexia with extensive depletion of muscle protein seems to set in at about the same time as this perinodal adipose tissue disappears. Thus it is possible that deficiencies in perinodal adipose tissue and its capacity to support immune function, rather than reduction in whole body energy supplies *per se*, are the mechanism by which nutritional "stress" impairs immune function.

Trypanosoma cruzi, the protoctistan parasite that causes Chagas' disease, can invade preadipocytes in vitro and alter their pattern of secretions *(185)*. Similar behavior in vivo could explain the long-term changes in metabolism and body composition associated with this disease.

3. CONCLUSIONS

Adipose tissue around lymph nodes is specialized and the tissues function together. Adipose tissue in the omentum and perinodal adipose tissue elsewhere, and probably "yellow" bone marrow, should be regarded as integral and essential parts of their adjacent lymphoid structures. As well as secreting and responding to cytokines, adipocytes associated with lymphoid structures selectively accumulate and store certain fatty acids, especially those that are essential precursors for eicosanoids and docosanoids, and release them in response to local lipolytic signals. Quality as well as quantity of fatty acids is important for efficient immune function. Local provisioning of lymphoid tissues partially emancipates immune function from changes in quantity and composition of food. Paracrine control of lipolysis by lymphoid cells reduces competition with other tissues for energy stores, thus enabling fever and other energetically expensive defenses against pathogens to take place simultaneously with immune responses and unrelated functions such as lactation and exercise. The partitioning of white adipose tissue into a few large and numerous small depots, many associated with lymphoid structures and other tissues (such as muscle), and its local, paracrine control evolved alongside the more elaborate and expensive immune system *(38)*.

ACKNOWLEDGMENTS

I thank Dr. R. H. Colby for helpful comments on the manuscript, and the Leverhulme Trust, Bristol-Myers Squibb (USA), the Open University Trustees' Fund, and the North West London Hospital Trust (St. Mark's Hospital, Northwick Park) for financial support.

REFERENCES

1. Rosen BS, Cook KS, Yaglom J, et al. Science 1989;244:1483–1487.
2. Choy LN, Rosen BS, Spiegelman BM. J Biol Chem 1992;267:12,736–12,741.
3. Cook KS, Min HY, Johnson D, et al. Science 1987;237:402–495.
4. Coppack SW. Proc Nutr Soc 2001;60:349–356.
5. Morelli AE, Thomson AW. Trends Immunol 2003;24:108–111.
6. Serhan CN, Hong S, Gronert K, et al. J Exp Med 2002;196:1025–1037.
7. Önfelt B, Nedvetzki S, Yanagi K, et al. J Immunol 2004;173:1511–1513.
8. Lee JY, Plakidas A, Lee WH, et al. J Lipid Res 2003;44:479–486.
9. Jump, DB. Crit Rev Clin Lab Sci 2004;41:41–78.
10. Dykstra M, Cherukuri A, Sohn HW, et al. Annu Rev Immunol 2003;21:457–481.
11. Pizzo P, Viola A. Curr Opin Immunol 2003;15:255–260.
12. Cohen AW, Combs TP, Scherer PE, et al. Am J Physiol Endocrinol Metab 2003;285:E1151–E1160.
13. Raclot T. Prog Lipid Res 2003;42:257–288.
14. Mattacks CA, Sadler D, Pond CM. Lymph Res Biol 2004;2:107–129.
15. Weisberg SP, McCann D, Desai M, et al. J Clin Invest 2003;112:1796–1808.
16. Khazen W, M'Bika JP, Tomkiewiez C, et al. FEBS Lett 2005;579:5631–5634.
17. Mills SC, Windsor AC, Knight SC. Clin Exp Immunol 2005;142:216–228.
18. Mattacks CA, Sadler D, Pond CM. Brit J Nutr 2004;91:883–891.
19. Hauner H. Proc Nutr Soc 2005;64:163–169.
20. Mattacks CA, Pond CM. Brit J Nutr 1997;77:621–643.
21. Cousin B, Munoz O, André M, et al. FASEB J 1999;13:305–331.
22. Cousin B, André M, Casteilla L, et al. J Cellul Physiol 2001;186:380–386.
23. Charriere G, Cousin B, Arnaud E, et al. J Biol Chem 2003;278:9850–9855.
24. Lin Y, Lee H, Berg AH, et al. J Biol Chem 2000;275:24,255–24,263.
25. Ajuwon KM, Jacobi SK, Kuske JL, et al. Am J Physiol–Regul Integr Comp Physiol 2004;286: R547–R553.
26. Villena JA, Cousin B, Pénicaud L, et al. Int J Obes 2001;25:1275–1280.
27. Mattacks CA, Sadler D, Pond CM. Adipocytes 2005;1:43–56.
28. Suganami T, Nishida J, Ogawa Y. Arterioscl Thromb Vasc Biol 2005;25:2062–2068.

29. Saillan-Barreau C, Cousin B, André M, et al. Biochem Biophys Res Commun 2003;309:502–505.
30. Bouloumie A, Curat CA, Sengenes C, et al. Curr Opin Clin Nutr Metab Care 2005;8:347–354.
31. Chehab FF, Qiu J, Ogus S. Rec Progr Horm Res 2004;59:245–266.
32. Orci L, Cook WS, Ravazzola M, et al. Proc Natl Acad Sci USA 2004;101:2058–2063.
33. Tansey JT, Sztalryd C, Gruia-Gray J, et al. Proc Natl Acad Sci USA 2001;98:6494–6499.
34. Xu HY, Barnes GT, Yang Q, et al. J Clin Invest 2003;112:1821–1830.
35. Yamauchi T, Kamon J, Minokoshi Y, et al. Nat Med 2002;8:1288–1295.
36. Yamauchi T, Kamon J, Waki H, et al. J Biol Chem 2003;278:2461–2468.
37. Pond CM. Prostagland Leukot Essent Fatty Acids 2005;73:17–30.
38. Pond CM. J Exp Zool 2003;295A:99–110.
39. Morin CL, Eckel RH, Marcel T, et al. Endocrinology 1997;138:4665–4671.
40. Mueller A, O'Rourke J, Chu P, et al. Proc Natl Acad Sci USA 2003;100:12,289–12,294.
41. Mattacks CA, Pond CM. Cytokine 1999;11:334–346.
42. Sopasakis VR, Sandqvist M, Gustafson B, et al. Obes Res 2004;12:454–460.
43. Grünfeld C, Zhao C, Fuller J, et al. J Clin Invest 1996;97:2152–2157.
44. Morin CL, Gayles EC, Podolin DA, et al. Endocrinology 1998;139:4998–5005.
45. Zhang YY, Guo KY, Diaz PA, et al. Am J Physiol–Regul Integr Comp Physiol 2002;282:R226–R234.
46. Laplante M, Sell H, MacNaul KL, et al. Diabetes 2003;52:291–299.
47. Rahmouni K, Mark AL, Haynes WG, et al. Am J Physiol–Endocrinol Metab 2004;286:E891–E895.
48. Zalatan F, Krause JA, Blask DE. Endocrinology 2001;142:3783–3790.
49. Bing C, Bao Y, Jenkins J, et al. Proc Natl Acad Sci USA 2004;101:2500–2505.
50. Cinti S. J Endocrinol Invest 2002;25:823–835.
51. Rosen ED. Ann NY Acad Sci 2002;979:143–158.
52. Gyllensten L. Acta Anat 1950;10:130–160.
53. Harvey NL, Srinivasan RS, Dillard ME, et al. Nature Genet 2005;37:1072–1081.
54. Pond CM. Proc Nutr Soc 1996;55:111–126.
55. Tripp RA, Topham DJ, Watson SR, et al. J Immunol 1997;158:3716–3720.
56. Rozman C, Feliu E, Rozman M, et al. Medicina Clinica 1993;101:441–445.
57. Yokota T, Meka CSR, Medina KL, et al. J Clin Invest 2002;109:1303–1310.
58. Yokota T, Meka CSR, Kouro T, et al. J Immunol 2003;171:5091–5099.
59. Suzawa M, Takada I, Yanagisawa J, et al. Nat Cell Biol 2003;5:224–230.
60. Maurin AC, Chavassieux PM, Vericel E, et al. Bone 2002;31:260–266.
61. Justesen J, Stenderup K, Eriksen EF, et al. Calcif Tissue Int 2002;71:36–44.
62. Justesen J, Mosekilde L, Holmes M, et al. Endocrinology 2004;145:1916–1925.
63. Loh M, Dubrovska G, Lauterbach B, et al. FASEB J 2002;16:1057–1063.
64. Verlohren S, Dubrovska G, Tsang SY, et al. Hypertension 2004;44:271–276.
65. Nawrocki AR, Scherer PE. Curr Opin Pharmacol 2004;4:281–289.
66. Iacobellis G, Corradi D, Sharma A. Nat Clin Pract Cardiovasc Med 2005;2:536–543.
67. Montani JP, Carroll JF, Dwyer TM, et al. Int J Obes 2004;28:S58–S65.
68. Pond CM, Mattacks CA. Brit J Nutr 2003;89:375–382.
69. Balazy M. Prostagland Lipid Mediat 2004;73:173–180.
70. Ivanova PT, Milne SB, Forrester JS, et al. Mol Interv 2004;4:86–96.
71. Stulnig TM. Int Arch Allergy Immunol 2003;132:310–321.
72. Calder PC. Proc Nutr Soc 2002;61:345–358.
73. Hausman DB, DiGirolamo M, Bartness TJ, et al. Obes Rev 2001;2:239–254.
74. Fukumura D, Ushiyama A, Duda DG, et al. Circ Res 2003;93:E88–E97.
75. Rupnick M, Panigrahy D, Zhang C, et al. Proc Natl Acad Sci USA 2002;99:10,730–10,735.
76. Rockson SG. Lymph Res Biol 2004;2:105–106.
77. Pond CM. Proc Nutr Soc 2001;60:365–374.
78. Prins JB, O'Rahilly S. Clin Sci 1997;92:3–11.
79. Prins JB, Walker NI, Winterford CM, et al. Biochem Biophys Res Comm 1994;201:500–507.
80. Pond CM. Biochem Soc Trans 1996;24:393–400.
81. Mattacks CA, Sadler D, Pond CM. J Anat Lond 2003;202:551–561.
82. Hausman DB, Lu J, Ryan DH, et al. Exp Biol Med 2004;229:512–520.

83. Sadler D, Mattacks CA, Pond CM. J Anat Lond 2005;207:761–789.
84. Toni R, Malaguti A, Castorina S, et al. J Endocrinol Invest 2004;27:182–186.
85. Wellen KE, Hotamisligil GS. J Clin Invest 2005;115:1111–1119.
86. Calle EE, Kaaks R. Nat Rev Cancer 2004;4:579–591.
87. Henrichot E, Juge-Aubry CE, Pernin AS, et al. Arterioscl Thromb Vasc Biol 2005;25:2594–2599.
88. Yudkin JS, Eringa E, Stehouwer CDA. Lancet 2005;365:1817–1820.
89. Moreno-Aliaga MJ, Campíon J, Milagro FI, et al. Adipocytes 2005;1:1–15.
90. Pond CM. *The Fats of Life*. Cambridge University Press, Cambridge: 1998.
91. Dhurandhar NV, Israel BA, Kolesar JM, et al. Int J Obesity 2000;24:989–996.
92. Dhurandhar NV. Drug News Perspect 2004;17:307–313.
93. Björntorp P. Int J Obesity 1996;20:291–302.
94. Seidell JC, Cigolini M, Deslypere J-P, et al. Am J Epidemiol 1991;133:257–265.
95. Singh D. J Personal Soc Psychol 1993;654:293–307.
96. Frayn KN. Br J Nutr 2000;83:S71–S77.
97. Grimble RF. Curr Opin Clin Nutr Metab Care 2002;5:551–559.
98. Dandona P, Aljada A, Bandyopadhyay A. Trends Immunol 2004;25:4–7.
99. Fain JN, Bahouth SW, Madan AK. Int J Obes 2004;28:616–622.
100. Brorson H. Acta Oncol 2000;39:407–420.
101. Ryan TJ. Clins Dermatol 1995;13:493–498.
102. Fogt F, Zimmerman RL, Daly T, et al. Int J Mol Med 2004;13:681–683.
103. Pond CM, Mattacks CA. J Zool Lond 1986;209:35–42.
104. Striker R, Conlin D, Marx M, et al. Clin Infect Dis 1998;27:218–220.
105. Shaw AJ, McLean KA, Evans BA. Int J STD AIDS 1998;9:595–599.
106. Carr A, Samaras K, Burton S, et al. AIDS 1998;12:F51–F58.
107. Saint-Marc T, Partisani M, Poizot-Martin I, et al. AIDS 1999;13:1659–1667.
108. Carr A, Emery S, Law M, et al. Lancet 2003;361:726–735.
109. Tien PC, Grunfeld C. Curr Opin Infect Dis 2004;17:27–32.
110. John M, Nolan D, Mallal S. Antivir Therap 2001;6:9–20.
111. Madge S, Kinloch-de-Loes S, Mercey D, et al. AIDS 1999;13:735–737.
112. Leonard EG, McComsey GA. Pediatr Infect Dis J 2003;22:77–84.
113. Pond CM. Trends Immunol 2003;24:13–18.
114. Hruz PW, Murata H, Mueckler M. Am J Physiol 2001;280:E549–E553.
115. Mattacks CA, Sadler D, Pond CM. Comp Biochem Physiol 2003;135C:11–29.
116. O'Neil SP, Mossman SP, Maul DH, et al. AIDS Res Hum Retrovir 1999;15:203–215.
117. Lopez MC, Huang DS, Watson RR. Lymphology 2002;35:76–86.
118. Pond CM, Mattacks CA. Cytokine 2002;17:131–139.
119. Chen DL, Misra A, Garg A. J Clin Endocrinol Metab 2002;87:4845–4856.
120. Brambilla P, Bricalli D, Sala N, et al. AIDS 2001;15:2415–2422.
121. Vigano A, Mora S, Testolin C, et al. J AIDS 2003;32:482–489.
122. Galli M, Cozzi-Lepri A, Ridolfo AL, et al. Arch Intern Med 2002;162:2621–2628.
123. Galli M, Gervasoni C, Ridolfo AL, et al. AIDS 2003;17 Suppl 1:S155–S161.
124. Path G, Bornstein SR, Gurniak M, et al. J Clin Endocrinol Metab 2001;86:2281–2288.
125. Bandyopadhyay GK, Lee LY, Guzman RC, et al. Lipids 1995;30:155–162.
126. Raclot T, Langin D, Lafontan M, et al. Biochem J 1997;324:911–915.
127. Dong KL, Bausserman LL, Flynn MM, et al. J AIDS 1999;21:107–113.
128. Gervasoni C, Ridolfo AL, Trifirò G, et al. AIDS 1999;13:465–471.
129. Evans DL, Pantanowitz L, Dezube BJ, et al. Clin Infect Dis 2002;35:1113–1119.
130. Gervasoni C, Ridolfo AL, Rovati L, et al. AIDS Patient Care STDS 2002;16:307–311.
131. Kahn JL, Wolfram-Gabel R, Bourjat P. Clin Anat 2000;13:373–382.
132. Heufelder AE. Acta Med Austr 2001;28:89–92.
133. Crohn BB, Ginzburg L, Oppenheimer GD. J Am Med Assoc 1932;99:1323–1329.
134. Sheehan AL, Warren BF, Gear MWL, et al. Brit J Surg 1992;79:955–958.
135. McGowan IM, Fairhurst RM, Shanahan F, et al. Neuroimmunomodulation 1997;4:70–76.
136. Lautenbach E, Lichtenstein GR. J Clin Gastroent 1997;25:456–459.

137. Desreumaux P, Ernst O, Geboes K, et al. Gastroenterology 1999;117:73–81.
138. Barbier M, Vidal H, Desreumaux P, et al. Gastroenterol Clin Biol 2003;27:983–985.
139. Schaffler A, Scholmerich J, Buchler C. Nat Clin Pract Gastroenterol Hepatol 2005;2:103–111.
140. Westcott EDA, Windsor ACJ, Mattacks CA, et al. Inflamm Bowel Dis 2005;11:820–827.
141. Pond CM. Écoscience 2003;10:1–9.
142. Harizi H, Gualde N. Prostaglandins Leukot Essent Fatty Acids 2002;66:459–466.
143. Trebble TM, Arden NK, Wootton SA, et al. Clin Nutr 2004;23:647–655.
144. Gassull MA, Fernández-Bañares F, Cabré E, et al. Gut 2002;51:164–168.
145. Geerling BJ, van Houwelingen AC, Badart-Smook A, et al. Am J Gastroenterol 1999;94:410–417.
146. Miura S, Tsuzuki Y, Hokari R, et al. J Gastroenterol Hepatol 1998;13:1183–1190.
147. Calder PC, Zurier RB. Curr Opin Clin Nutr Metab Care 2001;4:115–121.
148. Calder PC. Lipids 2003;38:343–352.
149. Mattacks CA, Sadler D, Pond CM. Brit J Nutr 2002;87:375–382.
150. Hyams J. J Pediatr Gastroenterol Nutr 1994;19:7–21.
151. Treem WR. J Pediatr Gastroenterol Nutr 1999;28:135–136.
152. Book DT, Smith TL, McNamar JP, et al. Am J Rhinol 2003;17:87–90.
153. Compston JE. J Endocrinol 2002;173:387–394.
154. Schoon EJ, Geerling BG, Van Dooren IMA, et al. Aliment Pharmacol Ther 2001;15:783–792.
155. Stockbrügger RW, Schoon EJ, Bollani S, et al. Gastroenterology 2001;120:3180.
156. Lord GM, Matarese G, Howar JK, et al. Nat Lond 1998;394:897–901.
157. Schmidt SM, Schag K, Muller MR, et al. Cancer Res 2004;64:1164–1170.
158. Dixit VD, Schaffer EM, Pyle RS, et al. J Clin Invest 2004;114:57–66.
159. La Cava A, Matarese G. Nat Rev Immunol 2004;4:371–379.
160. Matarese G, La Cava A. Trends Immunol 2004;25:193–200.
161. Matarese G, La Cava A, Sanna V, et al. Trends Immunol 2002;23:182–187.
162. Zhang HH, Kumar S, Barnett AH, et al. J Clin Endocrinol Metab 1999;84:2550–2556.
163. Schoof E, Stuppy A, Harig F, et al. Eur J Endocrinol 2004;150:579–584.
164. Montague CT, Prins JB, Sanders L, et al. Diabetes 1997;46:342–347.
165. Hube F, Lietz U, Igel M, et al. Horm Metab Res 1996;28:690–693.
166. Laharrague P, Larrouy D, Fontanilles AM, et al. FASEB J 1998;12:747–752.
167. Morroni M, De Matteis R, Palumbo C, et al. J Anat 2004;205:291–296.
168. Hamrick MW, Pennington C, Newton D, et al. Bone 2004;34:376–383.
169. Thomas T, Gori F, Khosla S, et al. Endocrinology 1999;140:1630–1638.
170. Jensen MD, Moller N, Nair KS, et al. Am J Clin Nutr 1999;69:18–21.
171. Himms-Hagen J. Crit Rev Clin Lab Sci 1999;36:575–655.
172. Murad A, Nath AK, Cha ST, et al. FASEB J 2003;17:U32–U46.
173. Park HY, Kwon HM, Lim HJ, et al. Exp Mol Med 2001;33:95–102.
174. Oliver P, Pico C, De Matteis R, et al. Dev Dyn 2002;223:148–154.
175. Haskó G, Cronstein BN. Trends Immunol 2004;25:33–39.
176. Johnson RW. Vet Immunol Immunopathol 2002;87:443–450.
177. Murray MJ, Murray AB. Am J Clin Nutr 1979;32:593–596.
178. Vangipuram SD, Sheele J, Atkinson RL, et al. Obes Res 2004;12:770–777.
179. Lin E, Kotani JG, Lowry SF. Nutrition 1998;14:545–550.
180. Lochmiller RL, Deerenberg C. Oikos 2000;88:87–99.
181. Pedersen BK. Immunol Cell Biol 2000;78:532–535.
182. Nova E, Samartin S, Gomez S, et al. Eur J Clin Nutr 2002;56:S34–S37.
183. Birmingham CL, Hodgson DM, Fung J, et al. Int J Eating Disord 2003;34:269–272.
184. Brichard SM, Delporte ML, Lambert M. Horm Metab Res 2003;35:337–342.
185. Combs TP, Nagajyothi F, Mukherjee NS, et al. J Biol Chem 2005;280:24,085–24,094.

12 Adipose Tissue and Mast Cells

Adipokines as Yin–Yang Modulators of Inflammation

George N. Chaldakov, Anton B. Tonchev, Nese Tuncel, Pepa Atanassova, and Luigi Aloe

Abstract

Recently, the endocrine activity of adipose tissue cells has been intensively studied. In effect, a wide range of exported secretory proteins, dubbed adipokines, have been identified as constituents of the adipose proteome (adipokinome). Besides their effects on glucose and energy metabolism, adipokines are potent modulators of inflammation. This chapter provides a state-of-the-science review of adipokine-mediated paracrine signaling that may be implicated in the pathogenesis of inflammation-related diseases such as atherosclerosis, thyroid-associated ophthalmopathy, and breast cancer. We also point out a possible contribution of adipose tissue-associated mast cell secretory activity to the development of these diseases. Finally, we provide arguments for yin-yang (protective vs pathogenic) roles of adipokines in inflammation. This hypothesis may provide further novel drug targets for the development of adipopharmacology of inflammatory diseases.

Key Words: Adipobiology; atherosclerosis; breast cancer; epicardial adipose tissue; ophthalmopathy.

1. INTRODUCTION

Today, increasing attention is being focused on the importance of adipose tissue in disease *(1)*, one of the most exciting examples being the rapidly growing interest in understanding the adipose tissue secretion of signaling proteins collectively designated adipokines *(2–5)*. These multifunctional molecules, via endocrine and paracrine pathways *(6,7)*, are potent modulators of inflammation (reviewed in refs. *5,8–11*).

This chapter reviews data of adipose tissue paracrine signaling in the pathogenesis of low-grade inflammation-related diseases such as atherosclerosis, thyroid-associated ophthalmopathy, and breast cancer. We also point out a possible contribution of adipose mast cell secretory activity to the development of these disorders. Finally, we provide arguments for differential, yin–yang (protective vs pathogenic) roles of adipokines in inflammation-related diseases. This may provide basis of adipose tissue-targeted pharmacology.

From: *Nutrition and Health: Adipose Tissue and Adipokines in Health and Disease*
Edited by: G. Fantuzzi and T. Mazzone © Humana Press Inc., Totowa, NJ

2. ADIPOSE TISSUE

Particularly well developed in humans is white adipose tissue (WAT), a major metabolic and secretory organ. Human WAT is partitioned into two large depots (visceral and subcutaneous), and many small depots associated with various organs, including heart, blood vessels, major lymph nodes, ovaries, mammary glands, eyes, and bone marrow. Another major adipose tissue subtype, brown adipose tissue, is present around kidneys, adrenals, and aorta, as well as within the mediastinum and neck. In adult humans, brown adipose tissue is very scarce and probably not functional.

2.1. Adipokines: Inhibitory (Yin) and Stimulatory (Yang) Signals in Inflammation

Celsus's description (first century AD) of inflammation signs includes *rubor et tumor cum calor et dolor*. Inflammation is an essential biological response aiming at recovering from injury, wound healing being a paradigm of such a homeostatic phenomenon. However, what begins as a protective response becomes a damaging process in excess; hence, inflammation is increasingly recognized as the underlying basis of a significant number of diseases. Recent studies based on a pangenomic approach in human subcutaneous WAT revealed that a panel of inflammatory molecules was upregulated in obese compared to lean subjects (ref. *12* and references therein). Of note, a calorie-restriction diet improved the anti-inflammatory profile of obese subjects via increase of anti-inflammatory and decrease of proinflammatory molecules *(12)*. Further, weight loss resulted in decrease of adipose macrophage number and an increased production of interleukin (IL)-10, a well-known anti-inflammatory cytokine *(13)*. These sophisticated analyses, as well as others *(8–11)*, support the hypothesis that adipose tissue-secreted factors may indeed be potent modulators of inflammation-related disorders such as obesity, type 2 diabetes, metabolic syndrome, atherosclerosis, inflammatory bowel disease, thyroid-associated (Graves') ophthalmopathy, breast cancer, and nonalcoholic fatty liver disease. Accordingly, the field of the role of adipose tissue in inflammation and metabolism has attracted great attention, exemplified by the rapidly growing interest in understanding adipose tissue protein secretion *(1,2,5–7,12–15)*. A paradigm-shifting discovery was that of leptin at the end of 1994 *(16)*. Although the birth hour of adipoendocrinology may be traced to the identification of the adipocyte-secreted enzyme lipoprotein lipase, followed in 1987 by adipsin *(17)*, leptin's discovery paved the way toward intensively studying adipose tissue endocrine function. As such, adipose tissue cells, represented by adipocytes, matrix cells, and stromovascular cells *(12,13,18)*, synthesize and release a diverse range of multifunctional molecules termed adipocytokines *(2,3)* or adipokines *(4,5)*, the latter terminology being more accurate than the former. Adipokines have been introduced as a term *(4)* that should be used exclusively to cover the secretory proteins (e.g., growth factors, cytokines, chemokines, enzymes, and matrix proteins) that are synthesized and released not solely by adipocytes, but also by matrix cells and stromovascular cells, including local macrophages *(13,19,20)* and, supposedly, mast cells. Because of recent advances in genomic and proteomic approaches, the secretory proteome of adipose cells (adipokinome) *(5)* is constantly being enriched with newly identified adipokines *(6,7,12,13,18–28)* (Table 1). Further, the whole spectrum of adipose secretory products (secretome) *(5)* is not limited to adipokines, but also

Table 1
Selected List of Adipokines

CYTOKINES
Leptin, IL-1[a], IL-1Ra, IL-6, IL-10, IL-18, TNF- α[a], LIF[a], oncostatin M
CHEMOKINES
MCP-1 (CCL2)[a], IL-8 (CXCL8)[a], Eotaxin (CCL11)[a], RANTES (CCL5)[a], IP-10
GROWTH FACTORS
FGF[a], TGF-β[a], NGF[a], CNTF, MCSF[a], BMP-2, HB-EGF, IGF
ANGIOGENIC FACTORS
VEGF[a], angiogenin, angipoietin-2, HGF[a]
RENIN–ANGIOTENSIN SYSTEM
Renin, angiotensinogen, angiotensin I, II, chymase[a], cathepsin D/G
ACUTE PHASE REACTANTS
SAA, PTX-3, lipocalin, ceruloplasmin, MIF, haptoglobin
HEMOSTATIC FACTORS
PAI-1[a], TF
ENZYMES
LPL, adipsin, MMP[a], tryptase[a]
OTHERS
Adiponectin, FIZZ1, resistin (FIZZ3), visfatin, vaspin, omentin, ASP, PEDF, prolactin, agouti protein, prohibitin, osteonectin (SPARC), TIMP-1, -2, adrenomedullin, CGRP, MT-1,-2, HIF-1α, Type VI collagen

[a]Secreted also by mast cells.
For a more extensive list of molecules comprised the adipose tissue secretome, *see* refs. *12,13*, and *28*.

includes a variety of nonproteins such as prostaglandins, fatty acids, monobutyrin, and steroid hormones.

In addition to their importance in lipid, glucose, and energy homeostasis, adipokines are pivotally involved in coordinating a variety of processes such as inflammation and immunity *(8,9–11,29)* and vascular biology-related processes including artery relaxation via perivascular adipose tissue-derived relaxing factor *(30)*, arteriolar constriction and insulin resistance *(31)*, and smooth muscle cell growth *(32)*.

3. ADIPOSE MAST CELLS

Mast cells were first described in 1878 by Paul Ehrlich (1854–1915) in his doctoral thesis, "Contribution to the Theory and Practice of Histological Staining" *(33)*. Ehrlich observed that mast cells were commonly located in connective tissue near blood vessels and nerves, as well as in inflammatory and tumor lesions. Mast cells are phenotypically and functionally versatile effector cells that have traditionally been associated with the immunoglobulin E-mediated allergic response. However, recent studies implicate these cells in the regulation of inflammation and fibrosis *(34–36)*, angiogenesis *(37)*, and neuroimmune interactions *(34,38)*, which could associate with various inflammatory diseases.

A wealth of evidence demonstrates that the mast cell is indeed "master" of protein secretion *(35)*. From a theoretical standpoint, adipose mast cell-secreted proteins may

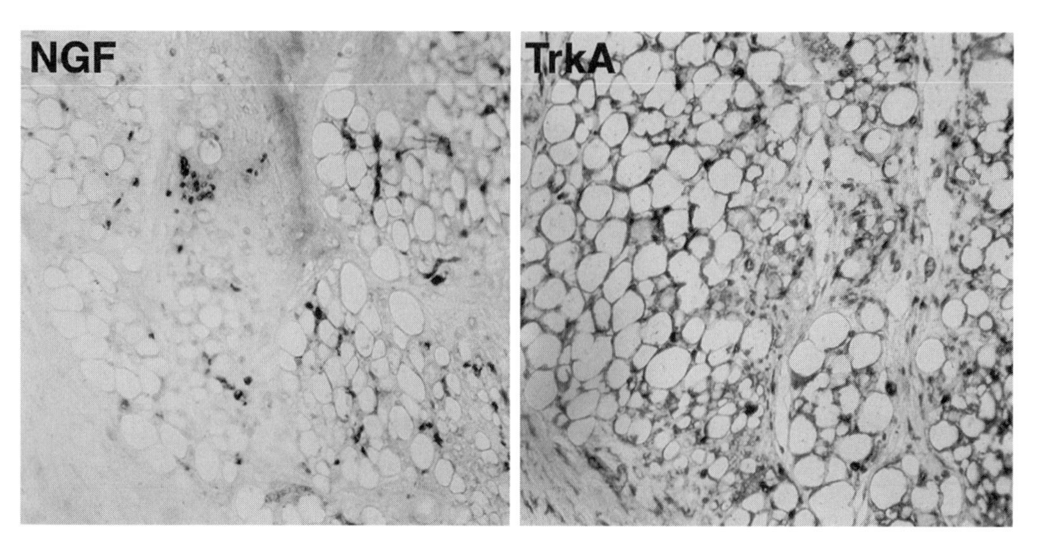

Fig. 1. Immunohistochemical localization of NGF and its high-affinity receptor TrkA in newborn human subcutaneous skin adipose tissue. Note the preferential stromal distribution of positive signal. TrkA protein is present also in adipocytes. Magnification, ×200.

potentially contribute to the whole body of adipokinome (*see* Table 1). At present, the knowledge of the biology of mast cells in adipose tissue is, however, limited as compared to that of macrophages *(13,19)*. Indeed, one has to go back more than 10 yr to find information, for example, about the role of brown adipose tissue-associated mast cell-secreted histamine in thermogenesis *(39)*. Likewise, whereas most studies deal with the effects of adipokines on macrophages or lymphocytes, only a single paper reported a stimulatory effect of leptin on mast cell growth, as demonstrated in biopsies of subcutaneous abdominal adipose tissue from patients with metabolic syndrome *(40)*. Also, our ongoing study on the involvement of neurotrophins in adipose tissue biology demonstrates a prominent immonoreactivity for nerve growth factor (NGF) and its high-affinity receptor tyrosine kinase-A (Trk-A) expressed in the stromal compartment of subcutaneous adipose tissue (Fig. 1). Some recent data about adipose mast cells in the pathobiology of diseases under the scope of the present chapter will be discussed in the following sections.

4. PARACRINE EFFECTS OF ADIPOSE TISSUE

The possibility that the endocrine secretory activity of large adipose depots may directly contribute to the altered blood plasma levels of certain adipokines has recently gained considerable attention *(1–11)*. Further, the paracrine secretory activity of the small adipose depots has, at long last, become a focus in the biology of disease. Similarly to endocrine products of large adipose depots reaching many organs through the bloodstream, paracrine products of organ-associated adipose depots can affect their neighboring tissues by a variety of adipokines (*see* subheading 4.1.3).

4.1. Perivascular Adipose Tissue and Cardiovascular Disease

In our previous papers *(4,41)*, we emphasized the importance of investigating the molecular composition of artery-associated adipose tissue, as it may yield clues to a possible paracrine transmission of protective and pathogenic signals derived from the perivascular adipose tissue toward the adjacent artery wall. Such an outside-to-inside signaling *(30,42)*, recently dubbed vasocrine signaling *(31)*, is implicated in the obesity-related insulin resistance phenotype *(31)* and various vascular disorders *(32)*. Moreover, inflammatory biomarkers measured in blood plasma may not adequately reflect local vascular inflammation.

An intriguing example of perivascular adipose tissue is the (sub)epicardial adipose tissue (EAT) that is conjunctioned to the adventitia of the most atherosclerosis-prone portions of the coronary artery—that is, the most proximal part of its left anterior descending branch. The possible involvement of EAT in coronary atherosclerosis and other cardiac pathologies has recently been addressed. Epicardial adipose tissue is a visceral fat depot around the heart, especially the right-ventricular free wall and left-ventricular apex. This neglected tissue is now recognized as a potent producer of various inflammation-related adipokines *(43–48)*. Specifically, recent findings demonstrate that the portion of the left anterior descending coronary artery running in the EAT develops atherosclerotic lesions, whereas the portion running in the myocardium is free of atherosclerotic lesions (ref. *41* and references therein). Further, the "atherosclerotic" EAT exhibits (1) reduced levels of adiponectin, an anti-inflammatory and antiatherosclerotic adipokine *(45)*, (2) elevated levels of monocyte chemoattractant protein-1, IL-1β, IL-6, tumor necrosis factor (TNF)-α *(44,46,47)*, and NGF *(43,49)*, and (3) the presence of inflammatory cell infiltrates, including mast cells *(43)*, lymphocytes *(44)*, and macrophages *(47)* (reviewed in refs. *32,48,49*).

4.2. Orbital Adipose Tissue and Thyroid-Associated Ophthalmopathy

Thyroid-associated (Graves') ophthalmopathy (TAO) has an autoimmune pathogenesis possibly related to the thyrotropin receptor *(50–53)*. The symptoms of TAO result from inflammation, fibrosis, and accumulation of orbital adipose tissues. Immunohistochemical analysis of orbital tissue biopsies from patients with TAO demonstrates that the thyrotropin receptor is expressed in fibroblast-like cells, accompanied by mast cell infiltrates *(50,51)*. Whether these mast cells, via their fibrogenic *(34–36)* and/or angiogenic *(37)* potential, may contribute to TAO-associated fibrosis and orbital adipose tissue hypertrophy, respectively, remains to be evaluated. Further, transforming growth factor-β inhibits, whereas IL-6 stimulates, thyrotropin receptor expression *(52)*, suggesting that the pathogenesis of TAO may be influenced by competing inhibitory (yin) and stimulatory (yang) adipokine effects within the orbit. One study examined 2686 genes, of which 25 known genes were upregulated in TAO orbital tissues, whereas 11 genes were downregulated *(53)*. Upregulated genes included secreted frizzled-related protein (sFRP)-1 and several adipocyte-related genes, including peroxisome proliferator-activated receptor (PPAR)γ and adiponectin. Treatment of TAO orbital preadipocytes in vitro with recombinant sFRP-1 significantly increased their adiponectin and leptin secretion *(53)*.

4.3. Mammary Adipose Tissue and Breast Cancer

It is known that inflammation can promote tumorigenesis. There is compelling evidence indicating that both normal mammary gland development and breast cancer growth depend, in part, on microenvironment, of which adipose tissue is a key component (ref. *28* and references therein). Interestingly, the mammary gland microenvironment during postlactational involution shares similarities with inflammation, which may be promotional for breast cancer development associated with pregnancy *(54)*. Recently, an elegant study by Celis et al. *(28)* provided the most extensive proteomic analysis of the mammary adipose secretome in high-risk breast cancer patients.

Adipose fibroblasts are another important cellular component of breast cancer microenvironment. These cells, being *bona fide* steroidogenic cells, are one of the major extragonadal sources of estrogen secretion. Estrogen synthesis is mediated by the enzyme aromatase cytochrome P450 (P450arom), which converts androgens to estrogens *(55)*. In breast cancer, one of the most aggressive human cancers, intratumoral proliferation of breast adipose fibroblasts is accompanied by increased P450arom expression by these cells, leading to proliferation of breast epithelial cells *(56)*. Notably, breast cancer commonly associates with a prominent immune, especially mast cell, response *(57–59)*. TNF-α and IL-6, which may potentially derive from both adipose cells and mast cells, upregulate aromatase expression *(60)*. Further, mast cell-secreted tryptase is a potent stimulator of fibroblast proliferation *(61)*, and adipocytes also produce tryptase *(12)*.

A novel piece to the puzzle of breast cancer is that NGF, a molecule known to be produced by adipocytes *(5,27,28,43,49)* and mast cells *(34,62)*, stimulates breast cancer cell proliferation *(63,64)*. Importantly, the antiestrogen drug tamoxifen inhibits NGF-mediated breast cancer cell proliferation through inhibition of the Trk-A receptor *(63)*. These data suggest a novel, NGF-mediated mechanism in the action of an old drug, tamoxifen, in breast cancer pharmacotherapy. Together these findings open possibilities for an adipose NGF-/mast cell-oriented therapy of breast cancer *(1)*, and pressingly call for studies on pharmacology of this neoplastic disorder.

5. CONCLUSIONS

Adipose tissue is a major source of and target for inflammatory signals. Although adipocyte–macrophage (*13,19,20,47*) and adipocyte–lymphocyte (*29*) interactions enjoy the researchers' appreciation, adipose mast cells have been relatively less studied until now. Nonetheless, adipocytes and mast cells share several biological features: (1) they are *bona fide* secretory cell types; (2) they cover almost the same spectrum of secretory proteins (*see* Table 1); and (3) they are co-implicated in the pathobiology of various inflammatory diseases. Despite these associations, further investigations will be required to illuminate the biology of mast cells in mast cells in health and disease. The following example might be a "role model" for such studies: activated human mast cells synthesize and release large amounts of plasminogen activator inhibitor type 1 through a nonconventional secretory pathway, using multivesicular endosome-mediated secretion of exosomes *(65)*. If this appears to be the case for adipose mast cells, it may further "inflame" adipose tissue. Also, comparing the biological responses of mast cells in wild-type mice with those of genetically engineered knock-in or knockout mice may provide new insights into adipose mast cell biology. Finally, a further suggesiton of a possible relation between mast cells and adipocytes is underscored by the observation that hyperlipidemia develops in mast cell-deficient W/WW mice *(66)*.

Because the actions of adipokines are complex and diverse, we need to design novel studies to determine how these molecules affect various inflammatory processes. Mechanistically, promotion of anti-inflammatory (yin) and suppression of proinflammatory (yang) adipokine-mediated signals may result in an improvement of inflammatory

Table 2
Examples of Adipokines as Possible Yin–Yang Modulators of Inflammation

Yin *Anti-inflammatory signals*	*Yang* *Proinflammatory signals*
Adiponectin *(1–3,5,6,8,45,67)*[a]	TNF-α *(5–9,44)*
IL-10 *(5,13,67)*	Interleukin-1, -6 *(14,18,44)*
Nerve growth factor *(5,27,43,49)*	Leptin *(8)*
Transforming growth factor-β *(52)*	Plasminogen activator inhibitor-1 *(5–9)*
receptor antagonist *(10)*	IL-18 *(71)*
Pigment epithelium-derived factor *(21,68)*	Resistin *(8,14,46)*
Calorie restriction *(12)*	Monocyte chemoattractant protein-1 *(20,46)*
Exercise-induced myokines *(70)*	IL-8 (CXCL8) *(8,14,19,44,46)*
Adrenomedullin *(69)*	Eotaxin (CCL11) *(19,46)*
Calcitonin gene-related peptide *(72)*	RANTES (CCL5) *(9,14,19,46)*
Metallothionein-1,-2 *(72)*	Hypoxia-inducible factor-1α *(13)*

[a]References are indicated in parentheses.

disease therapy (Table 2). The present challenge is thus to cultivate an adipocentric thinking about how we can make adipokines work for the benefits of patients. It is our belief that we should collaborate to more easily (and pleasantly) achieve that goal, as advised by the yin–yang philosophy also named "The Book of Ease."

REFERENCES

1. Chaldakov GN, Stankulov IS, Hristova M, et al. Curr Pharm Des 2003;9:1023–1031.
2. Funahashi T, Nakamura T, Shimomura I, et al. Inter Med 1999;38:202–206.
3. Matsuzawa Y. Nat Clin Pract Cardiovasc Med 2006;3:35–42.
4. Chaldakov GN, Fiore M, Ghenev PI, et al. Int Med J 2000;7:43–49.
5. Trayhurn P, Wood IS. Br J Nutr 2004;92:347–355.
6. Kershaw EE, Flier JS. J Clin Endocrinol Metab 2004;89:2548–2556.
7. Hauner H. Physiol Behav 2004;83:653–658.
8. Fantuzzi G. J Allergy Clin Immunol 2005;115:911–919.
9. Juge-Aubry CE, Henrichot E, Meier CA. Best Pract Res Clin Endocrinol Metab 2005;19:547–566.
10. Berg AH, Scherer PE. Circ Res 2005;96:939–949.
11. Yudkin JS. Int J Obes Relat Metab Disorb 2003;27(Suppl.3):S25–S28.
12. Viguerie N, Pottou C, Cancello R, et al. Biochemie 2005;87:117–123.
13. Cancello R, Henegar C, Viguerie N, et al. Diabetes 2005;54:2277–2286.
14. Fruhbeck G. Curr Med Chem–Cardiovasc Hematol Agents 2004;2:197–208.
15. Fruhbeck G, Nutr R, Salvador J. Nutr Res 2004;24:803–826.
16. Zhang Y, Proenca R, Maffei M, et al. Nature 1994;372:425–432.
17. Cook KS, Min HY, Johnson D, et al. Science 1987;237:402–405.
18. Fain JN, Madan AK, Hiler ML, et al. Endocrinology 2004;145:2273–2282.
19. Bouloumie A, Curat CA, Sengenes C, et al. Curr Opin Clin Nutr Metab Care 2005;8:347–354.
20. Weisberg SP, Hunter D, Huber R, et al. J Clin Invest 2006;116:115–124.
21. Drevon CA. Biochem Biophys Acta 2005;174:287–292.
22. Hugo ER, Brandebourg TD, Comstock CE, et al. Endocrinology 2006;147:306–313.
23. Fukuhara A, Matsuda M, Nishizawa M, et al. Science 2005;307:426–430.
24. Kloting N, Berndt J, Kralisch S, et al. Biochem Biophys Res Commun 2006;339:430–436.
25. Vasudevan AR, Wu H, Xydakis AM, et al. J Clin Endocrinol Metab 2006;91:256–261.

26. Rehman J, Traktuev D, Li J, et al. Circulation 2004;109:1292–1298.
27. Trayhurn P. Acta Physiol Scand 2005;184:285–293.
28. Celis JE, Moreira JM, Cabezon T, et al. Mol Cell Proteomics 2005;4:492–522.
29. Pond CM. Prostaglandins Leukot Essent Fatty Acids 2005;73:17–30.
30. Gollasch M, Dubrovska G. Trends Pharmacol Sci 2004;25:647–653.
31. Yudkin JS, Eringa E, Stehouwer CDA. Lancet 2005;365:1817–1820.
32. Montani J-P, Carroll JF, Dwyer TM, et al. Int J Obes 2004;28:S58–S65.
33. Vyas H, Krishnaswamy G. Methods Mol Biol 2006;315:3–11.
34. Chaldakov GN, Ghenev PI, Valchanov KP, et al. Biomed Rev 1995;4:1–6.
35. Galli SJ. N Engl J Med 1993;328:257–265.
36. Galli SJ. Int Arch Allergy Immunol 1997;113:14–22.
37. Norrby K. APMIS 2002;110:355–371.
38. Barbara G, Stanghellini V, De Giorgio R, et al. Neurogastroenterol Motil 2006;18:6–17.
39. Desautels M, Wollin A, Halvorson I, et al. Am J Physiol 1994;266(3Pt2):R831–837.
40. Chaldakov GN, Fiore M, Stankulov IS, et al. Arch Physiol Biochem 2001;109:357–360.
41. Chaldakov GN, Stankulov IS, Aloe L. Atherosclerosis 2001;154:237–238.
42. Gao YJ, Holloway AC, Zeng ZH, et al. Obes Res 2005;13:687–692.
43. Chaldakov GN, Stankulov IS, Fiore M, et al. Atherosclerosis 2001;159:57–66.
44. Mazurek T, Zhang LF, Zalewski A, et al. Circulation 2003;108:2460–2466.
45. Iacobellis G, Pistilli D, Gucciardo M, et al. Cytokine 2005;29:251–255.
46. Henrichot E, Juge-Aubry CE, Pernin A, et al. Arterioscler Thromb Vac Biol 2005;25:2594–2599.
47. Baker AR, da Silva NF, Quinn DW, et al. Cardiovasc Diabetol 2006;5:1.
48. Iacobellis G, Corradi D, Sharma AM. Nat Clin Pract Cardiovasc Med 2005;2:536–543.
49. Chaldakov GN, Fiore M, Stankulov IS, et al. Prog Brain Res 2004;146:279–289.
50. Ludgate M, Crisp M, Lane C, et al. Thyroid 1998;8:411–413.
51. Boschi A, Daumerie C, Spiritus M, et al. Br J Ophthalmol 2005;89:724–729.
52. Bahn RS. Thyroid 2002;12:193–195.
53. Kumar S, Leontovich A, Coenen MJ, et al. J Clin Endocrinol Metab 2005;90:4730–4735.
54. McDaniel SM, Rumer KK, Biroc SL, et al. Am J Pathol 2006;168:608–620.
55. Simpson ER. J Mammary Gland Biol Neoplasia 2000;5:251–258.
56. Meinhardt U, Mullis PE. Horm Res 2002;57:145–152.
57. Kankkunen JP, Harvima IT, Naukkarinen A. Int J Cancer 1997;72:385–388.
58. Kashiwase Y, Morioka J, Inamura H, et al. Int Arch Allergy Immunol 2004;134:199–205.
59. Samoszuk M, Kanakubo E, Chan JK. BMC Cancer 2005;5:121.
60. Irahara N, Miyoshi Y, Taguchi T, et al. Int J Cancer 2006;118:1915–1921.
61. Coussens IM, Raymond WW, Bergers G, et al. Genes Dev 1999;13:1382–1397.
62. Aloe L, Tirassa P, Bracci-Laudiero L. Curr Pharm Des 2001;7:113–123.
63. Chiarenza A, Lazarovic P, Lempeureur L, et al. Cancer Res 2001;61:3002–3008.
64. Dolle L, Adriaenssens E, El Yazidi-Belkoura I, et al. Curr Cancer Drug Targets 2004;4:463–470.
65. Al-Nedawi K, Szemraj J, Cierniewski CS. Arterioscler Thromb Vas Biol 2005;25:1744–1749.
66. Hatanaka K, Tanishita H, Ishibashi-Ueda H, et al. Biochem Biophys Acta 1986;878:440–445.
67. Wolf AM, Wolf D, Aliva MA, et al. J Hepatol 2006;44:537–543.
68. Zang SX, Wang JJ, Gao G, et al. FASEB J 2006;20:323–325.
69. Gonzalez-Rey E, Fernandez-Martin A, Chorny A, et al. Gut 2006;55:824–832.
70. Pedersen AM, Pedersen BK. J Appl Physiol 2005;98:1154–1162.
71. Wood IS, Wang B, Jenking JR, et al. Biochem Biophys Res Commun 2005;337:422–429.
72. Penkowa M. Biomed Rev 2002;13:1–15.

13 Bone Marrow Adipose Tissue

Patrick Laharrague and Louis Casteilla

Abstract

Bone marrow (BM) adipose tissue should no longer be considered simply as a filling material for bone cavities that is not needed for hematopoietic activity. In addition to its potential role as an energy store, BM adipose tissue exhibits a considerable adaptive plasticity and secretes a broad spectrum of hormones, cytokines and growth factors whose receptors are present on different cells of the stromal microenvironment. BM adipocytes, originating like osteoblasts from mesenchymal stem cells, display a marked metabolic and secretory activity. Among the various secreted adipokines, leptin, and adiponectin have opposite effects on hematopoiesis, immunity, inflammation, and bone remodeling. As a whole, a counterbalance exists between adipogenesis and erythropoiesis, and between adipose and bone formation. The better knowledge of the different paracrine and endocrine agents involved in the subtle and complex regulation of hematopoiesis and its osseous environment suggests that BM adipose tissue may represent a target for drugs in situations such as blood diseases or osteoporosis.

Key Words: Bone marrow; adipocytes; microenvironment; mesenchymal stem cells; hematopoiesis; osteogenesis; adipogenesis.

1. INTRODUCTION

Many studies have focused on brown and white fat in both rodents and humans. The organization and properties of bone marrow (BM) adipose tissue have received much less attention.

Although adipocytes are the most abundant cell type found in adult human bone marrow, their function is not fully understood. Several hypotheses—often conflicting—have been proposed to clarify their role. Most of them have not been rigorously verified *(1)*. For some authors, marrow fat simply fills the spaces free of hematopoietic cells in the closed and rigid cavity of bone, accommodating to hematopoietic demands by altering its volume—i.e., by contracting in the case of heightened hematopoiesis or by expanding in the event of decreased marrow activity *(2,3)*. Evans et al. considered marrow adipose tissue as an ordinary fat pad, the free fatty acids produced by lipolysis raising the circulation to participate in the general metabolism of the organism *(4)*. Others have suggested that the BM fat stores are important for local nutrition rather than for total energy supplies *(5)*. However, most recent studies indicate that BM adipose tissue possesses a significant endocrine function, considerable adaptive plasticity, and an

From: *Nutrition and Health: Adipose Tissue and Adipokines in Health and Disease*
Edited by: G. Fantuzzi and T. Mazzone © Humana Press Inc., Totowa, NJ

important *in situ* action, and is substantially involved in the regulation of hematopoiesis *(1,6–8)* and osteogenesis *(9,10)*.

2. PLASTICITY OF BM ADIPOSE TISSUE

2.1. Evolution Throughout the Life Span

In neonatal mammals no adipose cells are seen in any marrow cavities, which are primarily hematopoietic *(11)*, presumably because the elevated requirements for red cell production during neonatal life demand the resources of the entire potential of the marrow. With advancing age, the demands for red cell production recede, and the number of adipocytes in the bone marrow slowly and progressively increases, resulting in the appearance of fatty marrow.

In the rabbit, adipocytes begin to develop at 2 wk of age in both trunk and limb bones, the adult pattern being fully established by 4 mo *(12)*. In humans, during the first few years of life, the marrow of most bones is "red"—i.e., hematopoietic—but, with increasing age, the active marrow gradually recedes from the distal portions of the skeleton toward the trunk. This process actually commences before birth, and in the toes, by the age of 1 yr, the marrow is almost entirely "yellow" or fatty *(13)*. At about the age of 18, actively red hematopoietic marrow is found only in the vertebrae, ribs, sternum, skull bones, and proximal epiphyses of the femur and humerus *(14)*. It has been calculated that in the adult there are 0.56 g of marrow per gram of blood and that bone marrow accounts for 3.4 to 5.9% of body weight, or 1600 to 3700 g, roughly the weight of the liver *(14)*.

The preference of hematopoietic tissue for centrally located bones is still puzzling. Higher central tissue temperature with greater vascularity has been invoked to explain the hematopoietic distribution *(15,16)*. However, complete reactivation of peripheral fatty marrow in the rat demands more than simply increased environmental temperature, suggesting that there is an inherent determinant of the cellularity of marrow in different sites *(17,18)*.

Although total body fat may decrease in old age *(19)*, percent body fat declines very little, and may even remain constant or increase *(20)*. This occurs because fat is redistributed from fat depots to other organs, mainly muscle and liver *(21,22)*. BM fat is also increased in old age *(23–25)*. With aging, adipocytes increase in size and number, and the composition of BM fatty acids also changes, characterized by an overall increase in the unsaturation rate *(26)*. The polyunsaturated fatty acids released by adipocytes could lead to an inhibitory effect on osteoblastic proliferation *(27,28)*. In this way, the increase in marrow adipose tissue that occurs with aging may contribute to the age-related decrease in bone formation. We will further examine the close relationships between osteogenesis and adipogenesis.

2.2. Response to Starvation

Fat cell volume and the number of fat cells in the marrow remain essentially unchanged in the rabbit subjected to acute starvation, despite drastic weight loss and depletion of peripheral white fat stores *(29)*. BM adipose cells exhibit no ultrastructural change during short-term starvation when fat is mobilized from extramedullary adipose tissue *(30)*. In prolonged induced starvation or in patients with severe anorexia nervosa, BM fat atrophies and an accumulation of gelatinous extracellular mucopolysaccharide-type matrix is observed *(31,32)*.

2.3. Balance With Erythropoiesis

A reciprocal relationship exists between adipogenesis and erythropoiesis *(23)*. When rabbits are treated chronically with phenylhydrazine to induce hemolysis or are subjected to chronic bleeding, erythropoiesis is stimulated. In such situations, the mean volume of marrow fat falls, whereas that of nonmedullary fat pads is unchanged. BM adipocyte shrinking is accompanied by preferential release of unsaturated fatty acids *(4)*. This specific loss of unsaturated fatty acids suggests that increased hematopoiesis stimulates lipolysis of BM adipocytes *(33,34)*. However, the ultimate fate of these unsaturated fats is currently unknown.

In some situations, the reverse phenomenon is seen. In patients with aplastic anemia, an increased percentage of BM is occupied by adipocytes *(35)*. Keeping mice hypertransfused for up to 6 wk completely suppresses erythropoiesis. Sequential electron microscopic study performed during this sustained polycythemia reveals a shift from erythropoietic to granulopoietic tissue, destruction of BM macrophages, and accelerated development of marrow adipocytes. Accumulation of fat in stromal cells precedes the expansion of granulopoietic tissue, and adipocytes decline as granulopoiesis returns to normal *(36)*. The question of this association of marrow adipocytes with the induction of granulopoiesis, also suggested by in vitro studies *(37)*, will be further discussed.

2.4. Adiposity vs Bone Formation

Aging of the human skeleton is characterized by decreased bone formation and bone mass. The decrease in bone volume associated with age-related osteopenia is accompanied by an increase in marrow adipose tissue as determined by histomorphometry *(38,39)*. Adipogenesis is also observed in almost all conditions that lead to osteoporosis *(40,41)*, such as ovariectomy *(42)*, limb immobilization *(43)*, alcoholism *(44)*, and excessive treatment with glucocorticoids *(45)*. Conversely, adipogenesis is inhibited in conditions with increased bone formation *(46)*.

Studies in an age-related osteopenia animal model, SAMP6 mice, show that increased adipogenesis and myelopoiesis in the bone marrow are associated with a reduced number of osteoblast progenitors and decreased bone formation *(47)*.

Hindlimb suspension in the rat, a model of skeletal unloading, induces osteopenia by inhibiting bone formation in long bones *(48)*. This results from impaired recruitment of osteoblasts and decreased expression of bone matrix proteins. In addition to altering osteoblastogenesis, skeletal immobilization increases adipogenesis in human bone marrow.

Osteopenia and osteoporosis, which are associated with increased fracture rate and delayed fracture healing, are known long-term complications of insulin-dependent diabetes mellitus. In diabetes, both human and rodent studies find reduced bone volume, with decreased expression of osteoblastic markers together with an increase of adipocyte markers. Using a streptozotocin-induced diabetic mouse model, Botolin et al. observed suppressed osteoblastic maturation and increased BM adipose storage; unlike BM, peripheral adipose tissue was decreased in size, suggesting that the marrow environment and adipose stores are regulated differently from peripheral sites *(49)*.

Habitual consumption of significant quantities of ethanol is recognized as a major factor for osteopenia and increased fracture risk in both men and women. Bone mass is decreased in alcoholics and their osteopenic skeletons show increased bone marrow adiposity *(50,51)*.

Thus there is evidently a close relationship between adipocytes and osteoblasts. We will further develop this specific point.

3. MORPHOLOGICAL AND FUNCTIONAL CHARACTERIZATION OF BM ADIPOCYTES

Most of our knowledge has been obtained from histological and biochemical studies in the rabbit and the mouse. Cultures of either stromal-derived cell lines or primary human mesenchymal stem cells (MSCs) recently yielded additional information.

Morphologically, the marrow adipocyte is very similar to the extramedullary white adipose cell. Electron microscopic study of the rabbit marrow reveals extremely large adipocytes (140–160 μm), preferentially located close to the wall of BM sinuses *(52)*. Numerous organelles are located mainly in the perinuclear cytoplasm, compatible with high metabolic activity. They are rarely situated in the narrow cytoplasm adjacent to the central fat globule. Mitochondria are elongated and few in number. Variable amounts of glycogen are present *(53,54)*.

Histochemically, BM adipocytes essentially contain neutral triglycerides and free fatty acids, most of which are saturated and monounsaturated *(52)*. Bone marrow adipocytes thus stain well with Oil red O and Nile blue sulfate. The cytochemical staining pattern of freshly isolated marrow and extramedullary adipocytes *(54)* reveals differences in esterase activity (Table 1).

In the rabbit, palmitate turnover per gram triglyceride is fivefold greater in BM adipose tissue than in subcutaneous or perinephric adipose tissues; however, when expressed on the basis of individual cells, incorporation of the free fatty acid in marrow and in nonmedullary fat cells appears similar *(55)*. Gas chromatography reveals that marrow fat contains a higher concentration of unsaturated fatty acids. As a whole, these studies and those performed during stimulation of erythropoiesis indicate that there is greater lipolysis and lesser storage in BM fat than in nonmedullary fat pads *(55)*. Our opinion is that, contrary to white adipose tissues, fat storage and energy conservation are probably a secondary function of marrow fat.

4. BM FAT CELL PRODUCTS

Like their nonmedullary counterparts, BM adipocytes are now considered as active and potent secretory and endocrine cells. In addition to fatty acids, they secrete adipokines, which are mainly regulators of adipogenic, osteogenic, or hematopoietic development.

4.1. Leptin

The cloning of leptin, the *ob* gene product, emphasized the secretory function of adipocytes. It is well established that this hormone is secreted mainly by adipocytes and circulates in the blood. In addition to control of food intake and energy expenditure via its action on the hypothalamus, leptin plays an important role in numerous physiological functions such as adrenal, thyroid, and reproductive functions, glucose homeostasis, lipogenesis, and lipid oxidation, growth, and development during fetal and neonatal life *(56,57)*. Studies of obese rodents carrying the mutation (*ob/ob* mice) or with a mutation of the gene encoding for the ob receptor (*db/db* mice) reveal that leptin is also involved in immunity, inflammation, hematopoiesis *(58)*, and bone mass regulation *(59)*.

Table 1
Comparison Between Bone Marrow Adipocytes and Extramedullary Brown and White Adipocytes

Characteristics	*BM adipocytes*	*Brown adipocytes*	*White adipocytes*
Tissue localization	Bone marrow cavities	Mainly around heart and vessels	Dispersed
Morphology	Unilocular	Multilocular	Unilocular
Mitochondria	+	+++	+
Histochemistry:			
acetate esterases	+	–	–
chloroacetate esterases	+	–	–
butyrate esterases	Fluororesistant	Fluorolabile	Fluorolabile
alkaline phosphatases	±	?	–
Specific marker	–	UCP1	–
Development	Postnatal	Perinatal	Postnatal
Main physiological role	Hematopoiesis Osteogenesis Energy store?	Thermogenesis	Energy conservation
Response to starvation	–	–	+
Response to hemolysis	+	–	–
Insulin sensitivity	–	+	+
Differentiating agents:			
insulin	–	++++	+++
dexamethasone	+++	+++	+++
thiazolidinediones	+ (rodents)	+	+
catecholamines	–	++	–
T3	?	++	+

Leptin-deficient *db/db* mice exhibit an inherent deficit in lymphopoietic progenitors and are unable to completely recover their lymphopoietic populations after irradiation *(60)*. A decrease in the number of blood lymphocytes, phenotypic abnormalities in macrophages, and deficient expression of proinflammatory cytokines are observed in *ob/ob* mice *(58,61,62)*.

We first reported that human BM adipocytes secrete large amounts of leptin *(63)* and that leptin secretion may be directly regulated by the proinflammatory/hemopoietic cytokines interferon (IFN)-γ, interleukin (IL)-6, IL-1β, and tumor necrosis factor (TNF)-α *(64)*. The leptin receptor is expressed in hematopoietic progenitors *(60,65,66)*. Recombinant leptin acts on murine hematopoietic stem cells to stimulate multilineage expansion *(60,67)*, and significantly stimulates the development of granulocyte-macrophage colonies from human $CD34^+$ hematopoietic progenitors *(68)*.

Studies on human BM stromal cell lines show that leptin enhances osteoblastic and inhibits adipocytic differentiation *(69)*. These cells express both the short and long forms of the leptin receptors *(69,70)*. Moreover, cultures of human BM primary osteoblasts produce relatively large amounts of leptin *(70,71)*.

All these observations suggest that leptin may be an important paracrine signaling molecule between adipocytes, myeloid and lymphoid progenitor cells, and osteoblasts in the BM microenvironment.

4.2. Adiponectin

The protein designated Acrp30, adipoQ, or adiponectin represents a major fat cell-restricted product in the mouse and in humans *(72–74)*. It has also been isolated from human serum and termed GBP28 *(75)*. Adiponectin is attracting interest because of its potential involvement in obesity, diabetes, and cardiovascular diseases. BM adipocytes can highly and specifically express adiponectin. The biological effects of leptin and adiponectin overlap:

- Adiponectin may be implicated in the native and adaptive immune response. Recombinant adiponectin influences differentiation of early myeloid and lymphoid progenitor cells, probably through prostaglandin metabolism *(76)*. Adiponectin reduces the viability of monocyte cell lines, inhibits LPS-induced production of TNF-α, appears to use the C1qRp receptor on normal macrophages and blocks their ability to phagocytose particles *(77)*.
- Recombinant adiponectin inhibits fat cell formation by marrow-derived stromal cells through a paracrine negative feedback loop. The cyclooxygenase (COX)-2-dependent prostanoid pathway is important for this suppressive activity *(78)*.
- Adiponectin appears to be directly involved in osteogenesis, as the adiponectin receptor (AdipoR) is detected in human osteoblasts and recombinant adiponectin promotes their proliferation and differentiation. The proliferation response is mediated by the AdipoR/JNK pathway, whereas the differentiation response is mediated via the AdipoR/p38 pathway *(79)*.

4.3. Cytokines and Growth Factors

BM-derived stromal cell lines produce numerous factors implicated in lymphoid development and in granulocytic and monocyte/macrophage differentiation, such as IL-6, IL-7, transforming growth factor (TGF)-β, granulocyte-macrophage colony-stimulating factor, granulocyte colony-stimulating factor (G-CSF), and macrophage colony-stimunating factor (M-CSF) *(1)*. Human BM adipocytes in primary culture secrete trace amounts of IL-1β and TNF-α and significant and regulated levels of IL-6 *(64)*. These results indicate that bone marrow adipocytes may contribute to the complex network of cytokines involved in the control of hematopoiesis and immune response.

IL-6 may also be involved in the plasticity of the hematopoietic microenvironment, as BM-derived murine cell lines (LDA11 and MBA13.2), as well as murine (MC3T3) and human (MG-63) osteoblast-like cell lines, display the IL-6 receptor *(80)*. Therefore, myofibroblasts and osteoblasts appear as potential targets for the actions of IL-6. It is noteworthy that IL-6 stimulates stromal progenitor differentiation toward osteoblastic lineage *(81)*.

So, in addition to their potential role as an energy store, BM adipocytes are endocrine cells that secrete a broad spectrum of hormones, cytokines, and growth factors involved in bone remodeling, inflammation, and hematopoietic activity. Bone marrow adipose tissue should no longer be considered simply as a filling material but also must be looked at as an adaptable tissue with marked metabolic and secretory activity, involved in the subtle and complex regulation of hematopoiesis and its osseous environment. These intricate and balanced regulations will be further dealt with in the following sections.

5. BM ADIPOCYTES ORIGINATE FROM MESENCHYMAL STEM CELLS

There are two stem cell compartments in mammal bone marrow. Besides the well-known hematopoietic stem cells, BM has a multipotent population of cells capable of differentiating into adipocytes, osteoblasts, and other mesodermal pathways *(82,83)*. These MSCs, or BM stromal cells, have been cloned from humans, mice, and rats. They produce many cytokines and participate in the microenvironment that supports the proliferation and differentiation of hematopoietic stem cells *(1,84)*.

5.1. In Vitro Experiments

The nonhematopoietic MSCs of bone marrow were discovered by Friedenstein et al. *(85)*, who described clonal, plastic adherent cells from BM capable of differentiating into osteoblasts, adipocytes, and chondrocytes *(86–90)*. These cells are also stromal cells, structural components of the bone marrow that support ex vivo culture of hematopoiesis by providing extracellular matrix components, cytokines, and growth factors *(85,91,92)*.

5.1.1. Protocols for Isolation and Culture of MSCs

Protocols for isolation and culture of MSCs from different species vary, but human MSCs are typically isolated from the mononuclear layer of the bone marrow after separation by discontinuous density gradient centrifugation *(85,88,90,93)*. In some cases, the MSCs are further purified based on the expression of the primitive MSC marker, STRO-1 *(94)*. The mononuclear layer is simply cultured and the MSCs adhere to the culture plastic. Over time in culture, the nonadherent hematopoietic cells are washed away, resulting in small, adherent fibroblast-like cells. When cultured in a medium containing fetal calf serum or fresh human serum, after an initial lag phase the cells divide rapidly, with an average initial doubling time of 12 to 24 h. As the cultures approach high density, the MSCs enter a stationary phase and transform from a spindle-like morphology to a larger, flatter phenotype (Fig. 1). Typically, the MSCs recovered from a 2-mL bone marrow aspirate can be expanded 500-fold over about 3 wk, resulting in a theoretical yield of 12.5 to 35.5 billion cells. The cells generally retain their multipotentiality for at least 6 to 10 more passages *(95)*.

5.1.2. Adipogenic Differentiation

Adipogenic differentiation is classically induced by incubation of monolayers in a culture medium containing dexamethasone, isobutyl-methylxanthine (IBMX), and indomethacin. Insulin is not necessary *(63,96,97)*. IBMX is a phosphodiesterase inhibitor that blocks the conversion of cAMP to 5′AMP, resulting in upregulation of protein kinase A. Indomethacin is a known ligand for the peroxisome proliferator-activated receptor (PPAR)α/γ, a key early transcription factor in adipogenesis *(98)*.

Thiazolidinediones (TZDs) are antidiabetic compounds that were discovered to be potent stimulators of adipogenesis. TZDs bind and activate the nuclear receptor PPARγ *(99–101)*. They are routinely used in adipogenic treatment protocols to induce fat droplet accumulation within a few days in mesenchymal precursor cultures *(101,102)*. In vitro cell differentiation studies indicate that TZDs inhibit osteogenic differentiation

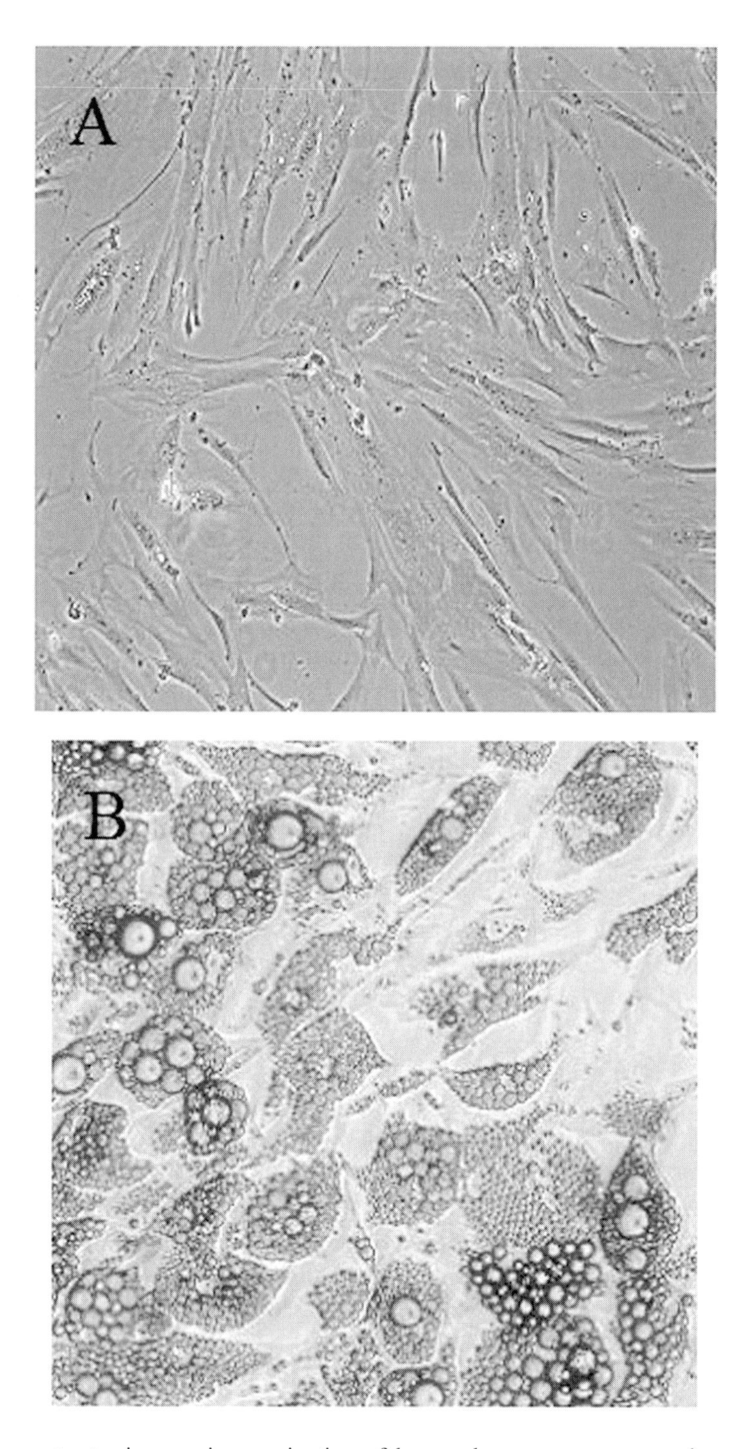

Fig. 1. Phase contrast microscopic examination of human bone marrow mesenchymal stem cells, before **(A)** and after **(B)** adipogenic differentiation.

while promoting adipogenic conversion of MSCs *(103,104)*. Studies of animals treated with PPARγ ligand TZDs have yielded conflicting results concerning BM fat cell differentiation and bone status: either significant loss of bone mineral density *(105,106)* and increased fat content in bones *(107,108)*, or no adverse effect *(109,110)*. Procedural differences may account for these discrepancies.

As MSCs in BM stroma undergo adipogenic differentiation, adipogenesis-related genes (i.e., CCAAT enhancer-binding protein [C/EBP], acylCoA synthetase, lipoprotein lipase, fatty acid binding protein 4, and PPARγ2) undergo upregulation, whereas osteogenesis-related genes (i.e., collagen I and osteocalcin) are downregulated *(98,111,112)*.

The cDNA microarray technique, confirmed by semiquantitative reverse transcription-polymerase chain reaction, has recently been applied to analyze gene expression profiles of human stromal cells incubated in the adipogenic medium. The temporal gene expression patterns indicate that genes differentially expressed during MSC adipogenesis are similar to those previously identified as "adipose-specific" in the differentiation processes of preadipocyte cell lines such as 3T3-L1 and 3T3-F442A *(113,114)*.

Human MSC-derived adipocytes (Fig. 1) display functional characteristics of mature adipocytes from adipose tissues, including specific intracellular signaling pathways for TNF-α and catecholamine-regulated lipolysis, as well as secretion of adiponectin and leptin. Similarly to differentiated preadipocytes, BM adipocytes display lipolytic effects mediated by β-adrenoceptors and antilipolytic effects mediated by the α2A-adrenoceptor, and they also express proteins with a pivotal role in human lipolysis, including β2-AR, α2A-AR, and hormone-sensitive lipase *(115)*.

5.1.3. Commitment of MSCs Among Phenotypes

Commitment of MSCs among phenotypes, as well as commitment to a particular lineage with suppression of alternative phenotypes, is governed by specific transcription factors. For instance, core binding factor a1 (Cbfa1/Runx2) is a transcription factor that is required for commitment of MSCs to the osteoblast lineage *(116,117)*. In contrast, PPARγ2 gene expression destines cells to adipocyte differentiation *(101,118)*. Transfection of stromal cells with PPARγ2, and subsequent activation with an appropriate ligand, causes the development of adipocytes, supporting the idea that PPARγ2 plays a critical role in the differentiation of mesenchymal cells to adipocytes *(119–121)*. This is confirmed by the work of Akune et al., who observed that embryonic stem cells from PPARγ-deficient mice spontaneously undergo osteogenesis but fail to undergo adipogenesis, and that their BM stromal cells exhibit a twofold reduction in adipogenesis and a threefold increase in osteogenesis *(122)*.

Additional pathways include the notch/delta/jagged ligands and receptors, related to the epidermal growth factor receptor family. Overexpression of notch in vitro inhibits osteogenesis and enhances adipogenesis in cell models *(123)*. Consistent with this is the observation that inhibition of the notch pathway interferes with adipogenesis in vitro *(124)*. Undifferentiated stromal cells and preadipocytes express Pref-1, a secreted notch/delta/jagged-like ligand that inhibits adipogenesis and is itself downregulated following activation of the adipogenic pathway *(125)*. Mice deficient in Pref-1 display skeletal malformations, growth retardation, and obesity *(126)*.

Overexpression of notch leads to downregulation of the Wingless (Wnt) receptor and its downstream mediator, β-catenin *(123)*. Wnt signaling blocks apoptosis and regulates differentiation of mesenchymal progenitors through inhibition of glycogen synthase kinase 3 and stabilization of β-catenin. The antiadipogenic effects of Wnt may be mediated through Frizzled (Fz)1 and/or Fz2, as these Wnt receptors are expressed in preadipocytes and their expression declines upon induction of differentiation. Activated Fz1 increases stability of β-catenin, inhibits apoptosis, induces osteoblastogenesis, and inhibits adipogenesis. Although activated Fz2 does not influence apoptosis or osteoblastogenesis, it inhibits adipogenesis through a mechanism independent of β-catenin. An important mediator of the β-catenin-independent pathway appears to be calcineurin, because inhibitors of this serine/threonine phosphatase partially rescue the block to adipogenesis caused by Wnt or activated Fz2. These data support a model in which Wnt signaling inhibits adipogenesis through both β-catenin-dependent and β-catenin-independent mechanisms *(127)*. Wnt signaling maintains preadipocytes in an undifferentiated state through inhibition of the adipogenic transcription factors C/EBPα and PPARγ *(128)*. Suppression of Wnt signaling is achieved by PPARγ through accelerating the degradation of β-catenin by the proteosome *(129)*. The necessary inhibition of Wnt signaling for the progression of adipogenesis provides an interesting insight into the regulation of osteogenic versus adipogenic commitment by MSCs, as PPARγ activation can inhibit osteogenesis *(130)*. This suggests that a fine balance between activated PPARγ and Wnt signaling controls the differentiation potential of MSCs to either bone or adipose tissue *(131)*.

Cadherins are a family of integral transmembrane proteins that mediate calcium-dependent cell–cell adhesion and that can also modulate intracellular signaling. Through their association with β-catenin, cadherins can also interfere with the Wnt signaling system. In transgenic mice expressing truncated cadherin mutant in osteoblasts, differentiation and function of osteoblasts are hampered and BM progenitor cell commitment to the alternative adipogenic lineage via interference with β-catenin signaling is favored. This results in decreased bone formation, delayed acquisition of peak bone mass, and increased body fat in young animals *(132)*.

Cytokines such as IL-1β, IL-11, TGF-β, and TNF-α inhibit adipogenesis in marrow MSC in vitro and in vivo *(133–139)*. The ligand-induced transactivation function of PPARγ is suppressed by IL-1, TGF-β, and TNF-α; this suppression is mediated through NFκB activated by the TAK1/TAB1/NFκB-inducing kinase cascade, a downstream cascade associated with IL-1 and TNF-α signaling *(140)*. Treatment of ST2 cells, a mesenchymal cell line derived from mouse BM, with low concentrations of the PPARγ ligand troglitazone, leads to lipid accumulation and expression of adipocyte-associated differentiation markers such as AP2 and LPL. This adipogenesis is prevented by IL-1 or TNF-α. ST2 cells treated with both troglitazone and cytokines differentiate into osteoblasts expressing both alkaline phosphatase protein and mRNA *(140)*. These results suggest that expression of IL-1 and TNF-α in bone marrow may alter the fate of pluripotent mesenchymal stem cells, directing cellular differentiation toward osteoblasts rather than adipocytes by suppressing PPARγ function.

Considering the cascade of transcription factors sequentially induced during early adipogenesis, some authors hypothesize that adipogenesis is the default pathway for MSCs that do not receive proper inductive signals to become osteoblasts, chondrocytes, myocytes, or other mesodermal cells *(131)*.

5.2. *In Vivo Observations*

In the mouse, mesenchymal progenitors with the potential to differentiate into cells of the osteogenic, adipogenic, and chondrogenic lineages are present in most of the sites harboring hematopoietic cells. They first appear in the aorta–gonad–mesonephros region at the time of the emergence of hematopoietic stem cells. They are found in the embryonic circulation. They increase numerically in hematopoietic sites during development to a plateau level found in adult BM. This colocalization of mesenchymal progenitor/stem cells to the major hematopoietic territories suggests that, as development proceeds, mesenchymal progenitors expand within these potent hematopoietic sites *(141)*.

In the 1970s, pioneering work by Friedenstein et al. demonstrated that rodent MSCs can be grown ex vivo and maintain their differentiation capacity in vivo upon reimplantation *(85)*. Following transplantation into irradiated animals, MSCs durably engraft in the bone, cartilage, and lungs of mice *(142)*, and produce fibroblasts or fibroblast-like cells that can be reisolated and cultured from the lungs, calvaria, cartilage, long bones, tails, and skin *(142)*.

Human MSCs, when transplanted *in utero* in sheep, contribute in a site-specific manner to chondrocytes, adipocytes, myocytes, cardiomyocytes, bone marrow stromal cells, and thymic stroma *(143)*. The human cells are found to persist in multiple tissues for as long as 13 mo, despite the development of immunological competence in the sheep, thus providing some of the first evidence that MSCs may possess unique immunological properties in addition to multiplicity of differentiation *(143)*.

5.3. *Mesenchymal Progenitors From Non-BM Adipose Tissues*

Recently, nonmedullary adipose tissue has been shown to have a population of pluripotent stem cells, exhibiting a fibroblast-like morphology and the ability to differentiate into multiple mesenchymal lineages including bone, cartilage, and fat *(144,145)*. Flow cytometry and immunohistochemistry show that human MSCs have a marker expression that is similar to that of these adipose tissue-derived cells. They express CD9, CD10, CD13, CD29, CD34, CD44, CD49d, CD49e, CD54, CD55, CD59, CD90, CD105, CD106, CD146, and CD166, and are absent for HLA-DR and c-kit expression. However unlike BM MSCs, adipose tissue-derived stromal cells express the hematopoietic and endothelial progenitor marker CD34 *(144,146)*. Microarray analysis of gene expression reveals that fewer than 1% of genes are differentially expressed between BM and adipose MSCs *(147,148)*.

6. ADIPOCYTES AND BONE-FORMING CELLS

Transdifferentiation exists between osteoblasts and adipocytes *(149,150)*. Even though these cells may have a phenotype characterized by late markers of differentiation, cell dedifferentiation followed by redifferentiation may still occur *(151,152)*. This plasticity between the differentiation of osteogenic and adipogenic human cells suggests that adipocytes are generated at the expense of osteoblasts from a common progenitor. As previously observed, this plasticity also has significant physiological and pharmacological implications because decrease in bone mass is accompanied by an increase in BM adipose tissue.

6.1. Animal Models of Osteoporosis or Increased Bone Formation

In the mouse, aging causes a decrease in the commitment of MSC to the osteoblast lineage and an increase in the commitment to the adipocyte lineage. The expression of osteoblast-specific transcription factors, Runx2 and Dlx5, and of osteoblast markers, collagen and osteocalcin, is decreased in aged MSC. Conversely, the expression of adipocyte-specific transcription factor PPARγ2 is increased, as is a gene marker of adipocyte phenotype, fatty acid binding protein aP2. In addition, expression of different components of TGF-β and BMP2/4 signaling pathways is altered, suggesting that the activities of these two cytokines essential for bone homeostasis change with aging *(136)*.

Studies by Hamrick et al. reveal that the bone phenotype of the *ob/ob* mouse is more complex than has previously been appreciated. It is characterized by short femora with reduced cortical thickness and reduced trabecular bone volume. In the spine, however, although cortical thickness is still reduced, vertebral length, bone mineral density, and trabecular bone volume are all increased. These results indicate that the effects of altered leptin signaling on bone differ significantly between axial and appendicular regions. Few adipocytes are observed in BM from lumbar vertebrae, whereas in the femur the number of marrow adipocytes per unit area is increased 235-fold. Leptin treatment induces loss of BM adipocytes and increases bone formation in these leptin-deficient *ob/ob* mice *(153)*.

Tornvig et al. increased marrow adipogenesis in mice experimentally via treatment with the PPARγ ligand troglitazone and found that trabecular bone volume did not change with an increase in marrow adipocytes *(109)*. These results suggest that bone mass and BM adipogenesis can be regulated independently.

Transgenic mice expressing Δ-FosB not only develop a severe and progressive osteosclerotic phenotype characterized by increased bone formation, but also show pronounced decrease in adipogenesis with decreased abdominal fat, low serum leptin levels, and a reduced number of adipocytes in the bone marrow *(46)*. The inhibitory effect of Δ-FosB on adipocyte differentiation appears to occur at early stages of stem cell commitment, affecting C/EBPβ functions *(154)*.

These animal studies make it clear that, at least in long bones, adipose and bone tissues develop at the expense of each other from mesenchymal progenitors. What are the endocrine cues for such determination? As well as leptin and adiponectin, already discussed, numerous hormones are implicated in the equilibrium between fat and bone.

6.2. Hormonal Control of Adipogenesis and Osteogenesis

6.2.1. Estrogen

It is well-known that estrogen affects the accumulation and distribution of peripheral fat during sexual maturation and menopause *(155,156)*. But there is also some evidence that oophorectomy-induced bone loss is accompanied by increased fat mass in BM *(42,157)*, suggesting that BM fat may also be a target for estrogen.

Heine et al. *(158)* showed that estrogen receptor (ER) knockout mice increase their adipose tissue. In humans, BM adipocytes express cytochrome P450 enzyme aromatase, which converts circulating androgens to estrogens *(159)*. Mice with aromatase deficiency enhance their adiposity *(160)*.

Estrogen stimulates bone metabolism directly via receptors on osteoblasts, stimulates both osteoclastogenesis and osteoblastogenesis *(161)*, and inhibits adipocyte differentiation in mouse BM stromal cell lines that express estrogen receptor (ER) α or β *(162)*. The phytoestrogen genistein can enhance the commitment and differentiation of human MSCs toward the osteoblast lineage, whereas adipogenic differentiation and maturation are hindered. This inhibition of adipogenic differentiation can be reversed by obstructing ER and TGF-β1 signaling *(163)*.

6.2.2. Growth Hormone

Peripheral fat depots are well-known targets of growth hormone (GH) action. In humans, GH deficiency results in adiposity and GH treatment reduces fat mass *(164,165)*. On the contrary, both body fat and BM fat increase with age *(166)* as GH secretion decreases *(167)*.

Dwarf rats (*dw/dw*) with isolated GH deficiency *(168)* have a markedly increased number of adipocytes in the BM; these adipocytes are also larger, implying increased fat storage in BM. GH treatment counteracts these changes, reducing adipocyte number and size and restoring the amount of BM fat to normal *(169)*.

Hypophysectomy also results in increased bone marrow adipogenesis and fat accumulation in rats, as demonstrated by the increased triglyceride content of bone marrow. Treatment of hypophysectomized rats with GH reverses these changes. The increased adipogenesis in hypophysectomized rats is also evident in primary BM stromal cell cultures, with not only a greater number of adipocytes but also increased expression of markers of adipocyte maturation: PPARγ2, adipsin, and leptin *(170)*.

Although many GH effects may be mediated via IGF-1, adipocytes express GH receptors, which mediate direct effects on lipolysis *(171–174)*. GH may also influence adipocyte numbers directly because it affects preadipocyte/adipocyte differentiation in vitro *(175)*.

GH also stimulates longitudinal growth and bone formation *(176)*, and GH deficiency is associated with decreased bone mass *(177)* that can be increased with GH treatment *(178)*.

Most in vitro studies on GH and bone formation have focused on mature osteoblasts, as they express GH receptors and can proliferate and differentiate in vitro in response to GH *(179,180)*. However, GH receptors are also present on BM stromal progenitor cells *(181)* and GH can increase the proliferation of stromal osteoblast-like precursors in vitro *(182)*. GH could therefore act on the progenitors of adipocytes and osteoblasts, affecting their proliferation and differentiation.

6.2.3. Thyroid Hormone

Thyroid hormone, T3, regulates a wide range of developmental and physiological processes including skeletal development, longitudinal bone growth, and adult bone metabolism *(183)*. T3 acts through nuclear hormone receptors encoded by two distinct but closely related genes, TRa and TRh. TRs are expressed in osteoblasts, osteoclasts, and chondrocytes *(184)*, and have also been reported in cultured adipocytes *(185)*.

Mice with targeted mutations in the TRa and TRh genes (Trα1–/–β–/–) have reduced body weight at birth and during postnatal development, a distinct skeletal phenotype with shortened long bones, as well as increased BM fat. The latter is a result of an increased

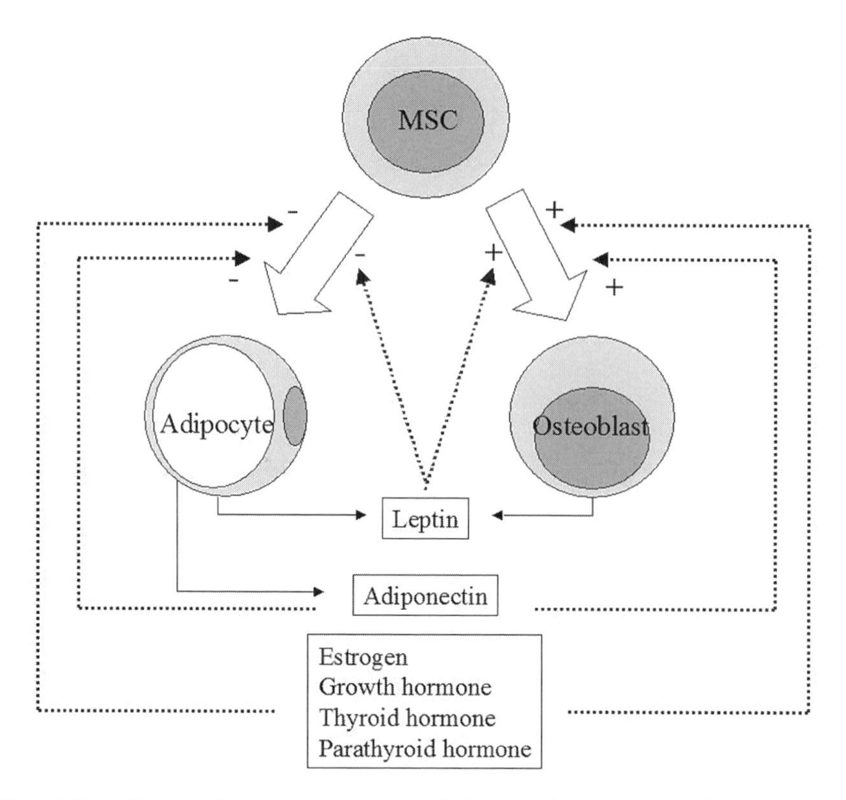

Fig. 2. Regulating effects of leptin, adiponectin, and the main hormones on adipogenesis and osteogenesis from mesenchymal stem cell (MSC) precursors.

number of adipocytes and is associated with increased expression of mature adipocyte markers *(186)*. Most studies on T3 and osteoblasts conclude that T3 promotes osteoblast differentiation *(187,188)*. T3 has also been shown to directly increase adipocyte differentiation in several different cultured cell lines *(185,189,190)*. These data of adipogenic actions of T3 are not in accordance with the observations in Trα1–/–β–/– mice. Differences between the in vivo and the in vitro situation or a different origin of the adipocytes in the various studies may contribute to these discrepancies.

6.2.4. Other Hormones

Parathyroid hormone and parathyroid hormone-like peptides potentiate osteoblastogenesis and inhibit adipogenesis at least in vitro *(191,192)*. Expression of prolactin receptor can be induced by adipocyte differentiation of BM stromal cells. However, it is currently unclear what induces this expression in the BM microenvironment *(193)*.

Therefore, the great majority of hormones exert a negative control on BM adipogenesis and stimulate bone-forming cells (Fig. 2). This point may be considered in conjunction with the idea previously put forward that, from a transcriptional point of view, adipogenesis could be the default pathway for MSCs not receiving inductive signals. As a whole, there is an obvious balance between adipose and bone formation within the BM microenvironment. This equilibrium is regulated in a subtle and complex manner, mainly via hormonal control, and could present a target for pharmacological intervention in disorders such as osteoporosis *(149,194)*.

7. BM ADIPOCYTES AND HEMATOPOIESIS

Bone marrow fat cells have long been suspected to have an influence on hematopoiesis *(111)*. Many arguments come from histological studies, in vitro studies using rodent-derived stromal cell lines, and, more recently, primary culture of human BM adipocytes.

7.1. Histological Observations

The location of the different hematopoietic cells in the bone marrow is not random: clumps of megakaryocytes are found adjacent to venous sinuses; red blood cells tend to be organized as colonies, or as erythroblastic islets, consisting of a central macrophage surrounded by differentiating erythroblasts; adipocytes are typically closely associated with granulocytes and monocytes *(195)*. This suggests preferential interactions of hematopoietic progenitors with different cells of the microenvironment.

7.2. Preadipocyte Stromal Cell Lines

BM-derived stromal cells have been cloned from mice and rats *(1)*. Most of the cells that support myelopoiesis and/or lymphopoiesis in long-term culture systems are preadipocytes. They secrete many cytokines and growth factors known to regulate proliferation or differentiation of hematopoietic progenitors. They also provide extracellular matrix components required for intracellular adhesion or cell recognition *(1)*. However, when these stromal cells develop into adipocytes, spontaneously or in response to adipogenic agents such as glucocorticoids, the expression of extracellular matrix and cytokines is frequently altered and this affects hematopoiesis. In that respect, it has been shown that fully differentiated fat cells from an adult-derived cell line produce less colony stimulating factor-1 than their precursors *(139)* and that expression of stem cell factor IL-6 and leukemia inhibitory factor declined with terminal adipocyte differentiation of an embryo-derived stromal line *(196)*. It is noteworthy that these cytokines act preferentially on the more immature hematopoietic progenitors. This could indicate that undifferentiated mesenchymal or stromal cells interact essentially with noncommitted hematopoietic progenitors, whereas adipocytes play a role in the blood cell differentiating process.

7.3. Primary Human BM Adipocytes

We compared the potential of human BM adipocytes and MSCs to support hematopoiesis in coculture systems. Adipocytes do not maintain self-renewal of $CD34^+$ hematopoietic progenitor cells, but these cells show full myeloid and B-lymphoid differentiation *(197)*. These data confirm that differentiated adipocytes do not maintain hematopoietic clonogenic progenitors, unlike the adipocyte precursors, MSCs, which support the self-renewal and proliferation of purified human $CD34^+$ progenitors and maintain myeloid cells for several weeks in culture *(198)*.

Inversely also, human osteoblasts supported the development of hematopoietic colonies from $CD34^+$ progenitors and maintained long-term culture-initiating cells, but could not support granulopoietic differentiation without added cytokines *(199)*. Osteoblasts could thus be critical for the maintenance and self-renewal of hematopoietic stem cells, whereas adipocytes may be implicated in differentiating processes.

These contrasting properties are all the more puzzling, as adipocytes and osteoblasts share a common progenitor, the MSC. These distinct characteristics (probably because of different secretory and/or surface factors) could be significant in some situations, such as increased BM adipogenesis during aging or osteoporosis.

What are the hematopoietic effects of the products more specifically secreted by adipocytes, namely cytokines, leptin and adiponectin?

- BM adipocytes in primary culture secrete only trace amounts of IL-1β and TNF-α. On the contrary, they produce significant levels of IL-6, a secretion stimulated by both IL-1β and TNF-α *(200)*. Besides having a proinflammatory effect, IL-6 is an important regulator of marrow hematopoiesis. The number of IL-6 receptors on hematopoietic progenitor cells increases significantly with maturation of these cells *(201)*, and IL-6 is a regulator of granulopoiesis in vivo *(202)*. IL-6 is also involved in normal B-cell differentiation, and is a key growth and survival factor for malignant B-cells in multiple myeloma *(203)*.
- We first reported that human BM adipocytes secrete large quantities of leptin *(63)*, which appears to play a part in the regulation of hematopoietic progenitors and their differentiation into granulocyte and monocyte precursors. The concentration of leptin required for this effect in vitro (50–100 ng/mL) is rather high, but is within the range of plasma leptin levels observed in obese subjects *(64)*. As leptin concentrations in bone marrow and plasma are highly correlated in humans *(64)*, is leptin involved in the leukocytosis associated with obesity and, more broadly, is there any correlation between leptin levels and blood cell counts? Wilson et al. observed that in obese Pima Indians most of the variance in the leukocyte count attributable to body fat could be accounted for by plasma leptin concentration *(204)*. We confirmed that leptin and leukocyte count are also correlated in French obese subjects *(64)*. Concerning nonobese subjects, Togo et al. *(205)* reported a negative correlation between leptin and hemoglobin levels in adult Japanese males, but no correlation between leptin levels and leukocyte counts. On the contrary, an association of serum leptin level with leukocyte and erythrocyte counts in adolescent Japanese males aged 15 to 16 yr was reported by Hirose et al. *(206)*. In a large population of European subjects, we observed that there was no statistically significant relationship between circulating leptin levels and blood cell parameters in healthy middle-aged men and women. However, a role for high leptin concentrations in situations such as obesity or sepsis cannot be excluded, as suggested by the weak correlation we observed in hospitalized patients *(207)*.
- In vitro, leptin modulates cytokine secretion from T-lymphocytes and macrophages, increases the proliferation of naive T-cells while reducing the proliferation of memory T-cells, and enhances the phagocytic activity of mature macrophages *(58)*. Mice with leptin deficiency or resistance have reduced T-cell function. The decrease in leptin that accompanies starvation or food restriction also induces immune suppression, probably through downregulation of the T-cell response *(208,209)*.
- In conclusion, leptin probably plays a direct role in the proliferation of BM hematopoietic progenitors and their differentiation along the granulocyte–macrophage and naive T-cell lineages. In situations with high levels of circulating leptin (inflammation, obesity) or with falling leptin levels (starvation), a regulatory role on immune response is likely, either directly or indirectly through its effects on other cells of the microenvironment and on cytokines.

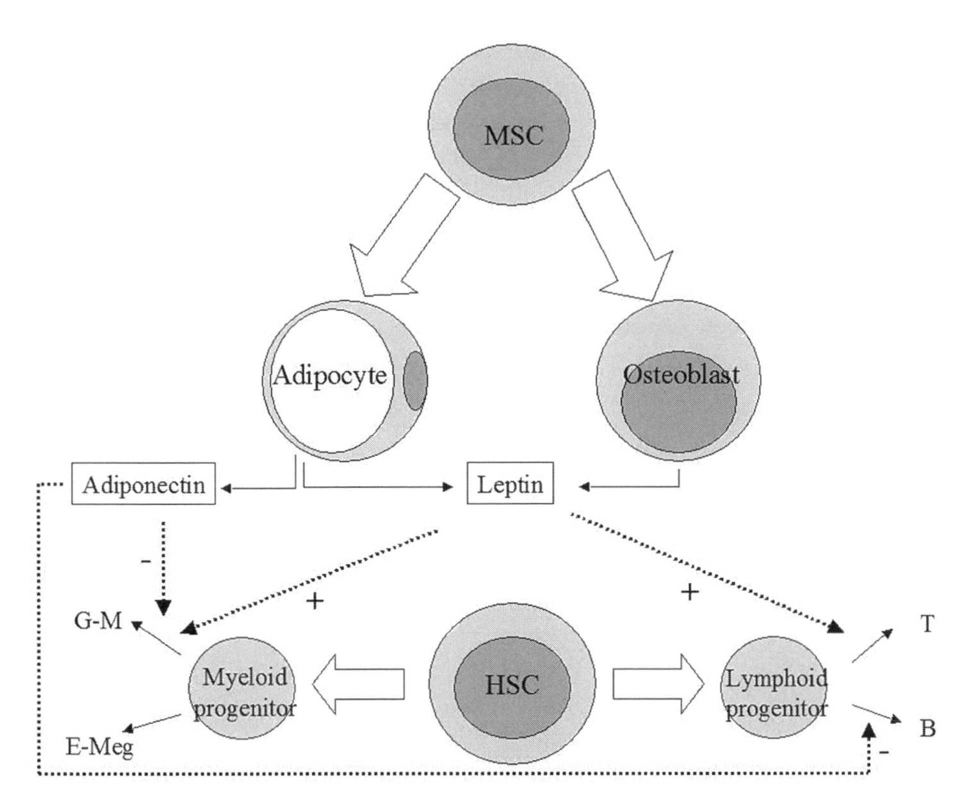

Fig. 3. Regulating effects of leptin and adiponectin on hematopoietic differentiation. MSC: mesenchymal stem cell; HSC: hematopoietic stem cell; G-M: granulocytes and monocytes/macrophages; E-Meg: erythroblasts and megakaryocytes.

As previously indicated, adiponectin serves as a negative regulator for myelomonocytic progenitor growth and inhibits macrophage functions *(77)*. In that respect, adiponectin is a negative regulator of the immune response at two levels: it suppresses the phagocytosis of mature macrophages and inhibits the growth of macrophage precursors. Addition of adiponectin to long-term bone marrow cultures influences the earliest lymphocyte precursors and strongly inhibits B-lymphopoiesis *(76)*.

Overall, leptin exerts a proinflammatory role and adiponectin appears to act as an anti-inflammatory molecule *(210)*. The effects of adiponectin and leptin on hematopoiesis, immunity, and inflammation appear to be diametrically opposite (Fig. 3).

8. CONCLUSIONS

Well over a decade after the paper published in 1990 by Gimble *(1)*, we are still asking questions about the function of adipocytes in the bone marrow stroma. Plastic BM adipose tissue does not simply fill bone cavities that are not needed for hematopoietic activity. It may serve as an energy store for local needs (i.e., hematopoiesis and bone modeling) or participate in the overall energy metabolism of the body. Curiously, since the pioneering works of Tavassoli et al. *(23,26)*, these hypotheses have not really been verified. Undoubtedly, preadipocytes and adipocytes secrete numerous cytokines and hormones whose receptors are present on different cells of the stromal microenvironment. They are direct and indirect regulators of granulopoiesis and T-lymphocyte commitment, and modulate macrophage function and secretion. In that respect, independent of nutritional status, BM adipose tissue is involved in the hematopoietic, immune, and inflammation systems. Some arguments indicate that adipogenesis may be the default differentiation pathway of BM mesenchymal progenitors not submitted to paracrine and endocrine agents. This is likely when considering the adipogenesis/osteogenesis balance. Such a counterbalance between adipogenesis and hematopoiesis remains to be demonstrated at the cell level. If transdifferentiation between cells of the mesodermal lineage depending on the current needs of the organism is confirmed, then BM adipose tissue may represent a target for drugs in situations such as aplastic anemia or osteoporosis.

REFERENCES

1. Gimble JM. New Biol 1990;2:304–312.
2. Erslev AJ. Clin Orthoped 1967;52:25–36.
3. Ascenzi A. In: Bourne G, ed. *Biochemistry and Physiology of Bone*, Vol. 4. Academic Press, New York: 1976, pp. 403–444.
4. Evans JD, Baker JM, Oppenheimer MJ. Am J Physiol 1955;181:504–508.
5. Tran MA, Lac DT, Berlan M. J Lipid Res 1981;22:1271–1276.
6. Tavassoli M. Exp Hematol 1984;12:139–146.
7. Pietrangeli CE, Hayashi S, Kincade PW. Eur J Immunol 1988;18:863–872.
8. Gimble JM, Youkhana K, Hua X, et al. J Cell Biochem 1992;50:73–82.
9. Beresford JN, Bennet JH, Devlin C, et al. J Cell Sci 1992;102:341–351.
10. Dodds RA, Gowen M, Bradbeer JN. J Histochem Cytochem 1994;42:599–606.
11. Charbord P, Tavian M, Humeau L, et al. Blood 1996;87:4109–4119.
12. Bigelow CL, Tavassoli M. Acta Anat 1984;118:60–64.
13. Emery JL, Follet GF. Br J Haematol 1964;10:485.
14. Wintrobe MM. In: *Clinical Hematology,* Lea & Febiger, Philadelphia: 1967, pp. 1–53.
15. Huggins C, Blocksom BH. J Exp Med 1936;64:253–274.
16. Huggins C, Blocksom BH, Noonan WJ. Am J Physiol 1936;115:395–401.
17. Maniatis A, Tavassoli M, Crosby WH. Blood 1971;37:581–586.
18. Crosby WH. In: Stohlman F, ed. *Hemopoietic Cellular Proliferation.* Grune & Stratton, New York: 1970, pp. 87–94.
19. Ravaglia G, Forti P, Maioli F, et al. J Gerontol 1999;54:M70–M76.
20. Kehayias JJ, Fiatarone MA, Zhuang H, et al. Am J Clin Nutr 1997;66:904–910.
21. Kirkland JL, Tchkonia T, Pirtskhalava T, et al. Exp Gerontol 2002;37:757–767.
22. Chumlea WC, Rhyne RL, Garry PJ, et al. Am J Hum Biol 1989;1:457–462.
23. Tavassoli M. In: Tavassoli M, ed. *Handbook of the Hemopoietic Microenvironment.* Humana Press, Totowa, NJ: 1989, pp. 157–187.
24. Robey PG, Bianco P. In: Rosen C, Bilezikian JP, eds *The Aging Skeleton..* Academic Press, San Diego, CA:1999, pp. 145–157.
25. Moore SG, Dawson KL. Radiology 1990;175:219–223.
26. Tavassoli M, Houchin DN, Jacobs P. Scand J Haematol 1977;18:47–53.
27. Maurin AC, Chavassieux PM, Frappart L, et al. Bone 2000;26:485–489.
28. Maurin AC, Chavassieux PM, Vericel E, et al. Bone 2002;31:260–266.
29. Bathija A, Davis S, Trubowitz S. Am J Hematol 1979;6:191–198.
30. Tavassoli M. Experientia 1974;30:424–425.
31. Tavassoli M, Eastlund DT, Yam LT, et al. Scand J Haematol 1976;16:311–319.
32. Seaman JP, Kjeldsbert CR, Linker A. Hum Pathol 1978;9:685–692.
33. Bathija A, Davis S, Trubowitz S. Am J Hematol 1978;5:315–321.
34. Dietz AA, Steinberg B. Arch Biochem 1953;45:1–9.
35. Islam A. Med Hypotheses 1988;25:209–217.
36. Brookoff D, Weiss L. Blood 1982;60:1337–1344.

37. Allen TD, Dexter TM, Simmons PJ. In: Dexter TM, Garland JM, Testa NG, eds. *Colony-Stimulating Factors. Molecular and Cellular Biology*. Marcel Dekker, New York: 1990, pp. 1–38.
38. Rozman C, Feliu E, Berga L, et al. Exp Hematol 1989;17:34–37.
39. Justesen J, Stenderup K, Ebbesen EN, et al. Biogerontology 2001;2:165–171.
40. Meunier P, Aaron J, Edouard C, et al. Clin Orthop 1971;80:147–154.
41. Verma S, Rajaratnam JH, Denton J, et al. J Clin Pathol 2002;55:693–698.
42. Martin RB, Zissimos SL. Bone 1991;12:123–131.
43. Ahdjoudj S, Lasmoles F, Holy X, et al. J Bone Miner Res 2002;17:668–677.
44. Wang YS, Li YB, Mao KY, et al. Clin Orthoped Rel Res 2003;41:213–224.
45. Wang GW, Sweet D, Reger S, et al. J Bone Joint Surg 1977;59A:729–735.
46. Sabatakos G, Sims NA, Chen J, et al. Nat Med 2000;6:985–990.
47. Kajkenova O, Lecka-Czernik B, Gubrij I, et al. J Bone Miner Res 1997;12:1772–1779.
48. Desplanches D, Mayet MH, Sempore B, et al. J Appl Physiol 1987;63:558–563.
49. Botolin S, Faugere M-C, Malluche H, et al. Endocrinology 2005;146:3622–3631.
50. Wezeman FH, Gong Z. Alcohol Clin Exp Res 2001;25:1515–1522.
51. Wezeman FH, Gong Z. Res Soc Alcoholism 2004;28:1091–1101.
52. Tavassoli M. Arch Pathol 1974;98:189–192.
53. Oberling F, Cazenave JP, Waitz R. Path Biol 1972;20:337–347.
54. Tavassoli M. Scand J Haematol 1978;20:330–334.
55. Trubowitz S, Bathija A. Blood 1977;49:599–605.
56. Flier JS. J Clin Endocrinol Metab 1998;83: 1407–1413.
57. Mantzoros CS, Moschos SJ. Clin Endocrinol 1998;49:551–567.
58. Fantuzzi G, Faggioni R. J Leukoc Biol 2000;68:437–446.
59. Hamrick MW, Pennington C, Newton D, et al. Bone 2004;34:376–383.
60. Bennett BD, Solar GP, Yuan JQ, et al. Curr Biol 1996;6:1170–1180.
61. Lee FY, Li Y, Yang EK, et al. Am J Physiol 1999;276:C386–C394.
62. Loffreda S, Yang SQ, Lin HZ, et al. FASEB J 1998;12:57–65.
63. Laharrague P, Larrouy D, Fontanilles AM, et al. FASEB J 1998;12:747–752.
64. Laharrague P, Truel N, Fontanille, AM, et al. Horm Metab. Res 2000;32:381–385.
65. Cioffi JA, Shafer AW, Zupancic TJ, et al. Nat Med 1996;2:585–589.
66. Gainsford T, Willson TA, Metcalf D, et al. Proc Natl Acad Sci USA 1996;93:14,564–14,568.
67. Umemoto Y, Tsuji K, Yang FC, et al. Blood 1997;90:3438–3443.
68. Laharrague P, Oppert JM, Brousset P, et al. Int J Obes Relat Metab Disord 2000;24:1212–1216.
69. Thomas T, Gori F, Khosla S, et al. Endocrinology 1999;140:1630–1638.
70. Reseland JE, Syversen U, Bakke I, et al. J Bone Miner Res 2001;16:1426–1433.
71. Gordeladze JO, Drevon CA, Syversen U, et al. J Cell Biochem 2002;85:825–836.
72. Scherer PE, Williams S, Fogliano M, et al. J Biol Chem 1995;270:26,746–26,749.
73. Hu E, Liang P, Spiegelman BM. J Biol Chem 1996;271:10,697–10,703.
74. Maeda K, Okubo K, Shimomura I, et al. Biochem Biophys Res Commun 1996;221:286–289.
75. Nakano Y, Tobe T, Choi-Miura NH, et al. J Biochem (Tokyo) 1996;120:803–812.
76. Yokota T, Meka CSR, Kouro T, et al. J Immunol 2003;171:5091–5099.
77. Yokota T, Oritani K, Takahashi A, et al. Blood 2000;96:1723–1732.
78. Yokota T, Meka CSR, Medina KL, et al. J Clin Invest 2002;109:1303–1310.
79. Luo XH, Guo LJ, Yuan LQ, et al. Exp Cell Res 2005;309:99–109.
80. Bellido T, Stahl N, Farruggelia T, et al. J Clin Invest 1996;97:431–437.
81. Tagushi Y, Yamamoto M, Yamate T, et al. Proc Assoc Am Physicians 1998;110:559–574.
82. Owen M. J Cell Sci 1988;10:63–76.
83. Bianco P, Riminucci M, Gronthos S, et al. Stem Cells 2001;19:180–192.
84. Kincade PW, Yamashita Y, Borghesi L, et al. Vox Sanguinis 1998;74:265–268.
85. Friedenstein AJ, Chailakhyan RK, Latsinik NV, et al. Stromal cells responsible for transferring the microenvironment of the hemopoietic tissues. Cloning in vitro and retransplantation in vivo. Transplantation 1974;17:331–340.
86. Friedenstein AJ, Gorskaja U, Kalugina NN. Exp Hematol 1976;4:267–274.
87. Caplan AI. J Orthopaed Res 1991;9:641–650.

88. Pereira RF, Halford KW, O'Hara MD, et al. Proc Natl Acad Sci USA 1995;92:4857–4861.
89. Pittenger MF, Mackay AM, Beck SC, et al. Science 1999;284:143–147.
90. Sekiya I, Larson BL, Smith JR, et al. Stem Cells 2002;20:530–554.
91. Dexter TM, Spooncer E, Simmon P, et al. In: Wright DG, Greenberger JS, eds. *Long Term Bone Marrow Culture*. Alan R. Liss, New York: 1984, pp. 57–96.
92. Austin TW, Solar GP, Ziegler FC, et al.Blood 1997;89:3624–3635.
93. Colter DC, Class R, DiGirolamo CM, et al. Proc Natl Acad Sci USA 2000;97:3213–3218.
94. Gronthos S, Graves SE, Ohta S, et al. Blood 1994;84:4164–4173.
95. Gregory CA, Prockop DJ, Spees JL. Exp Cell Res 2005;306:330–335.
96. Greenberger JS. Nature 1978;275:752–754.
97. Greenberger JS. In Vitro 1979;15:823–828.
98. Sekiya I, Larson BL, Vuoristo JT, et al. J Bone Miner Res 2004;19:256–264.
99. Picard F, Auwerx J. Annu Rev Nutr 2002;22:167–197.
100. Lehmann JM, Moore LB, Smith-Oliver TA, et al. J Biol Chem 1995;270:12,953–12,956.
101. Rosen ED, Walkey CJ, Puigserver P, et al. Genes Dev 2000;14:1293–1307.
102. Willson TM, Brown PJ, Sternbach DD, et al. J Med Chem 2000;43:527–550.
103. Gimble JM, Robinson CE, Wu XY, et al. Mol Pharmacol 1996;50:1087–1094.
104. Sottile V, Seuwen K. FEBS Lett 2000;475: 201–204.
105. Jennermann C, Triantafilou J, Cowan D, et al. J Bone Miner Res 1995;10:S361.
106. Rzonca SO, Suva D, Gaddy DC, et al. Endocrinology 2004:145:401–406.
107. Deldar A, Williams G, Stevens C. Diabetes 1993;42:179.
108. Ali AA, Weinstein RS, Stewart SA, et al. Endocrinology 2005:146:1226–1235.
109. Tornvig L, Mosekilde L, Justesen J, et al. Calcif Tissue Int 2001;69:46–50.
110. Sottile V, Seuwen K, Kneissel M. Calcif Tissue Int 2004;75:329–337.
111. Gimble JM, Robinson CE, Wu X, et al. Bone 1996;19:421–428.
112. Hata K, Nishimura R, Ueda M, et al. Mol Cell Biol 2005;25:1971–1979.
113. Gronthos S, Franklin DM, Leddy HA, et al. J Cell Physiol 2001;189:54–63.
114. Lee RH, Kim BC, Choi IS, et al. Cell Physiol Biochem 2004;14:311–324.
115. Ryden M, Dicker A, Götherström C, et al. Biochem Biophys Res Comm 2003;311:391–397.
116. Komori T, Yagi H, Nomura S, et al. Cell 1997;89:755–764.
117. Ducy P, Zhang R, Geoffroy V, et al. Cell 1997;89:747–754.
118. Shao D, Lazar MA. J Biol Chem 1997;272:21,473–21,478.
119. Shi X, Chang Z, Blair H, et al. Bone 1998;23:S454.
120. Lecka-Czernik B, Moerman EJ, Grant DF, et al. Endocrinology 2002;143:2376–2384.
121. Tontonoz P, Hu E, Spiegelman BM. Cell 1994;79:1147–1156.
122. Akune T, Ohba S, Kamekura S, et al. J Clin Invest 2004;113:846–855.
123. Sciaudone M, Gazzerro E, Priest L, et al. Endocrinology 2003;144:5631–5639.
124. Garces C, Ruiz-Hidalgo MJ, de Mora JF, et al. J Biol Chem 1997;272:29,729–29,734.
125. Smas CM, Sul HS. Int J Obes 1996;20:S65–S72.
126. Moon YS, Smas CM, Lee K, et al. Mol Cell Biol 2002;22:5585–5592.
127. Kennell JA, MacDougald OA. J Biol Chem 2005;280:24,004–24,010.
128. Ross SE, Hemati N, Longo KA, et al. Science 2000; 289:950–953.
129. Liu J, Farmer SR. Biol Chem 2004;279:45,020–45,027.
130. Khan E, Abu-Amer Y. J Lab Clin Med 2003;142:29–33.
131. Westendorf JJ, Kahler RA, Schroeder TM. Gene 2004;341:19–39.
132. Castro CH, Shin CS, Stains JP, et al. J Cell Science 2004;117:2853–2864.
133. Ron D, Brasier AR, McGehee RJ, et al. J Clin Invest 1992;89:223–233.
134. Torti FM, Torti SV, Larrick JW, et al. J Cell Biol 1989;108:1105–1113.
135. Locklin RM, Oreffo RO, Triffitt JT. Cell Biol Int 1999;23:185–194.
136. Moerman EJ, Teng K, Lipschitz DA, et al. Aging Cell 2004;3:379–389.
137. Delikat SE, Galvani DW, Zuzel M. Cytokine 1995;7:338–343.
138. Ohsumi J, Miyadai K, Kawashima I, et al. FEBS Lett 1991;288:13–16.
139. Umezawa A, Tachibana K, Harigaya K, et al. Mol Cell Biol 1991;11:920–927.
140. Suzawa M, Takada I, Yanagisawa J, et al. Nat Cell Biol 2003;5:224–230.
141. Mendes SC, Robin C, Dzierzak E. Development 2005;132:1127–1136.

142. Pereira RF, O'Hara MD, Laptev AV, et al. Proc Natl Acad Sci USA 1998;95:1142–1147.
143. Liechty KW, MacKenzie TC, Shaaban AF, et al. Nat Med 2000;6:1282–1286.
144. Zuk PA, Zhu M, Ashjian P, et al. Mol Biol Cell 2002;13:4279–4295.
145. Zuk PA, Zhu M, Mizuno H, et al. Tissue Eng 2001;7:211–228.
146. Dicker A, Le Blanc K, Astrom G, et al. Exp Cell Res 2005;308:283–290.
147. Hung SC, Chang CF, Ma HL, et al. Gene 2004;340:141–150.
148. Nakamura T, Shiojima S, Hirai Y, et al. Biochem Biophys Res Commun 2003;303:306–312.
149. Nuttall ME, Patton AJ, Olivera DL, et al. J Bone Miner Res 1998;13:371–382.
150. Park SR, Oreffo RO, Triffitt JT. Bone 1999;24:549–554.
151. Nuttall ME, Gimble JM. Bone 2000;27:177–184.
152. Bennett JH, Joyner CJ, Triffitt JT, et al. J Cell Sci 1991;99:131–139.
153. Hamrick MW, Della-Fera MA, Choi YH, et al. J Bone Miner Res 2005;20:994–1001.
154. Kveiborg M, Sabatakos G, Chiusaroli R, et al. Mol Cell Biol 2004;24:2820–2830.
155. Tchernof A, Poehlman ET, Despres JP. Diabetes Metab 2000;26:12–20.
156. O'Sullivan AJ, Martin A, Brown MA. J Clin Endocrinol Metab 2001;86:4951–4956.
157. Benayahu D, Shur I, Ben-Eliyahu S. J Cell Biochem 2000;79:407–415.
158. Heine PA, Taylor JA, Iwamoto GA, et al. Proc Natl Acad Sci USA 2000;97:12,729–12,734.
159. Frisch RE, Canick JA, Tulshinsky D. J Clin Endocrinol Metab 1980;51:394–396.
160. Jones ME, Thorburn AW, Britt KL, et al. Proc Natl Acad Sci USA 2000;97:12,735–12,740.
161. Compston JE. Physiol Rev 2001;81:419–447.
162. Okazaki R, Inoue D, Shibata M, et al. Endocrinology 2002;143:2349–2356.
163. Heim M, Frank O, Kampmann G, et al. Endocrinology 2004;145:848–859.
164. Snel YE, Brummer RJ, Doerga ME, et al. Am J Clin Nutr 1995;61:1290–1294.
165. Bengtsson BA, Johannsson G, Shalet SM, et al. J Clin Endocrinol Metab 2000;85:933–942.
166. Burkhardt R, Kettner G, Bohm W, et al. Bone 1987;8:157–164.
167. Corpas E, Harman SM, Blackman MR. Endocr Rev 1993;14:20–39.
168. Charlton HM, Clark RG, Robinson IC. et al. J Endocrinol 1988;119:51–58.
169. Gevers EF, Loveridge N, Endocrinology 2002;143:4065–4073.
170. Yaw A-D, Tapiador CD, Evans JF, et al. Am J Physiol Endocrinol Metab 2003;284:E566–E573.
171. Davidson MB. Endocr Rev 1987;8:115–131.
172. Vikman K, Carlsson B, Billig H, et al. Endocrinology 1991;129:1155–1161.
173. Yang S, Bjorntorp P, Liu X, et al. Obes Res 1996;4:471–478.
174. Ridderstrale M, Groop L. Mol Cell Endocrinol 2001;183:49–54.
175. Hansen LH, Madsen B, Teisner B, et al. Mol Endocrinol 1998;12:1140–1149.
176. Ohlsson C, Bengtsson BA, Isaksson OG, et al. Endocr Rev 1998;19:55–79.
177. de Boer H, Blok GJ, Van der Veen EA. Endocr Rev 1995;16:63–86.
178. Johansson AG, Engstrom BE, Ljunghall S, et al. J Clin Endocrinol Metab 1999; 84:2002–2007.
179. Barnard R, Ng KW, Martin TJ, et al. Endocrinology 1991;128:1459–1464.
180. Kassem M, Blum W, Ristelli J, et al. Calcif Tissue Int 1993;52:222–226.
181. Lincoln D, Sinowatz F, Gabius S, et al. Anat Histol Embryol 1997;26:11–28.
182. Kassem M, Mosekilde L, Eriksen EF. Growth Regul 1994;4:131–135.
183. Mundy GR. Bone 1991;12:S1–S6.
184. Abu EO, Bord S, Horner A, et al. Bone 1997;21:137–142.
185. Dace A, Sarkissian G, Schneider L, et al. Endocrinology 1999;140:2983–2990.
186. Kindblom JM, Gevers EF, Skrtic SM, et al. Bone 2005;36:607–616.
187. Varga F, Rumpler M, Luegmayr E, et al. Calcif Tissue Int 1997;61:404–411.
188. Klaushofer K, Varga F, Glantschnig H, et al. J Nutr 1995;125:1996S–2003S.
189. Darimont C, Gaillard D, Ailhaud G, et al. Mol Cell Endocrinol 1993;98:67–73.
190. Flores-Delgado G, Marsch-Moreno M, Kuri-Harcuch W. Mol Cell Biochem 1987;76:35–43.
191. Chan GK, Deckelbaum RA, Bolivar I, et al. Endocrinology 2001;142:4900–4909.
192. Chan GK, Miao D, Deckelbaum R, et al. Endocrinology 2003;144:5511–5520.
193. McAveney KM, Gimble JM, Yu-Lee L. Endocrinology 1996;137:5723–5726.
194. Nuttall ME, Gimble JM. Curr Opin Pharmacol 2004;4:290–294.
195. Weiss LP. In: Handin RI, Lux SE, Stossel TPJB, eds. *Blood. Principles & Practice of Hematology.* Lippincott, Philadelphia: 1995, pp. 155–169.

196. Nishikawa M, Ozawa K, Tojo A, et al. Blood 1993;81:1184–1192.
197. Corre J, Planat-Bénard V, Corberand JX, et al. Br J Haematol 2004;127:344–347.
198. Moreau I, Duvert V, Caux C, et al. Blood 1993;82:2396–2405.
199. Taichman RS. Blood 2005;105:2631–2639.
200. Laharrague P, Fontanilles AM, Tkaczuk J, et al. Eur Cytokine Netw 2000;11:634–639.
201. McKinstry WJ, Li CL, Rasko JEJ, et al. Blood 1997;89:65–71.
202. Liu F, Poursine-Laurent J, Wu HY, et al. Blood 1997;90:2583–2590.
203. Treon SP, Anderson KC. Curr Opin Hematol 1998;5:42–48.
204. Wilson CA, Bekele G, Nicolson M, et al. Br J Haematol 1997;99:447–451.
205. Togo M, Tsukamoto K, Satoh H, et al. Blood 1999;93:4444–4445.
206. Hirose H, Saito I, Kawai T, et al. Clin Sci 1998;94:633–636.
207. Laharrague P, Corberand J, Pénicaud L, et al. Haematologica 2000;85:993–994.
208. Flier JS. Nat Med 1998;4:1124–1125.
209. Lord GM, Matarese G, Howard LK, et al. Nature 1998;394:897–901.
210. Fantuzzi G. J Allergy Clin Immunol 2005;115:911–919.

IV

Weight Gain and Weight Loss

14 The Epidemiology of Obesity

Carol A. Braunschweig

Abstract

It is estimated that globally more than 1 billion adults are overweight and at least 300 million are obese. This epidemic evolved over many decades in the industrialized countries; however, today it is occurring in a little as 10 years in underdeveloped countries. In the United States, the prevalence of overweight increased significantly from 56% in 1994 to 65.2% in 1999–2000 with individuals at the higher end of the body mass index getting larger at a faster rate than the remaining population in both men and women and in all race/ethnic groups. Women in all age groups have higher obesity prevalence than men. Alarmingly, this epidemic is no longer restricted to adults. Childhood obesity has increased more than threefold in the United States in the past three decades with the majority of the increase occurring within the past 10 years. Initially, obesity in children was primarily isolated to those from lower economic households; however, recently this association has been weakening as greater numbers of middle and upper income children have become obese. Obesity is associated with many diseases and tremendously increases health care costs. Although there is consensus that this epidemic is increasing throughout the developed world, the causes for the surge remain unknown and pose challenges for clinicians, researchers and policy makers.

Key Words: Overweight; obesity; NHANES; surveys; trends; BMI.

1. INTRODUCTION

The fundamental basis for studying the epidemiology of obesity centers on the hypothesis that there are causal and preventative factors that can be identified through systematic investigation of different populations, or subgroups within a population, in different places or times. Epidemiological studies allow the quantification of the magnitude of the exposure–disease relationship in free living human populations and thus provides the possibility of altering this risk through individual or population-based interventions.

This chapter describes the current distribution patterns for adult and childhood obesity worldwide, with specific emphasis on populations in the United States. The modifying influences of gender, race/ethnicity, age, and socioeconomic status on obesity rates are discussed. A final section is devoted to central adiposity's role in obesity-related diseases and a comparison of obesity rates using waist circumference standards rather than the traditional measure of body mass index (BMI). Knowledge of current obesity distributions and the secular changes that have occurred better position clinicians to select the most appropriate populations for future research investigations and design interventions and governmental policies and programs tailored to meet the unique requirements of those individuals and populations with the greatest needs.

From: *Nutrition and Health: Adipose Tissue and Adipokines in Health and Disease*
Edited by: G. Fantuzzi and T. Mazzone © Humana Press Inc., Totowa, NJ

2. DEFINITION

Obesity is defined as an unhealthy excess of body fat that increases the risk of morbidity and premature mortality. Unfortunately, accurate measures of actual fat mass require sophisticated technologies that are not typically available in clinical settings. The BMI (weight in kg/height in m^2) has become the method of obesity classification used by the World Health Organization (WHO) and the National Institutes of Health (NIH) *(1)*. This index provides a simple measure of the relation between height and weight that correlates with percentage body fat in young and middle-aged adults *(2)*. Increased morbidity and mortality from hypertension, stroke, coronary artery disease, type 2 diabetes, sleep apnea, and some cancers increases as BMI rises above 25 kg/m^2, and a sharp increase occurs when BMI exceeds 30 kg/m^2 *(1)* (Fig. 1).

The obesity classification categories based on BMI are similar for men and women and are provided in Table 1.

3. WORLDWIDE EPIDEMIC

It is estimated that, globally, more than 1 billion adults are overweight and at least 300 million are obese *(3,4)*. Obesity rates range from 5% in rural China, Japan, and central Africa to over 75% in urban Samoa. In the nonindustrialized countries obesity generally follows an economic gradient, with the least economically developed having the least obesity (approx 1.8%). As economics expand, obesity prevalence increases from 4.8% in "developing" countries, to 17.1% in "economy in transition," to 20.4% in "developed market economy" countries *(4)*.

Improvement of a country's economy results in shifts from high fertility and mortality rates with high prevalence for malnutrition, poor sanitation, and infectious diseases to more urbanization and chronic diseases associated with urban–industrial lifestyles, including obesity, type 2 diabetes, and cardiovascular diseases *(5)*. Figure 2 illustrates the differences in the prevalence and rate of change for overweight and obesity for various countries at different stages of economic development.

As illustrated in Fig. 2B, the US prevalence for overweight and obesity is increasing by approx 0.5% annually, whereas rates in Mexico, Brazil, and Morocco are two to five times higher (e.g., Mexico 2.4%/yr, Brazil 1%/yr). In the developed countries the changes in diet and physical activity that lead to their current levels of obesity occurred over many decades, whereas in the developing countries of Asia, North Africa, the Middle East, Latin America, and in parts of sub-Saharan Africa, these changes are occurring in as little as 10 to 20 years, in economies that are at far lower levels than those of the industrialized countries. These changes have profound implications for their future health care needs *(6)*.

4. OBESITY EPIDEMIC IN THE UNITED STATES

Prevalence rates for overweight and obesity in the United States are generated from the National Health and Nutrition Examination Survey (NHANES) program. These cross-sectional surveys have been conducted since 1960 on a large representative sample of the US population and provide national estimates for overweight and obesity in children and adults from 2 to 74 yr of age. The National Health Examination Survey (NHES I,

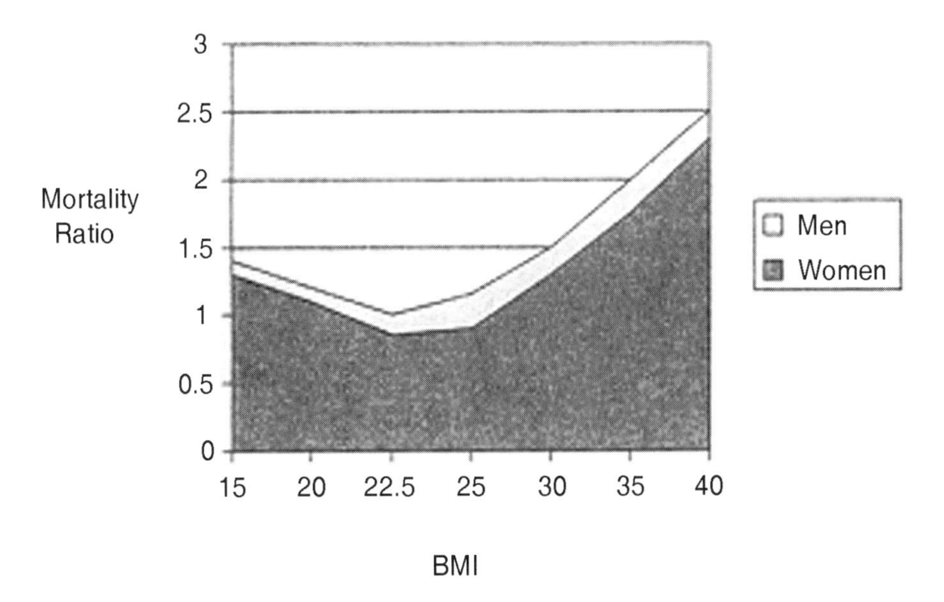

Fig. 1. J-shaped curve of all-cause mortality compared with body mass index. From ref. *1*.

Table 1
Classification of Overweight and Obesity by BMI

	Obesity class	*BMI (kg/m²)*
Underweight		< 18.5
Normal		18.5–24.9
Overweight		25.0–29.9
Obesity	I	30.0–34.9
	II	35.0–39.9
	III	≥ 40

1960–62) was followed by three NHANES surveys (NHANES I, 1971–74; NHANES II, 1976–80; NHANES II,I 1988–94). In 1999 NHANES became a continuous survey; the 1999–2000 data were first published in 2002.

Data for these surveys reveal dramatic increases in overweight or obesity between NHANES II and III, which were continued at an even greater rate in 1999–2000. The prevalence of overweight increased significantly, from 56% in NHANES III to 65.2% in 1999–2000. These changes are summarized in Table 2 *(7)*.

When changes in weight are partitioned into overweight (BMI 25–29.9) and obese (BMI ≥ 30), it can be see that the prevalence for *overweight* increased very little (<2%); however, the prevalence for obesity doubled, from 15.1 to 31.1%, between NHANES III and 1999–2002. Within the obesity category, the prevalence of *self-reported* BMI greater than 40 increased from 0.5 to 2% of the US population and those with a BMI greater than 50 quadrupled, from 0.5 to 2% between 1986 and 2000 *(8)*. Individuals at the higher end of the BMI spectrum appear to be getting larger at a faster rate than the remaining population.

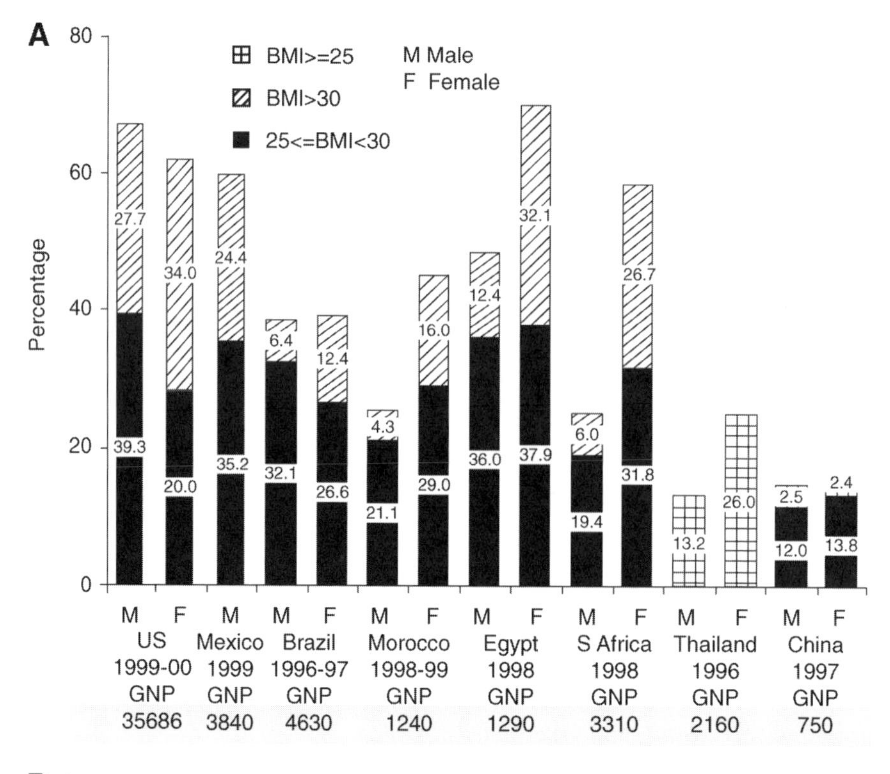

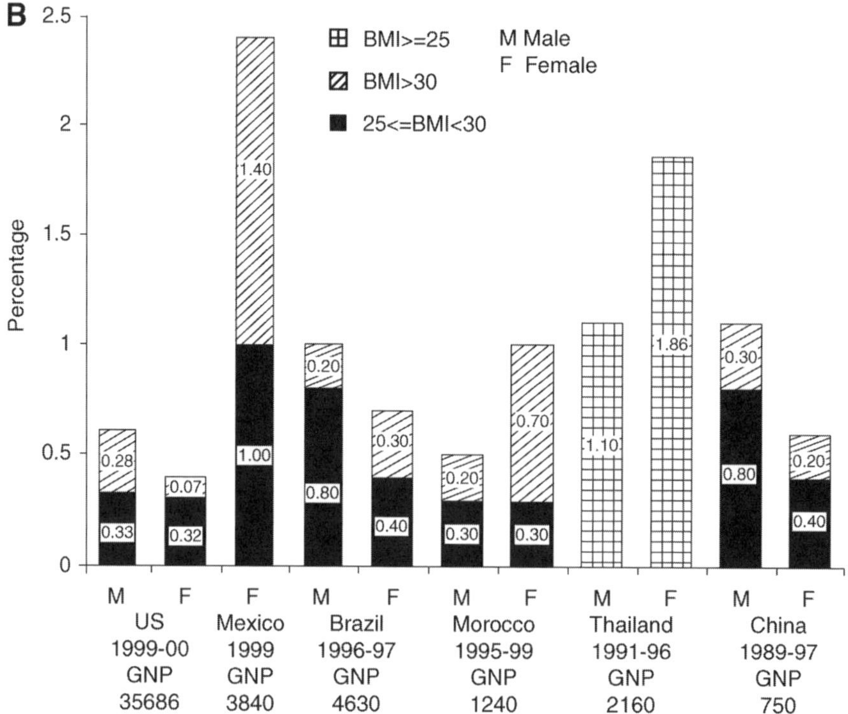

Fig. 2. Obesity patterns and trends across the world, adults aged 20 yr and older. **(A)** Prevalence rates; **(B)** obesity trends (the annual percentage point increase in prevalence). With permission from ref. *6*.

Table 2
Prevalence and Trends of Overweight and Obesity Among Adults in the United States, 1960–2002

Survey year	*Overweight or obese % (SE)*	*Overweight but not obese % (SE)*	*Obese % (SE)*
1960–1962	44.8 (1.0)	31.5 (0.5)	13.3 (0.6)
1971–1974	47.7 (0.7)	33.1 (0.6)	14.6 (0.5)
1976–1980	47.4 (0.8)	32.3 (0.6)	15.1 (0.5)
1988–1994	56.0 (0.9)	32.7 (0.6)	23.3 (0.7)
1999–2002	65.2 (0.8)	34.1 (0.8)	31.1 (1.0)

5. GENDER AND OBESITY

Obesity classification and treatment recommendations are the same for men and women (Table 1) *(1)*; however, at any given BMI women throughout the world are more likely to have greater amounts of adipose tissue than men *(9)* and have higher rates of obesity owing to biological differences *(10)*. In the United States, proportionate increases in obesity prevalence have occurred for both men and women 20 to 74 yr of age since 1960, and women have had higher prevalence rates in all age groups in each of the national surveys (Table 3).

Overall, in both men and women, there has been a general shift of the top half of the BMI distribution toward higher BMI levels, with the heaviest subgroup of the population becoming much heavier in NHANES III than II *(11)*. For 20- to 49-yr-old men changes in the lower portion of the BMI distribution was less than 1 BMI unit, whereas there was a fairly large shift at the upper end among 50- to 74-yr-old men. For women a gradual increase across the entire BMI distribution occurred. Collectively these findings suggest that the entire spectrum is changing, with the most marked changes occurring at the upper end of the distribution.

6. ETHNICITY/RACE AND OBESITY

Obesity and extreme obesity prevalence increased in all sex and racial/ethnic groups between NHNAES III and 1999–2000. Differences in obesity rates by ethnicity/race were found to vary by gender *(12)*. Specifically, no difference in obesity rates were found among African American, Hispanic, or Caucasian men; however, among women, African Americans had a higher prevalence of overweight and obesity (77 and 50%, respectively) than Caucasians (57 and 30%, respectively), and Mexican American women had a prevalence that was intermediate between these two (72% overweight and 40% obese). Significant increases in extreme obesity occurred between NHANES III and 1999–2000 in all men and women and for non-Hispanic black women, with similar trends in all other groups (Table 4).

Interestingly, although there has been a significant increase in the prevalence of obesity in the past 15 yr, the mean BMI measurements have changed much less. Between NHANES III and 1999–2002 the mean BMI increased 1.7 and 2.5 BMI units in non-Hispanic white and black women respectively, representing a 6.6% and 8.8% increase in rate of change, whereas overall obesity prevalence for women during that same time period increased 7.8 and 10.8% respectively, representing a 34% and 38.2% increase in

Table 3
Prevalence of Obesity in Men and Women in the United States, 1960–2002[a]

	Men				*Women*			
Survey yr	*20–74 yr (%)*	*20–39 yr (%)*	*40–59 yr (%)*	*60–74 yr (%)*	*20–74 yr (%)*	*20–39 yr (%)*	*40–59 yr (%)*	*60–74 yr (%)*
1960–1962	10.7	9.8	12.6	8.4	15.8	9.3	18.5	26.2
1971–1974	12.1	10.2	14.7	10.5	16.6	11.2	19.7	23.4
1976–1980	12.7	9.8	15.4	13.5	17.0	12.3	20.4	21.3
1988–1994	20.6	14.9	25.4	23.8	25.0	20.6	30.4	28.6
1999–2002	27.7	23.7	28.8	35.8	34.0	28.4	37.8	39.6

[a]Adapted from ref. *7*.

Table 4
Prevalence of Obesity (BMI > 30) and Extreme Obesity (BMI > 40) in Different Racial Groups in the United States, 1988–2002[a]

	Prevalence of obesity (BMI > 30)							
	Men				*Women*			
Survey yr	*All*	*Non-Hispanic white (%)*	*Non-Hispanic black (%)*	*Mexican American (%)*	*All (%)*	*Non-Hispanic white (%)*	*Non-Hispanic black (%)*	*Mexican American (%)*
1988–1994	20.2	20.3	21.1	23.9	25.4	22.9	38.2	25.3
1999–2002	27.6	28.2	27.9	27.3	33.2	30.7	49.0	38.4
	Prevalence of extreme obesity (BMI > 40)							
	Men				*Women*			
1988–1994	1.7	1.8	2.4	1.1	4.0	3.4	7.9	4.8
1999–2002	3.3	3.3	3.4	2.9	6.4	5.5	13.5	5.7

[a]Adapted from refs. *12,13*.

rate of change, respectively (Tables 4 and 5). These findings further support the observation that overall shifts in obesity prevalence are occurring among the heavier end of the BMI spectrum.

7. OBESITY AND SOCIOECONOMIC STATUS

Obesity has frequently been reported in the past to occur disproportionately among economically disadvantaged populations in the United States *(15–17)*; however, more recently, substantial variation in obesity rates between income and ethnic groups over time have emerged *(14,18)*. Using level of education attainment to categorize socioeconomic status (SES), Zhang and Wang found a gradual decrease in the association between SES (determined according to level of education) and obesity between 1971 and 2000, particularly among women, and a disproportionate increase in the obesity prevalence in the high-SES groups *(18)* (Fig. 3).

Table 5
BMI in 18- to 64-Yr-Old Non-Hispanic White and Black Men and Women Between NHANES I (1960–1962) and 1999–2002[a]

	Women		*Men*	
	Non-Hispanic white mean BMI	*Non-Hispanic black mean BMI*	*Non-Hispanic white mean BMI*	*Non-Hispanic black mean BMI*
NHANES I	24.3	26.9	25.4	25.4
NHANES II	24.5	26.9	25.3	25.3
NHANES III	25.9	28.5	28.5	26.4
1999-2002	27.6	31.0	27.9	27.3

[a]Adapted from ref. *14*.

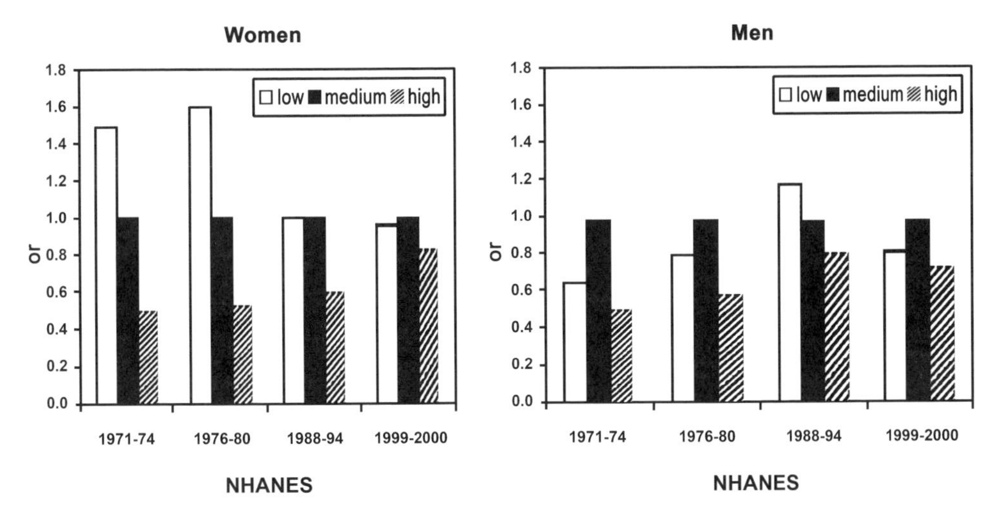

Fig. 3. Secular trends in the association between SES and obesity among US adults, 1971–2000. With permission from ref. *18*.

Chang and Lauderdale *(14)* compared obesity rates in non-Hispanic blacks, whites, and Mexican Americans by income using the poverty–income ratio (PIR) (grouped into four categories, 1 low, 4 high) between NHANES I and the 1999–2000 NHANES. They found that obesity increased over this interval at all levels of income in all ethnic/ racial groups. In African American and Caucasian women the lowest obesity prevalence occurred in the highest income category; however, among African American and Mexican American men, obesity prevalence was highest in the highest income category (28.3 and 36.2%, respectively) and lowest in the lowest two income categories (category 1 in African Americans, 22.8%; and category 2 for Mexican Americans, 19.3%). Additionally, the *rate of increase* in obesity was greater among the more affluent than the poor. Among African American women the rate of increase for obesity was lowest among the lowest PIR (14.5%) and highest in the middle-income category (27%); among non-Hispanic white women the lowest increase (13.0%) was found in the highest income category (PIR category 4) and the highest increase was found in PIR category 2. The significant "catch-up" in obesity prevalence that has occurred in high-SES groups implies that obesity is a much more equal-opportunity condition than previously suspected. These findings suggest that social–environmental factors that affect all SES groups may be the primary contributors to the increase in obesity that has occurred over the past 30 yr.

8. CHILDHOOD OBESITY

There has been a threefold increase in childhood obesity in the United States in the past three decades *(12)*. Coinciding with this increase has been an increase in prevalence for type 2 diabetes, hypertension, gallbladder disease, hyperlipidemia, orthopedic complications, sleep apnea, and nonalcoholic steatohepatitis in children. Obese children are predisposed to adult obesity and have increased risks for adult obesity-related diseases *(19)*. Approximately 30% of obese adults became obese during childhood, and 80% of obese adolescents become obese adults *(20)*.

No single cut-point for obesity classification can be used for children because between birth and approx 20 yr of age, height and weight—and thus, BMI—vary by age. To address this children (age 2–11) and adolescents (age 12–19) are classified as at risk of overweight or overweight if they are at or above the 85th or the 95th percentile, respectively, for their age and gender using the 2000 Centers for Disease Control and Prevention Growth Charts *(21)*. The prevalence and trends between 1971 and 2002 of overweight in 6- to 11-yr-olds and 12- to 19-yr-olds are presented in Tables 6 and 7.

The majority of the increase in overweight prevalence among children has occurred within the past 10 yr, and the most recent data indicate that this trend continues. In 1999– 2000, 29.9% of 6- to 19-yr-old children were at risk of overweight and 15% were overweight; in 2000–2001 these levels increased to 30.5% at risk of overweight and 16.5% overweight *(13)*.

Gender and racial–ethnic differences are presented in Table 7. Distinct from adults, where women at all ages have higher rates for obesity, very little difference in overweight prevalence exists between boys and girls. Ethnic differences for overweight do exist in children. African American and Mexican American children are significantly heavier than Caucasian boys and girls. Similar to observations in adults, a pronounced upward shift (at least 2 BMI units) in the upper end of the BMI distributions in all age/gender groups occurred during this time period; however, very little change occurred in the lower end of the BMI distributions, and median BMI remained virtually unchanged. The heaviest groups of children are much heavier in NHANES III than they were in previous surveys *(11)*.

The secular trends of the impact of SES on obesity in children have received very limited study. In Canada the prevalence of overweight has been reported to vary inversely by SES *(22)*; however, the data used in their analysis spanned 1981 to 1996 and thus do not reflect the most recent decade. A recent report using the 1971–2002 NHANES surveys examined trends in obesity prevalence and SES status categorized by level of parental education attained (less than high school, high school, college, or higher) (Table 8) *(23)*. The report found an overall weakening of association between obesity risk and SES category that varied by racial/ethnic groups.

Table 6
Prevalence of Overweight (≥ 95th Percentile) in Children and Adolescents in the United States, 1960–2002[a]

Survey yr	*Children 6–11 (%)*	*Adolescents 12–19 (%)*
1971–1974	4.0	6.1
1976–1980	6.5	5.0
1988–1994	11.3	10.5
1999–2002	15.8	16.1

[a]Adapted from ref. *7*.

Table 7
Prevalence of Overweight (≥95th Percentile) by Sex, Age, and Racial/Ethnic Group in the United States, 1999–2002[a]

Age	*All*	*Non-Hispanic white*	*Non-Hispanic black*	*Mexican American*
Boys				
6–19	16.8	14.3	17.9	25.5
2–5	9.9	8.2	8.0	14.1
6–11	16.9	14.0	17.0	26.5
12–19	16.7	14.6	18.7	24.7
Girls				
6–19	15.1	12.9	23.2	18.5
2–5	10.7	9.1	9.6	12.2
6–11	14.7	13.1	22.8	17.1
12–19	15.4	12.7	23.6	19.9

[a]Adapted from ref. *13*.

In NHANES I the overweight prevalence was 7.1, 6.3, and 3.8% in low-, medium-, and high-SES non-Hispanic white girls, compared with 8.2, 14.8, and 1.9% in non-Hispanic black girls; in 1999–2002 these figures became 17.9, 10.6, and 10.6% vs 24.5, 18.7, and 38.0%, respectively. Although the prevalence for overweight remains higher among African American than among Caucasian children, the highest rates of increase in overweight were found in the highest SES groups for both racial/ethnic groups. The disproportionate protection that once was conferred by being within the higher SES category is eroding among all racial/ethnic groups. Although the cause for this weakening secular trend remains uncertain, these findings should be considered when predicting future trends in obesity prevalence and guide the development of programs and governmental policies designed for obesity prevention and management.

9. ADIPOSE TISSUE DISTRIBUTION

The previous sections have presented obesity data based on BMI classifications. The primary assumption of BMI is that it is an independent index of body fat—i.e., it

Table 8
Odds Ratios for Overweight in 2- to 19-Yr-Old Children in Low and High SES Compared With Middle SES Categories

	NHANES I *OR (95% CI)*	*NHANES II* *OR (95% CI)*	*NHANES III* *OR (95% CI)*	*1999–2000* *OR (95% CI)*
Low SES	0.82 (0.48–1.41)	1.15 (0.79–1.69)	1.24 (0.73–2.09)	1.04 (0.82–1.33)
High SES	0.66 (0.43–1.0)	0.58 (0.34–1.0)	0.42 (0.23–0.76)	0.99 (0.68–1.43)

assumes that after adjusting for body weight-for-stature, all individuals with the same BMI will have the same fatness, regardless of their age, gender, ethnicity, or disease history. Many investigations have demonstrated that this assumption is not correct for various populations. For example, it is widely recognized that body fatness increases with age, is greater in females than in males, and differs among ethnic groups (Asian Indians and Chinese have greater percentage of fat at any BMI). Thus, categorization of an individual's overweight or obesity status should not be done without consideration of other factors that could have influenced their BMI measurement.

Centrally located adipose tissue as assessed by waist circumference (WC) was first associated with heightened risk for obesity-related disease risks approx 50 yr ago *(24)*. Since that time numerous studies have confirmed the effect of excessive central fat distribution on increased risks for insulin resistance *(25)*, type 2 diabetes *(26)*, cardiovascular disease *(27,28)*, hypertension *(29)*, and inflammation *(30)*.

Women tend to gain adipose tissue peripherally, and men centrally. Because of the more peripheral distribution in women, they tend to have lower disease risks and less disease for a given weight than men. Visceral fat accumulation is thought to increase risks more than subcutaneous fat. Differentiation between subcutaneous and visceral fat depots requires analysis with computerized tomography or magnetic resonance imaging. These are expensive tests and are not readily available in many clinical settings.

The NIH and WHO recommend waist circumference (measured midway between the lower rib and the iliac crest) cut-points for central obesity classification at ≥88 cm for women and ≥102 cm for men *(1,3)*. Age-, gender-, and race/ethnicity-specific median WC for men and women in NHANES I, II, and III were reported by Okosun et al. *(31)*. Changes in WC distributions that occurred between 1960 and 1962 and 1999 and 2000 for men and women are illustrated in Fig. 4.

Waist circumference shifted to the right (higher values) in both men and women between 1960 and 2000, although women tended to have greater shifts then men. This trend parallels the changes that occurred for BMI over this time period. No specific differences in WC were found between African-American men or women in NHANES I (1960–62), followed by an increasing gradient in both genders (Fig. 5). In subsequent surveys Caucasian men had significantly larger WC than African-American men, whereas African-American women had larger WC than Caucasian women.

The prevalence of abdominal obesity stratified by BMI across 1960–2000 is provided in Table 9. By 1999–2000 more than 90% of all obese individuals were abdominally obese, and more than two-thirds of overweight women also had central obesity. At BMIs greater than 35 the additional risk conferred by measuring WC is lost, as virtually all of

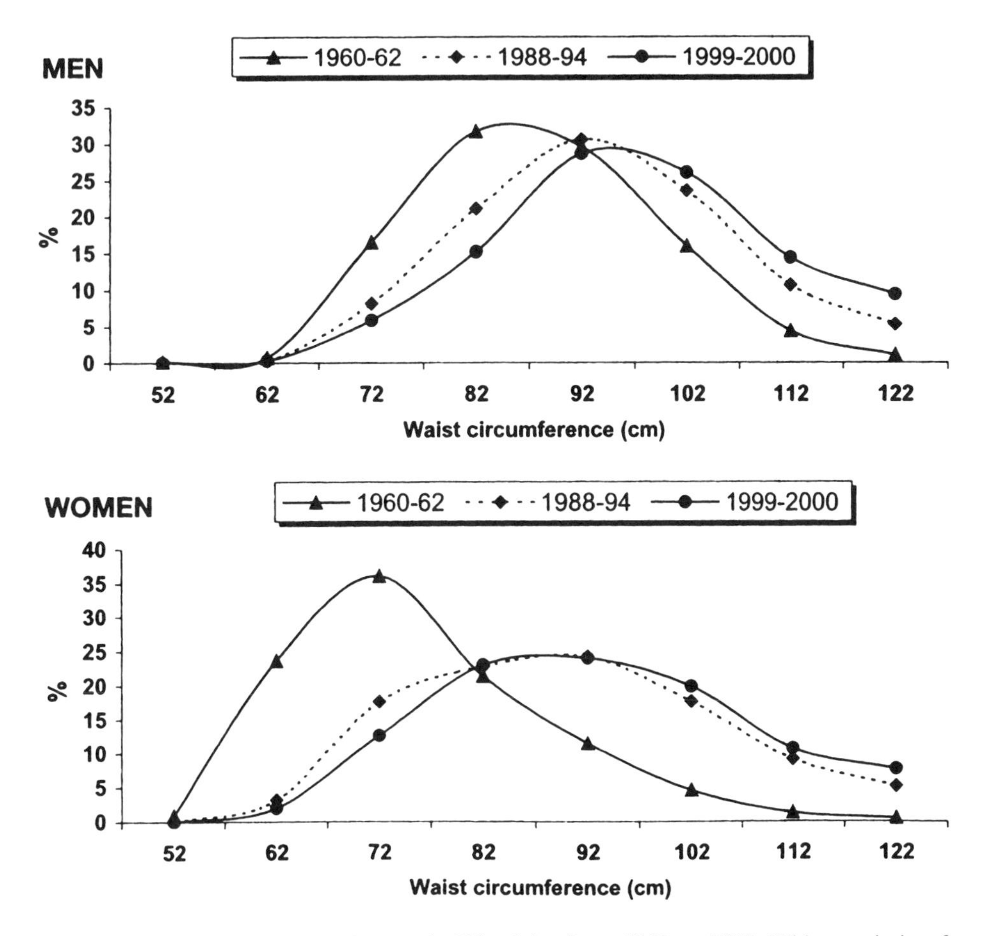

Fig. 4. Distribution of waist circumference in US adults from 1960 to 2000. With permission from ref. *32*.

these individuals exceed the WC cut-points. The overall age-adjusted prevalences of abdominal obesity in men were 12.7, 29, and 38.3% in 1960–62, 1988–94, and 1999–2000 respectively; among women, these rates were 19.4, 38.8, and 59.9%, respectively. These secular trends in abdominal obesity are similar in direction but greater in magnitude than those for generalized obesity (men, 13.1, 23.4, and 31.0%; women, 15.6, 25.8, and 34.3%). The NIH clinical recommendations for the need to institute weight-loss therapy have been divided into two levels. Action level I (men WC ≤ 94 cm, women ≤ 80 cm) indicates the need for lifestyle modifications and level II (men WC > 102 cm, women > 88 cm) indicates that professional help is needed.

Overall there has been a shift in abdominal obesity across the entire population, including those with normal BMI, since 1960, and abdominal obesity rates have increased faster than generalized obesity. The higher health risk for diabetes and cardiovascular disease from excessive central fat, independent of generalized obesity, suggests that public health strategies that address this issue are urgently needed. Studies focused on the identification of factors that are causing increases in the entire population are needed to guide interventions focused on the entire population rather than solely on the heaviest individuals.

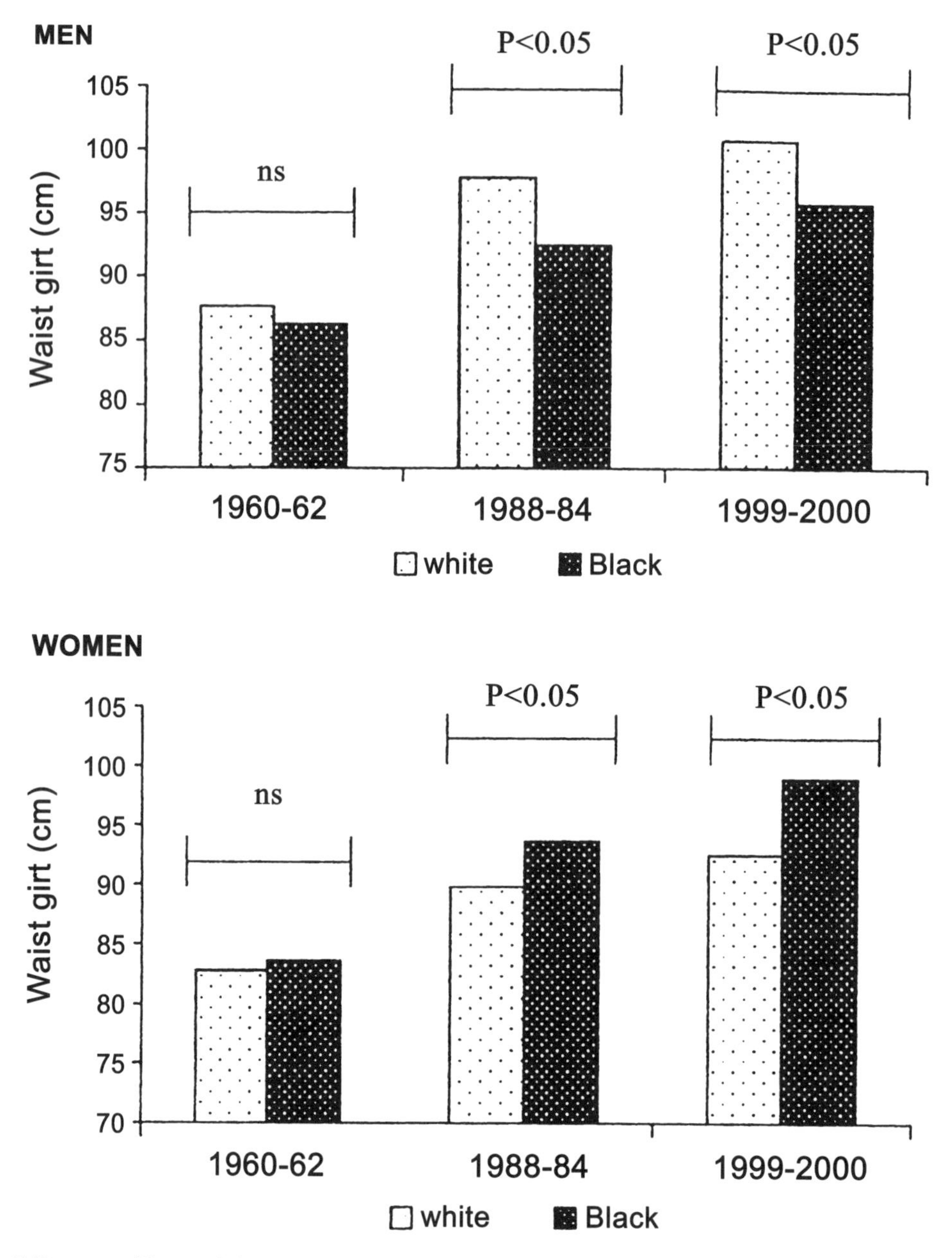

Fig. 5. Race-specific trends in waist circumference in adults from 1960 to 2000. With permission from ref. *32*.

10. CONCLUSIONS

Obesity becomes a disease when excess body fat reaches levels associated with increased morbidity and mortality. In 2000 the WHO designated obesity as the major unmet public health problem worldwide *(3)*. It is probable that the increase in obesity that has occurred is the result of the "modern lifestyle" that encourages increased caloric consumption and discourages physical activity. All recent studies suggest that

Table 9
Prevalence of Abdominal Obesity Stratified by Gender and BMI Across NHANES Studies[a]

	1960–1962	*1988–1994*	*1999–2000*	*p-value*
BMI category men	Prev (%)	Prev (%)	Prev (%)	
<25	0.3	0.8	1.5	<0.001
25–29.0	12.1	25.4	29.6	<0.001
>30	65.4	88.3	93.5	<0.001
BMI category women				
<25	0.8	11.9	11.5	<0.001
25–29.9	15.9	70.3	69.0	<0.001
>30	75.4	98.1	98.7	<0.001

[a]Adapted from ref. *32*.

this epidemic continues to expand into younger, more diverse populations that were previously thought to be somewhat impervious to its development. A complex web of problems can be expected to emerge when this epidemic is superimposed on the aging US population, ultimately predisposing large segments of society to reduced independence, increased health care costs, prolonged medical care, and ultimately shorter, less productive lives. Partnerships between public and private sectors are needed to develop both population-based environmental approaches and individual strategies for successful prevention and treatment.

REFERENCES

1. National Institutes of Health/National Heart, Lung and Blood Institute. Obes Res 1998:6 (Suppl 2):51S–209S.
2. Willett W, Dietz W, Colditz G. N Engl J Med 1999;341:427–434.
3. World Health Organization. *Obesity: Preventing and Managing the Global Epidemic*. WHO Technical Report, Series 894, World Health Organization, Geneva: 2000.
4. World Health Organization. *Global Strategy on Diet, Physical Activity and Health*, 2003. http://www.who.int/dietphysicalactivity/publications/facts/obesity/en.
5. Olshansky S, Ault A. Milbank Mem Fund Q 1986;64:355–391.
6. Popkin B, Gorgon-Larsen P. Int J Obes 2004;28:52–59.
7. Flegal K. Physiol Behav 2005;86:599–602.
8. Sturm R. Arch Intern Med 2003;163:21:146–148.
9. Norgan N. Int J Obes Relat Metab Disord 1997;21:738–746.
10. James W, Jackson-Leach R, Mhurchu N. In: Ezzati M, Lopez A, Rodgers A, et al., eds. *Comparative Quantification of Health Risks: Global and Regional Burden of Disease Attributable to Selected Major Risk Factors,* Vol. 1. WHO, Geneva: 2004, pp. 497–596.
11. Flegal K, Troiano R. Int J Obes Relat Metab Disord 2000;24:807–818.
12. Flegal K, Carroll M, Ogden C, et al. JAMA 2002;288:1723–1727.
13. Hedley A, Ogden C, Johnson C, et al. JAMA 2004;291:2847–2850.
14. Chang V, Lauderdale D. Arch Intern Med 2005;165:2122–2128.
15. Paeratakul S, Lovejoy J, Syan D, et al. Int J Obes 2002;26:1205–1210.
16. Adler NE, Boyce T, Chesney MA, et al. AM J Psychol 1994;49:15–24.
17. Drewnowski A. Am J Prev Med 2004;27:154–162.
18. Zhang Q, Wang Y. Obes Res 2004;12:1622–1632.
19. Guo S, Huang C, Maynard L, et al. Int J Obes Relat Metab Disord 2000;24:1628–1635.

20. Pi Sunyer F. Obesity. In: Shils M, Olson J, Shike M, eds. *Modern Nutrition in Health and Disease,* 9th ed. Lippincott Williams & Wilkins, Baltimore: 1999, pp. 1395–1418.
21. Kuczmarski R, Ogden C, Guo S, et al. Vital Health Stat 11, 2002;246:1–190.
22. Willms J, Tremblay M, Katzmarzyk P. Obes Res, 2003;11:668–673.
23. Wang Y, Zhang Q. Obes Res 2005;13:A7.
24. Vague J. Am J Clin Nutr 1956;4:20–34.
25. Bonora E. Int J Obes 2000;20:S32–S35.
26. Lundgren H, Bengtsson C, Blohme G, et al. Int J Obes 1989;13:413–423.
27. Krotkiewski M, Bjorntorp P, Sjostrom L, et al. J Clin Invest 1983;72:1150–1162.
28. Kissenbah A, Vydelingum N, Murray R, et al. J Clin Endocrinol Metab 1982;54:254–260.
29. Van Lenthe F, Van Mechelen W, Kemper H, et al. Am J Epidemiol 1998;147:686–692.
30. Lemieux I, Pascot A, Prud'homme D, et al. Arterioscler Thromb Vasc Biol 2001;21:881–883.
31. Okosun I, Chandra K, Boev A, et al. Prev Med 2004;39:197–206.

15 Environmental Aspects of Obesity

Lisa Diewald, Meredith S. Dolan, and Myles S. Faith

Abstract

The dramatic increase in obesity in both children and adults over the past two to three decades represents a national and global concern because obesity can significantly increase the risk of a myriad of chronic health conditions. Macroenvironmental influences on food consumption, including changes in food supply trends, fast food consumption, increases in portion size and food availability as well as changes in physical activity patterns in schools and workplaces have been implicated. Societal efforts undertaken to stem the surge of obesity in children and adults must address this "toxic environment," which encourages food over-consumption and discourages physical activity. Some community-based efforts aimed at reversing this trend are meeting with promising results; however, additional research, policy changes, and a committed partnership between industry, schools, individuals and the media may be necessary to achieve lasting impact.

Key Words: Macroenvironmental; obesity; toxic environment; portion size; energy intake; physical activity; television viewing.

1. INTRODUCTION

The dramatic rise in overweight and obesity in the United States and worldwide has prompted global concern, as the prevalence statistics are staggering. Since 1998, the prevalence of overweight in the United States has increased by 9% and 5% in adults and children, respectively *(1–5)*. The rate of obesity in childhood has doubled over the past 20 to 30 yr. Because obesity tracks into adulthood, especially in adolescence, this concerning trend in childhood overweight is predictive of future health concerns *(6)*.

A number of factors, both genetic and environmental, play a role in the development of obesity. However, this chapter will focus on several key macroenvironmental factors that have been identified as probable causes for the recent dramatic increases in obesity prevalence *(7)*. For the purposes of this review, "macroenvironmental" influences are defined as community-level aspects of the environment that reach beyond single individuals or families, encompassing groups of individuals or families. These factors reflect the broader social environment, beyond clinical or treatment settings that reach a more select group of individuals.

"Macro-level" environmental influences addressed in this chapter include increased food portion sizes, increased availability of energy-dense and highly palatable foods,

From: *Nutrition and Health: Adipose Tissue and Adipokines in Health and Disease*
Edited by: G. Fantuzzi and T. Mazzone © Humana Press Inc., Totowa, NJ

greater public demand for convenience foods, an increased number of meals eaten away from home, limited access to healthy foods, decreased daily lifestyle activities, and shifts in the structure of and participation in school-based physical education. No single factor is responsible for exacerbating the problem of obesity; more likely, the synergistic effect of these trends has created an environment that discourages physical activity and encourages excess food consumption *(8–10)*.

2. THE TOXIC ENVIRONMENT

Brownell coined the term "toxic environment," referring to a series of social and economic changes that have occurred in the United States and other industrialized nations during the past several decades *(8,11)*. Brownell argues that progressive societal changes set the stage for the rising obesity prevalence, even though strong causal inferences cannot be determined from these observational trends. These changes include increased portion sizes, proliferation of fast-food restaurants, the perceived value of fast-food products, increased availability of energy-dense foods in schools and workplaces, the accelerated use of labor-saving devices, and fewer opportunities for physical activity at schools, parks, and playgrounds (Table 1). A brief discussion of each of these points is provided below.

2.1. Food Supply Trends

There have been changes in the national food supply that correspond to the increasing obesity prevalence in the United States. For example, data from the US Marketing System indicate that per capita calorie availability increased by 15% from 1970 to 1994, from 3300 kcal per capita in 1970 to 3800 kcal per capita in 1994 *(10)*. This trend was coupled with a 22% increase in fats and oils added to the food supply, resulting in a doubling of added fat consumption, from 32 lb per capita in 1909 to 63 lb per capita in 1998 *(12,13)*. The increased dietary fat intake was accompanied by a large increase in snack and confectionery foods, a dramatic 131% rise in soft drink consumption, and a marked decrease in milk intake *(10,11)*. National surveys on snacking habits of children, ages 2 to 18, also determined that, despite the fact that the macronutrient and energy content of snacks did not change, snacking *frequency* increased from 1977 to 1998 across all age groups. Thus, the increased snacking frequency resulted in an increased total snack energy intake *(14)*. The largest relative increase in daily energy intake was attributed to snacks rather than meals *(15)*. In 1977, snacks made up 11.3% of total energy intake, increasing to 17.7% in 1996 *(15)*. Snacks made up a larger segment of the dietary intakes of 2- to 18-yr-olds than in any other age group *(16)*. These food consumption trends may enhance the risk of excess weight gain in susceptible individuals and populations.

2.2. Portion Size

The current environment provides an overabundance of high-fat, high-sugar, energy-dense foods. The global trend toward large portion sizes began at least in the 1980s and continues today in fast-food restaurants and other similar dining establishments around the world, but appears especially pronounced in the United States. Indeed, most foods are available in larger portions compared with portion size in the 1970s *(17,18)*. A report of 63,380 individuals participating in the Nationwide Food Consumption Survey and

Table 1
Environmental Factors Contributing to Toxic Environment and Obesity Risk

Food consumption trends	*Physical activity trends*
↑ Meals away from home	↑ TV viewing
↑ Portion size	↑ Computer use
↑ Vending machines/soft drink access	↓ Access to parks/trails
↑ Fast food restaurants	↓ School physical education time
↑ Food advertising	↑ Labor-saving devices
↓ Healthy school cafeteria options	↓ Leisure-time physical activity

Continuing Survey of Food Intake by Individuals found that the largest portions were consumed at fast-food restaurants, although portion size clearly varied by food type. Between 1977 and 1996, energy intake from a variety of foods including salty snacks, soft drinks, hamburgers, french fries, and Mexican foods increased between 49 and 133 kcal per serving *(16,18)*.

Young and Nestle's study of marketplace foods found that portion sizes of virtually all foods they sampled increased significantly since the 1970s, consistently exceeding USDA standard portion sizes *(17)*.

Two studies linked large portion size with an increase in total energy intake among children *(18–21)*. Between 1989 and 1996, the mean increase in children's calorie intake ranged from 80 to 230 kcal per day *(22,23)*. Large portion sizes, along with an increase in sugar-sweetened beverage consumption, were associated with excess energy intake *(22,23)*. In a study of 3- to 5-yr-old preschoolers' eating patterns, Rolls and colleagues served three different portion sizes at an experimental lunch meal. Results indicated that the older preschool children ate significantly more calories when provided with the larger-size meal compared with the smaller-size meal. This phenomenon did not hold true for the younger children. The authors proposed that younger children may more effectively self-regulate, or compensate, energy intake than older children, who may be more vulnerable to overeating when provided with larger portions *(20)*.

Fisher et al. extended this study by testing the effects of portion-size provision on preschoolers' self-selected portion size and food intake *(24)*. They found that repeated exposure to meals with large portion sizes consistently increased preschoolers' bite size and entrée size at meals. Moreover, after consuming the larger portion sizes, children exhibited a compensatory decrease in the intake of other foods served during the same meal *(24)*. These experimental data demonstrate how large portion sizes can promote positive energy balance, and potentially contribute to overweight in children. Protecting and preserving compensation skills may be of paramount importance in reducing childhood overeating.

In the presence of large portion sizes, adults also tend to overeat *(2,9)*. Rolls and colleagues studied eating behavior of adults by serving four different portions of an entrée item. Results showed that participants consumed 30% more calories when served the largest portion compared with when they were served the smallest portion *(25)*. Similarly, when served portions that were 50% larger than baseline portions, participants in another study consumed 16% more calories (i.e., 328 kcal for women and 522 kcal for men) over baseline caloric intake *(2,26)*.

Portion sizes provided at home and in restaurants have increased significantly, and intake of these foods has also increased over the past two to three decades *(9,18,27)*.

2.3. Sweetened Beverage Intake

There has been a clear trend in recent years toward higher per capita consumption of sugar-sweetened beverages, including fruit drinks, carbonated beverages, and sports drinks. Across all age groups, sweetened beverage intake increased by 135%, with a concomitant 38% decrease in milk consumption *(28–30)*. The largest increases in sweetened beverage intake were noted in young adults (ages 19–39) and children (ages 2–18) *(27,28)*. As a result, added sugar now comprises 20% of the total caloric intake in a child's diet *(22)*. Nearly 75% of all children drink at least one 8-oz serving of a sugar-sweetened beverage daily, and the average child consumes 1.4 servings daily. This corresponds to an average annual sugar intake of 14.05 kg/yr *(22)*. In addition, one-third of teen males consume more than three soft drinks daily, providing an additional 30 teaspoons of added sugar, equivalent to 120 calories *(22,31)*.

The seemingly innocuous consumption of a "healthy beverage," such as 100% fruit juice, also may predispose children to excess caloric intake *(32)*. Because 100% fruit juice is often perceived as "healthy" owing to its high vitamin and mineral content, many parents do not limit their children's juice intake. In a study of more than 10,000 families enrolled in the Missouri Special Supplemental Nutrition Program for Women, Infants, and Children (WIC), at-risk for overweight and overweight preschoolers who consumed one or more servings per day of juice were significantly more likely to be overweight 1 yr later compared with preschoolers who consumed less than one serving per day. Interestingly, no associations were found in children who were below the 85th percentile for body mass index (BMI) at initial assessment *(33)*. That heavier children may be more vulnerable to excess weight gain in response to food or beverages is consistent with the notion of a gene–environment interaction *(34,35)*.

Children generally maintain a consistent level of energy intake *(36,37)* although there is variability in "caloric compensation" ability *(38,39)*. Moreover, beverage intakes appear to be less precisely regulated than solid foods in humans and animals *(40–43)*. In young children, energy imbalances as subtle as 30 kcal/d can promote overweight, which could be easily achieved by juice intake. Because a standard 6- to 8-oz fruit juice box has approx 100 kcal per serving, children consuming multiple boxes each day might ingest as much as about one-fourth of their daily energy requirements just from juice *(44)*. Indeed, overconsumption of sugar-sweetened beverages has been associated with excess weight gain in studies with adults and children *(45,46)*.

2.4. Dining Out Trends/Fast-Food Consumption

There is increasing popularity for meals eaten away from home, especially at fast-food restaurants. Increased frequency of consumption of fast food meals has been associated with increased weight status. The average American's fast-food meal consumption frequency rose from 9.6% in 1977 to 23.5% in 1996 *(15)*. Compared with 1977, when meals eaten at home represented 76.9% of daily caloric intake, Americans now typically consume less than 65% of total energy intake at home *(15)*. In 1955, restaurant meal sales accounted for 25.5% of every food dollar spent, increasing to 46.1% in 1999 *(15)*. Nearly one-half of adults eat out in a restaurant daily, and studies have demonstrated an increase in energy intake when individuals are presented with large portions *(47,48)*.

Among 891 adults enrolled in the "Pound of Prevention Study," increased frequency of fast-food consumption was associated with significantly greater total energy intake, higher percent fat intake, more frequent consumption of hamburgers, french fries, soft drinks, and less consumption of fiber and fruit. Over 3 yr, each additional fast-food meal consumed per week was associated with an excess weight gain of 0.72 kg. In a prospective study of more than 3000 young adults enrolled in the Coronary Artery Risk Development in Young Adults (CARDIA) study, frequency of fast-food restaurant visits at baseline (visits/wk) predicted excess 15-yr weight gain and worsening of insulin resistance in Caucasian and African American respondents *(49)*.

Visiting fast food restaurants may promote obesity by promoting increased consumption of energy dense foods. Prentice and Jebb reviewed the nutritional content of the foods sold at three popular fast-food outlets *(50)*. The average energy densities for the three menus were 1.7-fold greater than the average British diet. Bowman et al. reported that 30.3% of children ate fast food every day, providing them with 187 additional calories per dining visit *(51)*. One study exploring the dining habits of young adults reported that those who dined out twice a week in fast-food restaurants experienced an 86% increase in obesity risk compared with those visiting a fast-food establishment once weekly *(52)*. In many cases, frequent visits to fast-food restaurants were associated with reduced fruit and vegetable intake, as energy-dense foods took the place of fiber-rich fruit and vegetable sources *(53)*.

2.5. Food Advertising

Increased consumer demands for convenient and expanded food selections have paralleled increases in food advertisement funds. The fast food industry spends more than 95% of its advertising budget on television advertisements, capitalizing on the fact that televisions are present in 98% of households *(10,54)*. Foods that tend to be most heavily advertised are those with the highest calorie content, including convenience foods, baked items, snacks, soft drinks, and alcohol *(55,56)*. Conversely, fruits and vegetables are the least frequently advertised items. Moreover, greater monetary resources appear to be allocated toward foods of lower nutritional quality. Whereas two of the largest fast-food companies spent nearly $500 million annually on advertising campaigns, a national "five a day for better health" campaign targeting fruit and vegetable consumption had only $1 million in advertising funds *(57–59)*.

2.6. Physical Activity Trends

Improvements in technology have led to a gradual transition from labor-intensive occupations to labor-saving occupations. Changes in physical education requirements, leisure time activity patterns, accessible park and recreation space, and home computer access have also affected daily energy expenditure levels in the population. The mechanisms for these associations are complex. For instance, increased television viewing may displace physical activity time, may be accompanied by snacking, and provides exposure to food advertising, all of which can promote positive energy balance.

Both children and adults spend a significant percentage of their leisure time engaged in sedentary activities such as television viewing, home computer tasks, and video games. Home computer ownership in the United States is expanding, increasing from 7.9% in 1984 to 36.6% in 1997 *(10)*. By 1997, 50% of all children ages 3 to 17 and 40% of all adults had access to a computer at home *(10)*. A survey of 10,000 households conducted by Media Metrix in 1999 found that simultaneous use of both the computer and television was reported by 49% of respondents. Such sedentary activities may be contributing to obesity in the United States *(60)*.

2.7. School Physical Education Trends

Declines in physical education participation and structure have further reduced the amount of time children spend in moderate to vigorous physical activity, especially as they get older. The National Children and Youth Fitness Study (NCYFS) II found that nearly all first- through fourth-grade students were registered in physical education classes at school and received physical education 3.1 times per week on average *(61)*. Thirty-six percent received daily physical education. In some cases, however, recess time substituted for instruction by a physical education teacher. Traditionally, physical education enrollment rates and participation drop dramatically in older grades. According to 2003 Youth Risk Behavioral Surveillance System data, for example, only 55.7% of adolescents in grades 9 through 12 were enrolled in physical education classes and, nationwide, 44.9% of students reported inadequate or virtually no moderate or physical activity in the previous 7 d *(62)*.

2.8. Access to Health Clubs, Parks, and Recreational Areas

An active lifestyle depends to some extent on access to safe, adequate recreational and exercise facilities, yet a recent survey indicated that only 46% of municipal and county parks and recreation departments had fitness trails. Hiking and biking trails were present in only 29 and 21% of city and county recreation departments nationally *(10,62,63)*. However, access to park facilities does not guarantee regular participation in physical activity. Although 51% of adults reported that easier access to exercise facilities would help them stay active, two studies demonstrated that physical activity did not improve despite access to free exercise facilities *(64,65)*. Conversely, health club membership and usage are increasing, especially in higher income individuals. Memberships rose between 1988 and 1998 by 51% *(10,66)*. Members reported using the facilities an average of 13.3 d more annually in 1998 than in 1988 *(10,66)*.

2.9. Television Viewing

Increased television viewing has been associated with increased energy intake and body weight *(67)*. The mechanisms for this association are multifold. Increased television viewing may displace more active leisure time pursuits, be accompanied by eating, and/or promote excess energy intake through exposure to food advertising. According to Neilson Media Research, individuals ages 12 and older watch an average of 28 h of television weekly, representing a significant increase over the past two decades *(54)*. Data from the 2003 Youth Risk Behavioral Surveillance System indicated that nationwide, 38.2% of students in grades 9 through 12 watch television 3 or more hours daily *(62)*. Younger children spend an average of at least 2 h daily watching television or playing video games and one-quarter of all US children watch more than 4 h of television each day *(68–70)*.

Television viewing may be associated with food consumption in some individuals. A study conducted by Francis et al. showed that television viewing was associated with increased snack food consumption in girls who were 5, 7, and 9 yr old, which in turn predicted girls' increase in BMI from ages 5 through 9 yr *(67)*. Another study of ethnically diverse third- and fifth-grade children compared food consumption during television viewing with food consumption when the television was off. Children consumed 17 to 26% of total energy intake while watching television, representing a significant percentage of total calorie intake *(71)*. However, the consumption of high-fat foods was linked with BMI only in the younger children.

Time spent watching television may also be accompanied by food intake in other select individuals and groups. A study of 861 women and 198 men in a community-based sample found an association between television viewing and caloric intake, with wide differences existing between socioeconomic groups. In higher income women, every additional hour of television viewing was associated with an excess 50 kcal energy intake and an excess 136 kcal for lower income women. No association between television viewing and energy intake was found in male participants *(10,72)*. Thus, television viewing has been shown to be a risk factor for snack consumption and, in turn, increased weight status, especially in those individuals who are predisposed to obesity. Increased television viewing has been associated with increased energy intake and body weight *(67)* and decreased physical activity. Thus, decreased energy expenditure has been negatively associated with BMI *(73–75)* and maintenance of weight loss *(76,77)*.

3. IMPLICATIONS FOR CHANGE: CHANGING ENVIRONMENTS VS INDIVIDUALS

In response to growing concern over the obesity problem in the United States, in 2001 the US Surgeon General issued a report, "A Call to Action to Prevent and Decrease Overweight and Obesity," advocating healthier school and workplace environments. Other position statements and guidelines issued by the Centers for Disease Control and Prevention, the Institute of Medicine (*see* Table 2), and the American Academy of Pediatrics have provided an initial framework for formulating and implementing public health policies focusing on child and adult overweight *(4,22,58,78,79)*.

Several ideas have been proposed to alter the environment to make it easier for consumers to access and select healthier food choices. Public health interventions such as issuing a surcharge to high-fat items ("fat tax"), subsidizing healthy food choices in schools and workplaces, price reductions on healthy snacks, and limiting or prohibiting vending machine sales in schools have been proposed.

Vending machines are positioned in approx 1.5 million schools and workplaces nationwide *(80–82)*, and present a potentially effective medium for testing and implementing pricing strategies to promote healthier food choices for adults and children. A study by French et al. suggested that when low-fat vending machine item prices were reduced by 50%, sales of these healthy snacks increased by 80% without a significant change in profitability *(57,81)*. A comparable study found that a similar price reduction in fresh fruit and vegetable selections in high school and workplace cafeterias was accompanied by a two- to fourfold and a threefold increase in sales, respectively *(80,81,83)*. Combining a social marketing component to promote healthy food choices,

Table 2
Summary of Institute of Medicine Recommendations for Schools to Address Childhood Obesity, September 2004

- Establish nutritional standards for all "competitive foods"
- Establish a minimum of 30 min of activity during the school day
- Enhance school health curricula
- Ensure that all school meals meet the Dietary Guidelines for Americans
- Ensure that all schools are as "advertising free" as possible
- Conduct annual assessments of student weight, height, BMI, and make data available to parents

Adapted from ref. *4*.

as done in the CHIPS (Changing Individuals' Purchase of Snacks) study with 10, 25, and 50% price reductions, resulted in a small but significant positive effect. Low-fat sales rose by 9, 39, and 93% with the aforementioned pricing strategy *(80,81)*.

When a 10% surcharge was added to the cost of 7 high-fat menu and snack selections concurrently with a 25% price reduction for low-fat items, the estimated revenue generated from the sales of high-fat items was within 5% of expected *(80,81)*. The use of such "fat taxes" to help subsidize healthier food choices in community settings and government assistance programs warrants further exploration, testing a wider array of food selections initiated in a more extensive variety of sites.

In 2004, Congress adopted legislation requiring the development of a "Wellness Policy" addressing the sales of competitive foods in all public schools and improving the school environment to promote physical education opportunities *(84)*. A large survey conducted in the 51 largest school districts in the United States demonstrated that some progress in improving the school environment to promote health is being made. Of the 51 school districts studied, 39% had adopted competitive food policies that exceeded federal and state requirements and nearly all of these districts had implemented policies banning the sale of soda in schools and placing restrictions on vending machine sales *(78)*. None of the school districts had met all the Institute of Medicine recommendations, although a significant shift toward a healthier school environment was observed.

4. CONCLUSIONS

In conclusion, the obesity epidemic is influenced by several macroenvironmental factors. No single environmental influence is responsible for the growing prevalence of overweight and obesity in the United States and other countries. Strategies for reversing the trend must include efforts to change our "toxic environment," which encourages excess energy consumption and a sedentary lifestyle. Examples of environmental manipulations to help combat the obesity epidemic include limiting television viewing, reducing children's access to high-fat, high-sugar foods, the development of safe and convenient areas for physical activity, educating individuals on appropriate serving sizes, and reducing the cost of healthy snack foods in schools and workplaces.

REFERENCES

1. Flegal KM, Carroll MD, Ogden CL, Johnson CL. JAMA 2002;288(14):1723–1727.
2. Rolls B. Nutrition Today 2003;38(2):42–53.

3. Gillis LJ, Oded B-O. J Am Coll Nutr 2003;22(6):539–545.
4. Institute of Medicine. *Progress in Preventing Childhood Obesity: Focus on Schools*. The National Academic Press, Washington DC: 2005.
5. Ogden CL, Flegal KM, Carroll MD, Johnson CL. JAMA 2002;288(14):1728–1732.
6. Deckelbaum RJ, Williams CL. Obes Res 2001;9 Suppl 4: 239S–243S.
7. Hill JO, Peters JC. Science 1998;280(5368):1371–1374.
8. Battle EK, Brownell KD. Addictive Behav 1996;21(6):755–765.
9. Brownell KD. Pediatrics 2004;113(1 Pt 1):132.
10. French SA, Story M, Jeffery RW. Ann Rev Pub Health 2001;22:309–335.
11. Brownell K. *Food Fight*. Contemporary Books, New York: 2004.
12. Putnam J. Food Rev 2000;23:13.
13. Jacobson MF. Brownell KD. Am J Pub Health 2000;90(6):854–857.
14. Jahns L, Siega-Riz AM, Popkin BM. J of Pediatr 2001;138(4):493–498.
15. Nielsen SJ, Siega-Riz AM, Popkin BM. Obes Res 2002;10(5):370–378.
16. Tippett K, Cypel YS. Design and Operation. The Continuing Survey of Food Intakes by Individuals and the Diet and Health Knowledge Survey 1994–96. US Department of Agriculture, Service AR, ed. Washington, DC: 1997.
17. Young LR, Nestle M. Am J Pub Health 2002;2(2):246–249.
18. Nielsen SJ, Popkin BM. JAMA 2003;289(4):450–453.
19. McConahy KL, Smiciklas-Wright H, Birch LL, Mitchell DC, Picciano MF. J Pediatr 2002;140(3): 340–347.
20. Rolls BJ, Engell D, Birch LL. J Am Dietetic Assn 2000;100(2):232–234.
21. Edelman B, Engell D, Bronstein P, Hirsch E. Appetite 1986;7(1):71–83.
22. Fox S, Meinen A, Pesik M, Landis M, Remington PL. WMJ 2005;104(5):38–43.
23. Gleason P, Suitor C. *Changes in children's diets: 1989–1991 to 1994–1996*, US Department of Agriculture, Food and Nutrition. Service Alexandria, VA. 2001.
24. Fisher JO, Rolls BJ, Birch LL. Am J Clin Nutr 2003;77:1164–1170.
25. Rolls BJ, Morris EL, Roe LS. Am J Clin Nutr 2002;76(6):1207–1213.
26. Kral TVE, Meengs JS, Wall DE, et al. FASEB J, In press.
27. Nicklas TA, Baranowski T, Cullen KW, Berenson G. J Am Coll Nutr 2001;20(6):599–608.
28. Nielsen SJ, Popkin BM. Am J Prev Med 2004;27(3):205–210.
29. Morton JF, Guthrie J. Fam Econ Nutr Rev 1999;11:44–47.
30. Harnack L, Stang J, Story M. J Am Dietetic Assn 1999;99(4):436–441.
31. National School Lunch Program. *Foods sold in competition with USDA school meal programs: A report to Congress*. United States Department of Agriculture, Washington DC. 2001.
32. Dennison BA, Rockwell HL, Baker SL. Pediatrics 1997;99(1):15–22.
33. Welsh JA, Cogswell ME, Rogers S, et al. Pediatrics 2005;115(2):223–229.
34. Ebbeling CB, Sinclair KB, Pereira MA, Garcia Logo E, Feldman HA, Ludwig DS. JAMA 2004;291(23): 2828–2833.
35. Bouchard C, Tremblay A, Despres JP, et al. N Engl J Med, 1990;322(21):1477–1482.
36. Birch LL, Deysher M. Learn Motiv 1985;16:341–355.
37. Faith MS, Keller KL, Johnson SL, et al. Am J Clin Nutr 2004;79(5):844–850.
38. Faith MS, Pietrobelli A, Nunez C, Heo M, Heymsfield SB, Allison DB. Pediatrics 1999;104(1 Pt 1): 61–67.
39. Mattes RD. Physiol Behav 1996;59(1):179–187.
40. Rolls BJ, Kim S, Fedoroff IC. Physiol Behav 1990;48(1):19–26.
41. Beridot-Therond ME, Arts I, Fantino M, De La Gueronniere V. Appetite 1998;31(1):67–81.
42. De Castro JM. Physiol Behav 1993;53(6):1133–1144.
43. Goran MI. Am J Clin Nutr 2001;73(2):158–171.
44. Ludwig DS, Peterson KE, Gortmaker SL. Lancet 2001;357(9255):505–508.
45. Tordoff MG, Alleva AM. Am J Clinl Nutr 1990;51(6):963–969.
46. Lin BIL, Guthrie J, Blaylock JR. Agric Econ Report 1996;746:1–37.
47. Lin BH, Frazao E, Guthrie J. Fam Econ Nutr Rev 1999;12:85–89.
48. Pereira MA, Kartashov AI, Ebbeling CB, et al. Lancet 2005;365(9453):36–42.

49. Prentice AM, Jebb SA. Obes Rev 2003;4(4):187–194.
50. Bowman SA, Gortmaker SL, Ebbeling CB, Pereira MA, Ludwig DS. Pediatrics 2004;113(1 Pt 1): 112–118.
51. Pereira MA, Kartashov AI, Ebbeling CB, et al. Circulation 2003:107.
52. French SA, Story M, Neumark-Sztainer D, Fulkerson JA, Hannan P. Int J Obes Rel Metab Disord 2001;25(12):1823–1833.
53. 2000 Report on television. In: *Neilsen Media Research 2000*, A.N. Co, Editor. New York: 2000.
54. Kotz K, Story M. J Am Dietetic Assn 1994;94(11):1296–1300.
55. Story M, Faulkner P. Am J Pub Health 1990;80(6):738–740.
56. French SA, Story M, Jeffery RW, et al. J Am Dietetic Assn 1997;97(9):1008–1010.
57. Krebs NF, Jacobson MS. Pediatrics 2003;112(2):424–430.
58. Leading National Advertisers, Advertising Age 1999.
59. *Simultaneous Use of the PC and Television Growing Rapidly*, Metrix M, Editor. New York: 1999.
60. Ross JG, Pate RR. J Phys Educ Rec Dance 1987;58:51–56.
61. Centers for Disease Control and Prevention. Surveillance Summaries, May 21, 2004. MMWR 2004;22–27.
62. Healthy People 2000: National Health Promotion and Disease Prevention Objectives. US Dept of Health and Human Services, Washington, DC: 1988.
63. French SA, Jeffery RW, Oliphant JA. Am J Health Promot 1994;8(4):257–262.
64. Sherwood NE, Morton N, Jeffery RW, et al. Am J Health Promot 1998;13:12–18.
65. American Sports Data Health Club Trends Report, I.R.S.C. Assoc, Editor. 1999.
66. Francis LA, Lee Y, Birch LL. Obes Res 2003;11(1):143–151.
67. Andersen RE, Crespo CJ, Bartlett SJ, Cheskin LJ, Pratt M. JAMA 1998;279(12):938–942.
68. Larson RW, Verma S. Psychol Bull 1999;125(6):701–736.
69. Johnson RK. Proc Nutr Soc 2000;59(2):295–301.
70. Matheson DM, Killen JD, Wang Y, Varady A, Robinson TN. Am J Clin Nutr 2004;79(6):1088–1094.
71. Jeffery RW, French SA. Am J Pub Health 1998;88(2):277–280.
72. Jakicic JM, Otto AD. Am J Clin Nutr 2005;82(1 Suppl):226S–229S.
73. Parsons TJ, Power C, Manor O. Int J Obes (London), 2005;29(10):1212–1221.
74. Wareham NJ, van Sluijs EM, Ekelund U. Proc Nutr Soc 2005;64(2):229–247.
75. McGuire MT, Wing RR, Klem ML, Lang W, Hill JO. J Consult Clin Psychol 1999;67(2):177–185.
76. Wing RR, Phelan S. Am J Clin Nutr 2005;82(1 Suppl):222S–225S.
77. Greves MH, RF. Inter J Behav Nutr Phys Act 2004;3(1):1–10.
78. The Surgeon General's call to action to prevent and decrease overweight and obesity. US Dept of Health and Human Services. Rockville, MD: 2001.
79. French SA. J Nutr 2003;133(3):841S–843S.
80. French SA, Jeffery RW, Story M, et al. Am J Pub Health 2001;91(1):112–117.
81. 1999 state of the vending industry report, A. Merchandiser, Editor. 1999.
82. Jeffery RW, French SA, Raether C, Baxter JE. Prev Med 1994;23(6):788–792.
83. Child Nutrition and WIC Reauthorization Act of 2004. 118 Stat 729, 2004.

16 Developmental Perspectives on the Origins of Obesity

Christopher W. Kuzawa, Peter D. Gluckman, and Mark A. Hanson

Abstract

This chapter reviews the developmental pathways contributing to the origin of obesity. Evolutionary considerations are emphasized. At birth more than half of a human baby's metabolism is devoted to the brain and it is suggested that the extreme neonatal and early childhood adiposity of humans is an adaptation to provide an energy reserve during periods of nutritional stress arising from infections and the process of weaning. This chapter also reviews the substantial experimental and clinical evidence for prenatal and early postnatal factors in the development of obesity. Developmental pathways that may lead to obesity include fetal undernutrition caused by an impaired intrauterine environment, fetal overnutrition and macrosomia caused by maternal diabetes, and infant overnutrition caused by excessive early feeding. There is evidence for interactions between these pathways and for intergenerational influences. Finally, this chapter discusses the implications for the global obesity epidemic of mismatch between the genotype, environment, and lifestyle, and underlines the potential role of inappropriate adaptive responses during development in populations undergoing rapid nutritional transition.

Key Words: Obesity; evolution; development; environment; fetal nutrition; developmental plasticity; adaptive responses; prediction; mismatch.

1. UNDERSTANDING OBESITY REQUIRES A DEVELOPMENTAL AND EVOLUTIONARY PERSPECTIVE

Ever since Neel's proposition of the "thrifty gene" hypothesis more than 40 yr ago *(1)*, evolutionary explanations for the origins of human obesity have assumed that our tendency to put on weight in a modern environment is the vestige of a trait that was beneficial under the more austere nutritional conditions of the past. Neel proposed that, for millions of years during the Paleolithic, humans and our hominin ancestors survived as roaming bands of foragers who faced an unpredictable food supply. Given this, an ability to capitalize on any excess energy by efficiently depositing it as fat during periods of "feast" would have boosted the chances of surviving the inevitable future "famine." We inherited our genes from ancestors who survived these recurrent ecological crises, which now leave us prone to obesity and diabetes in a contemporary environment of nutritional abundance.

From: *Nutrition and Health: Adipose Tissue and Adipokines in Health and Disease*
Edited by: G. Fantuzzi and T. Mazzone

Neel's hypothesis was invaluable for stimulating interest in the evolutionary origins of human obesity. In particular, the thrifty gene model heightened awareness that the environment can change more rapidly than the genome, potentially leading to novel diseases through "mismatch" between genes and environment. Despite these important contributions, the hypothesis has difficulty explaining more recent advances in our understanding of the obesity epidemic and its health sequelae. As crosscultural data accumulate, the heterogeneity in the prevalence and health consequences of obesity are not easily reconciled with a purely gene-based model of obesity risk *(2)*. Observations among societies experiencing ongoing nutritional transitions in Asia, Latin America, and elsewhere document a heightened metabolic disease risk for a given level of body mass index or adiposity *(3–5)*. The hypothesis would need to be modified to explain this. Additionally, there is now evidence that famine was less common among our foraging ancestors than presumed by Neel's model *(6)*, raising doubts about its central assumptions *(2)*. Further complexities arise as etiological insights change; for example, current understanding of the pathogenesis of obesity-related disease now includes a significant role for immune activation, and this is not easily encompassed within the model *(7)*.

That human metabolism is not primarily crafted to survive famine is suggested by a closer examination of the developmental trends in body composition that characterize the human lifecycle. As illustrated in Fig. 1, body fat in humans constitutes a larger percentage of weight at birth than in any other mammal so far studied *(8)*. This is followed by a continued fast pace of fat deposition during the early postnatal months. In well-nourished populations, adiposity reaches peak levels at around the age of weaning before gradually declining to a nadir in childhood, when humans reach their lowest level of adiposity in the lifecycle before again increasing in the prepubertal phase. If the threat of famine is what molded the human metabolic propensity to deposit and maintain extra body fat, it is not obvious why children's bodies should do so little to prepare for these difficult periods. These developmental changes in body composition suggest that the evolutionary forces that selected for the size of the energy buffer during early life were primarily aimed at defending against nutritional stress that has largely subsided by mid-childhood. Indeed, the low priority placed on maintaining an energy reserve during childhood suggests that the background risk of starvation faced by our ancestors may have been smaller than often thought.

This chapter first reviews the nutritional ecology of the period spanning mid-gestation through early childhood and consider the influence that natural selection operating at this age may have had on the modern human genome and the risk of metabolic disease. We adopt a developmental perspective to move beyond Neel's model and take into account the sources and age-specific intensity of nutritional stress and natural selection. There is now substantial evidence that developmental responses to early nutritional environments can modify our genetic pattern of ontogenesis (developmental plasticity), with lasting effects on our physiology and metabolism. Secondly, we discuss evidence for several developmental pathways now known to link early environmental experiences, including nutrition, to later metabolism, weight gain, and disease. Just as Neel emphasized the past adaptive significance of genes, these newly discovered and potentially adaptive modes of developmental response may provide greater flexibility in the face of ecological change than can be achieved through the slow process of genetic change.

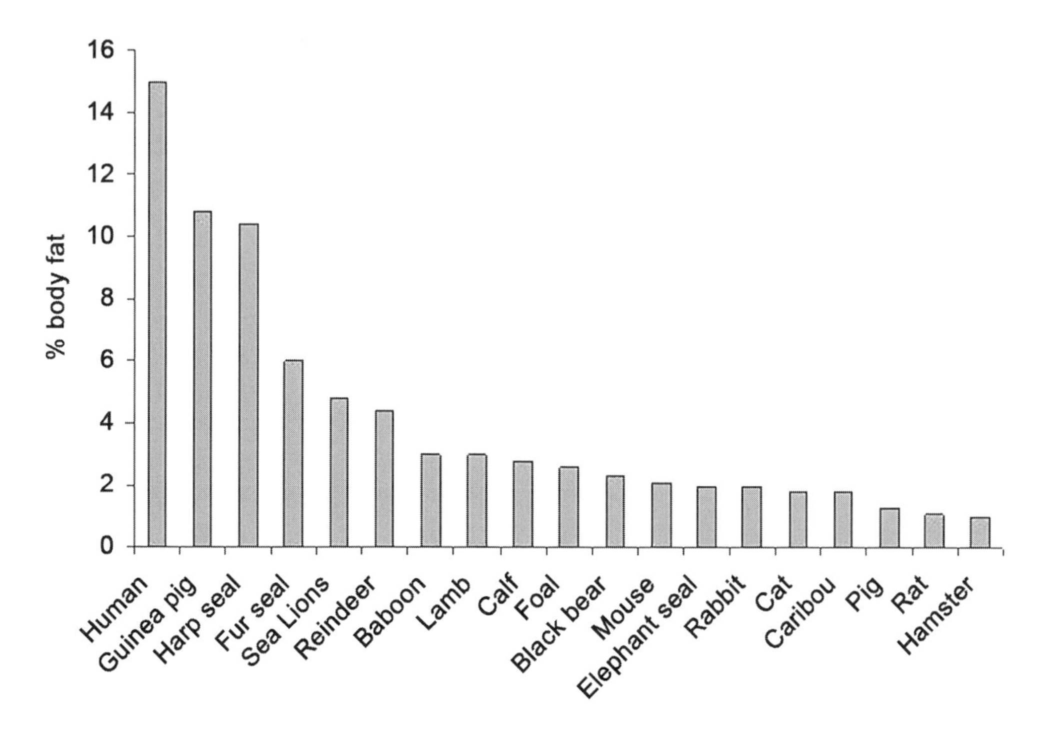

Fig. 1. Percentage of body fat at birth in mammals. Adapted from ref. *8*.

These considerations underscore the need to take development seriously in studies of metabolic disease. We conclude by speculating that the accelerating pace of the global obesity epidemic and its rising disease burden may be the result of two related forms of mismatch: that between the human genome and the novel lifestyle of contemporary human populations and, in more rapidly changing environments, that between the constraints imposed by developmental processes together with early nutrition and the environment and lifestyle subsequently experienced in adulthood.

2. DEVELOPMENTAL PERSPECTIVE ON ENERGY BALANCE IN HUMAN EVOLUTION

Humans are unusual with respect to their high level of natural "obesity" *(9)*. This trait is particularly obvious at birth, when humans have more body fat than any other species *(8)*. Explanations for our distinctive "baby fat" have traditionally proposed that it is related to loss of body hair, and it is widely assumed that natural selection compensated for this by increasing the layer of insulative body fat *(10,11)*. A competing perspective notes that this excess adipose tissue is well suited to serve as a backup energy supply for another distinctive human trait—our large brains *(8)*. Humans are exceptional in the size of their brains, and at birth more than half of the body's metabolism is devoted to this organ *(12)*. Unlike energy expended on other tissues or systems, brain metabolism is less flexible during a period of nutrient shortage, and must be maintained within narrow limits. Thus, our large brains impose a double burden on metabolism during infancy: they increase demand for energy while restricting the flexibility of metabolic requirements when that demand is not met.

Other factors common during infancy can impede the supply of nutrients, ensuring that negative energy balance is a frequent occurrence at this age in most populations. We are born with a naïve adaptive immune system and must become exposed to specific pathogens to acquire the repertoire of antibodies necessary to protect us from future infection. Exclusively breastfed infants are initially shielded from exposure to pathogens and gain some passive immunity through their milk, so that they are often quite healthy in the early postnatal months. But, as energy requirements outstrip the supply capacity of breast milk by roughly 6 mo of age, less sterile supplemental foods must be introduced and infectious disease becomes unavoidable in all but the most sanitary environments. These childhood infections, in turn, are a source of nutritional stress, and indeed it is primarily through their effects on nutritional status that they compromise health and contribute to early-life mortality and poor childhood growth (*see* ref. *13* for a review). The ensuing nutritional depletion has the effect of suppressing immune function, leaving the infant more prone to future infection and thus a compounding cycle of nutritional stress.

Human infants thus face a profound energetic dilemma: at the age when they are most dependent on provisioning by caretakers to maintain the high and inflexible metabolic requirements of their large brains, they are also at risk of separation from that supply chain as a result of illness and the nutritional stresses of weaning. It is this confluence of factors, and the link between nutritional stress and immature immune function, that accounts for much of the high infant mortality in many societies *(14)*. And this is where evolution has likely favored neonatal adiposity as a strategy. It is not difficult to imagine how infants with a predisposition to deposit fat prior to weaning might be better represented among the subset who survive to adulthood to reproduce and pass on their genes *(8)*. Infants typically experience not a single infection, but cycles of infection followed by recovery. Thus, it is those infants who efficiently replenish their energy reserves during recovery intervals who will be better equipped energetically to survive any future illness. From this perspective, the infant experiencing cycles of illness faces a metabolic challenge not unlike Neel's proposed recurrent episodes of feast and famine.

The evolutionary imprint of this energetic stress is most conspicuous in our distinctive adiposity at birth, indicating that the process of fat deposition begins much earlier, during gestation itself—as does the challenge of protecting a relatively large brain. The size of the brain is large relative to body size throughout late gestation and the fetal brain is relatively protected under conditions of fetal undernutrition. Recent studies also show visceral fat to be conserved in the face of fetal undernutrition *(15,16)*. Thus, early human development, beginning *in utero* and continuing into early childhood, is an energetically precarious stage in the lifecycle, even though the supply line of nutrients and common causes of nutritional insufficiency shift at parturition and at weaning.

These sources of energy stress have largely receded by mid-childhood: the fraction of total metabolism required by the brain has declined substantially and children have already acquired immunity against the major pathogens that they are likely to face. It thus makes sense that the human body places little priority on maintaining sizable body fat stores by this age. This would of course *not* be true if famine were the main source of selection on human body composition and metabolism, in which case we should expect no decline in adiposity in childhood.

3. DEVELOPMENTAL PATHWAYS TO OBESITY

Given that infant and child mortality is highest as a result of developmentally mediated undernutrition during the earliest stages of the lifecycle, natural selection operating at this age may have had an important influence on the evolution of human metabolism. It is notable, for instance, that the challenge of surviving recurrent infections shares similarities with Neel's vision of a feast–famine scenario, and might be expected to favor the rise of similar metabolic-disease predisposing genes. In this sense, early life might be likened to an "ontogenetic bottleneck" through which any adult metabolic traits must first pass *(17)*. Although all humans experience this age of heightened energetic vulnerability, there is much variation in the environments that individuals experience, which determines whether they will be forced to rely on such contingencies as brain sparing, lipolysis, or rapid replenishment of body fat. There is now considerable experimental evidence that one's early nutritional experiences, both *in utero* and during infancy, may act through developmental plasticity to permanently influence traits such as appetite, tissue-specific insulin sensitivity, and weight gain *(18)*. As will be discussed, there is much interest in establishing whether these responses have an adaptive basis, and if so, whether they are primarily designed to serve a short- or long-term adaptive function *(17–21)*. What is certain is that these responses can have profound long-term implications for the risk of gaining excess body weight and the metabolic consequences of that weight gain. There is increasing evidence for at least two and possibly three developmental pathways that can lead to obesity; these will be discussed separately.

3.1. Fetal Undernutrition Pathway

There is a considerable body of epidemiological data relating an impaired intrauterine environment to the development of later central adiposity. For instance, studies of elderly populations show greater visceral adiposity among individuals who were born small *(22)*. Children of smokers, who experience prenatal hypoxia and are likely to be born small, are more likely to develop obesity *(23)*. Studies of the Dutch winter famine of 1944–1945, where previously well-nourished women were exposed to Nazi-imposed severe rationing, show that those who were pregnant during the famine gave birth to offspring who subsequently became obese *(24)*.

Experimentally, it can be demonstrated that animals born to undernourished mothers are relatively more obese; this is particularly evident if the offspring are placed on a high-fat diet after birth *(25)*. The obesity induced by these early life exposures has both central and peripheral components. In both the rat *(26)* and sheep *(27,28)*, changes in the neuroendocrine anatomy in the hypothalamus are described following nutritional limitations induced *in utero*. Animals born after adverse fetal manipulation also tend to be hyperphagic and to have a preference for fatty foods. They are also sarcopenic and mature to have peripheral insulin resistance, fatty liver, and truncal obesity *(29)*.

Such observations have generally been made in the context of studies of the early life antecedents of maturity-onset diabetes and heart disease—the so-called developmental (or fetal) origins of adult disease paradigm. There is now a large body of epidemiological, clinical, and experimental data showing that an adverse intrauterine environment predicts a variety of physiological and metabolic traits that are consistent with a greater risk of lifestyle diseases in later life, including but not limited to obesity. For instance, both impaired pancreatic islet cell function and greater insulin resistance are reported among individuals born small *(30)*, or among rats or sheep exposed to undernutrition as fetuses *(31)*. There have been major reviews of the developmental origins model recently *(21,32,33)*.

Within this literature, there has been debate as to whether it is the initial impaired nutritional environment or the delayed response of rapid postnatal catch-up growth that is causally related to these later-life health consequences *(34)*. The experimental data show that the rate of weight gain is related to both the prenatal and the postnatal nutritional exposure of the growing rat *(25,29)* and a number of experiments or clinical observations, such as the Dutch famine data, suggest that there is a clear fetal component. The clinical data also suggest that the propensity for obesity is in part prenatally determined. For instance, whereas subcutaneous fat at birth is linearly related to later body weight, visceral fat is not reduced in smaller babies *(15,16)*, suggesting that the tendency for visceral obesity can be induced *in utero*, or that such individuals preferentially protect deposition and maintenance of visceral depots. In perhaps the most studied cohort, from Helsinki, adults who developed diabetes or insulin resistance were born small and did not show catch-up in weight during the early postnatal years. However, they did experience an earlier childhood adiposity rebound and put on weight faster in late childhood *(35)*. Not dissimilar but more limited data are reported from an Indian cohort *(36)*.

Genetic polymorphisms clearly impact on the interaction between the developmental environment and long-term phenotypic outcomes. For example, a polymorphism in the PPARγ gene is associated with an elevated risk of insulin resistance in adults but only if birthweight was reduced. Presence of the polymorphism in the absence of fetal growth impairment is not associated with an increased risk of insulin resistance *(37)*. Recent studies have shown similar interactions between polymorphisms in a number of loci, birth size phenotype, and later disease risk. These include the angiotensin converting enzyme gene *(38)*, plasma cell glycoprotein-1 *(39)*, glucokinase *(40)*, and the vitamin D receptor *(41)*. In such analyses, birthweight is likely to generally be a surrogate for an adverse intrauterine environment rather than causally related, but in some cases—for example, the glucokinase mutation *(40)*—the polymorphism may affect insulin sensitivity or action, adding directly to the strength of the association as insulin is an important regulator of fetal growth *(42)*.

But although there are clearly genetic influences, the thrifty genotype model cannot explain the experimental or clinical data linking early undernutrition to elevated adult metabolic disease risk, or findings such as research on the Dutch winter cohort discussed above. An adaptive explanation for such observations was first proposed by Hales and Barker *(19)*, who suggested that smaller babies make a number of adaptations *in utero* to survive, including inducing insulin resistance, which leave them more able to cope in a poor postnatal environment but which also made them more at risk of disease in later life. As with the thrifty gene hypothesis, this "thrifty phenotype" model has limitations. For instance, it cannot explain the continuous relationship between birth size and later disease risk that can be seen even in infants above the mean birthweight, nor can it explain why growth-retarded infants tend to protect visceral adiposity at birth. It also assumes that insulin resistance is present at birth, and both clinically and experimentally this appears not to be the case *(43,44)*. In some cases growth-retarded neonates indeed appear to have heightened insulin sensitivity *(43)*. It also assumes that slowed growth rate is a central adaptation to a compromised prenatal nutritional environment, yet many of the experimental associations do not rely on birthweight; this is also the case in some clinical observations such as the Dutch famine data *(45,46)* and some recent cohort studies *(35,47)*. Finally, the hypothesis cannot explain developmental induction in the face of physiological challenges other than nutrition. The thrifty phenotype concerned itself only with glucose/insulin metabolism, whereas subsequent work has shown effects on myriad other systems *(21,48)*. A recent study of children conceived by in vitro fertilization shows them to be taller, leaner and to have heightened insulin sensitivity *(49)*, suggesting that fetal responses to environmental manipulation need not act only in a direction toward "thriftiness." Thus, a more comprehensive model is needed if we are going to explain the broader pattern of responses, of which obesity is one manifestation.

A more general adaptive model based on the concept of predictive adaptive responses and developmental mismatch has recently been proposed *(50,51)*. The mismatch model suggests that all fetuses (and embryos) across the full range of environments sense their environment and, based on such cues, set their life course strategy, mediated by epigenetic processes, according to the environments they predict that they will meet postnatally *(21,52,53)*. In particular, if the fetus predicts an adverse nutritional environment it will induce strategies such as preserving fat mass at the expense of metabolically costly muscle (sarcopenia) while reducing expenditures on other structures and costly physiological functions, including immune function, investment in reproduction, and cellular maintenance functions that normally delay the onset of aging *(18,48,54)*. The model can provide an explanation of why a population undergoing rapid nutritional transition is at greater risk *(18)*.

The underlying mechanisms in this case would require a capacity to predict future conditions on the basis of early life cues, which may be nutritional but could also be, for example, endocrine or oxidative stress, and to respond developmentally in an appropriate fashion. Modeling shows that such predictive responses need not be completely accurate to provide a fitness advantage *(51,55)*. Moreover, it has been argued that the fidelity of the prenatal nutritional cue may be enhanced by maternal buffering mechanisms and intergenerational effects on fetal nutrition *(17)*. For instance, fetal growth rate—and, by implication, fetal nutrition—is predicted not only by the mother's current nutrition but also by her nutritional status prior to pregnancy *(56)*, which reflects her cumulative nutritional experiences in the years prior to conception. Birthweight is also predicted by the mother's (but not the father's) own growth rate during both childhood and during her fetal life, and by extension, the nutritional conditions experienced by the grandmother as conveyed *in utero* to the mother as a fetus. Such intergenerational influences on fetal nutrition and growth are well established *(57)*. By integrating nutritional information across several generations, fetal nutrition may provide a higher-fidelity cue of typical conditions than might otherwise be possible *(17)*.

It is also important to note in this context that the fetal environment is, with the exception of gestational diabetes, always nutritionally constrained to varying degrees *(21,58,59)*. The processes of maternal constraint have evolved particularly in monotocous species to limit fetal growth to match the maternal size phenotype, so that vaginal delivery is possible in a species with a uniquely large head but a pelvis narrowed by the adoption of an upright posture. The mechanisms of fetal growth limitation are poorly understood, but may involve both utero–placental function and interactions between the maternal and paternal genomes via the imprinted IGF-2 system *(60)*. It has been suggested that maternal constraint has had the additional evolutionary advantage of moving fetal development toward a set-point favoring predictive adaptation matched to a nutritionally deprived environment *(51)*. This might provide an additional fail-safe protection for the infant. As constraint operates even in normal pregnancies, this limits the range of nutritional environments that the infant can be matched to postnatally, and as a result of improved access to food and more sedentary lifestyles more children and adults are now placed in environments above this limit. Disease risk is enhanced as a result.

Constraint and developmental induction can well extend into the neonatal period, as the neonate is entirely dependent on the mother for nutrition. Recent studies provide some support for this model. For example, if neonatal rat pups born to undernourished mothers are treated in the neonatal period with leptin, they do not become obese even if placed on a high-fat diet *(29)*, suggesting that the predictive trajectory can be altered by neonatal manipulation.

In this model, then, factors that reduce the fidelity of the predictive cue can lead to a developmentally based mismatch between biology and environment, not unlike Neel's proposal of a gene–environment mismatch *(59,61)*. Indeed, the processes mediating the predictive mechanisms must have been selected and preserved through evolution, rather as have Neel's thrifty genes. The fidelity of the predictive cues may be reduced by factors such as faulty maternal transduction (for example, maternal smoking inhibiting transplacental nutrient transfer) or severe maternal constraint limiting nutrient supply to the fetus (for example, young or primiparous mothers) or by genetic polymorphisms affecting mechanisms regulating nutrient delivery to fetal tissues *(39)*. The signal may also have low fidelity as a result of rapid environmental or nutritional change, as exemplified by populations experiencing a "nutrition transition" to higher intake of dietary energy and fats coupled with reduced physical expenditure *(3,61)*. Under these circumstances, the nutritional cue conveyed to the fetus may become inaccurate as the environment changes, thereby heightening risk of metabolic derangements as individuals poorly nourished early in life gain weight. This proposal may help explain why the BMI is a particularly strong predictor of hypertension, diabetes, or heart disease in nations like China, Brazil, and the Philippines, all of which are experiencing an ongoing, rapid nutritional transition of recent onset *(3,62)*. Conversely, the finding that compromised birth outcome does not predict elevated adult risk for cardiovascular disease in The Gambia, where poor nutritional conditions have not improved, is also consistent with this model *(63)*.

3.2. Fetal Overnutrition Pathway

In addition to these effects of prenatal undernutrition, it is well recognized that the fetuses of diabetic mothers are born with relative obesity, suggesting a separate pathway linking prenatal nutrition with later risk of metabolic disease. Maternal hyperglycemia leads to fetal hyperglycemia and fetal hyperinsulinemia, which promotes excessive fat deposition during the third trimester. The degree of subcutaneous adiposity is increased even under conditions of subclinical maternal hyperglycemia. The increased adiposity in fetal life may then be magnified by postnatal overnutrition, and the obesity that is generated often, in turn, leads to type 2 diabetes *(64)*.

3.3. Multigenerational Influences

The situation across generations is more complex, and available data suggest that it is possible to begin with either prenatal undernutrition or overnutrition and to stabilize on a pattern of intergenerational macrosomia and diabetes. Mothers who were born as macrosomic babies are themselves at risk of gestational diabetes and thus of "transmitting" macrosomia to the next generation. Alternatively, mothers born in an impaired intrauterine environment are at greater risk, because of their altered insulin sensitivity, of developing subclinical or gestational diabetes, as discussed above, and may subsequently give birth to a macrosomic infant at heightened risk of developing obesity as a result *(65)*. This intergenerational sequence has been suggested as an explanation for the extremely high prevalence of diabetes among some Native American groups *(2)*. In India, where mothers are small owing to nutritional limitations in childhood and intergenerational stunting, women who give birth to infants in the upper tertile of the birthweight range (mean 3300 g, mean for whole sample 2700 g) are more likely to develop type 2 diabetes within 8 yr than are women who deliver smaller babies *(66)*. Thus it would appear that when the mother is small, she may give birth to a relatively small baby, although relatively large for that population, and that this baby then follows the overnutrition pathway.

3.4. Infant Overnutrition Pathway

Although there are conflicting data, recent systematic reviews concur that breastfeeding confers some protection against the development of childhood and adult obesity *(67,68)*. In turn, these studies imply that feeding infant formula or cow's milk results in a greater risk of obesity. This could reflect the greater caloric and protein load of cow's milk, which can lead to overnutrition in infancy, or as yet unidentified beneficial effects of other breast milk components, such as growth factors or hormones such as leptin. In rats, high nutrition in infancy can induce both peripheral and central components of obesity, involving changes in both local depots and hypothalamic neuroendocrine pathways *(69,70)*. The effect is apparent in both premature and term infants, suggesting that it is early feeding that is particularly sensitizing to the development of later obesity *(71)*. A recent long-term cohort study identifying an association between adult obesity and rapid weight gain in the first week of life of formula-fed infants tends to confirm a deleterious effect of early overnutrition rather than an intrinsic protective effect of breastfeeding *(72)*.

Although the studies are not comparable, either in terms of populations or their historical and geographical situations, there is an unexplained paradox: one group of studies suggests the importance of infant overnutrition as predisposing to obesity *(34)*, whereas a separate group of studies points to lower birthweight and underweight in infancy followed by earlier adiposity rebound in childhood as the causative factors *(35,36)*. This suggests that there are likely to be two or more independent developmental pathways that may link infant nutrition with later weight gain. Alternatively, they may be both reflections of a common pathway informed by pattern of growth in which a constrained nutritional environment (fetal or infant) is followed by a nutritionally enriched environment. The degree of constraint varies, some pregnancies are more constrained than others *(58)*, and the speed, nature, and degree of nutritional transition can vary; all influence the pattern of developmental plasticity and the degree of mismatch and thus the disease risk *(18)*. Additionally there is clearly genetic variation influencing the outcome of environmental interactions during development *(37)*. Even though the specifics of the pathways involved remain unclear, this does not diminish their substantial contribution to disease risk, a contribution that is still not widely realized. Calculations performed on the basis of birthweight, weight at 1 yr, and weight at 11 yr suggest that abnormal patterns of fetal and infant growth could account for roughly 50% of the risk of heart disease and non-insulin-dependent diabetes *(73)*. This is a fertile area for ongoing research.

4. CONCLUSIONS

In the more than four decades since Neel's publication of the thrifty gene hypothesis, obesity has been viewed as the result of an ancient metabolism adapted to feast–famine conditions, now thrust into a modern world of chronic nutritional excess. Famine or other ecological crises may indeed have been among the most important sources of nutritional mortality among our adult ancestors, but this should not be extrapolated uncritically to the human lifecycle as a whole. Before reaching adulthood, humans first must survive the energetic turmoil and high mortality associated with gestation, parturition, and weaning, exacerbated by a relatively large and energetically demanding brain and compounded by the inevitable infections of childhood. If human metabolism is "designed" to survive nutritional stress, we have argued that the nutritional stress of early life is likely to have left a more prominent evolutionary imprint on our metabolic homeostasis and modes of adaptive response than has famine. As one example of the potential utility of the model discussed above, the intimate ties between infectious disease and nutritional stress during early life provide a useful starting point for considering the interconnections between metabolic and inflammatory processes that are now recognized as central to the metabolic syndrome.

The extensive research documenting the developmental origins of health and disease (DOHaD) paradigm makes it clear that susceptibility to obesity and metabolic disease is not dictated solely by one's inherited genome and adult lifestyle, but is also powerfully influenced by developmental processes initiated by nutrition *in utero* and continuing into the postnatal period. As discussed, one proposal we favor to explain these findings states that fetal, and perhaps infant, nutrition acts as a cue for predicting future nutrition, thus allowing the organism to adaptively modify its metabolism, physiology, and life history characteristics in a predictive fashion. That fetal undernutrition can lead in certain circumstances to a greater predisposition to weight gain, or a greater sensitivity to the adverse health effects of weight gain, intuitively supports this proposition of early life prediction, for it shows that undernutrition contributes strongly to heightened future disease risk only when conditions have changed significantly between birth and adulthood. Also supporting this model is the finding that, when environments do not change and individuals poorly nourished early in life remain marginally nourished as adults, the risk of obesity-related conditions is not elevated *(63)*.

If prediction were the rule across the full range of early environments, we might expect prenatal overnutrition to protect against the adverse health effects of overnutrition later in life. There is some evidence that this might be the case in the absence of pathological causes of fetal overnutrition. Whereas infants born thin and who become fat as children are at greater risk of later disease, children born with a high ponderal index and who have a high ponderal index in childhood are at no particular enhanced risk *(74)*. Experimental exposure to high-cholesterol diets *in utero* induces greater postnatal cholesterol tolerance in pigs *(75)*. High prenatal nutrition is associated with longevity in mice *(76)*, whereas after birth it is undernutrition that is associated with longevity.

The contrary situation of macrosomia caused by maternal diabetes is likely to be a recent pathological development reflecting the very different nutritional environments of modern humans compared with conditions prevalent in the Paleolithic and Neolithic. Maternal undernutrition was likely to be far more common during hominin evolution than was gestational diabetes, which—like obesity itself—has emerged as a health problem only in recent generations. As is true for all biological systems, developmental responses to early environments must have limits within which they are designed to operate, reflecting the range of expected conditions likely to have been experienced by our ancestors. This system may now be pushed by chronic intergenerational excess to a state rarely expressed phenotypically in the past.

Although questions remain regarding the function of early-life developmental plasticity, the available data are sufficient to propose a tentative revised model for the origins of obesity and related diseases. The model includes two forms of mismatch, each reflecting adaptive processes operating on different temporal scales. The first is a process of gene–environment mismatch akin to that proposed by Neel. This could help explain the general human tendency to gain weight under modern conditions of reduced energy expenditure and increased intake, and for metabolic conditions such as insulin resistance to be triggered inappropriately by factors such as novel proinflammatory features of the environment. The second form of mismatch is based on developmental plasticity, as documented by the DOHaD literature, which may help to explain the heterogeneity in the obesity epidemic and its consequences. When conditions change markedly within a single generation, or when early-life predictive signals have low fidelity, this may lead to additional mismatch and metabolic disturbance. As emphasized, these latter effects could help explain some of the important features of disease transitions in populations experiencing a particularly rapid pace of dietary and lifestyle change *(77)*.

This mismatch model potentially helps to explain some features of the modern obesity epidemic that are not addressed by a gene-centered model and is also consistent with the growing appreciation of the role of developmental processes, including plasticity, in adaptation and evolution among a wide range of organisms *(78–80)*. Natural selection has favored a range of strategies to help our ancestors manage the vagaries of ecology and nutrition, including a genetic architecture with an appropriate set of metabolic priorities and flexibility in developmental processes capable of responding more quickly to changing conditions than would be possible via genetic change alone *(52)*. Although these processes may have allowed efficient tracking of gradual changes in past environments, especially the more threatening situations of reduced nutrition, their adaptive capacities are now overtaxed, leading to metabolic disease, when they are confronted with the evolutionarily unprecedented pace of change in modern environments.

REFERENCES

1. Neel JV. Am J Hum Genet 1962;14:353–362.
2. Benyshek DC, Martin JF, Johnston CS. Med Anthropol 2001;20:25–64.
3. Popkin BM. Asia Pac J Clin Nutr 2001;10:S13–S18.

4. Bell AC, Ge K, Popkin BM. Obes Res 2002;10:277–283.
5. Kuzawa CW, Adair LS, Avila JL, et al. Am J Hum Biol 2003;15:688–696.
6. Cohen MN, Armelagos GJ. *Paleopathology at the Origins of Agriculture.* Academic Press, New York: 1984.
7. Dandona P, Aljada A, Bandyopadhyay A. Trends Immunol 2004;25:4–7.
8. Kuzawa CW. Yearbook Phys Anthropol 1998;41:177–209.
9. Pond C. In: Morbeck M, Galloway A, Zihlman A, eds. *The Evolving Female: a Life History Perspective.* Princeton University Press, Princeton, NJ: 1997, pp. 147–162.
10. Hardy A. New Scientist 1960;7:642–645.
11. Morris D. *The Naked Ape.* McGraw-Hill, New York: 1967.
12. Holliday M. In: Falkner F, Tanner JM, eds. *Human Growth: A Comprehensive Treatise.* Plenum, New York: 1986, pp. 117–139.
13. Scrimshaw NS. J Nutr 2003;133:316S–321S.
14. Pelletier DL, Frongillo EAJ, Habicht JP. Am J Pub Health 1993;83:1130–1133.
15. Yajnik CS, Fall CHD, Coyaji KJ, et al. Int J Obes 2003;27:173–180.
16. Harrington TAM, Thomas EL, Frost G, et al. Pediatr Res 2004;55:437–441.
17. Kuzawa CW. Am J Hum Biol 2005;17:5–21.
18. Gluckman P, Hanson M. Science 2004;305:1733–1736.
19. Hales CN, Barker DJ. Diabetologia 1992;35:595–601.
20. Bateson P, Barker D, Clutton-Brock T, et al. Nature 2004;430:419–421.
21. Gluckman PD, Hanson MA. *The Fetal Matrix: Evolution, Development, and Disease.* Cambridge University Press, Cambridge: 2005.
22. Kensara O, Wootton S, Phillips D, et al. The Hertfordshire Study Group. Am J Clin Nutr 2005;82:980–987.
23. Toschke AM, Montgomery SM, Pfeiffer U, et al. Am J Epidemiol 2003;158:1068–1074.
24. Susser M, Stein Z. Nutr Rev 1994;52:84–94.
25. Vickers MH, Breier BH, Cutfield WS, et al. Am J Physiol 2000;279:E83–E87.
26. Plagemann A, Harder T, Rake A, et al. J Nutr 2000;130:2582–2589.
27. Bloomfield FH, Oliver MH, Hawkins P, et al. Endocrinology 2004;145:4278–4285.
28. Bloomfield FH, Oliver MH, Giannoulias CD, et al. Endocrinology 2003;144:2933–2940.
29. Vickers MH, Gluckman PD, Coveny AH, et al. Endocrinology 2005;146:4211–4216.
30. Hofman PL, Cutfield WS, Robinson EM, et al. J Clin Endocrinol Metab 1997;82:402–406.
31. Armitage JA, Khan IY, Taylor PD, et al. J Physiol 2004;561:355–377.
32. Wild SH, Byrne CD. Nutr Res Rev 2004;17:153–162.
33. McMillen IC, Robinson JS. Physiol Rev 2005;85:571–633.
34. Singhal A, Lucas A. Lancet 2004;363:1642–1645.
35. Eriksson JG, Forsen T, Tuomilehto J, et al. Diabetologia 2003;46:190–194.
36. Bhargava SK, Sachdev HS, Fall CHD, et al. N Engl J Med 2004;350:865–875.
37. Eriksson J, Lindi V, Uusitupa M, et al. Clin Genet 2003;64:366–370.
38. Kajantie E, Rautanen A, Kere J, et al. J Clin Endocrinol Metab 2004;89:5738–5741.
39. Kubaszek A, Markkanen A, Eriksson JG, et al. J Clin Endocrinol Metab 2004;89:2044–2047.
40. Weedon MN, Frayling TM, Shields B, et al. Diabetes 2005;54:576–581.
41. Jordan KM, Syddall H, Dennison EM, et al. J Rheumatol 2005;32:678–683.
42. Gluckman PD, Pinal CS. Endocrine 2002;19:81–89.
43. Mericq V, Ong KK, Bazaes RA, et al. Diabetologia 2005;48:2609–2614.
44. Ozanne SE, Hales CN. Proc Nutr Soc 1999;58:615–619.
45. Ravelli G-P, Stein ZA, Susser MW. N Engl J Med 1976;295:349–353.
46. Ravelli AC, van der Meulen JH, Osmond C, et al. Am J Clin Nutr 1999;70:811–816.
47. Ong KK, Ahmed ML, Emmett PM, et al. BMJ 2000;320:967–971.
48. Kuzawa CW, Pike IL. Am J Hum Biol 2005;17:1–118.
49. Cutfield W. IVF children are taller with increased IGF-I, IGF-II and IGFBP-3 levels suggesting altered genetic imprinting. Proceedings of the 3rd International Congress on Developmental Origins of Health and Disease. Toronto, Nov. 16–19, 2005.
50. Gluckman PD, Hanson MA, Spencer HG, et al. Proc R Soc Lond B 2004;272:671–677.
51. Gluckman PD, Hanson MA, Spencer HG. Trends Ecol Evol 2005;20:527–533.
52. Bateson P. Int J Epidemiol 2001;30:928–934.

53. Gluckman PD, Hanson MA. Trends Endocrinol Metab 2004;15:183–187.
54. Gluckman PD, Hanson MA. Trends Endocrinol Metab 2006;17:7–12.
55. Jablonka E, Oborny B, Molnar I, et al. Phil Trans R Soc Lond B 1995;350:133–141.
56. Morton SMB. In: Gluckman PD, Hanson MA, eds. *Developmental Origins of Health and Disease*. Cambridge University Press, Cambridge, UK: 2006, pp. 98–129.
57. Ounsted M, Scott A, Moar VA. Ann Hum Biol 1988;15:119–129.
58. Gluckman PD, Hanson MA. Semin Fetal Neonatal Med 2004;9:419–425.
59. Gluckman PD, Hanson MA. Pediatr Res 2004;56:311–317.
60. Haig D. Q Rev Biol 1993;68:495–532.
61. Gluckman PD, Hanson MA, Morton SM, et al. Biol Neonate 2004;87:127–139.
62. Colin Bell A, Adair LS, Popkin BM. Am J Epidemiol 2002;155:346–353.
63. Moore SE, Halsall I, Howarth D, et al. Diabet Med 2001;18:646–653.
64. Eidelman AI, Samueloff A. Semin Perinatol 2002;26:232–236.
65. Gluckman PD, Hanson MA. In: Hornstra G, Uauy R, Yang X, eds. *The Impact of Maternal Nutrition on the Offspring* (Nestle Nutrition Workshop Series Pediatric Program Vol. 55). Karger AG, Basel: 2005, pp. 17–27.
66. Yajnik CS, Joglekar CV, Pandit AN, et al. Diabetes 2003;52:2090–2096.
67. Arenz S, Ruckerl R, Koletzko B, et al. Int J Obes 2004;28:1247–1256.
68. Owen CG, Martin RM, Whincup PH, et al. Pediatrics 2005;115:1367–1377.
69. Plagemann A, Heidrich I, Gotz F, et al. Exp Clin Endocrinol 1992;99:154–158.
70. Davidowa H, Plagemann A. Neurosci Lett 2004;371:64–68.
71. Singhal A, Fewtrel M, Cole TJ, et al. Lancet 2003;361:1089–1097.
72. Stettler N, Stallings VA, Troxel AB, et al. Circulation 2005;111:1897–1903.
73. Barker DJP, Eriksson JG, Forsén T, et al. Int J Epidemiol 2002;31:1235–1239.
74. Osmond C, Barker DJP. Environ Health Perspect 2000;108:545–553.
75. Norman JF, LeVeen RF. Atherosclerosis 2001;157:41–47.
76. Ozanne SE, Hales CN. Nature 2004;427:411–412.
77. Prentice AM, Moore SE. Arch Dis Child 2005;90:429–432.
78. West-Eberhard MJ. *Developmental Plasticity and Evolution*. Oxford University Press, New York: 2003.
79. Kirschner MW, Gerhart J. *The Plausibility of Life: Resolving Darwin's Dilemma*. Yale University Press, New Haven: 2005.
80. Gluckman PD, Hanson MA. *Mismatch: Why Our World No Longer Fits Our Bodies*. Oxford University Press, Oxford: 2006.

17 Genetics of Obesity

Five Fundamental Problems With the Famine Hypothesis

John R. Speakman

Abstract

The last 50 yr have witnessed a major epidemic of obesity in Western societies. The development of obesity has a strong genetic component, yet the timescale of its increase cannot have come about because of population genetic changes. Consequently, the most accepted model is that obesity is a consequence of a gene–environment interaction. This current model suggests that we have an ancient genetic predisposition to deposit fat that is particularly expressed in the modern environment. Why we have this genetic predisposition has been a matter of much speculation. There is currently a broad consensus that over evolutionary time we have been exposed to regular periods of extreme food shortage (famines), during which time fatter individuals would have had a selective advantage. Consequently, individuals with genes promoting the efficient deposition of fat during periods between famines (so-called "thrifty genes") would be favored. In the modern environment this genetic predisposition prepares us for a famine that never comes, and an epidemic of obesity follows. This chapter presents five fundamental flaws with the famine hypothesis for the genetic origins of obesity. The flaws are that (1) modern hunter–gatherer and subsistence farming communities show no fat deposition between famines; (2) famines occur only about once every 100–150 yr and involve increases in total mortality that only rarely exceed 10% of the population; (3) most people in famines die of disease rather than starvation, the latter factor accounting for between 5 and 25% of the famine mortality; (4) famine is a modern phenomenon and most human populations have probably not experienced more than 100 famine events in their history; and (5) the age distribution of mortality during famine would not result in differential mortality between lean and obese people. Even ignoring this latter factor, a simple genetic model presented in this chapter shows that famines provide insufficient selective advantage over an insufficient time period for a thrifty gene to have any penetration in the modern human population.

Key Words: Famine; hunger; starvation; mortality; selection; thrifty gene; genetics; obesity.

1. INTRODUCTION

Western societies are in the throes of an obesity epidemic that has grown over the past 50 yr or so to become classed as the greatest health threat facing the Western world *(1)*. Estimates of the prevalence of obesity suggest that in the United States obesity affects 20 to 35% of the population, with an additional 35 to 40% of individuals being classed as overweight, depending on the source of the data *(2,3)*. Recent evidence points to

From: *Nutrition and Health: Adipose Tissue and Adipokines in Health and Disease*
Edited by: G. Fantuzzi and T. Mazzone © Humana Press Inc., Totowa, NJ

levels of obesity rising among populations throughout Europe, southeast Asia, and the third world, so that the epidemic has taken on a global dimension *(4–7)*. The rapidity of the rise of the prevalence of obesity, combined with our knowledge of modern birth and death rates, indicates that the epidemic cannot be caused by the genetic restructuring of the population. Consequently, there must be an environmental cause. Yet, when studies have explored the contribution of genetic and shared environmental factors on individual susceptibility to obesity, the dominant effect always emerges as genetic *(8–10)*. Moreover, many gene mapping studies have been performed identifying more than 200 candidate genes that may influence our susceptibility to develop obesity *(11–13)*. The inevitable conclusion, then, is that obesity results from a gene by environmental interaction *(14,15)*: some individuals have a genetic predisposition to become obese, which is exposed only in the modern environment.

Faced with this scenario, several previous researchers have speculated about the evolutionary processes that may have led to this genetic predisposition. The scenarios presented to date are fundamentally similar, in that they postulate that the modern epidemic results from the natural selection of traits that in our ancient history were advantageous, but when the resultant adapted genome is immersed in modern society it confers disadvantages. Among the first to make this suggestion was Neel *(16)*, who proposed that diabetes and obesity stemmed from natural selection on our ancient ancestors, favoring a "thrifty genotype" that enabled highly efficient storage of fat during periods of food abundance. Neel argued that such a genotype would have been extremely advantageous for primitive humans, who were exposed to periods of food shortage, because it would allow them to efficiently deposit fat stores and thus survive any subsequent period of food shortage *(16)*. In modern society, however, where food supply is almost always available, this thrifty genotype proves deleterious because it promotes efficient storage of fat in preparation for a period of shortage that never arrives. The development of insulin resistance was seen as part of this adaptive thrifty genotype, helping early humans with the process of efficient fat deposition.

Since Neel there have been many papers broadly reiterating the theme that obesity, and its sequela of metabolic syndrome, are fundamentally a consequence of a genotype that was at some historical stage ideally adjusted to an ancient environment, characterized by unpredictable energy resources, that fails to cope in an environment characterized by constant and highly available supplies of energy. In the past decade this original idea has been heavily promoted in many papers as a plausible evolutionary scenario, underpinning not only the obesity epidemic but epidemics of other chronic diseases *(10,17–28)* and aging *(29)*. In particular, the critical role that has been played by historical periods of famine in the process of selection has been strongly reiterated *(18,19,23,26–28)*. It is the purpose of this chapter to challenge the orthodox opinion that famine has been a significant force in the process molding our genetic predisposition to obesity.

2. SUPPORTIVE EVIDENCE FOR THE FAMINE HYPOTHESIS

There is no doubt that famine has been a common feature of human history for a long time *(25,30–32)*, and that during famine individuals face enormous deprivation and increased mortality *(30,33)*. There are numerous biblical references to famine periods, which attest to the great antiquity of the process. The best known of these—the seven years of drought predicted by Joseph in Egypt, "This seven-year famine fell upon Egypt and all the world" (Genesis 41:57)—was estimated to have occurred between 1878 and 1871 BCE. Prentice *(25)* lists many other records of famine that also date back thousands of years. The impression is that famine is a universal feature of human existence that has always been with us, even in prehistoric times. Chakravarthy and Booth *(23)* indicated that the major period that resulted in selection of 95% of our modern biology was the late Paleolithic period, from 50,000 to 10,000 yr ago, and that this period was also characterized by cycles of feast and famine.

The process of selection depends on the intensity of selection (i.e., differential mortality with respect to the trait in question) and the frequency at which such selective events takes place. The frequency of famines depends to an extent on definitions, but periods of starvation due to natural shortages of food supply may be relatively frequent. In Britain periods of severe food deprivation occur roughly every 10 yr or so (190 documented events in 2000 yr) *(30)*. In 50,000 yr since the dawn of the Paleolithic period, therefore, individuals may have encountered approx 5000 famine events that have molded the modern genome. Indeed, the importance of famine as a primary force in demography was emphasized by Malthus in his seminal publication, "Essay on the Principle of Population," published in 1798. This essay was read by Darwin and Wallace and appears to have been a primary stimulus to both of them for the idea of evolution by natural selection. In fact, the role of famine as a contributor to evolution in higher animals was emphasized by Darwin *(34)*.

That individuals involved in famines undergo extreme privation is reflected in the many anecdotes that document individuals resorting to cannibalization of their own children during such events *(26,27)*. Because individuals are unlikely to sacrifice their own offspring—who are a direct component of their fitness—unless they are under extreme risk of mortality themselves, the argument is advanced that not only is famine a common event, but the level of selection during famine must be intense. Moreover, this increased mortality is combined with a decline in fertility—a combination, it is argued, that must have had a profound effect on the evolution of "thrifty genes" predisposing us to deposit fat efficiently whenever famines were not present.

A cornerstone of the "famine hypothesis" is the idea that individuals who have large deposits of fat in their bodies survive periods of famine, whereas their lean counterparts do not. Consequently, the genes that predispose to efficient deposition of fat during postfamine good times are favored. It is certainly the case that individuals who are obese store much greater amounts of energy. However, by virtue of their much larger body size, obese people also burn up energy at a faster rate. The relationship between energy expenditure and body size has a much shallower gradient than the relationship between energy storage and body size. Consequently, the time period that a fatter person can survive in the total absence of food is, in theory, longer. Data concerning the time that people of different body weights survive under conditions of total food abstinence are available from measurements made on people making political gestures by engaging in hunger strikes where they voluntarily starve themselves to death *(35)*. These studies confirm that individuals who are fatter at the start of a complete fast live longer than those who start out thinner.

To date the idea that famine predisposes us to inherit thrifty genes has been a completely qualitative argument. How realistic is it to expect the evolution of thrifty genes during only 5000 famine events? Given the above scenario we can model the population genetics for the penetration of a thrifty gene into a population undergoing periodic famine. Consider therefore a population of 5 million individuals (an approximate estimate for the world population in the Paleolithic period) that consists entirely of individuals that have a dominant allele "A." Initially all individuals have the genotype AA, which generates a lean phenotype inefficient at depositing fat. A mutation happens in a single individual to generate a recessive allele "a," which is a "thrifty" gene. Hence the population level probabilities of finding these alleles immediately after the mutation happens are 0.9999999 for A and 0.0000001 for a. We will assume that the thrifty gene confers on its owner a selective advantage during famines because of reduced mortality. Let us set survival during a famine at 0.68 for carriers of the AA genotype (i.e., 32% annual mortality). In reality the efficiency of fat storage will be a polygenic trait influenced by several genes, each having a relatively small effect. Let us assume that the mutation increases fat storage during the period prior to any famine such that the risk of mortality is reduced in the homozygotic aa genotype to 29% (survival = 0.71). We will assume that mortality of the heterozygote lies midway between these at 30.5% (survival = 0.29). The patterns of representation of the three genotypes AA, Aa, and aa over a sequence of 1000 famine events are shown in Fig. 1. Over a period of around 500 famine events following the mutation, little happens. The thrifty gene remains a very rare allele and the population is completely dominated (>99.9% of individuals) by the lean AA genotype. However, between 500 and 750 famine events this situation changes. The Aa genotype and then the aa genotype increase in frequency. With more famine events the aa genotype then replaces the Aa genotype until, after 950 famine events, all the 5 million individuals in the population are homozygous for the thrifty gene. This model shows that if a thrifty gene that results in more efficient fat deposition confers only a 3% improvement in survival, it would spread to dominate the population in only one-fifth of the hypothesized number of famine events that may have occurred since the start of the Paleolithic. Several independent mutations affecting fat storage would not influence this pattern of penetration of this thrifty aa genotype. However, if two genes were to interact synergistically—for example, if there were two genes, A and B, where the effect of the aa and bb genotypes both improved survival during famine by 3% but the aa × bb genotype resulted in an improvement of 9% instead of 6%, the penetration of the aa genotype would be slightly faster (Fig. 1B).

3. SOME FUNDAMENTAL PROBLEMS WITH THE FAMINE HYPOTHESIS

The second part of this chapter outlines what are considered to be five fundamental problems with the famine hypothesis as an explanation for the genetic contribution to the modern obesity epidemic.

3.1. Prevalence of Obesity Too Low During Periods Between Famines to Be a Strongly Selected Trait

If the argument is correct that in hunter–gatherer and subsistence agriculture communities there is strong selection for thrifty genes that enable individuals to deposit fat efficiently in times between famines, then during periods when these communities do not experience famine the individuals should become obese. If they do not become obese

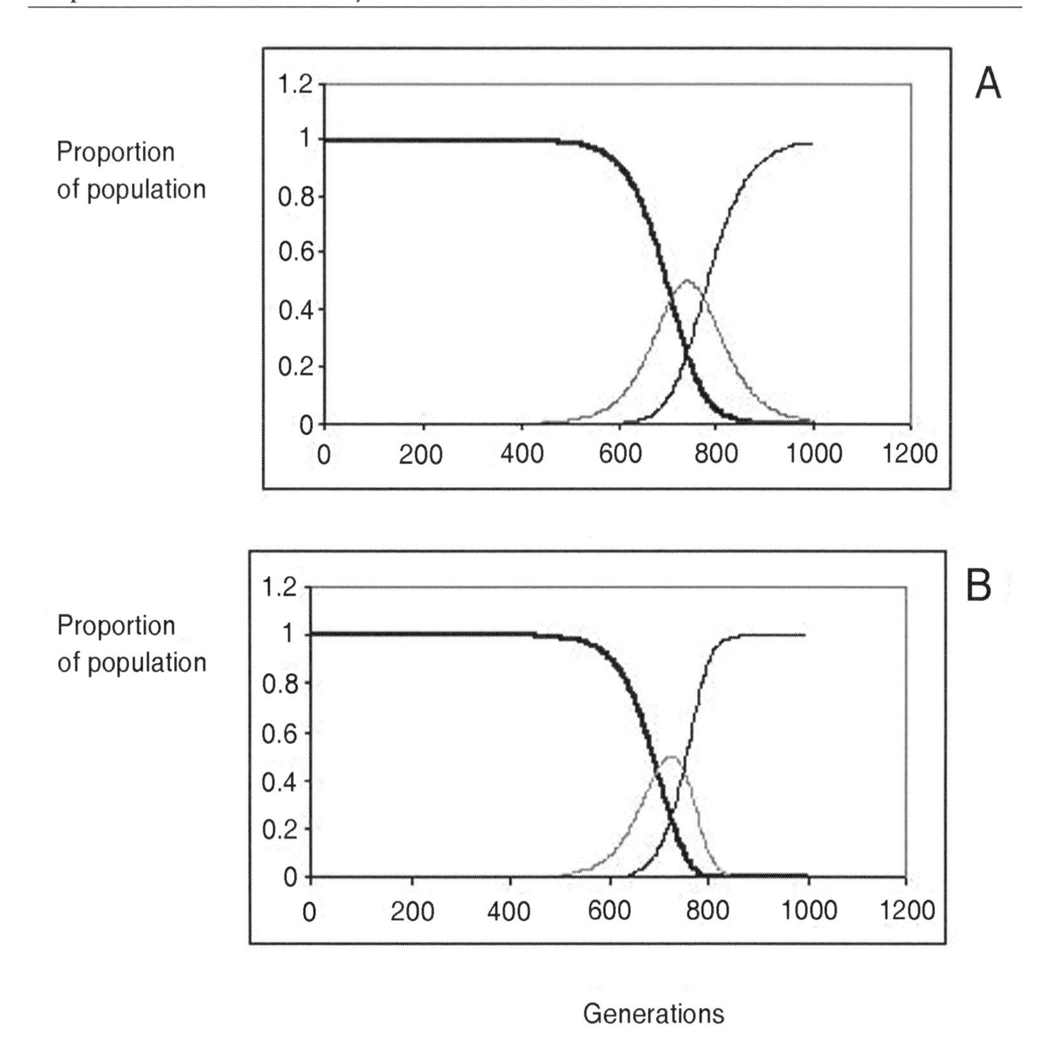

Fig. 1. Propotions of three different genotypes in relation to the number of famine events. There are two differrnt alleles. A lean allele (A) and a thrifty allele (a). The plots show the changing proportion of the population with AA (thick black line) Aa (grey) and aa (thin black line) genotypes. The population is set at 5 million individuals and the survival of the three genotypes during a famine is set at 0.68 for AA, 0.695 for Aa and 0.71 for aa in (**A**). Under this scenario of selection, the thrifty genotype (aa) completely replaces the lean phenotypes (AA) within 1000 famine events. In (**B**) there are two genes. The impact on survival of AA, Aa, and aa types is the same as in **A**, except the effect of the aa homozygous is enhanced in the presence of bb (survival = 0.74 not 0.71). In this situation the non-additive effect on survival of the aa × bb genotypes leads to a faster spread of the aa genotype in the population.

(or at least overweight) between famines then it is difficult to see how they would derive any survival advantage during the next ensuing period of famine.

Although hunting–gathering has largely died out as a lifestyle, there are still many tribal communities in the third world subsisting on agriculture, using practices effectively unchanged for thousands of years. If the model in Fig. 1 is realistic, all the individuals in these populations should carry thrifty genes conferring efficient fat storage capabilities. Table 1 summarizes some estimates of body mass index (BMI) in hunting–gathering and subsistence-farming populations. Only one of the papers from which

Table 1
Body Mass Indices of Hunter–Gatherer (HG) and Subsistence Agriculture (SA) Communities During Periods Between Famines

Location	*Population*	*Activity*	*Sex*	*BMI*	n	*Ref*
Namibia	!kung San	HG	M	19.4	238	69
			F	19.1		
	Kavango	SA	M	19.4	156	
			F	20.3		
Camaroon	Pygmy	HG	M, F	19.9–20.9		70
	Bantu	SA	M, F			
Australia	Aboriginal	HG	M, F	<20.0		71
Parguay	Ache	SA				72
Kenya	Ariaal	HG & SA	M, F	17.8	56	36
Ethiopia	Elka	SA	M	19.7	226	73
			F	20.0		

these measurements are taken indicates that during the collection of the anthropological measurements the communities were under any form of food restriction resembling a famine. Yet, despite being in generally nonfamine conditions, the individuals are universally characterized by being in the BMI range from 17.5 to 21.0—at the very lean end of normal. None of the communities even approaches the cutoff point for being overweight, never mind being obese. In the one study in which a population was studied in drought conditions (the Ariaal in Kenya *[36]*) it is interesting to note that the investigators detected no significant differences in BMI between a population of subsistence agriculturalists undergoing a drought, and a population of nomads that were not experiencing such conditions. It is hard to reconcile the data in Table 1 with any other conclusion than these individuals have not inherited the supposed thrifty genes that predispose them to weight gain and obesity during periods between famines.

3.2. Extent of Mortality During Famine and Frequency of Famine are Insufficient to Act as Driving Force for Evolution

Although details of the deprivations that people undergo during famine are impressive, and stories of eating one's own children indications of the desperation that famine victims endure, this information is fundamentally anecdotal. The key aspect of importance in terms of the evolutionary and demographic significance of famine is the effect it has on rates of mortality and fertility. Estimates of mortality in ancient famines are generally vague and imprecise and include phrases such as "countless multitudes" or "vast numbers" *(33)*. During the Middle Ages several famines were suggested to involve enormous mortality rates. For example, the famine of 1376 in Italy is said to have wiped out two-thirds of the population *(30)*, but evidence supporting this numerical estimate is limited. It is only since the 1800s that well-supported estimates of the number of people dying during famines have been generated. However, translating even these more reliable figures into meaningful mortality statistics is hampered by the problem that such numbers generally do not specify the size of the population to which they refer, nor the baseline rates of mortality in the same populations when famine was not occurring. These problems were illustrated by Watkins and Menken *(33)* in their discussion of mortality during the Chinese famine of 1878 to 1879, which has been described as one of the worst famines of modern times (13 million people died). As a percentage of the contemporary population of the five provinces in which the famine was centered (107 million), this gives a crude death rate of about 15%. This represents an absolute maximum rate attributable to the famine, as some mortality would have occurred anyway, and, as is very often the case, "mortality" statistics frequently fail to distinguish mortality from emigration. If the "mortality" is expressed relative to the wider population size of the whole of China (estimated at 430 million at the time), the estimated death rate is much lower, at only 3.5%. In contrast, records for specific towns in the affected region report populations before the famine of 145,000 inhabitants that crashed to postfamine levels of 45,000—implying a death rate of almost 60%. As a general trend, therefore, as the area involved in the estimates gets smaller, the estimated mortality gets larger. This is because larger areas are less likely to confound mortality with emigration, and also because famines generally are limited in their geographical extent.

Clearly, setting some bounds on the affected area makes most sense when attempting to estimate mortality rates. When this is done and comparisons are made between periods involving famine and periods of intervening feast, then the total rates of mortality for most famines are generally under 10% per annum. For example, mortality in England during the famine of 1597 was about 52% higher than the previous years when no famine occurred *(37)*. Similar estimates apply to the Bangladeshi famines in the 1970s, which involved increases in the crude death rates of 38% and 58%, and the Chinese "Great Leap Forward" famine of 1959–61, where crude death rates increased by 72 to 104% over the previous famine-free period *(38,39)*. Because crude death rates in the absence of famine in these instances were around 2% per annum, the estimated total mortalities during the famines was all less than 5% of the population per annum. Moreover, about half of this was mortality that would have occurred anyway, and of the remainder, for some famines, at least some of this extra mortality was actually emigration. The extent of mortality during periods of famine is much lower than is implied by the severity of individual accounts of the level of suffering.

The frequency with which famines occur depends to an extent on the inclusion criteria for what constitutes a famine. Keys et al. *(30)* identified 190 "famines" that occurred in Britain over 2000 yr, giving a rough occurrence rate of one famine per 10 yr. It is clear that not all of these involved the same duration of food shortage or levels of mortality. Indeed, Wrigley and Schofield *(40)* documented the occurrence of food supply crises in England over a 330-yr period between 1541 and 1871 and found that the mean duration of a supply crisis was, on average, only 2 mo. Given that mortality during really severe famine events lasting multiple years seldom exceeds 10% per annum, it is likely that mortality increases during such short events would be barely detectable. In fact, estimates of the frequency of occurrence for severe mortality crises in Europe (defined as crises leading to significant increases in mortality in a given year relative to the baseline over the previous 10 yr) were an order of magnitude less frequent *(41)*, and these mortality events include all causes, so the occurrence of significant famine must be even less frequent. This calculation suggests that rather than occurring every 10 yr, famines involving significant mortality occur less than once every century *(41)*. Data from China paint a very similar picture. The province of Hubei consists of 71 counties. Over the 63 yr between 1850 and 1912 the records of famine events across the province during this period reveal that famines occurred at a county level on 29 occasions. This is from a total of $71 \times 63 = 4473$ county-years of observation. In other words, on average, each county could expect to see a famine event once every 154 yr (i.e., 4473/29) *(42)*.

These data show that, when a more quantitative approach is taken, famine is actually a rare demographic event, and when it does occur the mortality impact is relatively low. In addition to direct mortality a major impact in terms of evolution and selection favoring "thrifty obesity-inducing genes" is the reduction in fertility that also occurs during periods of famine, which has been suggested to be important *(26)*. There is no question that declines in fertility during famine can be large. In the 1974 Bangladeshi famine, fertility rates for females aged between 15 and 40 declined by, on average, 30% relative to nonfamine years *(43)*. Similar data are available for other famines *(25,33,44)*. The causes of the famine-induced decline in fertility are multifactorial, including loss of libido and spousal separation (when one person in a couple moves to find food), but it is at least plausible to suggest that part of the effect may reflect amenorrhea caused by reduced body condition. In this case the burden of lost fertility would fall more on the leaner individuals; this could be a selective pressure favoring the evolution of metabolic efficiency and obesity. However, these speculations about the role of fertility have ignored the evidence that following famine there is often a fertility rebound once the disaster is over. In 1962, for example, following the end of the Great Leap Forward famine in China, fertility rates were elevated by 25% relative to the period prior to the famine *(38)*. Although such a rebound is not always observed, the fact that females who survive the famine are not permanently impaired and may enjoy catch-up fertility will certainly minimize any impact of the reduction during the famine period. The modern consensus, despite the huge role avowed for famine in demographic processes by Malthus *(45)*, is that famine is trivial in demographic terms. Detailed models of the effect of famine suggest that even extremely severe events (such as might occur once each century) involving increased death rates by 10% per annum and reduced fertility by 30%, both lasting for a period of 2 yr, would have a barely detectable impact on a population over the time frame of 100 yr *(46)*.

3.3. Historical Pattern of Famine Occurrence is Incompatible With Other Aspects of Hypothesis

Although it has been suggested that famine is a feature of our history stretching back to the earliest part of the Paleolithic 50,000 yr ago *(17,23)*, famine is actually a phenomenon born of the age of agriculture. It is widely accepted that pre-Neolithic hunter–gathering communities were probably far less prone to famine because they exploited a range of food types and had the mobility to follow food resources. Once agriculture and stable communities developed, however, large populations became dependent on agricultural crop production *(27,47)*. As the yield of such crop production is potentially sensitive to weather patterns, adverse weather events could cause severe problems. Weather and natural causes certainly take their toll in terms of precipitating some famine events *(48,49)*. However, there is a broad school of thought, following the work of Sen *(50)*, which suggests that most famine events have in fact very little to do with natural factors interfering with food production patterns and much more to do with societal factors leading to inequable distribution of resources—the so-called "entitlement hypothesis" for famine causality. An obvious case in point is the Chinese "Great Leap Forward" famine, which was precipitated primarily by communist China exporting its food production in return for arms and technology, rather than distributing it to the indigenous population *(51–53)* (but *see* ref. *54* for additional important factors). This entitlement hypothesis, at its extreme, considers all famines to result from human action and should be considered acts of genocide *(55)*.

Such causes for famine can occur not only after the dawn of agriculture but also after the emergence of quite sophisticated societies. Prentice *(27)* therefore considered that famine has basically been an important evolutionary phenomenon for only the last 6000 yr or so. Using an estimate that famines occur on average every 10 yr *(30)* he suggested that this still leaves 600 selection events working over time to remodel the genetic makeup of the population. Using the model in Fig. 1, this would be insufficient to favor complete penetration of a thrifty gene mutation. As we have seen, however, the estimate of one severe famine event per decade is probably at least an order of magnitude too great. Using the estimate of 6000 yr ago for the origin of famine susceptibility suggests that most human populations have probably experienced only 60 or so famine events in their history. Even if we take a generous estimate of the susceptibility dating back to the dawn of agriculture and increasingly developed social systems at 10,000 yr ago *(56)*, this still means most populations of humans have, in their entire history, experienced only 100 severe famine events, and even in these severe events mortality seldom exceeded 10% of the population per annum.

Whether the size and frequency of these events are enough to drive the natural selection of thrifty genes depends on the extent to which the mortality is distributed between lean and obese subjects (*see* Subheading 3.4.). However, the modern timing of the events (over the past 6000 to 10,000 yr) is fundamentally incompatible with claims made by others advocating the famine hypothesis that the human genetic constitution has remained essentially unaltered over this time period *(17,23)* and that any significant evolutionary forces must predate the Neolithic.

3.4. People in Famines Generally Do Not Die of Starvation

Although fat people live longer under conditions of voluntary complete starvation, in accord with the predictions of the balance of energy expenditure relative to storage (fasting endurance), these conditions are not necessarily a good model for what happens during natural (or man-made) conditions of famine. In fact, it may come as a surprise to most people to discover that during most famines relatively few of the people that die perish from frank starvation—that is, depleting their body reserves to the point where they run out of energy. Perhaps the most comprehensive data available on causes of mortality during famine comes from the Irish potato famine, which hit the island of Ireland between 1845 and 1850. The cause of the famine was a combination of societal and natural disasters. During the period in question the staple diet of the population was almost entirely restricted to a single crop: the potato. During the famine years the potato crop was devastated by an infection of the potato blight fungus. In 1841 the population of Ireland was assessed by a national census to be 8,175,124 individuals. In 1851, after the famine, the population was 6,552,385—a direct loss of 1.623 million individuals. By factoring in the likely population increase between 1841 and the start of the famine in 1846, and also accounting for the 925,000 people who are estimated to have emigrated out of Ireland during the famine years, the mortality has been estimated to be slightly more than 1 million people. This matches closely the records of 985,000 deaths documented in 686 pages of mortality statistics meticulously compiled in the immediate postfamine period by Wilde *(57)*—mostly on the basis of personal recall by survivors of deaths within their own families. It has been pointed out (from as far back as Wilde himself) that personal testimony may be biased by poor memory and inaccurate recall. Yet the death of one's family is something seldom forgotten, even in times when mortality is high, so this seems unlikely to be a serious bias. A more serious issue, however, is that the causes of death that wiped out entire families would be under-reported simply because no survivors would be around to make the report *(58)*. Paradoxically, therefore, the most serious causes of death would be less likely to be represented.

Nevertheless, despite these caveats the mortality statistics for the Irish famine still provide a comprehensive picture of what the major causes of mortality were during one of the most serious famine events ever known. In fact, mortality during the worst famine years of 1847, 1848, and 1849 were 18.5, 15.4, and 17.9% (mean = 17.3%) of the population—considerably greater than reported in most other famines (above), and much higher than the background mortality rates in the prefamine years of 1842, 1843, and 1844, which averaged 5.1, 5.2, and 5.6% (mean = 5.3%). In 1841 there were 140,000 deaths recorded in Ireland by the census commissioners, of which deaths only 17 were attributable to starvation. During 1847, at the height of the famine, the reported total deaths numbered 250,000. Yet only 6000 of these were deaths attributed to starvation. Consequently, in this major famine the mortality directly attributable to the famine as excess deaths above background amounted to about 12% of the population annually, but of these deaths only 5% could be directly attributed to starvation. Starvation-induced mortality during the peak famine years was therefore only about 0.6% of the population per annum. Even if this mortality was biased completely to the leanest members of the population, this differential mortality would be very unlikely to provide a selective force favoring obesity-related genes.

So if people don't die of starvation during famine, why do they die? The causes of mortality during major famine events are predictably complex. The major factors, however, are infectious diseases and diarrhea. Of the 985,000 deaths in Ireland during the famine, 23% were attributed to fever (probably typhoid). Approximately 35% were caused by diarrhea and dysentery, and the major cause of mortality during 1849 included an epidemic of cholera. The Irish famine is probably unique in the meticulous detail in which mortality has been recorded—even if there are some flaws inherent in relying on individual family recall. However, data are available from many other famines and the patterns of cause of mortality are all broadly similar. Starvation is generally a cause of less than 25% of the observed mortality. Infectious disease and diarrhea make up the vast majority of the effect. Data from the Ethiopian famine in 1997–2000 are typical. During this period in the Gode area, there was a major famine in part caused by drought. During this period approx 23% of famine mortality was caused by starvation, 22% by a measles outbreak, and 37% by diarrhea. The balance was attributable to other infectious diseases. In the Madras famine of 1840, 40% of deaths were attributed to cholera and smallpox outbreaks. In Bangladesh, which experienced two famines during the 1970s, infectious disease was also listed as the most significant cause of death *(59)*. Overall, during famine conditions, 60 to 95% of mortality is due to infectious disease *(60)*. Although these general principles hold, there are records of some famines where most of the mortality was caused by starvation (e.g., ref. *61*); these, however, appear to be rare exceptions.

Infectious diseases take hold in famine conditions for two reasons. These reflect a radical difference between those people voluntarily starving themselves to death to make a political statement and people who face externally imposed food shortages (natural or man-made) but do not wish to die. In the latter conditions people try to find food; in the former they do not. The first problem with this strategy is that people who are facing starvation become relatively unselective in their food choices. People eat weeds, tree bark, and various other plants that they would normally not select. Moreover, people will readily eat decomposing carrion and even eat the corpses of other individuals who have died. This shift in food choice has several major consequences. First, people die during famine of direct poisoning from eating poisonous plants *(62)*. Second, eating decomposing carrion or cannibalizing corpses can lead to intestinal infections and diarrhea. Third, the decline in food nutritional quality, in terms of vitamin and micronutrient quantities rather than energy quantities, may lead to a decrease in immunocompetence and hence the ability to fight off infectious disease. In fact, by the end of the Irish famine, there was a reappearance in Ireland of scurvy, generated by the virtual absence of any fresh fruit intake *(63)*. Although this was not a mortality factor, it points to the general decline in nutritional status of the population.

Individual immunosuppression, however, may be less significant in the spread of disease during famine than a general breakdown in sanitary conditions *(64,65)*. This occurs in part because people become mobile in the search for food and, in so doing, become disconnected from fixed assets that are a necessary part of good hygiene, such as the ability to wash oneself and one's clothes (eliminating lice and transmission of typhus) in clean water. Individual mobility compounds transfer of disease, particularly when people start to assemble in areas where they think they may get food. Finally, medical practices (such as the practices in the Irish famine of giving castor oil to cure diarrhea and bloodletting from already weakened individuals) may have aimed to resolve the medical problems of famine but in reality may have exacerbated them.

Although most people in famine do not die because of starvation, it might be argued that lean people would be more likely to succumb to these problems because their need to feed would be greater because of their lower energy reserves. Lean people might therefore be more prone to accidentally poison themselves, or to resort to eating food that was microbiologically suspect, giving them a greater susceptibility to intestinal problems such as diarrhea. Hence, although mortality is not caused by starvation *per se*, there might still be an overall bias in the direction of mortality favoring selection for obese genes. Unfortunately we have no direct evidence concerning the actual bias in mortality patterns with respect to the BMI of individuals at the start of any famine period. However, some evidence suggests that a bias in mortality toward lean people because they are "more hungry" during starvation and thus more likely to make poor food choices that compromise their health seems improbable. There is no evidence that perceptions of hunger that drive people to feed are greater in the lean than the obese. In fact, absolute energy demands and food intake are greater the larger a person is. If hunger is driven by a shortfall in intake relative to demands, then one would predict that hunger would actually be greater in the obese and consequently it would be the obese rather than the lean who would make more poor food choices.

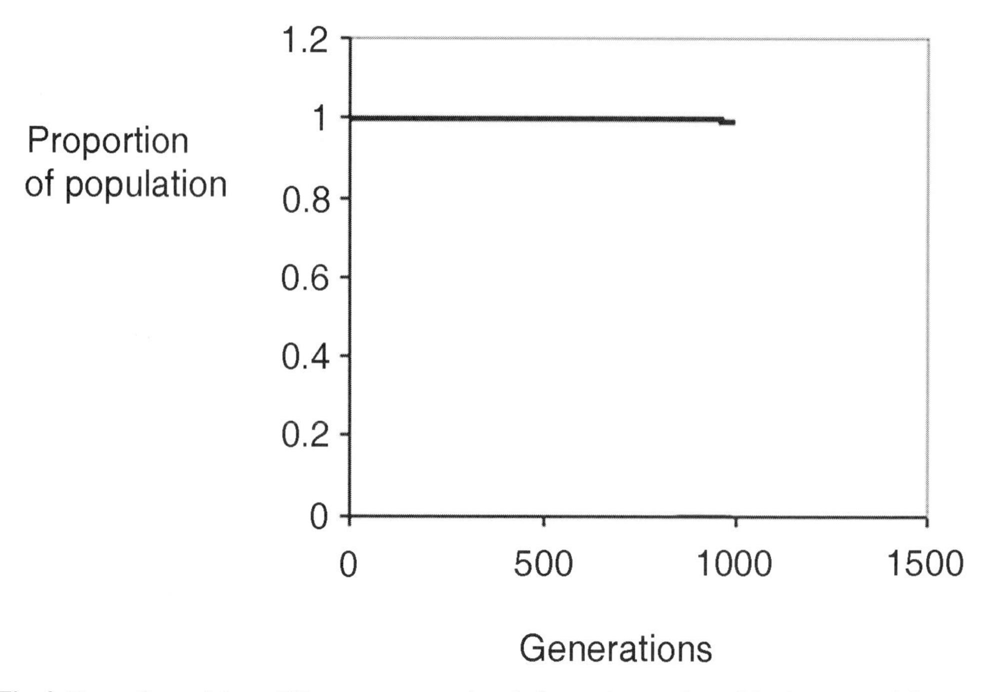

Fig. 2. Proportions of three different genotypes in relation to the number of famine events. There are two different alleles. A lean allele A and a thrifty allele a. The plot shows the changing proportion of the population with AA (thick black) Aa (grey) and aa (thin black) genotypes (note the aa and Aa genotypes have values along the *x*-axis for almost the entire 1000 events). The population is set at 5 million individuals and the survival of the three genotypes during a famine is 0.88 for AA, 0.89 for Aa and 0.90 for aa. Under this scenario of selection the thrifty genotype does not spread in the population even after 1000 famine events.

The role of individual immunosuppression in the spread of infectious disease is another area where it might be argued that the immune system would become more rapidly compromised in a lean person compared with an obese one. Yet the association of immune status with body weight is confounded by the covariation of low body weight and malnutrition (low nutrient intake).

How do these re-evaluations of the salient parameters affect the theoretical penetration into a population of a thrifty gene? I reparameterized the model presented in Fig. 1 by assuming the same starting conditions—a population of 5 million, but this time I assumed that the total mortality during the famine events was elevated by, on average, 10% and that of this 10% one-fifth (i.e., $0.2 \times 10\% = 2\%$ of total mortality) was caused by actual starvation. I also assumed that this starvation mortality occurred only among the lean phenotypes. Hence during the famine events the survival of the aa genotype was 0.90 (i.e., 10% mortality, 8% higher than background mortality of 2%) but for the AA genotypes the survival was 0.88 and for the Aa genotype it was 0.89. Under this selection scenario something completely different happens (Fig. 2). Instead of the thrifty gene completely replacing the lean phenotype in around 950 famine events, there is absolutely no penetration of the thrifty gene at all. In fact by 1000 famine events the homozygous lean phenotype has declined to only 98.5% of the population, with the balance

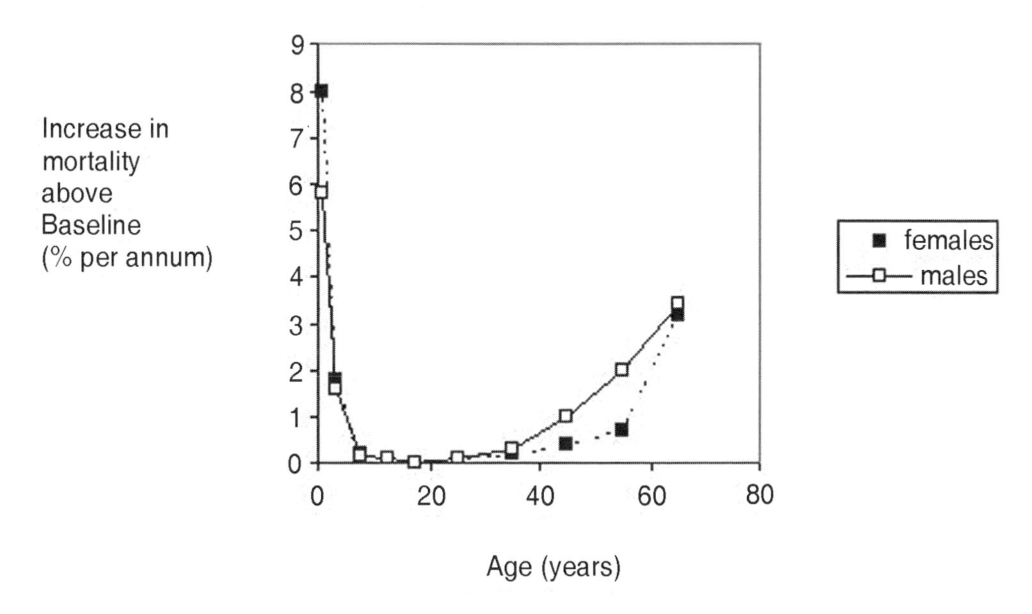

Fig. 3. Pattern of age-related mortality in the 1974 famine in Bangladesh. Mortality is concentrated in the youngest and oldest cohorts in both sexes. Among people aged 15 to 25 there was virtually no increase in mortality relative to non-famine years. (Drawn from data in ref. *68*.)

comprising Aa. This population genetic modeling indicates that the low frequency of famine, combined with the quantitative estimates of its severity and the low percentage of this mortality that reflects true starvation, makes famine as a driving force for the evolution of thrifty genes highly unlikely. Any mutant gene conferring a thrifty phenotype would never become established in the population. The last problem with the famine hypothesis is that even this scenario may be optimistic.

3.5. Burden of Mortality in Famines Affects Wrong People to Have Selection for Energy Efficiency

Although we have no direct information about the variation in mortality during famines with respect to the BMI of individuals prior to the period of food shortage, there are abundant data documenting the differential susceptibility of individuals with respect to other parameters. The most commonly reported of these is age. Time and again, studies of mortality in famines report that the burden of mortality falls disproportionately on two particular groups: the very young (<5 yr old) and the very old (>60 yr old) *(33,60,66,67)*. The Ethiopian famine in Gode between 1997 and 2000 provides a typical example. Of 293 deaths reported during the last 6 mo of the famine period, 158 (54%) were in children under 5, and a further 73 (25%) were in children aged between 5 and 14 (CDC data). In the Irish famine 60% of mortality was for children under 10 and the elderly aged over 60 *(57)*. Increases in mortality in relation to age during the Bangladeshi famines in the 1970s are shown in Fig. 3 *(68)*. The very distinctive pattern revealed in this plot is how little the mortality rate of the young adult population is elevated. For the youngest age group mortality increased by 6 to 8%, and in older individuals by 3 to 5%, but for individuals in their late teens and early 20s mortality was virtually unaffected. This variation in mortality reflects the fact that most people during famine conditions die of infectious diseases and it is the very young and the old who are most susceptible to these causes of mortality.

This pattern of mortality is a serious problem for the famine hypothesis for the adaptive selection of obesity-related traits. Any mortality in the elderly is completely unimportant in terms of its impact on selection because these individuals have already made their reproductive investment and passed their genes on to the next generation. If fat old people die before lean old people it makes absolutely no impact on the population genetics. Mortality in the very young might be more significant. However, it is very unlikely to be biased with respect to obesity simply because, until recently, obesity in this age class was virtually unheard of. The key group in which differential mortality might have some impact because the individuals involved may (1) exhibit a substantial variation in their phenotype and (2) still have reproductive investments to make (i.e., young adults prior to and during their reproductive years) are exactly the people who remain virtually unaffected by the impact of famine mortality. Consequently the assumption in the model presented in Figs. 1 and 2 that starvation during famine is biased toward individuals that have the lean (AA and Aa) phenotypes is unrealistically biased in favor of the penetration of the a allele. Yet even under this extremely optimistic situation the thrifty genes did not penetrate the population after 1000 famine events (more than an order of magnitude more events than most human populations have probably experienced).

4. CONCLUSIONS

In summary, I have highlighted five fundamental difficulties with the widely accepted view that the modern obesity epidemic has, at its core, a mismatch between an ancient metabolism finely tuned to cope with an environment characterized by periods of feast and famine, and a modern environment where opportunities to feast are plenty, but famine never comes. The five problems are that severe famines involving significant mortality are rare events and that even in the worst famines mortality seldom exceeds 10% per annum. Of that figure probably only 5 to 20% is caused by actual starvation, most of which falls on the very young and old. Most human populations have experienced such events probably less than 60 times. Genetic models suggest that this scenario would not favor the spread of a thrifty gene mutation. The prediction that thrifty genes will not evolve is compatible with empirical observations that hunter–gatherer and subsistence agriculture communities during periods between famines have very lean phenotypes—indicating they do not have thrifty genes helping them deposit fat stores in preparation for any forthcoming period of famine.

ACKNOWLEDGMENTS

I am grateful to Peter Grant and Ela Krol for interesting discussions on the issues raised in this chapter.

REFERENCES

1. World Health Organization. Obesity: preventing and managing the global epidemic. WHO Report 894. Geneva: 1998.
2. Flegal KM. Physiol Behav 2005;86:599–602.

3. Yun S, Zhu BP, Black W, et al. Int J Obes 2006;30:164–170.
4. Arroyo P, Loria A, Fernandez V, et al. Obes Res 2000;8:179–185.
5. Dorosty AR, Reilly JJ. Obes Res 2000;8:D18.
6. Kain J, Uauy R, Vio F, et al. Eur J Clin Nutr 2002;56:200–204.
7. Likitmasku S, et al. J Pediatr Endocrinol Metab 2003;16:71–77.
8. Barsh GS, Farooqi IS, O'Rahilly S. Nature 2000;404:644–651.
9. Perusse L, Chagnon YC, Rice T, et al. Med Sci 1998;14:914—924.
10. Hebebrand J. Ernahrungs-Umschau 2005;52:90.
11. Perusse L, Rankinen T, Zuberi A, et al. Obes Res 2005;13:381–490.
12. Snyder EE, Walts B, Perusse L, et al. Obes Res 2004;12:369–439.
13. Chagnon YC, Rankinen T, Snyder EE, Weisnagel SJ, Perusse L, Bouchard C. Obes Res 2003;11: 313–367.
14. de Castro JM. Br J Nutr 2004;92:S59–S62.
15. Speakman JR. J Nutr 2004;134:2090S–2105S
16. Neel JV. Am J Hum Genet 1962;14:352–353.
17. Eaton SB, Konner M, Shostak M. Am J Med 1988;84:739–749.
18. Lev-Ran A. Diabetes Metab Res Rev 2001;17:347–362.
19. Lev-Ran A. Diabetes Rev 1999;7:1–22.
20. Campbell BC, Cajigal A. Med Hypotheses 2001;57:64–67.
21. Ravussin E. J Clin Invest 2002;109:1537–1540.
22. Diamond J. Nature 2003;423:599–602.
23. Chakravarthy MV, Booth FW. J Appl Physiol 2004;96:3–10.
24. Scott EM, Grant PJ. Diabetologia 2006;49:1462–1466.
25. Prentice AM. Mech Ageing Dev 2005;126:976–981.
26. Prentice AM, Rayco-Solon P, Moore SE. Proc Nutr Soc 2005;64:153–161.
27. Prentice AM. Physiol Behav 2005;86:640–645.
28. Prentice AM. Br Med Bull 2001;60:51–67.
29. Kirkwood TBL, Shanley DP. Mech Ageing Dev 2005;126:1011–1016.
30. Keys AJ, Brozek J, Henschel O, et al. *The Biology of Human Starvation*. University of Minnesota Press, Minneapolis, MN: 1950.
31. McCance RA. Proc Nutr Soc 1975;34:161–166.
32. Elia M. Clinl Nutr 2000;19:379–386.
33. Watkins SC, Menken J. Pop Dev Rev 1985;11:647–675.
34. Darwin C. *On the Origin of Species*. Ward and Lock, London: 1859; reprint from 1910.
35. Elia M. In: Scrimshaw N and Schurch B. *Protein-Energy Interactions,* International Dietary Energy Consultancy Group, Proceedings of an IDECG workshop. Lausanne, Nestle Foundation, Switzerland: 1991.
36. Campbell B, O'Rourke MT, Lipson SF. Am J Hum Biol 2003;15:697–708.
37. Appleby AB. *Famine in Tudor and Stuart England*. Stanford University Press, Stanford, CA: 1978.
38. Ashton B, Hill K, Piazza A, et al. Pop Dev Rev 1984;10:613–645.
39. Coale AJ. Rapid population change in China, 1952–1982. Committee on Population and Demography. 27. National Academy Press, Washington, DC: 1984.
40. Wrigley EA, Schofield R. *The Population History of England 1541–1871*. Harvard University Press, Cambridge, MA: 1981.
41. Dupaquier J. In: Charbonneau H, LaRose A, eds. *The Great Mortalities*. Ordina, Liege: 1979, pp. 83–112.
42. Ho PT. *Studies on the Population of China, 1368–1953*. Harvard University Press, Cambridge, MA: 1959.
43. Ruzicka LT, Chowdhury AKMA. Vital events and migration—1975. Demographic surveillance system—MATLAB. Vol. 4. 12. Cholera Research Laboratory, Dacca: 1978.
44. Cai Y, Feng W. Demography 2005;42:301–322.
45. Malthus TR. First Essay on Population. 1798.
46. Watkins SC. A skeptical view of the demography of famines: results from a simultion model. Florence International Population Conference. Vol. 4, International Union for the Scientific Study of Population, Liege: 1985, pp. 339–348.

47. Diamond J. *The Rise and Fall of the Third Chimpanzee*. Harper, New York: 1993.
48. Therrell MD, Stahle DW, Soto RA. Bull Am Meteorol Soc 2004;85:1263.
49. Tauger MB. J Peasant Stud 2003;31:45–72.
50. Sen A. *Poverty and Famines: an Essay on Entitlement and Deprivation*. Clarendon Press, Oxford: 1981.
51. Lin JY, Yang DT. China Econ Rev 1998;9:125–140.
52. Li W, Yang DT. J Polit Econ 2005;113:840–877.
53. Lin JY, Yang DT. Econ J 2000;110:136–158.
54. Chang GH, Wen GJ. China Econ Rev 1998;9:157–165.
55. Edkins J. IDS Bull (Inst of Dev Stud) 2002;33:12.
56. Diamond J. *Guns, Germs and Steel. The Fates of Human Societies*. W. W. Norton, New York: 1997.
57. British Parliamentary Papers. The census of Ireland for the year 1851. Part V, Vols. 1 and 2. Tables of Death. 1856.
58. Mokyr J, Grada CO. Famine disease and famine mortality: lessons form Ireland, 1845–1850. Centre for Economic Research Working paper 99/12. University College Dublin: 1999.
59. Chen LC, Chowdhury AKMA. The dynamics of contemporary famine. Mexico International Population Conference (Vol. 1) International Union for the Scientific Study of Population, Liege: 1977, pp. 409–426.
60. Toole MJ, Waldman RJ. Bull WHO 1988;66:237–247.
61. Hionidou V. Pop Stud 2002;56:65–80.
62. Addis G, Urga K, Dikasso D. Hum Ecol 2005;33:83–118.
63. Crawford M. J Soc Social Hist Med 1988;1:281–300.
64. Carmichael A. In: Rotberg RI, Rabb TK. eds. *Hunger and History: The Impact of Changing Food Production and Consumption Patterns on Society*. Cambridge University Press, Cambridge: 1983, pp. 51–66.
65. Dirks R. In: Kiple KF, ed., *The Cambridge World History of Human Disease*. Cambridge University Press, Cambridge: 1993, pp. 157–163.
66. Murray M, Murray A, Murray N, Murray M. Lancet 1976;332:1283–1285.
67. Lindtjorn B. Br Med J 1990:301:1123–1127.
68. Menken J, Campbell C. Health Trans Rev 1992:2:91–108.
69. Kirchengast S. Ann Hum Biol 1998;25:541–551.
70. Kesteloot H, et al. Nutr Metab Cardiovasc Dis 1997;7:383–387.
71. Odea K. Clin Exp Pharmacol Physiol 1991;18:85–88.
72. Bribiescas RG. Am J Phys Anthropol 2001;115:297–303.
73. Alemu T, Lindtjorn B. Int J Epidemiol 1995;24:977–983.

18 Inherited and Acquired Lipodystrophies

Disorders of Adipose Tissue Development, Differentiation, and Death

Vinaya Simha and Anil K. Agarwal

Abstract

In humans, lipodystrophies constitute a heterogeneous group of inherited or acquired disorders characterized by the selective loss of adipose tissue. There is considerable variation in the extent of adipose tissue loss and the severity of metabolic complications resulting from insulin resistance, such as hyperinsulinemia, acanthosis nigricans, hypertriglyceridemia, diabetes mellitus, and hepatic steatosis. Among genetic lipodystrophies, the fat loss is observed either since birth as in congenital generalized lipodystrophy, or it occurs later in life as in familial partial lipodystrophy. Defects in several genes such as those encoding an enzyme (AGPAT2), a nuclear receptor (PPAR-γ), a nuclear lamina protein (LMNA) and its processing endoprotease (ZMPSTE24), a kinase (AKT2), and a protein of unknown function (BSCL2) have been found in patients with genetic lipodystrophies. Additional loci still remain to be discovered. Acquired lipodystrophies mainly occur due to autoimmune mediated fat loss or because of HIV-1 protease inhibitors induced adipose tissue atrophy. This chapter discusses various features of inherited and acquired lipodystrophies and their pathogenesis. We further discuss various possible mechanisms for the loss of adipose tissue based on our current understanding of adipocyte biology.

Key Words: Lipodystrophies; acyltransferase; endoprotease (ZMPSTE24); lamin A/C; seipin (BSCL2); PPARG; insulin resistance; diabetes mellitus; hepatic steatosis.

1. INTRODUCTION

Lipodystrophies are a heterogeneous group of inherited or acquired disorders, characterized by selective loss of adipose tissue (AT). The pattern and extent of AT loss vary among different types of lipodystrophies *(1)*. For example, in congenital generalized lipodystrophy (CGL), the loss of AT is near total and is observed from birth, whereas in familial partial lipodystrophy (FPL) the loss of AT is observed later in life, and is generally confined to peripheral regions, such as the extremities, or hips, often sparing the trunk. There is also a genetic heterogeneity associated with lipodystrophies. Currently, mutations in six genes—*AGPAT2*, *BSCL2*, *LMNA*, *PPARG*, *ZMPSTE24,* and *AKT2*—have been found in patients affected with lipodystrophies *(2–8)*.

From: *Nutrition and Health: Adipose Tissue and Adipokines in Health and Disease*
Edited by: G. Fantuzzi and T. Mazzone © Humana Press Inc., Totowa, NJ

Table 1
Classification of Lipodystrophies

Inherited (genetic) lipodystrophies
Autosomal-recessive
Congenital generalized lipodystrophy, types 1 and 2
Lipodystrophy associated with mandibuloacral dysplasia, types A and B
Lipodystrophy associated with SHORT syndrome
Lipodystrophy associated with neonatal progeroid syndrome
Autosomal-dominant
Familial partial lipodystrophy
Lipodystrophy associated with SHORT syndrome
Lipodystrophy associated with Hutchinson-Gilford progeria syndrome
Pubertal-onset generalized lipodystrophy
Acquired lipodystrophies
Lipodystrophy in HIV-infected patients
Acquired partial lipodystrophy
Acquired generalized lipodystrophy
Localized lipodystrophy

The initial description of lipodystrophic syndrome was provided by Mitchell and colleagues who, in the late 1800s, described a 12-yr-old girl who had lost AT from the upper half of the body and face since the age of 8 yr. This most likely was the first documented case of acquired partial lipodystrophy *(9)*. Not much progress was made until the mid-1900s, when Berardinelli and Seip reported patients with complete loss of AT since birth *(10,11)*. In the past few years, there has been a renewed interest in studying lipodystrophies, mainly because the affected patients have severe insulin resistance, diabetes mellitus (DM), and other associated features of metabolic syndrome. In addition, there has been a rapid increase in the incidence of lipodystrophy in HIV-infected subjects receiving highly active antiretroviral therapy (HAART) containing HIV-1 protease inhibitors. Consequently, pathological and genetic mechanisms of lipodystrophies might provide clues to the pathogenesis of other common disorders.

Based on clinical features, lipodystrophies can be broadly classified as either acquired or inherited (genetic) varieties. Depending on the extent of fat loss (localized, partial, or generalized), mode of inheritance and other associated features, each variety can be further subclassified as shown in Table 1.

In this review, we describe various genetic forms of lipodystrophies and their associated clinical and biochemical defects. We will also describe lipodystrophies associated with various other syndromes where the pathophysiology is not well understood, as well as acquired lipodystrophies. Finally, we will present our working hypothesis for the mechanism of AT loss, based on our current understanding of adipocyte commitment from mesenchymal cells and further differentiation.

2. CONGENITAL GENERALIZED LIPODYSTROPHY FROM MUTATIONS IN *BSCL2* AND *AGPAT2*

Since the first case description of CGL, an autosomal recessive disorder, by Berardinelli and Seip *(10,11)*, some 300 cases have been reported so far. Although described in patients of various ethnic origins, significant clusters of patients seem to be localized to some regions in Brazil and Lebanon. Affected patients have near-complete absence of subcutaneous AT from birth, leading to marked prominence of muscles and veins. During childhood, they are noted to have a voracious appetite, accelerated growth, and advanced bone age. Umbilical hernia or prominence of the periumbilical skin, and an acromegaloid appearance because of enlargement of hands, feet, and mandible, are other common features. Hepatomegaly from fatty infiltration may be seen at birth or later in life and may be accompanied by splenomegaly. Acanthosis nigricans is often observed over the neck, axilla, groin, and trunk. Postpubertal girls develop clitromegaly and features of polycystic ovarian syndrome. Less commonly, some patients develop multiple focal lytic lesions in the appendicular skeleton, hypertrophic cardiomyopathy, and mild mental retardation *(12–15)*.

Patients also develop significant metabolic abnormalities very early in life. Extreme insulin resistance is a striking feature of this syndrome. Most patients have high fasting and postprandial insulin levels. Onset of DM is usually seen during pubertal years, but, occasionally, neonates may also develop hyperglycemia. Severe hypertriglyceridemia leading to recurrent pancreatitis, low HDL cholesterol levels, and chronic steatohepatitis are other associated features. These patients have very low levels of serum leptin and adiponectin *(16)*. Administering recombinant leptin to these patients as a replacement therapy helped to control hyperglycemia, hypertriglyceridemia, and hepatic steatosis *(17)*, suggesting that hypoleptinemia may be partly responsible for the metabolic abnormalities.

2.1. Molecular Genetics

Genome-wide linkage analysis of several pedigrees and positional cloning strategies have led to the identification of two genetic loci for CGL: the 1-acylglycerol 3-phosphate-O-acyltransferase 2 (*AGPAT2*) gene on chromosome 9q34 *(2,18)*, which is associated with CGL1, and the Berardinelli Seip congenital lipodystrophy 2 (*BSCL2*) gene located on chromosome 11q13, which is associated with CGL2 *(3)*. AGPAT2 belongs to the superfamily of acyltransferase enzymes *(19)*. This enzyme is involved in the biosynthesis of triglycerides and glycerophospholipids by esterifying the sn-2 position of lysophosphatidic acid with a fatty acid to form phosphatidic acid (Fig. 1) *(20)*. AGPAT2 has a tissue-restricted expression pattern, such that its cognate mRNA is highly expressed in human omental adipose tissue and less in the liver and skeletal muscle *(2)*. This pattern suggests that defective AGPAT2 function is likely to impair triglyceride and glycerophospholipid synthesis in adipocytes, which might cause lipodystrophy (Fig. 1). When expressed in heterologous cultured cells, many of these mutations, either missense or amino acid deletions, have reduced enzymatic activity (Fig. 2) *(21)*. Although not directly determined, it is possible that these mutations also reduce intracellular glycerophospholipid content, which might also interfere with adipocyte function, as these phospholipids are important for signaling pathways and membrane structure and functions.

The second locus, *BSCL2,* encodes a 398-amino-acid protein, seipin, whose physiological function still remains undefined. Based on homology with other species, it appears that the predicted protein contains an additional 64 amino acids in its amino terminus, although no mutation has been observed in this region *(22)*. Intriguingly, the highest expression for *BSCL2* is detected in the brain and testis *(3)*, but seipin is only

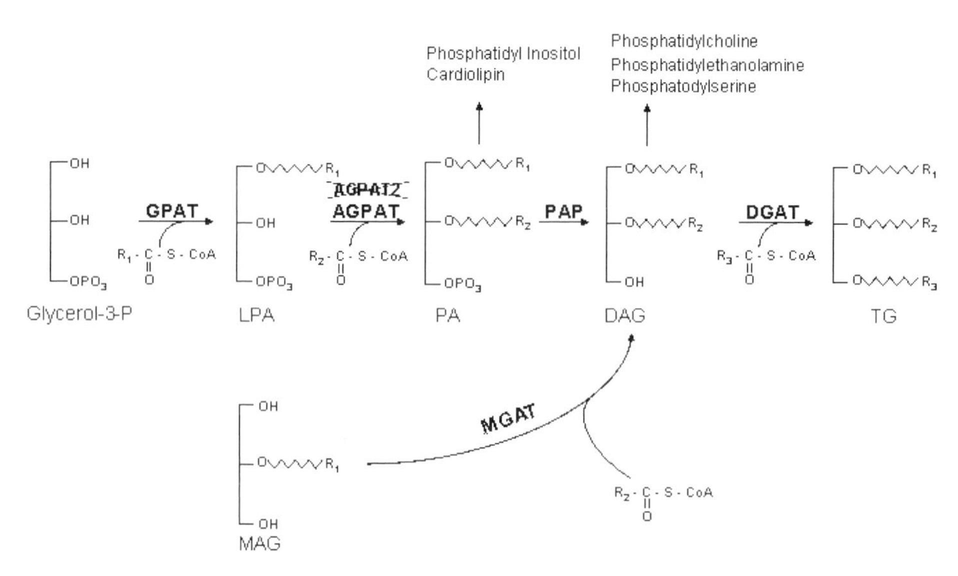

Fig. 1. Pathways for the biosynthesis of triacylglycerol. Glycerol-3-phosphate is the initial substrate for acylation at the sn-1 position by glycerol-3-phosphate acyltransferase (GPAT), to form 1-acylglycerol-3-phosphate or lysophosphatidic acid (LPA). LPA is further acylated at the sn-2 position by 1-acyl-glycerol-3-phosphate acyltransferase (AGPAT) to form phosphatidic acid (PA). In the next step, the phosphate group is removed by phosphatidate phosphohydrolase (PAP) to produce diacylglycerol (DAG). DAG is further acylated at the sn-3 position by diacylglycerol acyltransferase (DGAT) to produce triacylglycerol (TG). In addition, TG can be synthesized via the acylation of 2-monoacyl-glycerol by the enzyme monoacylglycerol-acyltransferase (MGAT), which is highly expressed in the small intestine. PA and DAG are also substrate for the synthesis of glycerophospholipids. PA is the substrate for synthesis of phosphatidylinositol and cardiolipin. Phosphatidylcholine, phosphatidylethanolamine, and phosphatidylserine are synthesized from DAG. LPA and related lipid molecules are also involved in signal transduction and are ligands for G protein-coupled receptors. R_1, R_2, and R_3 could be any long-chain fatty acid.

weakly expressed in adipose tissue *(23)*. This pattern suggests that the loss of AT in CGL2 might be mediated by a central mechanism involving the brain; however, a direct role of *BSCL2* in adipocyte development and differentiation may still be possible. Recently, a homology search revealed that seipin protein has weak but significant homology to the protein midasin *(22)*. Midasin belongs to the family of AAA$^+$-containing proteins, which have myriad functions in the cells. Thus, clues as to its role in causing lipodystrophy in CGL2 are beginning to appear. However, its precise role in AT loss remains far from clear.

2.2. Phenotypic Variation in CGL1 and CGL2

Identification of two different genetic loci, with genes associated with two different pathways, suggested that there might be phenotypic differences in these patients as well. Indeed, on further inspection, patients with CGL2 have increased prevalence of cardiomyopathy and mental retardation, whereas focal lytic lesions in the appendicular skeleton are mostly seen in patients with CGL1 *(14,15)*. Differences in body fat distribution have also been noted between the two genotypes *(24)*. Clinical examination

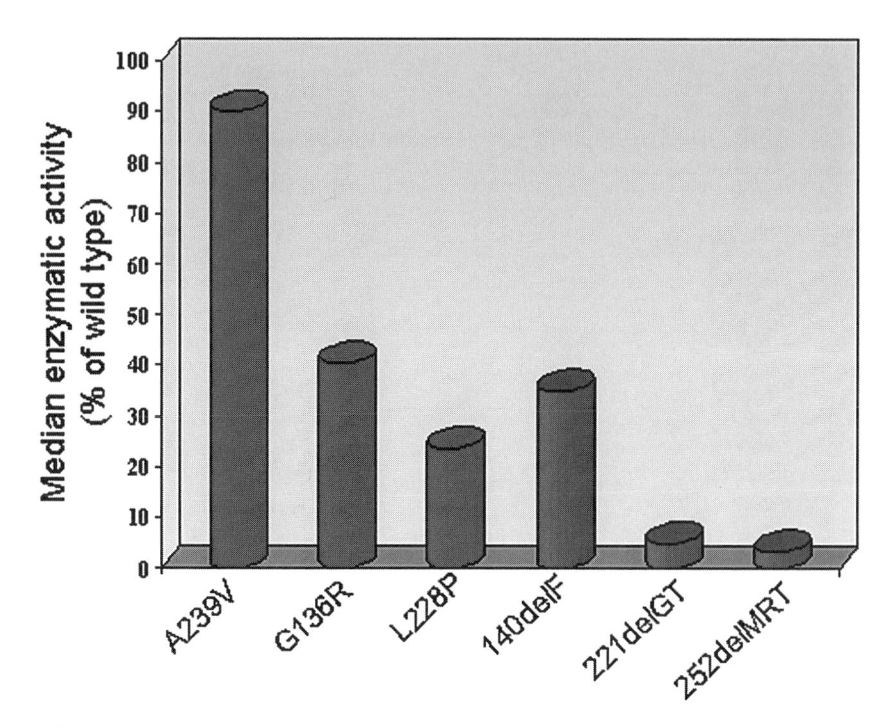

Fig. 2. AGPAT2 mutant enzymatic activities. Shown are the enzymatic activities of missence and amino acid deletion mutants. Enzymatic activities were determined in overexpressing CHO cells by the conversion of ^{3}H-lysophosphatidic acid (^{3}H-LPA) to ^{3}H-phosphatidic acid (^{3}H-PA). Shown are the mean values plotted as percentage of wild-type.

and magnetic resonance imaging (MRI) studies reveal that CGL2 patients have more severe fat loss involving the palms, soles, orbits, scalp, and periarticular regions, not seen in patients with CGL1 (Fig. 3). AT in these regions has been postulated to have a mechanical function, whereas AT in other regions, such as subcutaneous, intermuscular, bone marrow, intraabdominal, and intrathoracic regions, is metabolically active. The metabolic AT is lost with both the *AGPAT2* and *BSCL2* genotypes. The role of seipin in causing the additional loss of mechanical AT remains to be determined. Finally, it must be noted that certain patients with CGL do not have mutations in either *AGPAT2* or *BSCL2* and their pedigrees do not show linkage to these loci, thus suggesting the presence of additional CGL loci *(15)*.

3. FPL CAUSED BY MUTATIONS IN *LMNA, PPARG,* OR *AKT2*

Familial partial lipodystrophies are autosomal dominantly inherited disorders of AT. The most distinguishing feature, compared with CGL, is a normal body AT distribution at birth with progressive and variable loss of AT, mostly from the extremities and truncal region of the body. The candidate gene approach has now identified mutations in three genes—*LMNA (5,6)*, *PPARG (4,25)*, and *AKT2 (8)*—in these affected patients with FPL, but additional loci are also likely, as some patients do not harbor mutations in these three genes.

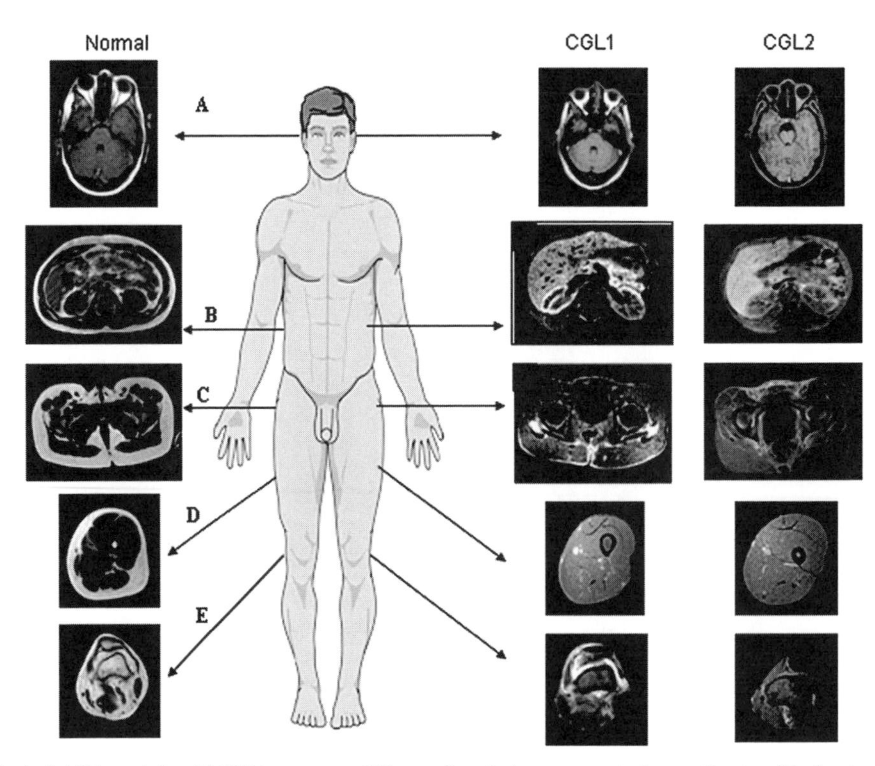

Fig. 3. Axial T1-weighted MRI images at different levels in representative patients with the two types of congenital generalized lipodystrophy (CGL) (right) and a normal subject (left) as indicated. Adipose tissue has a short T1 and a long T2 proton relaxation time compared with other tissues, and therefore appears bright because of high signal intensity on T1-weighted images. **(A)** Section through the orbits show preservation of retro-orbital fat and subcutaneous fat under the skull in patient with CGL1 (*AGPAT2* mutation) but not in patient with CGL2 (*BSCL2* mutation). **(B)** Section through the abdomen showing marked loss of both subcutaneous and intraabdominal adipose tissue in both the patients with CGL. Increased signal intensity in the liver suggests hepatic steatosis. **(C)** Section through the hip joint, again showing loss of subcutaneous AT in patients with CGL compared with normal. Note the preservation of fat around the femoral head (periarticular fat) in the patients with CGL1 compared with the patients with CGL2. **(D,E)** Sections through the thigh and knee joint demonstrating loss of subcutaneous, bone marrow, and intermuscular fat in both patients with CGL, whereas periarticular fat around the knee joint is preserved in patient with CGL1.

3.1. Familial Partial Lipodystrophy, Dunnigan Variety From LMNA Mutations

Familial partial lipodystrophy, Dunnigan Variety (FPLD) is an autosomal-dominant disorder first described in 1974 by Dunnigan and colleagues *(26)*. More than 300 patients, mostly of European descent, have since been reported, making this the most prevalent form of FPL *(1)*. As mentioned above, distribution of body fat is normal during childhood, and gradual loss of subcutaneous AT from the arms and legs is noted following puberty. In these patients there appears to be a redistribution of AT (Fig. 4). Although variable AT loss from the chest and anterior abdominal wall is observed, an excess fat deposition over the chin, supraclavicular area, face, dorsocervical, and intra-abdominal regions may occur. Women with FPLD are more easily diagnosed than men, owing to the unusual muscular appearance of the extremities. Metabolic abnormalities

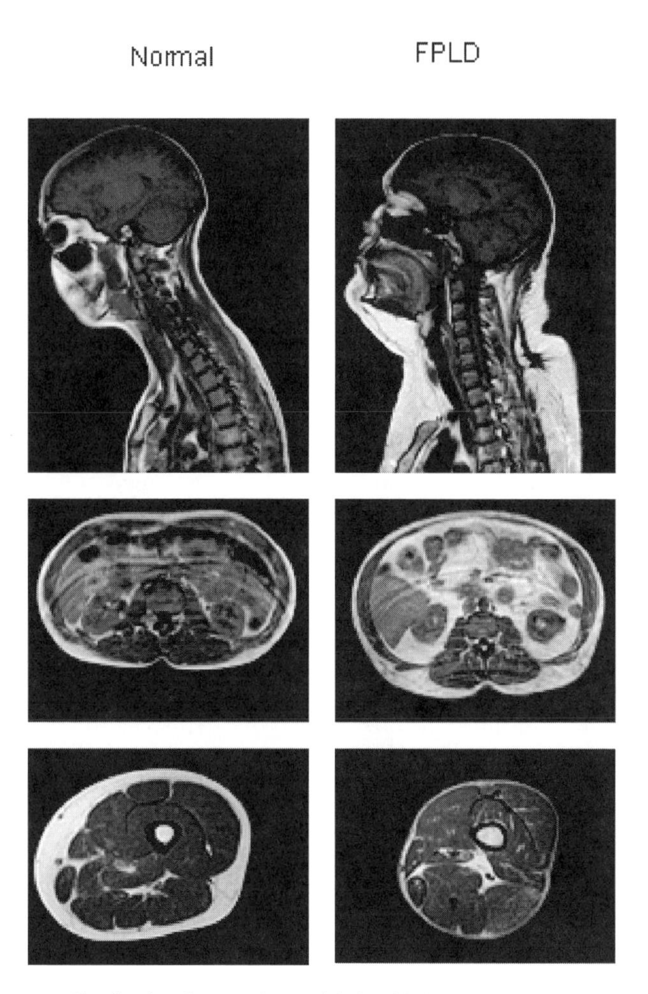

Fig. 4. Adipose tissue redistribution in a patient with familial partial lipodystrophy, Dunnigan variety (FPLD) (right) compared with a normal volunteer (left) as revealed by T1-weighted MRI imaging. **Top panel:** Sagittal section through the head and neck region showing excessive fat accumulation in the submental region, dorsocervical region, and the neck. This abnormal fat deposition is responsible for the "double chin" appearance and "buffalo hump" noted in patients with FPLD. **Middle panel:** Transaxial section through the abdomen demonstrating mild loss of subcutaneous fat anteriorly, and increased accumulation of intra-abdominal fat in the patient with FPLD compared to normal. **Bottom panel:** Transaxial section through the mid-thigh region showing loss of subcutaneous fat, but preservation of bone marrow and intermuscular fat in the patient with FPLD compared with normal.

such as DM and hyperlipidemia are also more prevalent in women *(27–29)*. Acanthosis nigricans and features of polycystic ovarian syndrome are seen in one out of four affected women. In some patients, cardiac conduction system disturbances resulting in atrial fibrillation and congestive heart failure have been noted *(30)*.

The genetic locus for FPLD has been traced to the *LMNA* gene located on chromosome 1q21-22 *(31)*. The *LMNA* gene encodes lamins A and C, which belong to the intermediate filament family of proteins that are involved in the nuclear lamina, providing structural integrity to the nuclear envelope *(32,33)*. Mutant lamins A and C may disrupt normal interaction with chromatin and other nuclear lamina proteins, resulting in apoptosis and premature death of adipocytes. However, this mechanism does not explain selective loss of fat that is confined to particular regions, such as extremities only. Furthermore, specific mutations lead to distinct phenotypes. Most of the missense mutations in FPLD affect the arginine residue at position 482, which resides in the globular domain of the protein that assumes the IgG structural fold *(34)*. Less severe lipodystrophy is associated with substitution of histidine to arginine at codon 582, which lies outside the IgG domain in the carboxyl terminus of the lamin A protein *(35)*. Mutations in exon 1 cause an overlap syndrome of FPLD and cardiomyopathy *(30)*. It is also interesting that some rare patients with *LMNA* mutations develop an entirely different phenotype of pubertal-onset generalized lipodystrophy associated with progeroid features *(36,37)*.

3.2. FPL From PPARG *Mutations*

Heterozygous missense mutations in the *PPARG* gene, encoding the nuclear receptor PPARγ, have been identified in 10 patients with FPL *(4,25,38–40)*. No case of homozygous mutations in the *PPARG* gene have been reported, consistent with the embryonic lethality of homozygous gene deletion in a murine model. All affected subjects had normal AT distribution at birth, and developed loss of AT from distal extremities after the second decade of life. Fat loss was less prominent from the arms and thighs, and facial fat was often spared. Features of insulin resistance, such as DM, hypertension, and hypertriglyceridemia, were noted. *PPARγ* is highly expressed in adipocytes and has a critical role in regulating adipocyte development and differentiation *(41)*. It is therefore possible that *PPARG* mutations cause lipodystrophy by affecting adipogenesis. However, as with *LMNA* mutations, the regional susceptibility of certain fat depots and the temporal progression of the disease remain perplexing.

3.3. FPL From AKT2 *Gene Mutation*

Recently, a heterozygous missense mutation, arginine 274 histidine, in the *AKT2* gene has been reported in a family in which affected subjects developed partial lipodystrophy of the extremities, insulin resistance, DM, and hypertension *(8)*. AKT2 is a serine/threonine kinase expressed in insulin-sensitive tissues and is involved in postreceptor insulin signaling. Data from homozygous *AKT2* gene deletion in mice suggests that AKT2 also regulates the expression of *PPARγ (74)*. Abnormal AKT2 functioning may cause lipodystrophy by decreasing adipocyte differentiation via *PPARγ* regulation or by dysfunctional postreceptor insulin signaling.

4. SYNDROMIC LIPODYSTROPHIES

Lipodystrophies associated with various syndromes with multiple clinical defects have also been identified. Most prominent among them are syndromes of mandibuloacral dysplasia (MAD) and SHORT, an acronym for **s**hort stature, **h**yperextensibility of joints and/or inguinal hernia, **o**cular depression, **R**eiger anomaly, and **t**eething delays. Although two genetic loci, *LMNA (42)* and *ZMPSTE24 (7)* have been identified in MAD, identification of a genetic derangement for SHORT syndrome remains lacking.

Worldwide, only 40 patients with MAD, an autosomal-recessive disorder, have been observed so far. The characteristic features of this syndrome include postnatal resorption of bone in the mandible, clavicles, and terminal phalanges (acro-osteolysis) in addition to AT loss, suggesting a common genetic or metabolic defect affecting both the skeleton and AT *(43)*. In addition, these patients also have short stature, delayed closure of cranial sutures, joint contractures, mottled skin pigmentation, and, more important, features of premature aging.

Two patterns of lipodystrophy have been described in patients with MAD *(43)*. Type A pattern is characterized by partial loss of subcutaneous AT from the extremities, with normal or excess fat over the face and neck. This pattern is similar to the fat loss seen in FPLD patients and, not surprisingly, *LMNA* mutations have been described in MAD patients with Type A lipodystrophy *(42)*. Type B pattern involves a more generalized loss of subcutaneous fat; these patients have compound, heterozygous mutations in *ZMPSTE24*, a gene encoding a zinc metalloproteinase, a microsomal protein, that is essential in post-translational modification of prelamin A *(7)*. The detrimental effects of *ZMPSTE24* mutations may be caused by either accumulation of prenylated prelamin A or lack of mature lamin A. Some MAD patients have no mutations in either *LMNA* or *ZMPSTE24*, suggesting additional, as yet unmapped, loci.

The genetic locus for SHORT and some of the neonatal progeroid syndromes has not been identified. The changes in body AT distribution are also not well defined, mainly because of the rarity of these syndromes. Patients with the autosomal dominant form of SHORT syndrome are described to have lipodystrophy affecting only the face, gluteal region, and elbows *(44)*. Patients with neonatal progeroid syndrome have been described to have generalized lipodystrophy, with sparing of fat over the sacral and gluteal areas *(45)*. There have been no reports of either skinfold measurements or MRI studies in patients with these rare syndromes.

5. ACQUIRED PARTIAL LIPODYSTROPHY

Also known as Barraquer-Simmons syndrome, more than 250 patients with acquired partial lipodystrophy (APL) have been reported in the literature *(46)*. APL is characterized by the gradual onset of bilaterally symmetrical loss of subcutaneous fat from the face, neck, upper extremities, thorax, and abdomen, in a cephalocaudal sequence, sparing the lower extremities. Often, excess fat accumulation is seen over the lower abdomen, gluteal region, thighs, and calves. The onset of fat loss usually occurs during childhood or adolescence, at a median age of 7 yr, whereas the excess fat accumulation is noted to occur at the onset of puberty or with weight gain and glucocorticoid therapy. Fat loss occurs in a gradual process over a period of few months up to 2 yr, and 75% of patients have discernible fat loss before 13 yr of age. Women are reportedly affected four times more often than men *(46)*. Whole-body MRI studies in affected patients confirmed the loss of subcutaneous fat from face, neck, trunk, and upper extremities, including the palms. Intra-abdominal fat (intraperitoneal and retroperitoneal) was also reduced, whereas bone marrow and retro-orbital fat were well preserved. These observations were corroborated by autopsy findings. Unlike other lipodystrophies, patients with APL rarely develop metabolic abnormalities related to insulin resistance. However, this condition is often associated with membranoproliferative glomerulonephritis and other autoimmune disorders, such as systemic lupus erythematosus, dermatomyositis/polymyosits, vasculitis, and undifferentiated connective tissue diseases.

6. ACQUIRED GENERALIZED LIPODYSTROPHY

Acquired generalized lipodystrophy (AGL) is another rare disorder and has been reported in approx 80 patients *(47)*. It is characterized by selective loss of AT from large regions of the body occurring after birth. AT loss usually involves face, trunk, and extremities, and sometimes also spreads to the palms and soles. Intra-abdominal fat may also be lost, although retro-orbital and bone marrow fat are generally well preserved. The loss of AT may occur precipitously, within a few weeks, or may be more insidious over several months. In some patients, fat loss is preceded by the appearance of tender, inflamed subcutaneous nodules caused by panniculitis, but in others the disease is associated with autoimmune diseases such as juvenile dermatomyositis, Sjögren's syndrome, juvenile rheumatoid arthritis, chronic active hepatitis, and autoimmune hemolytic anemia. However, in more than half the reported cases of AGL, neither panniculitis nor autoimmune diseases have been reported. Although autoimmune destruction of adipocytes may explain fat loss in the panniculitis and autoimmune variety, little is known about the pathogenesis of the idiopathic variety of AGL.

Similar to CGL, patients with AGL also show marked insulin resistance, including hyperinsulinemia, DM, hypertriglyceridemia, and low serum HDL cholesterol levels. DM usually presents a few years after the onset of lipodystrophy, but in some instances DM appeared almost simultaneously or even preceded the onset of lipodystrophy. Severe hypertriglyceridemia may be associated with eruptive xanthomas and pancreatitis. Other abnormalities include nonalcoholic steatohepatitis, acanthosis nigricans, and menstrual irregularities. Low levels of serum adipokines such as leptin and adiponectin have also been reported *(16)* and leptin replacement therapy appears to be a promising option to control the metabolic abnormalities *(17,48)*.

7. LIPODYSTROPHY IN HIV-INFECTED SUBJECTS

Lipodystrophy in HIV-infected subjects (LDHIV) is a newly identified variety of acquired lipodystrophy, but is by far the most prevalent form of any lipodystrophy. HIV-1 protease inhibitor (PI) therapy appears to be the major culprit, but some studies suggest that nucleoside reverse transcriptase inhibitors might contribute to it as well *(49–51)*. On average, 40 to 50% of ambulatory HIV-infected patients in the United States have demonstrated abnormalities in body fat distribution *(52)*, and more than 100,000 patients in the United States are estimated to be afflicted by this condition *(1)*. LDHIV may present with diverse morphological features, and it is a matter of debate whether LDHIV represents a composite of different syndromes. The majority of patients demonstrate loss of subcutaneous AT from peripheral regions, including the face, extremities, and gluteal region, a phenomenon sometimes referred to as lipoatrophy *(53)*. In most of these patients, subcutaneous AT over the trunk and visceral AT are reported to be either preserved or increased (lipohypertrophy). Prominence of the dorsocervical fat pad (buffalo hump) and breasts may also be seen. In most patients, features of fat loss and fat excess coexist, suggesting AT redistribution. However, a recent cross-sectional study in a large cohort of HIV-infected men suggests that fat accumulation is not reciprocally associated with peripheral fat loss *(54)*. Further, MRI studies showed that in HIV-positive men with clinically apparent peripheral lipodystrophy/lipoatrophy, less subcutaneous AT is present in both peripheral and central sites, as well as visceral AT, than in HIV-positive men without peripheral lipoatrophy. The overall prevalence of central lipohypertrophy in HIV-positive men was less than in controls. It would, therefore, appear that peripheral lipoatrophy is the cardinal feature of LDHIV, with lipohypertrophy as either a distinct phenomenon or a compensatory change. Further clarification by well-controlled longitudinal studies is needed.

Patients with LDHIV often have dyslipidemia, insulin resistance, and hepatic steatosis. Dyslipidemia in the form of hypertriglyceridemia, hypercholesterolemia, and low HDL cholesterol is much more common (50–70%) than hyperglycemia (0–20%) *(55,56)*. Metabolic abnormalities may sometimes precede changes in body fat distribution.

8. ADIPOSE TISSUE DISTRIBUTION IN HUMANS AND ADIPOCYTE DIFFERENTIATION

In humans, unlike other organs such as the liver, heart, or lung, the AT does not have a well-defined tissue demarcation. Fat under the skin or dermis is referred to as subcutaneous AT, whether it lies under the skin of the abdomen, extremities, or other parts of the body. Fat found in the visceral cavity of the body is largely referred to as intra-abdominal or omental AT. In humans, the subcutaneous (sc) and the omental forms are the two large AT depots that have been studied in detail, whereas smaller AT depots, such as those behind the eyes (retro-orbital), knees (periarticular), around the hip joints, or beneath the skull, have received little attention. It is unclear whether, in addition to these differences in the anatomical location of AT, there are physiological differences as well between these AT depots. Assuming that AT in the different anatomical sites has similar functions would be an oversimplification. Unfortunately, little is known about the physiological differences between the various anatomically distinct AT depots. Some investigators consider the two large depots, sc and intra-abdominal, as metabolically active, whereas the smaller depots, such as retro-orbital, periarticular, or beneath the skull, are seen as relatively inert or mechanical AT *(57)*. Others propose that these so-called mechanical ATs, which are intertwined with the lymphatic system, participate in the immune response via paracrine mechanisms *(58,59)*. Deletion of the mouse *Prox-1* gene, which is expressed only in the lymphatic system, revealed the connection between lymphatic system and AT when these animals developed obesity *(60,61)*.

Much of our understanding of AT biology has been derived by analyzing the mouse 3T3-L1 fibroblast cell line, which has been used extensively to dissect the mechanisms of adipocyte differentiation and maturation using an in vitro cell culture model system. As shown in Fig. 5A, the initial phase begins with the recruitment of mesenchymal stem cells to committed cell lineages: osteoblasts, preadipocytes, chondrocytes, and myoblasts. Although not much is known about the differentiation of these cell lineages, recent studies show that transcription factors such as TAZ (transcriptional coactivator with PDZ-binding motif) are among the very first transcription factors contributing toward the genesis of preadipocytes from undifferentiated mesenchymal cells *(62)*, with the acquisition of the cell surface markers $CD34^{+}/CD31^{-}$ *(63)*. Based on the deletion of *AGPAT2* and *BSCL2* genes in lipodystrophy, these genes could also affect the formation of preadipocytes.

Much is now known regarding the differentiation of preadipocytes to adipocytes, as described below. The capacity of mature adipocytes for enhanced lipogenesis is predominantly mediated by the activation of the master transcription factor SREBP-1c

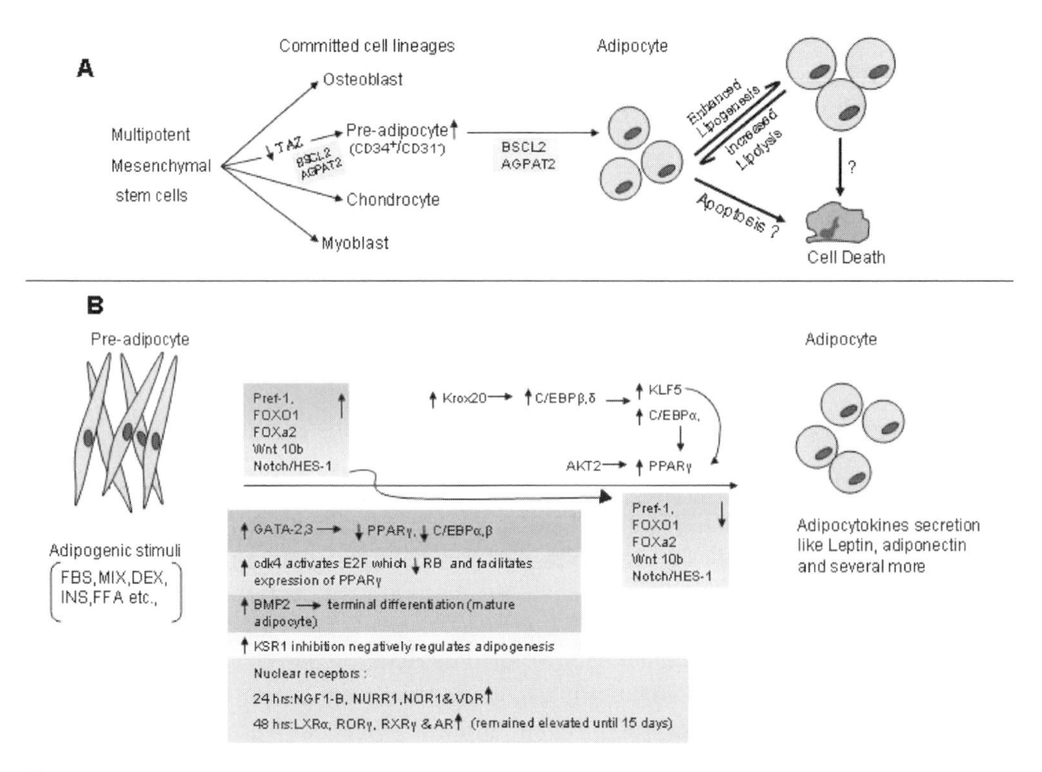

Fig. 5. Schematics for the differentiation and maturation of adipocytes from mesenchymal stem cells. **(A)** Mesenchymal stem cells, by a mechanism that is still not well understood, differentiate into cell lineages, including preadipocytes. A transcriptional coactivator with PDZ binding motif (TAZ) when downregulated increases the number of committed preadipocytes and acquire the cell surface markers such as $CD34^{+}/CD31^{-}$ *(63)*. Shown also are *AGPAT2* and *BSCL2* which could also affect this step of pre-adipocyte differentiation. These are then further differentiated into mature adipocytes by serial activation of the gene program, as shown in **(B)**. Genetic evidence obtained from lipodystrophic patients, *AGPAT2* and *BSCL2*, could affect this step as well. Mature adipocytes accumulate lipids owing to activation of transcription factors such as SEBRP-1c, which induces expression of all lipogenic enzymes and the activation of triglyceride-synthesizing genes. Mature adipocytes are engorged with lipid accumulation, enhanced lipogenesis, or return to the original cell size because of increased lipolysis, mainly owing to activation of hormone-sensitive lipases. The least understood step is adipocyte disintegration, most likely via activation of caspases-8 (apoptosis) *(93)* or additional unknown factor(s). **(B)** The main transcriptional events essential for differentiation of the preadipocytes to adipocytes are described in the text. Shown in the box are adipocyte inhibiton factors including Pref-1, FOXO1, FOXA2, members of Wnt (Wnt10b) and notch (Notch/HES-1) signaling pathway, which must be downregulated for the differentiation of preadipocytes to adipocytes to occur. Recent experiments have discovered additional factors necessary for the differentiation and maturation of the adipocytes. Shown are the transcription factor, Krox20, which is upregulated and, in turn, increases the expression level of C/EBP-β and -δ. Transcription factor KLF5 is shown to up regulate PPARγ. The level of PPARγ is also regulated by AKT2. The transcription factors GATA-2 and -3 downregulate PPARγ and C/EBP-α and -β. Proteins such as KSR1 inhibition negatively regulate adipogenesis. Expression of cdk4 activates E2F, a transcription factor, which facilitates expression of PPARγ via downregulation of retinoblastoma (RB) protein. Bone morphogenetic protein 2 (BMP2), a member of the transforming growth factor β superfamily, has a role in terminal differentiation of preadipocytes to adipocytes. Although a nuclear receptor

(64), which induces the expression of all lipogenic enzymes. Mature adipocytes cycle between stages of lipid accumulation during the fed state and lipolysis during fasting. It is unclear if aged adipocytes ultimately disintegrate via apoptosis or remain quiescent in a lipid-depleted state.

Many studies have established specific hormones, transcriptional factors, and intracellular signals that participate in the differentiation of preadipocytes to adipocytes. Treatment of 3T3-L1 cells with adipogenic factors, including dexamethasone, methylisobutylxanthine, insulin, and fetal bovine serum (Fig. 5B), sets in motion a cascade of events resulting in the upregulation of transcription factors such as the CCAAT/enhancer binding protein (C/EBP-β and -δ) *(65)*, which then induce expression of the C/EBP-α and PPARγ transcription factors *(66)*. Simultaneously, several transcription factors are downregulated as well, shown in the box in Fig. 5B. These include the well-known adipogenesis inhibition factor, preadipocyte factor (Pref)-1 *(67)*, FOXO1, FOXA2, members of Wnt (Wnt10b) and notch (Notch/HES-1) *(68,69)* signaling pathways, which are downregulated during the differentiation of preadipocytes to adipocytes *(70)*. The importance of the nuclear receptor PPARγ for adipogenesis is well documented, as overexpression in 3T3-L1 cells is sufficient for differentiation to adipocytes. In addition, other nuclear receptors, including nerve growth factor-induced gene B (NGF1-B), nuclear receptor-related factor 1 (NURR1), neuron-derived orphan receptor 1 (NOR1), and vitamin D receptor (VDR), are upregulated within the first 24 h of initiation of adipogenesis. After 48 h, the expression of other nuclear receptors including PPARγ, liver X receptor (LXR) α, retinoic acid-related orphan receptor (ROR)γ, retinoic acid X receptor γ (RXRγ), and androgen receptor (AR) rise; these factors remain elevated for 15 d *(71)*.

Recent experiments have discovered additional factors necessary for the differentiation and maturation of adipocytes. The increased expression of a zinc-finger transcription factor, Krox20 *(72)*, in turn increases the levels of C/EBP-β and -δ. C/EBP-β and -δ induce the expression of C/EBP-α, another zinc-finger transcription factor, Kruppel-like transcription factor (KLF5) *(73)*, and PPARγ. Mouse embryonic fibroblast cells obtained from homozygous deletion of Akt2 had impaired adipocyte differentiation and downregulation of PPARγ, suggesting that Akt2, a serine/threonine kinase, also activates expression of PPARγ *(74)*. Other zinc-finger transcription factors, such as GATA-2 and -3, are shown to downregulate PPARγ *(75)*, as well as C/EBP-α and -β, which could help explain lack of AT development in those depots that have increased expression of these transcription factors. Cell cycle regulators such as cyclin-dependent kinase 4 (cdk4) *(76)* may also facilitate the expression of PPARγ. Bone morphogenetic protein 2 (BMP2) *(77)*, a member of the transforming growth factor-β superfamily, contributes to terminal differentiation of preadipocytes to adipocytes. Scaffolding proteins such as kinase suppressor of Ras 1 (KSR1) negatively regulate adipogenesis *(78)*.

Fig. 5. (*Continued*) such as PPARγ is well studied for adipogenesis, other nuclear receptors such as NGF1-B, NURR1, NOR1, and VDR are also upregulated within the first 24 h of initiation of adipogenesis. The increase in expression of nuclear receptors, including PPARγ, LXRα, RORγ, RXRγ, and AR, occurs around 48 h after stimulation and remain elevated until 15 d. The interaction of many of these transcription factors, kinases, ligands, and metabolites may provide a better explanation for differential loss of adipose tissue.

9. POSSIBLE MECHANISMS FOR LOSS OF ADIPOSE TISSUE IN LIPODYSTROPHIES

The generalized loss of AT in patients with CGL seems to be caused by either a defect in the cell fate determination from the mesenchymal cells or the inability of the adipocytes to synthesize and accumulate lipids. The seipin protein, encoded by *BSCL2,* may act at both differentiation and lipogenesis. As discussed above, patients carrying the *BSCL2* mutations lose AT from all the anatomic regions of the body. This pattern suggests defects in both the determination of the meschenymal cells to preadipocytes and subsequent cellular differentiation to adipocytes as well. Seipin has homology to proteins such as midasin, which is associated with the nuclear membrane *(22)*, suggesting that alterations in seipin might affect nuclear function, possibly the export of mature RNA or import of proteins into the nucleus. Why seipin defects preferentially affects adipocytes is unclear.

In contrast, the mechanism of lipodystrophy in CGL1 seems more straight forward. Defective AGPAT2 enzyme precludes triglyceride (TG) synthesis from lysophosphatidic acid (LPA) (Fig. 1), eliminating a dominant pathway of TG formation. At the same time, AGPAT2 deficiency leads to accumulation of LPA and decreased synthesis of phosphatidic acid (PA). As PA is a substrate for important lipid signaling molecules, including phosphatidylinositol (PI), and affects the synthesis of phosphatidylethanolamine (PE) and phosphatidylcholine (PC), which are important for integrity of cellular membrane structures, it remains unclear how the defective phospholipids produced will affect normal functioning of the adipocytes.

Whereas the loss of AT in CGL is easy to comprehend, the partial loss of ATs in FPL patients is more difficult to explain. Several genes, such as *PPARγ, LMNA, ZMPSTE24,* and *AKT2,* are now associated with the partial loss of AT. Most intriguing among these are the heterozygous mutations in *PPARγ* associated with FPL. As discussed above, the role of PPARγ in adipocyte differentiation is undisputed. However, it is not clear how the heterozygous mutations in *PPARG* lead only to partial lipodystrophy in humans. It is possible that a cellular and molecular heterogeneity in AT from various anatomic regions might correlate with variable requirements of PPARγ in preadipocyte differentiation to adipocytes. In chicken preadipocytes, when the culture medium is supplemented with the fatty acid, oleic acid, there was an increased accumulation of lipid as determined by Oil-red-O staining, with no simultaneous increase in PPARγ expression *(79)*. This result suggests that there are subtle variations in the need for ligand-activated transcription factor for adipocyte differentiation. Like all the nuclear receptors, the PPARγ receptor has several domains, which might function independently. The amino-terminal ligand-independent activating domain and a carboxyl terminal ligand-dependent binding and activation domain are usually separated by the DNA binding and receptor dimerization region *(80)*. All the *PPARG* gene mutations reported so far, except one occurring in the promoter region, map to the ligand-binding domain of the PPARγ receptor. In FPL patients, the ligand-binding domain of the receptor may be nonfunctional, yet some function of the ligand-independent domain may still be active. This possibility is suggested by in vitro mutagenesis studies *(81)*, which show that PPARγ receptors were still able to recruit coactivators and provide basal receptor activity despite mutations in the ligand-binding domain. Some AT depots might require robust

PPARγ activity for proper development, such as adipocytes in extremities, but other depots might be less dependent on PPARγ and remain unaffected despite these mutations. Although mouse models and extensive cell culture experiments have not been conducted to determine the ligand-independent activation with PPARγ, as with glucocorticoid receptor, clearly there are ligand-dependent and -independent roles, which are being elucidated *(82,83)*.

Mutations in lamin A/C protein and the loss of AT have focused attention on the significant role of nuclear lamina protein and nuclear architecture in adipocyte biology. Most of the mutations found associated with lipodystrophy are found localized to the immunoglobulin fold domain of lamin *(6,29)*, which are important for protein–protein interactions. Recent experiments using left ventricular myocytes from *Lmna*$^{-/-}$ mice revealed defective import of SREBP-1c, a lipogenic transcription factor, into the nucleus, leading to a reduction in intranuclear SREBP-1c protein *(84)*. Defective import of SREBP-1c might indicate that additional defects in the nuclear import machinery for other protein might also exist.

Mutation in the ZMPSTE24 is presumed to cause loss of AT via defective processing of pre-lamin A protein. However, because ZMPSTE24 is a CaaX motif protease, it is likely that defective processing of other CaaX motif proteins might also be involved in alteration of adipocyte biolgy. In both *LMNA* and *ZMPSTE24* deficiencies, a defect in nuclear import of the proteins is observed, yet the nuclear export of mature RNA has not been well-studied. It is also likely, because of molecular heterogeneity of adipocytes in various AT depots, that additional defects in mRNA export occur in sc adipocytes compared to intra-abdominal adipocytes.

AKT2 is a serine/threonine kinase that has an important function in insulin signaling, downstream of PI3-kinase. AKT2 has many functions in the cell, including regulation of cell survival and apoptosis. This role of AKT2 appears to be cell-specific as well as dependent on the expression of various downstream target proteins. It is possible that some AT depots, such as those in the extremities, lose their AT in conjunction with additional localized defects yet to be identified.

9.1. Possible Mechanism for Loss of AT in Acquired Lipodystrophies

As mentioned before, some varieties of AGL are associated with autoimmune diseases, suggesting autoimmune destruction of AT. Some autoimmune diseases show a strong association with the HLA complex, but because of the rarity of these patients, HLA typing in AGL has not been explored. Fat loss in APL is believed to involve complement-mediated lysis of adipocytes, which might be initiated by viral infections *(46)*. There is evidence to suggest the presence of circulating autoantibodies, some of which may be directed against adipocytes in some patients with AGL and in many with APL *(46,85)*. The adipocyte-specific antigens against which these autoantibodies are directed remain to be identified.

LDHIV is strongly related to the use of HIV-protease inhibitors. Several in vivo and in vitro studies have shown the ability of PIs to impair adipocyte development by decreasing expression of SREBP-1c, PPARγ, and C/EBP-α *(86–89)*. Increased adipocyte apoptosis mediated by proinflammatory cytokines, such as TNF-α, and mitochondrial toxicity from nucleoside analogs are other potential mechanisms for fat loss in HIV-infected patients *(90–92)*.

10. CONCLUSIONS

Our understanding of the initial commitment of mesenchymal stem cells to preadipocytes, and the roles of the surrounding cellular milieu, are woefully inadequate to explain adipocyte biology and mechanisms of lipodystrophy. Adipocytes obtained either from sc or visceral depots secrete several factors known as adipocytokines, including leptin, adiponectin, and resistin. Whether each of these, individually or in combination, has any depot-specific influence on AT has not been studied. Studies with lipodystrophic patients have revealed, for the first time, that nuclear lamina proteins are also critical for adipocyte development. Future studies should include investigations of alterations in import and export of RNA, proteins, and small molecules, which affect the development of adipocytes. Equally important will be the role of acyltransferases in generating lipid signaling molecules critical for adipocyte maturation. As shown in Fig. 5, an increasing number of transcription factors or other proteins known to participate in adipogenesis have been identified; however, we know a great deal less about the endogenous ligands for many of these transcription factors. The function of endogenous ligands, and whether their synthesis is regulated by PPARγ, is still debated. The metabolites generated during adipogenesis—which may include, but are not limited to, lipids, steroids, or other small molecules—might have interesting and important contribution to adipogenesis. Lastly, there is a need to develop coculture systems to study the paracrine effects of the nervous system, surrounding cells, and even the lymphatic system on the development and proliferation of adipocytes.

ACKNOWLEDGMENTS

The authors thank A. Garg, MD, and R. J. Auchus, MD, PhD, for reviewing this manuscript. The study was supported in part by the National Institute of Health grants R01-DK54387 and M01-RR00633 and by the Southwestern Medical Foundation. The authors thank RG Huet and Vidya Rai for the graphics.

REFERENCES

1. Garg A. N Engl J Med 2004;350:1220–1234.
2. Agarwal AK, Arioglu E, De Almeida S, et al. Nat Genet 2002;31:21–23.
3. Magre J, Delepine M, Khallou E, et al. Nat Genet 2001;28:365–370.
4. Agarwal AK, Garg A. J Clin Endocrinol Metab 2002;87:408–411.
5. Shackleton S, Lloyd DJ, Jackson SN, et al. Nat Genet 2000;24:153–156.
6. Cao H, Hegele RA. Hum Mol Genet 2000;9:109–112.
7. Agarwal AK, Fryns JP, Auchus RJ, et al. Hum Mol Genet 2003;12:1995–2001.
8. George S, Rochford JJ, Wolfrum C, et al. Science 2004;304:1325–1328.
9. Mitchell S. Am J Med Sci 1885;90:105–106.
10. Berardinelli W. J Clin Endocrinol Metab 1954;14:193–204.
11. Seip M. Acta Paediatr 1959;48:555–574.
12. Fleckenstein JL, Garg A, Bonte FJ, et al. Skeletal Radiol 1992;21:381–386.
13. Seip M, Trygstad O. Acta Paediatr Suppl 1996;413:2–28.
14. Van Maldergem L, Magre J, Khallouf TE, et al. J Med Genet 2002;39:722–733.
15. Agarwal AK, Simha V, Oral EA, et al. J Clin Endocrinol Metab 2003;88:4840–4847.
16. Haque WA, Shimomura I, Matsuzawa Y, et al. J Clin Endocrinol Metab 2002;87:2395.
17. Oral EA, Simha V, Ruiz E, et al. N Engl J Med 2002;346:570–578.

18. Garg A, Wilson R, Barnes R, et al. J Clin Endocrinol Metab 1999;84:3390–3394.
19. Leung DW. Front Biosci 2001;6:D944–D953.
20. Agarwal AK, Garg A. Trends Endocrinol Metab 2003;14:214–221.
21. Haque W, Garg A, Agarwal AK. Biochem Biophys Res Commun 2005;327:446–453.
22. Agarwal AK, Garg A. Trends Mol Med 2004;10:440–444.
23. Agarwal AK, Barnes RI, Garg A. Int J Obes Relat Metab Disord 2004;28:336–339.
24. Simha V, Garg A. J Clin Endocrinol Metab 2003;88:5433–5437.
25. Hegele RA, Cao H, Frankowski C, et al. Diabetes 2002;51:3586–3590.
26. Dunnigan MG, Cochrane MA, Kelly A, et al. Q J Med 1974;43:33–48.
27. Garg A. J Clin Endocrinol Metab 2000;85:1776–1782.
28. Hegele RA, Kraw ME, Ban MR, et al. Arterioscler Thromb Vasc Biol 2003;23:111–116.
29. Haque WA, Oral EA, Dietz K, et al. Diabetes Care 2003;26:1350–1355.
30. Garg A, Speckman RA, Bowcock AM. Am J Med 2002;112:549–555.
31. Peters JM, Barnes R, Bennett L, et al. Nat Genet 1998;18:292–295.
32. Gruenbaum Y, Margalit A, Goldman RD, et al. Nat Rev Mol Cell Biol 2005;6:21–31.
33. Hutchison CJ, Worman HJ. Nat Cell Biol 2004;6:1062–1067.
34. Dhe-Paganon S, Werner ED, Chi YI, et al. J Biol Chem 2002;277:17,381–17,384.
35. Garg A, Vinaitheerthan M, Weatherall PT, et al. J Clin Endocrinol Metab 2001;86:59–65.
36. Caux F, Dubosclard E, Lascols O, et al. J Clin Endocrinol Metab 2003;88:1006–1013.
37. Chen L, Lee L, Kudlow BA, et al. Lancet 2003;362:440–445.
38. Barroso I, Gurnell M, Crowley VE, et al. Nature 1999;402:880–883.
39. Savage DB, Tan GD, Acerini CL, et al. Diabetes 2003;52:910–917.
40. Al-Shali K, Cao H, Knoers N, et al. J Clin Endocrinol Metab 2004;89:5655–5660.
41. Rosen ED, Sarraf P, Troy AE, et al. Mol Cell 1999;4:611–617.
42. Novelli G, Muchir A, Sangiuolo F, et al. Am J Hum Genet 2002;71:426–431.
43. Simha V, Garg A. J Clin Endocrinol Metab 2002;87:776–785.
44. Aarskog D, Ose L, Pande H, et al. Am J Med Genet 1983;15:29–38.
45. Pivnick EK, Angle B, Kaufman RA, et al. Am J Med Genet 2000;90:131–140.
46. Misra A, Peethambaram A, Garg A. Medicine (Baltimore) 2004;83:18–34.
47. Misra A, Garg A. Medicine (Baltimore) 2003;82:129–146.
48. Javor ED, Cochran EK, Musso C, et al. Diabetes 2005;54:1994–2002.
49. Mallal SA, John M, Moore CB, et al. AIDS 2000;14:1309–1316.
50. Bernasconi E, Boubaker K, Junghans C, et al. J Acquir Immune Defic Syndr 2002;31:50–55.
51. Saint-Marc T, Partisani M, Poizot-Martin I, et al. AIDS 1999;13:1659–1667.
52. Grinspoon S, Carr A. N Engl J Med 2005;352:48–62.
53. Carr A, Samaras K, Burton S, et al. AIDS 1998;12:F51–F58.
54. Bacchetti P, Gripshover B, Grunfeld C, et al. J Acquir Immune Defic Syndr 2005;40:121–131.
55. Dube MP, Johnson DL, Currier JS, et al. Lancet 1997;350:713–714.
56. Carr A, Samaras K, Thorisdottir A, et al. Lancet 1999;353:2093–2099.
57. Garg A, Fleckenstein JL, Peshock RM, et al. J Clin Endocrinol Metab 1992;75:358–361.
58. Pond CM. Prog Lipid Res 1999;38:225–248.
59. Pond CM. J Exp Zoolog A Comp Exp Biol 2003;295:99–110.
60. Harvey NL, Srinivasan RS, Dillard ME, et al. Nat Genet 2005;37:1072–1081.
61. Alitalo K, Tammela T, Petrova TV. Nature 2005;438:946–953.
62. Hong JH, Hwang ES, McManus MT, et al. Science 2005;309:1074–1078.
63. Sengenes C, Lolmede K, Zakaroff-Girard A, et al. J Cell Physiol 2005;205:114–122.
64. Shimomura I, Hammer RE, Richardson JA, et al. Genes Dev 1998;12:3182–3194.
65. Yeh WC, Cao Z, Classon M, et al. Genes Dev 1995;9:168–181.
66. Wu Z, Rosen ED, Brun R, et al. Mol Cell 1999;3:151–158.
67. Gregoire FM, Smas CM, Sul HS. Physiol Rev 1998;78:783–809.
68. Ross DA, Hannenhalli S, Tobias JW, et al. Mol Endocrinol 2006;20:698–705.
69. Ross DA, Rao PK, Kadesch T. Mol Cell Biol 2004;24:3505–3513.
70. Harp JB. Curr Opin Lipidol 2004;15:303–307.
71. Fu M, Sun T, Bookout AL, et al. Mol Endocrinol 2005;19:2437–2450.

72. Chen Z, Torrens JI, Anand A, et al. Cell Metab 2005;1:93–106.
73. Oishi Y, Manabe I, Tobe K, et al. Cell Metab 2005;1:27–39.
74. Peng XD, Xu PZ, Chen ML, et al. Genes Dev 2003;17:1352–1365.
75. Tong Q, Tsai J, Tan G, et al. Mol Cell Biol 2005;25:706–715.
76. Abella A, Dubus P, Malumbres M, et al. Cell Metab 2005;2:239–249.
77. Tang QQ, Otto TC, Lane MD. Proc Natl Acad Sci USA 2004;101:9607–9611.
78. Kortum RL, Costanzo DL, Haferbier J, et al. Mol Cell Biol 2005;25:7592–7604.
79. Matsubara Y, Sato K, Ishii H, et al. Comp Biochem Physiol A Mol Integr Physiol 2005;141:108–115.
80. Michalik L, Desvergne A, Dreyer C, et al. Int J Dev Biol 2002;46:105–114.
81. Molnar F, Matilainen M, Carlberg C. J Biol Chem 2005;280:26,543–26,556.
82. Eickelberg O, Roth M, Lorx R, et al. J Biol Chem 1999;274:1005–1010.
83. Schmitt J, Stunnenberg HG. Nucleic Acids Res 1993;21:2673–2681.
84. Nikolova V, Leimena C, McMahon AC, et al. J Clin Invest 2004;113:357–369.
85. Hubler A, Abendroth K, Keiner T, et al. Exp Clin Endocrinol Diabetes 1998;106:79–84.
86. Kannisto K, Sutinen J, Korsheninnikova E, et al. AIDS 2003;17:1753–1762.
87. Dowell P, Flexner C, Kwiterovich PO, et al. J Biol Chem 2000;275:41,325–41,332.
88. Bastard JP, Caron M, Vidal H, et al. Lancet 2002;359:1026–1031.
89. Caron M, Auclair M, Vigouroux C, et al. Diabetes 2001;50:1378–1388.
90. Villarroya F, Domingo P, Giralt M. Trends Pharmacol Sci 2005;26:88–93.
91. Galluzzi L, Pinti M, Guaraldi G, et al. Antivir Ther 2005;10(Suppl 2):M91–M99.
92. Lagathu C, Kim M, Maachi M, et al. Biochimie 2005;87:65–71.
93. Pajvani UA, Trujillo ME, Combs TP, et al. Nat Med 2005;11:797–803.

19 Mechanisms of Cachexia

Robert H. Mak and Wai W. Cheung

Abstract

Cachexia is brought about by a synergistic combination of a dramatic decrease in appetite and an increase in metabolism of fat and lean body mass. This combination is found in a number of disorders including cancer, chronic kidney disease (CKD), AIDS, cystic fibrosis, chronic heart failure, rheumatoid arthritis, and Alzheimer's disease. Cachexia has a stronger correlation with survival than any other current measure of diseases such as AIDS, cancer, and CKD. The underlying mechanism of increased resting metabolic rate may involve the increased activity of mitochondrial uncoupling proteins. Chronic overproduction of cytokines, such as interleukin (IL)-1β, IL-6, and tumor necrosis factor-α, may lead to cachexia in various chronic illness models, through the nuclear factor-κB and ATP–ubiquitin-dependent proteolytic pathways. Inhibition of these cytokines in experimental models ameliorated cachexia. Cytokines may also act on the central nervous system to alter the release and function of neuropeptides and melanocortin receptors in the hypothalamus, thereby altering both appetite and metabolic rate. Further research into the molecular pathways leading to cachexia may lead to novel therapeutic therapy for this devastating and potentially fatal complication of chronic disease.

Key Words: Cachexia; cancer; chronic kidney disease; AIDS; cytokines; neuropeptides; hypothalamus; melanocortin receptors.

1. INTRODUCTION

The role of energy homeostasis—the balance of nutrition and metabolism—is well known in normal growth, development, and health maintenance. The first description of cachexia came from Hippocrates, who wrote, "the flesh is consumed and becomes water, the abdomen fills with water, the feet and legs swell, the shoulders, clavicle, chest, and thigh melt away. This illness is fatal" *(1)*. Cachexia is derived from the Greek *kakos hexis*, meaning "bad condition," and is brought about by a synergistic combination of a dramatic decrease in appetite and an increase in metabolism of fat and lean body mass. This combination is found in a number of disorders including cancer, chronic kidney disease (CKD), acquired immune deficiency syndrome (AIDS), cystic fibrosis, chronic heart failure, rheumatoid arthritis, and Alzheimer's disease. Loss of adipose tissue reaches 85% and loss of skeletal muscle reaches 75% when cancer patients have lost 30% of body weight, a situation that often leads to death fairly quickly *(2)*. Cachexia could account for at least 20% of cancer deaths *(3)*. Poor nutrition is reported to have a prevalence of 30 to 60% and is an important risk factor for mortality

From: *Nutrition and Health: Adipose Tissue and Adipokines in Health and Disease*
Edited by: G. Fantuzzi and T. Mazzone © Humana Press Inc., Totowa, NJ

in patients with CKD *(4)*. Patients between the ages of 25 and 34 with end-stage renal disease have a 100- to 200-fold higher mortality than the general population *(5)*. Body mass retention in patients with AIDS has a stronger correlation with survival than any other current measure of the disease *(6)*. Loss of proteins from skeletal muscle is probably the most important factor regulating survival, as at this level of lean tissue loss, physiological functions, such as respiratory muscle function, are significantly impaired *(7)*. Asthenia is directly related to the substantial muscle atrophy and reduction in physical activity level and quality of life. Loss of protein from skeletal muscle could reduce the performance status and physical activity level of cancer patients *(8)*.

2. MALNUTRITION VS CACHEXIA

Malnutrition, in conditions such as kwashiorkor and pyloric stenosis, results from inadequate intake of nutrients despite a good appetite, and manifests as weight loss associated with protective metabolic responses such as decreased basic metabolic rate and preservation of lean body mass at the expense of fat mass. Cachexia differs from malnutrition in several key ways. First, despite the fact that the cachexic person is "starving," he or she is anorexic. Second, in normal starvation the metabolic rate decreases as a protective mechanism. This protective reduction in metabolic rate is not observed in cachexia. Resting energy expenditure is high in patients with cachexia from renal failure *(9,10)*. Third, in simple starvation fats are preferentially lost and there is preservation of lean body mass. In cachexia, lean tissues are wasted and fat stores are relatively underutilized *(11)*. Finally, the abnormalities in malnutrition can usually be overcome simply by supplying more food or altering the composition of the diet. To date this approach has not proven to be successful in cachexic patients. Stimulation of nutritional intake with megestrol acetate fails to restore the loss of lean body mass in cancer patients, whereas weight gain achieved is the result of an accumulation of adipose tissue and water *(12)*. There are fundamental metabolic abnormalities in cachexic patients that prevent them from utilizing dietary nutrients effectively. One important point about cancer cachexia is that the presence and severity of cachexia has no correlation to the size of the tumor. The tumor does not represent a "metabolic sink" for ingested nutrients. If this were the case, increasing nutrient intake would result in more rapid tumor growth, an outcome that is not observed clinically *(11)*. Furthermore, parabiosis experiments in animals demonstrated that cachexia is observed in the nontumor-bearing partner, indicating that circulating factors are largely responsible for this phenomenon *(11)*.

3. ANOREXIA AND CACHEXIA

Anorexia, defined as the loss of appetite and early satiety, often accompanies cachexia and has been suggested to play a role in the loss of body weight. However, loss of skeletal muscle is not prominent in primary anorexic states, as the brain adapts to using ketone bodies derived from metabolism of fat, reducing the requirement for gluconeogensis from amino acids derived from muscle proteins. This suggests that the metabolic changes in anorexia and cachexia are different. Indeed, appetite stimulants such as megestrol acetate failed to restore the loss of lean body mass, and weight gain was shown to be caused by an accumulation of adipose tissue and water *(12)*. Anorexia can arise from (1) decreased taste and smell of food, (2) early satiety, (3) dysfunctional hypothalamic membrane adenylate cyclase, (4) increased brain tryptophan, and (5) cytokine production *(11)*.

The etiology of anorexia in cachexic states is not well understood. CKD patients with anorexia regain appetite soon after starting dialysis treatment, presumably because of removal of one or more toxic factors that suppress appetite. Fractions in the middle molecular weight range isolated from normal urine and uremic plasma ultrafiltrate inhibited ingestive behavior in the rat. To investigate their site of action and specificity, rats were injected intraperitoneally, intravenously, or intracerebroventricularly with concentrated fractions of uremic plasma ultrafiltrate or normal urine and tested for ingestive and sexual behavior. An intraperitoneal injection of a urine fraction or a uremic plasma ultrafiltrate fraction inhibited carbohydrate intake by 76.3 and 45.9%, respectively, but an intravenous injection had no effect. An intracerebroventricular injection of urine middle molecular fraction or uremic plasma ultrafiltrate inhibited carbohydrate intake similarly. Injections of the corresponding fraction from normal plasma ultrafiltrate had no effect. Injection of urine or uremic plasma ultrafiltrate middle molecular fractions did not affect the display of sexual behavior. These results suggest that middle molecular fractions from uremic plasma ultrafiltrate or normal urine act in the splanchnic region and/or brain to inhibit food intake and that the effect is specific for ingestive behavior *(13)*.

4. ENERGY EXPENDITURE IN CACHEXIA

Energy expenditure is increased in cachexic patients with cancer and CKD. In patients with advanced pancreatic cancer, resting energy expenditure is increased, whereas total energy expenditure and physical activity levels are reduced *(14)*. In patients with CKD, energy expenditure is related to inflammation. Resting energy expenditure correlated with surrogate markers of inflammation, such as C-reactive protein, and treatment of infection and subsequent resolution of elevated C-reactive protein were associated with normalization of elevated resting energy expenditure in patients with CKD *(9)*. Furthermore, resting energy expenditure was associated with increased mortality and cardiovascular death in patients on peritoneal dialysis and was related to cachexia and inflammation in these patients *(10)*.

The underlying mechanism of increased resting metabolic rate may involve the increased activity of mitochondrial uncoupling proteins (UCPs). These proteins translocate protons across the inner mitochondrial membrane in a process not coupled to phosphorylation of ADP, so that energy is lost as heat. The principal UCPs are UCP1, which is found only in brown adipose tissue, and UCP3, which is found in both brown adipose tissue and skeletal muscle. Brown adipose tissue was found in the periadrenal tissues in 80% of cancer patients, compared with 13% of age-matched controls *(15)*. In mice bearing the cachexia-inducing MAC16 adenocarcinoma, UCP1 mRNA levels in brown adipose tissue and UCP3 mRNA levels in skeletal muscles were increased *(16)*. UCP3 mRNA levels were significantly higher in skeletal muscles of cancer patients with weight loss than in those who had not lost weight and in patients without cancer *(17)*. The increase in UCP expression in cancer patients would increase energy expenditure and contribute to weight loss. The increase in UCP3 mRNA in skeletal muscle may be caused by hydrolysis of triglycerides in adipose tissue, as treatment of tumor-bearing animals with nicotinic acid stimulated both hyperlipidemia and an increase in UCP3 mRNA in muscle *(18)*. Furthermore, cytokines and tumor products have been shown to directly regulate UCP expression, suggesting a further mechanism for the control of energy expenditure in cachexia *(11)*.

Futile cycles are often increased in cancer cachexia. Nonesterified fatty acids released from adipose tissue can immediately be re-esterified in the triacylglycerol/fatty acid substrate cycle. This process was increased threefold in tumor-bearing mice, although there were no differences in animals with and without cachexia *(19)*. An additional futile cycle that may account for energy loss in cancer cachexia is the Cori cycle *(20)*. In the Cori cycle, tumors consume large amounts of glucose and convert it to lactate, which circulates to the liver and is converted back to glucose. Gluconeogenesis uses six ATP molecules for each lactate–glucose cycle and is energy-inefficient. The Cori cycle is normally responsible for 20% of glucose turnover, but is increased to 50% in cachexic cancer patients *(20)*.

5. ADIPOSE TISSUE AND LIPID METABOLISM IN CACHEXIA

There is evidence of increased lipolysis in cachexic cancer patients. Fasting glycerol is high and there is an increased turnover of both glycerol and fatty acids *(19,21)*. The fatty acids released are rapidly oxidized, with a 20% increase in oxidation. Furthermore, triglycerol hydrolysis may be increased in cancer patients. Although there is no evidence for decreased level of lipoprotein lipase in adipose tissue, there is a twofold increase in the relative level of mRNA for hormone-sensitive lipase *(22)*. The fatty acids released serve as an energy source to drive futile metabolic cycles. A lipid-mobilizing factor isolated from the urine of cachexic cancer patients causes a significant increase in glucose oxidation and decreased blood glucose levels in experimental animals. This lipid-mobilizing factor also increases overall lipid oxidation. These results suggest that changes in carbohydrate metabolism and loss of adipose tissue, together with increased whole-body fatty acid oxidation in cachexic cancer patients, may arise from tumor production of a lipid-mobilizing factor *(23)*. Recently this lipid-mobilizing factor was identified to be Zinc-α2-glycoprotein (ZAG), which is a 43-kDa protein. It is overexpressed in certain human malignant tumors and acts as a lipid-mobilizing factor to stimulate lipolysis in adipocytes, leading to cachexia in mice implanted with ZAG-producing tumors. ZAG is overexpressed in white adipose tissue of tumor-bearing mice and may have a local role for in the substantial reduction of adiposity of cancer cachexia *(24)*. Loss of adipose tissue is coupled with an increase in UCP1 in brown adipose tissue *(23)*. ZAG increases UCP1 expression in brown adipose tissue, as well as the expression of UCP2 and UCP3 in myotubes *(25)*. The effects on UCP1 and UCP2 have been shown to be mediated through a β3-adrenergic receptor *(11)*.

6. SKELETAL MUSCLE IN CACHEXIA

Cachexia conditions are associated with a progressive loss of lean body mass. Skeletal muscle, which accounts for almost half the whole-body protein mass, is severely affected in cachexia secondary to cancer *(11)* and CKD *(26)*. There is evidence of decreased protein synthesis *(27)* and increased protein degradation *(28)*. An increase in the synthesis of acute-phase proteins in the liver may alter the balance of amino acids for protein synthesis, as acute-phase proteins contain high levels of sulfur amino acids *(29)*. Serum branched-chain amino acids are decreased in both cancer *(11)* and CKD patients *(30)*, reducing the stimulus for protein synthesis. The fact that appetite stimulation by megestrol acetate could not replenish lean body mass suggests that increase in protein catabolism outweighs the decrease in anabolism *(12)*. A biphasic model of protein degradation in cancer cachexia has been proposed, in which an early calpain-dependent release of myofilaments from myofibrils would be followed by ubiquitylation of myofilaments and subsequent degradation by the proteosome. Increased activity of the ubiquitin-proteasome proteolytic pathway has been reported in cachexia from cancer *(9)* or CKD *(31)*. Furthermore, the presence of metabolic acidosis exacerbates protein catabolism in CKD. Glucocorticoids are involved in accelerating protein degradation in muscle, which results in loss of lean body mass, whereas a low insulin level appears to play a permissive role in accelerating increased catabolism. Cellular mechanisms mediating these changes again include upregulation of the ubiquitin-proteasome pathway and branched-chain ketoacid dehydrogenase enzyme activity in muscle *(31)*.

7. CYTOKINES AND MUSCLE PROTEIN DEGRADATION IN CACHEXIA

Chronic overproduction of cytokines may lead to cachexia in various animal chronic illness models. Cytokines such as interleukin (IL)-1β, IL-6, and tumor necrosis factor (TNF)-α are upregulated in various animal cachexia models. The pathogenetic role of the cytokine network in upregulating muscle protein degradation has been documented as the result of a complex interplay among TNF-α, IL-6, IL-1β, interferon (IFN)-γ, and a variety of humoral mediators. Neutralization of these factors by genetic or pharmacological methods leads to attenuation of cachexia. Chronic infusion of IL-1β or TNF-α causes anorexia, rapid weight loss, and catabolism of body protein stores, analogous to the state observed with chronic illness *(31–34)*. Anticytokine approaches have been proven at least partially effective in attenuating muscle weight loss in cancer. Mice bearing the Lewis lung carcinoma are protected from cachexia when treated with antibodies against IFN-γ *(35)*. Implantation of the Lewis lung cancer into transgenic mice lacking the TNF-α receptor I or overexpressing its soluble form does not result in muscle depletion *(36,37)*. The release of IL-6 seems to be crucial for the onset of cachexia in mice bearing the C-26 adenocarcinoma *(38)*. Inhibition of IL-6 release by intratumor injection of decoy oligonucleotides directed against the transcription factor NFκB can prevent cachexia *(39)*. Pentoxifylline, an inhibitor of TNF-α synthesis, partially prevented muscle hypercatabolism in sepsis *(40)* and AH-130 tumor-bearing rats by down-regulating the expression of components of the ATP-ubiquitin-dependent proteolytic system *(41)*. Suramin, capable of inhibiting the interaction of TNF-α and IL-6 with their receptors by inducing ligand deoligomerization *(42)* significantly reduced the severity of cancer cachexia *(43)*. Both pentoxifylline and suramin, either alone or in combination, prevented the depletion of muscle mass and significantly reduced the activity of muscle ATP-ubiquitin- and calpain-dependent proteolytic systems *(44)*. IL-15 treatment partly inhibited skeletal muscle wasting in AH-130 tumor-bearing rats by decreasing protein degradation rates to values even lower than those observed in nontumor-bearing control animals. This dramatic reduction in protein breakdown rates was associated with an inhibition of the ATP-ubiquitin-dependent proteolytic pathways *(45)*. Furthermore, β2-adrenergic agonists, such as clenbuterol, which are known to favor skeletal muscle hypertrophy, largely prevented skeletal muscle wasting in AH-130 tumor-bearing rats by restoring protein degradation rates close to control values. Again, this was associated with a decrease in the inhibition of the ATP-ubiquitin-dependent proteolytic pathways *(46)*.

8. PROTEOLYSIS-INDUCING FACTOR

The cytokines involved in the induction of cachexia appear to vary with tumor type. Mice bearing the MAC16 tumor are cachexic, whereas mice bearing the MAC13 tumor are not *(47)*. Proteolysis-inducing factor (PIF) was purified from the serum of cachexic mice bearing the MAC16 tumor *(48)*. PIF was shown to be present in the urine of patients with cancer cachexia from a range of solid tumors but absent if the tumor did not induce cachexia. PIF is detected in 80% of patients with pancreatic cancer; these patients have a significantly greater weight loss than those in whom PIF was not detected. In patients with advanced gastrointestinal cancer, PIF-positive patients lost weight, whereas PIF-negative patients did not. PIF caused significant weight loss in mice without a reduction in food and water intake. This weight loss was mainly a result of loss in lean body mass, which was associated with a 50% decrease in protein synthesis and a 50% increase in protein degradation. Skeletal muscle from mice treated with PIF showed an increased activity and expression of key components of the ubiquitin–proteasome proteolytic pathway. Proteasome inhibitors attenuated the enhanced protein degradation. Like TNF-α, PIF action was associated with rapid degradation of IκB and nuclear migration of NFκB. Inhibition of this process attenuated loss of body weight and protein degradation in mice with cancer cachexia. PIF was shown to activate both NFκB and the transcription factor STAT3, which leads to increased production of both IL-6 and IL-8 and increased production of C-reactive protein. Thus PIF is likely to be involved in the proinflammatory response observed in cachexia *(11)*.

9. LEPTIN AND GHRELIN IN CACHEXIA

Leptin is secreted by adipocytes and regulates adiposity and metabolic rate by reducing food intake and increasing energy expenditure. Leptin is also a member of the IL-6 superfamily of cytokines. Experimental elevation of leptin within the physiological range produces weight loss and relative anorexia. Leptin secretion is increased by both central and systemic immunological challenge and has been proposed as a potential mediator of inflammation-induced anorexia. The mechanism of how leptin expression and secretion is enhanced during inflammation is complex, but there is evidence for mediation by both IL-1β and TNF-α. Conversely, leptin induces production and release of IL-1β in the brains of normal rats and the release of both IL-1β and TNF-α from mouse macrophages. Collectively, these observations suggest a complex interplay between leptin and other cytokines in the regulation of metabolism and appetite during acute and chronic illness *(49)*.

Leptin is cleared from the circulation primarily by the kidneys *(50)*. Leptin levels are significantly increased in dialysis patients, even after correction for body mass index.

The percentage of body fat was strongly correlated with leptin levels in these patients. However, the ratio of leptin levels to body fat is significantly greater for dialysis patients than for control subjects. Increased leptin levels are associated with markers of poor nutritional status, such as low serum albumin and high protein catabolic rate in dialysis patients. In children with CKD, leptin levels increase with declining renal function, presumably by reduced renal clearance. Leptin levels are inappropriately elevated in these children in relation to the percentage of body fat and inversely correlate with dietary nutrient intake. Thus leptin may be an important factor in the pathogenesis of anorexia and cachexia in CKD *(51)*. Furthermore, serum leptin concentrations correlate with plasma insulin concentrations in patients with CKD independent of body fat content, suggesting that insulin resistance and hyperinsulinemia may contribute to elevated serum leptin concentrations or vice versa. The role of leptin resistance due to receptor insufficiency and saturable transport via the blood–brain barrier in limiting the potential impact of elevated leptin levels in patients with CKD is controversial and not well understood *(49)*.

A hormonal signal involved in the short-term control of appetitie is the gastric peptide ghrelin. Levels of this appetite-stimulating hormone rise in the blood before meals (when the stomach is empty), then fall quickly after food is consumed. Ghrelin is implicated as a factor that triggers the onset of eating, participating in a meal-to-meal control system that is itself sensitive to changes in insulin and leptin levels. In this way, the size and frequency of individual meals can be adjusted so as to minimize changes in body fat content. Preliminary observations show that markedly elevated plasma ghrelin concentrations are found in advanced CKD *(52,53)* and correlate with body mass index, fat mass, plasma insulin, and serum leptin levels. There is no difference in plasma ghrelin between CKD patients with and without signs of wasting. Changes in plasma ghrelin concentrations over time during peritoneal dialysis treatment are associated with changes in body composition *(53)*. Further studies are needed to examine relative changes in ghrelin versus inactive ghrelin metabolites during the course of progression of CKD and the putative associations with changes in food intake and body composition.

10. NEUROPEPTIDES IN CACHEXIA

A current hypothesis of cachexia in chronic illness is that cytokines released during cancer, CKD, or chronic inflammation act on the central nervous system to alter the release and function of a number of key neurotransmitters, thereby altering both appetite and metabolic rate *(54,55)*. The melanocortin system is critical in mediating the effect of cytokines, such as leptin, on metabolism. There are distinct local counterparts of the pro-opiomelanocortin (POMC) cells: agouti-related protein (AgRP) and neuropeptide Y (NPY) producing cells in the arcuate nucleus. Activation of POMC neurons by leptin triggers the release of α-melanocyte-stimulating hormone (α-MSH) from POMC axon terminals, which in turn activates the type 4 melanocortin receptor-4 (MC-4R), leading to suppressed food intake and increased energy expenditure. Simultaneously, leptin suppresses the activity of arcuate nucleus NPY/AgRP neurons, which otherwise would antagonize the effect of α-MSH on MC-4Rs through the release of AgRP. Not only does the NPY/AgRP system antagonize anorexigenic melanocortin cells at their target sites, where MC-4Rs are located, but it also very robustly and directly inhibits POMC perikarya through both NPY and the inhibitory neurotransmitter GABA, which acts through basketlike synaptic innervation of POMC cells by NPY/AgRP cell terminals. This apparent unidirectional anatomical interaction between the NPY/AgRP and POMC perikarya is of potential significance, as it provides a tonic inhibition of the melanocortin cells whenever the NPY/AgRP neurons are active *(56)*.

Marks et al. studied the role of melanocortin receptors in transducing the prolonged metabolic derangement observed in experimental cancer. These investigators demonstrated that the cachexia (poor appetite and weight loss) induced by cancer can be both reversed and prevented by administration of AgRP. Prevention of tumor-induced hypophagia with early and repeated AgRP injections resulted in maintenance of normal food intake. To further demonstrate that central melanocortin blockade attenuates cancer cachexia, they investigated this metabolic syndrome in MC4R knockout (KO) mice. The MC4-RKO animals had normal feeding and growth even when bearing a carcinoma that produced classic cachexia in wild-type (WT) control animals. These data clearly indicate that hypothalamic MC4-R plays a role in transducing cachexigenic stimuli from the periphery *(54)*.

We recently tested the hypothesis that leptin was an important cause of uremic cachexia via signaling through its receptor. Our results showed that uremic cachexia was attenuated in *db*/*db* mice, a model of leptin receptor deficiency. Nephrectomy in *db*/*db* mice did not result in any change in weight gain, body composition, resting metabolic rate, and efficiency of food consumption *(55)*. Recent studies suggested that *db*/*db* mice resisted LPS-induced anorexia by reducing TNF-α secretion. Thus, leptin may have an important role in the regulation of appetite, body composition, and metabolic rate in uremia. Indeed, elevated serum leptin was associated with lower dietary intake and higher catabolic rate in uremic children *(51)*. In another set of experiments, we demonstrated that uremic cachexia in experimental animals is attenuated by central MC4-R blockade via a genetic approach. Both homozygous and heterozygous MC4-RKO mice had no decrease in appetite after nephrectomy compared with WT animals. The most striking difference was that both the homozygous and heterozygous MC4-RKO animals continued to gain lean body mass and fat mass with no change in food consumption efficiency despite the cachexic effects of uremia, as demonstrated in the WT nephrectomized controls (N). The effects of nephrectomy on increasing resting metabolic rate were also much attenuated in both the homozygous group and the heterozygous group. These results are consistent with previous observations that MC4-RKO animals maintained normal metabolic rate and body composition even when bearing a carcinoma that produced classic cachexia in WT control animals *(54)*. These data strongly suggest that the hypothalamic MC4-R plays a significant role in transducing cachexigenic signals in uremia. We then tested the effect of central melanocortin receptor antagonism in the experimental uremic cachexia models using a pharmacological approach. These data clearly demonstrated that uremic cachexia is ameliorated by central administration of AgRP in WT-N mice. Repeated intracranial infusion of AgRP significantly regulates food intake, weight gain, body composition, resting metabolic rate, efficiency of food consumption, and circulating leptin concentrations in WT-N/AgRP mice as compared with WT-N mice *(55)*. AgRP is an antagonist for MC4-R. Tumor-induced and sepsis-induced cachexia are attenuated by MC4-R blockade with AgRP *(54)*.

11. CONCLUSIONS

The etiology of cachexia is complex and multifactorial and likely to involve different pathways for different diseases. Inflammation secondary to cytokines and other circulating factors such as PIF may play an important role. Despite their different origins and chemical composition, these cachexia-mediating factors may activate a common final intracellular pathway leading to activation of NFκB. There is also recent evidence that hypothalamic neuropeptides may play an important role. Further research into the molecular pathways leading to cachexia may lead to novel therapeutic therapy for this devastating and potentially fatal complication of chronic disease.

REFERENCES

1. Katz AM, Katz PB. Br Heart J 1962;24:257–264.
2. Fearon KCH. Proc Nutr Soc 1992;51:231–265.
3. Inagaki J, Rodriguez V, Bodey GP. Cancer 1974;33:568–571.
4. Lowrie EG, Lew NL. Am J Kidney Dis 1990;15:458–482.
5. Gruppen MP, Groothoff JW, Prins M, et al. Kidney Int 2003;63:1058–1065.
6. Suttmann U, Ockenga J, Selberg O, et al. J Acquired Immune Defic Syndr Hum Retrovirol 1995;8: 239–246.
7. Windsor JA, Hill Gl. Arch Surg 1988;208:209–217.
8. Dew WD, Begg C, Lavin PT, et al. Am J Med 1980;69:491–497.
9. Utaka S, Avesani CM, Draibe SA, et al. Am J Clin Nutr 2005;82:801–805.
10. Wang AY, Sea MM, Tang N, et al. J Am Soc Nephrol 2004;15:3134–3143.
11. Tisdale MJ. Physiology 2005;20:340–348.
12. Loprinzi CL, Schaid DJ, Dose AM, et al. J Clin Oncol 1993;11:152–154.
13. Mamoun AH, Soderstein P, Anderstam B, et al. J Am Soc Nephrol 1999;10:309–314.
14. Moses AGW, Slater C, Preston T, et al. Br J Cancer 2004;90:991–1002.
15. Shellock FG, Riedinger MS, Fishbein MC. J Cancer Res Clin Oncol 1986;111:82–85.
16. Bing C, Brown M, King P, et al. Cancer Res 2000;60:2405–2410.
17. Collins P, Bing C, McCullock P, et al. Br J Cancer 2002;86:372–375.
18. Busquets S, Carbo N, Almendro V, et al. FEBS Lett 2001;505:255–258.
19. Beck SA, Tisdale MJ. Lipids 2004;39:1187–1189.
20. Eden E, Edstrom S, Bennegard K, et al. Cancer Res 1984;44:1718–1724.
21. Shaw JH, Wolfe RR. Ann Surg 1987;205:368–375.
22. Thompson MP, Cooper ST, Parry PR, et al. Biochim Biophys Acta 1993;1180:238–242.
23. Russell ST, Tisdale MJ. Br J Cancer 2002;87:580–584.
24. Bing C, Bao Y, Jenkins J, et al Proc Natl Acad Sci USA 2004;101:2500–2505.
25. Sanders PM, Tisdale MJ. Cancer Lett 2004;212:71–81.
26. Mak RH, Cheung W, Cone RD, et al. Nature Clin Pract Nephrol 2006;2.
27. Emery PW, Edwards RH, Rennie MJ, et al. Br Med J Clin Res Ed 1984;289:584–586.
28. Lundholm K, Bylund AC, Holm J, et al. Eur J Cancer 1976;12:465–473.
29. Reeds PJ, Field CR, Jahoon F. J Nutr 1994;124:906–910.
30. Mak RHK. Pediatr Nephrol 1998;12:637–642.
31. Du J, Hu Z, Mitch WE. Eur J Clin Invest 2005;35:157–163.
32. Sherry BA, Gelin J, Fong Y, et al. FASEB J 1989;3:1956–1962.
33. Fong Y, Moldawer LL, Marano M, et al. Am J Physiol 1989;2563(Pt 2):R659–R665.
34. Plata-Salaman CR, Sonti G, Borkoski JP, et al. Physiol Behav 1996;603:867–875.
35. Matthys P, Heremans H, Opdenakker G, et al. Eur J Cancer 1991;27:182–187.
36. Llovera M, Garcia-Martinez C, Lopez-Soriano J, et al. Mol Cell Endocrinol 1998;142:183–189.
37. Llovera M, Garcia-Martinez C, Lopez-Soriano J, et al. Cancer Lett 1998;130:19–27.
38. Strassman G, Kambayashi T. Cytokines Mol Ther 1995;1:107–113.
39. Kawamura I, Morishita R, Tomita N, et al. Gene Ther 1999;6:91–97.

40. Breuille D, Farges MC, Rose F, et al. Am J Physiol 1993;265:E660–E666.
41. Combaret L, Rallier C, Taillandier D, et al. Mol Biol Rep 1999;26:95–101.
42. Alzani R, Corti A, Grazioli L, et al. J Biol Chem 1993;268:12,526–12,529.
43. Strassman G, Fong M, Freter CE, et al. J Clin Invest 1993;92:2152–2159.
44. Costelli P, Bossola M, Muscaritoli G, et al. Cytokine 2002;19:1–5.
45. Carbo N, Lopez-Soriano J, Costelli P, et al. Brit J Cancer 2000;83:526–531.
46. Costelli P, Garcia-Martinez C, Llovera M, et al. J Clin Invest 1995;95:2367–2372.
47. Cariuk P, Lorite MJ, Tordorov PT, et al. Br J Cancer 1997;76:606–613.
48. Tordorov P, Cariuk P, McDevitt T, et al. Nature 1996;379:739–742.
49. Mak RH, Cheung W, Cone RD, et al. Kidney Int 2006;69:794–797.
50. Sharma K, Considine RV, Michael B, et al. Kidney Int 1997;51:1980–1985.
51. Daschner M, Tonshoff B, Blum WF, et al. J Am Soc Nephrol 1998;9:1074–1079.
52. Perez-Fontan N, Cordido F, Rodriguez-Camona A, et al. Nephrol Dial Transplant 2004;19:2095–2100.
53. Rodriguez Ayala E, Pecoits-Filho R, Heimburger O, et al. Nephrol Dial Transplant 2004;19:421–426.
54. Marks DL, Ling N, Cone RD. Cancer Research 2001;61:1432–1438.
55. Cheung W, Yu PX, Little BM, et al. J Clin Invest 2005;115:1659–1665.
56. Cone RD. Nat Neurosci 2005;8:571–578.

20 Effect of Weight Loss on Disease

Sergio Josè Bardaro, Dennis Hong,
and Lee Swanström

Abstract

Obesity has a negative effect on the health of patients and is becoming an increasing problem worldwide. Morbid obesity in particular affects most major organ systems and is implicated in several of the most lethal medical conditions: diabetes, myocardial ischemia, hypertension, and many cancers. Weight loss has been shown to reverse many of the comorbidities associated with obesity, with a resulting positive benefit to the patient's health and quality of life. Achieving substantial and persistent weight loss is difficult and, to date, surgery has been the only truly effective option for treatment of obesity. Obesity has effects on multiple organ systems. Whereas some effects are permanent, many can be reversed if the patient loses a substantial amount of weight. This includes the endocrine system (diabetes), the musculoskeletal system (degenerative joint disease), and the cardiovascular system (hypertension, congestive heart failure, hypercoagulability and stroke). Dieting and exercise can be effective strategies for the moderately overweight but have a high failure rate (>90%) in the morbidly obese. Medical treatments have a mixed history and, in general, have only moderate efficacy and high side effect profiles. Surgery has been shown to be the most effective long term weight-loss strategy but is a relatively high-risk endeavor in this patient population. The laparoscopic gastric bypass is currently the most widely used bariatric surgery, achieving around 75% excess weight loss at 5-yr follow-up.

Obesity has a major detrimental effect on a patient's health. Many of the diseases and comorbidities related to obesity can be reversed or stabilized with substantial weight loss, which currently is best achieved by surgery.

Key Words: Weight loss; bariatric surgery; morbid obesity; gastric bypass.

1. INTRODUCTION

Obesity is epidemic in the United States and the Western world. Approximately 97 million adults are obese and the number continues to rise. Body mass index (BMI) has become the standard of measurement for obesity and is a relation between body weight (in kg) and body surface (square of the height in meters). Obesity is defined as a BMI greater than 30 kg/m^2 and morbid obesity as a BMI greater than 40 kg/m^2. The National Health and Nutrition Examination Survey (NHANES) reported that between 1988 and 1994, 22.9% of adults were obese, 55.9% were overweight, and 2.9% were extremely obese *(1)*.

Excess body fat predisposes an individual to obesity-related comorbidities, such as coronary heart disease, hypertension, dyslipidemia, type 2 diabetes, gallstones, cancer,

From: *Nutrition and Health: Adipose Tissue and Adipokines in Health and Disease*
Edited by: G. Fantuzzi and T. Mazzone © Humana Press Inc., Totowa, NJ

sleep apnea, osteorthritis, joint degeneration, reflux disease, nonalcoholic steatohepatitis, and others.

Weight loss is the most effective treatment for obesity-related physical and psychological comorbidities. The extent of weight loss and its effect on obesity-related comorbidities have been of great interest. Many studies have shown that there are major improvements in comorbidities even with a modest weight loss, and total resolution of many obesity-related diseases can occur without attaining a completely normal weight *(2–3)*. The implication that modest weight loss has disproportionate benefit is interesting and implies that the weight loss state *per se* has advantages in the obese *(4)*.

Many treatments are available for weight loss, including dieting, drugs, and weight loss surgery. Dieting and drugs seem to be effective for overweight or mildly obese individuals, but have a high rate of failure in the morbidly obese. Outcome studies show bariatric surgery to be the most effective treatment to achieve the desirable long-term results of substantial weight loss. The purpose of this chapter is to describe the alternatives of weight loss treatments and their effectiveness in resolution of comorbidities.

2. METHODS OF INDUCING WEIGHT LOSS IN THE OBESE PATIENT

2.1. Diet, Pharmacological, and Behavioral Treatment

The US Food and Drug Administration (FDA), as well as an expert panel convened by the National Heart, Lung, and Blood Institute, have recommended that weight loss medications be used only as an adjunct to a comprehensive program of lifestyle modification that includes diet, physical activity, and behavior therapy *(16)*. A common hypothesis is that medication should help facilitate adherence to lifestyle modification. By reducing appetite or nutrient absorption, medications may make it easier for patients to adhere to a low-calorie diet. Surprisingly little, however, is known about the specific benefits of combining therapies, or how and when they should be combined. The well-documented problem with lifestyle modification is weight regain after treatment termination. On average, in the year following treatment, patients regain 30 to 35% of their lost weight. Approximately 3 to 5 yr after therapy, 50% or more of participants have returned to their baseline weight *(17–18)*. These results are not entirely discouraging considering that most obese people, left untreated for 3 to 5 yr, would probably gain 0.5 to 1 kg per year *(19)*.

Orlistat and sibutramine are the only two drugs currently approved for long-term use to achieve and maintain weight loss *(20)*. A recent meta-analysis of pharmacological treatments for obesity showed mean weight loss of 2.9 kg after 1 yr with orlistat, and 4.5 kg after 1 yr with sibutramine *(21)*. For obese individuals who need to lose more than 10 kg, these results are suboptimal. On the horizon, rimonabant is a potential alternative. In a recent multicenter trial, this cannabinoid-1 receptor antagonist produced a mean weight loss of 6.6 kg after 1 yr *(22)*.

The Atkins and Protein Power diets are very high in total and saturated fat compared with current dietary guidelines. Long-term use of these diets for weight maintenance is likely to significantly increase serum cholesterol and risk for coronary heart disease. The Sugar Busters and Zone diets would lower serum cholesterol concentrations and likely reduce risk for coronary heart disease. High-carbohydrate, high-fiber, and low-fat diets would have the greatest effect in decreasing serum cholesterol concentrations and,

Table 1
Mechanism of Action of Various Contemporary Bariatric Surgeries

Mechanism	*Procedure*
Restrictive	Vertical banded gastroplasty
	Gastric banding
Malabsorptive	Jejunoileal bypass
	Biliopancreatic diversion
Combination	Roux-*en*-Y gastric bypass

thus, the risk for coronary heart disease. Even though high-fat diets may promote short-term weight loss, the potential hazards for worsening risk for progression of arteriosclerosis or atherosclerotic events override the short-term benefits *(23)*.

2.2. Bariatric Surgery

Bariatric surgery dates back to the 1950s, when intestinal bypasses were first performed. Total weight loss achieved correlated to the length of small bowel bypassed. However, because of severe complications, including diarrhea, electrolyte abnormalities, dehydration, liver dysfunction, protein malnutrition, and renal disease, these procedures were abandoned. Currently, there are several different types of bariatric procedures, which are classified as restrictive, malabsorptive, or a combination of both (Table 1). Each surgical procedure, depending on its mechanism of action, has benefits, side effects, and complications associated with it. The success rate for weight loss varies among them as well (Table 2).

Gastric bypass, either laparoscopic or open, is currently the most common weight loss surgery performed in the United States (Fig. 1). The mortality rate for this surgery is approx 0.5%, and perioperative complications occur in 10% of cases. Long-term considerations in gastric bypass include vitamin and mineral deficiencies, anemia, dumping syndrome, nausea, and vomiting.

Currently the most common purely restrictive procedure is adjustable gastric banding. These silicone bands are inflatable via a small reservoir that is placed under the skin of the anterior aspect of the abdominal wall, percutaneously injecting the port allows the diameter of the band to be adjusted. Adjustable bands can be inserted laparoscopically, thereby reducing the complications and discomfort of the open approach (Fig. 2A,B).

Easy laparoscopic placement, reversibility of the procedure, and lack of gastric or intestinal anastomosis are clear advantages of the procedure, although complications such as band slippage, pouch dilation, band/port infection, band erosion, esophageal dilation, failure to lose weight, nausea, and vomiting can occur. Although some studies have documented weight loss equal to gastric bypass, with fewer complications, many other groups have had disappointing outcomes. Recently, Jan et al. reported a 3-yr excess weight loss of 60 vs 57% ($p = 0.85$) in gastric bypass and lap-band, respectively *(9)*.

Biliopancreatic diversion (BPD), with or without duodenal switch, has some of the highest rates of weight loss, but also the highest rates of nutritional complications *(15)*. Unlike the gastric bypass, in which no stomach is removed, the BPD includes the resection

Table 2
Advantages and Disadvantages of Different Bariatric Procedures

Attributes	*LAGB*	*VBG*	*RYGB*	*BPD*	*BPD/DS*
Safe	+++	++	+++	+	+
Durable	+	++	++	+++	+++
Controlable/adjustable	YES	NO	NO	NO	NO
Easily reversible	++	+	+	NO	NO
Fully reversible	+++	+	+	NO	NO
Minimal invasiveness	+++	++	++	+/–	+
Low reoperation rate	+	+	+	+/–	+
Quality of life	++	++	+++	+/–	+
Nausea and vomiting	+++	++	+	+	+
Protein deficiency	NO	NO	+/–	+++	++
Micronutrients deficiency	+	+	++	+++	+++
Gallstone formation	+/–	+/–	+	+++	++

LAGB, laparoscopic adjustable gastric banding; VBG, vertical banded gastroplasty; RYGB, Roux-*en*-Y gastric bypass; BPD, biliopancreatic diversion; BPD/DS, biliopancreatic diversion with duodenal switch.

of 70% of the distal stomach transversally. The residual upper stomach is far larger than the small pouch performed for gastric bypass (Fig. 3). The patient can still consume large volumes of food, because malabsorption will induce weight loss. An extremely reduced common channel (50–100 cm) and alimentary limb (100 cm) ensure a high rate of malabsorption, but also a great number of complications (with additional dumping syndrome) result from this situation. Currently, BPD is mostly used as a treatment for hyperobese patients where more weight loss is desired. Many specialists debate whether this surgery and its complications outweigh the benefits *(13)*.

In summary, although bariatric surgery is the best solution for substantial long-term weight loss, there is no single perfect operation for all obese persons. At this time, laparoscopic gastric bypass and laparoscopic adjustable gastric banding seem to be the procedures that show desirable weight loss with an acceptable incidence of morbidity and mortality. Each patient should be individually evaluated and counseled regarding the advantages and disadvantages of all procedures in order to match the type of procedure to the patient's expectancies and needs.

3. SPECIFIC DISEASES AFFECTED BY SUBSTANTIAL WEIGHT LOSS

3.1. Infertility

The fertility of obese women compared with normal-weight women is lower in natural cycles and infertility treatment cycles *(24–27)*. The role of obesity in menstrual disorders, as well as the beneficial effect of weight reduction on ovulatory function in overweight anovulatory women, have been described *(28,29)*. As obesity is a known factor in infertility *(30)*, weight reduction is frequently incorporated as a part of infertility treatment *(31)*. The physiological mechanisms of fertility and weight loss are just beginning to be understood.

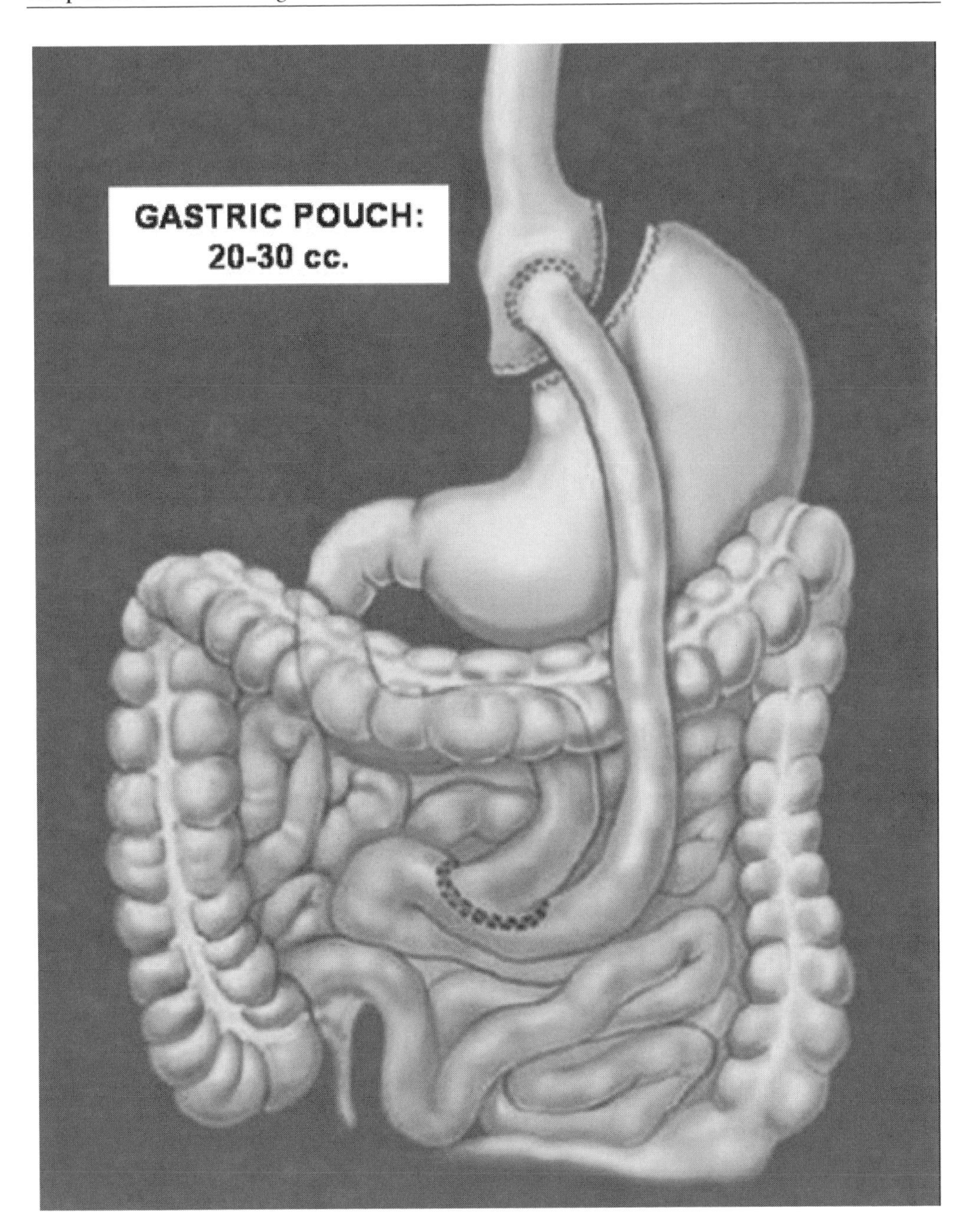

Fig. 1. Roux-*en*-Y gastric bypass.

Hollmann et al., from Germany, reported the effects of weight loss in obese, infertile women, describing the changes in hormone levels, menstrual function, and pregnancy rate. Of the 58 women in the study, 35 took part in a weight- reducing program lasting 32 ± 14 wk with a weight loss of 10.2 ± 7.9 kg (therapy group). At the time of first oral glucose tolerance testing, insulin resistance was a feature in 85% of the women in the therapy group, and 22% were hyperandrogenemic. Weight loss resulted in a significant

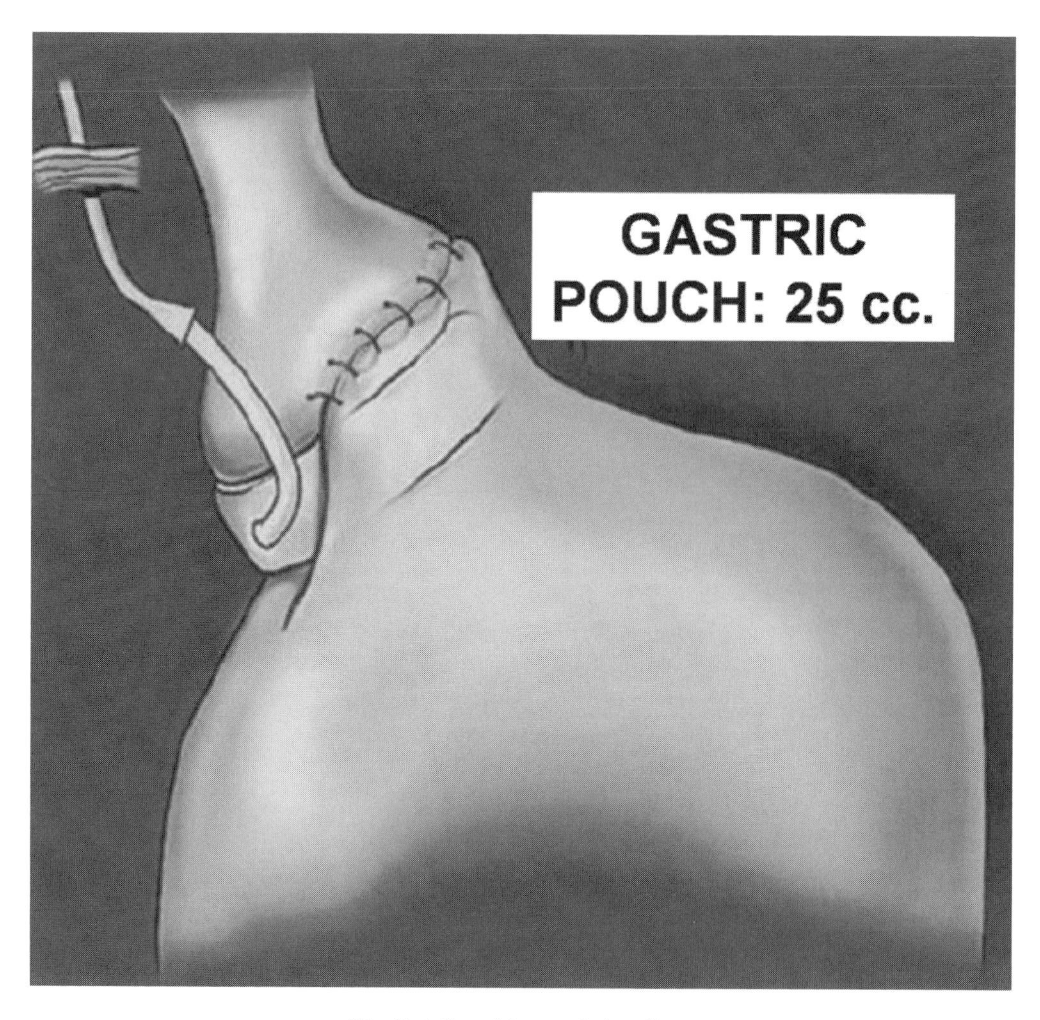

Fig. 2. Adjustable gastric banding.

reduction in blood glucose, insulin, androstenedione, dihydrotestosterone, and estradiol concentrations. The pregnancy rate was 29% in this group, and 80% showed an improvement of their menstrual function. Thus, weight reduction is an appropriate treatment for women with obesity-related endocrine derangement, menstrual irregularity, and infertility *(32)*.

Clark et al. reported that an average weight loss of 10 kg restored spontaneous ovulation in 90% in a cohort of previously anovulatory patients. In this group, 77.6% subsequently became pregnant and 67% achieved a live birth. This also showed the cost-effectiveness of weight loss as an infertility treatment. Prior to the study, 67 women had had infertility treatment for $550,000 for two live births, a cost of $275,000 per baby. After participation in this study and completing the weight loss program (diet and exercise for 6 mo), the same 67 women had treatment costs of $210,000 for 45 live births, a cost of $4600 per baby *(33)*.

These results support the view that weight loss should be a prerequisite for obese women prior to any assisted reproduction program.

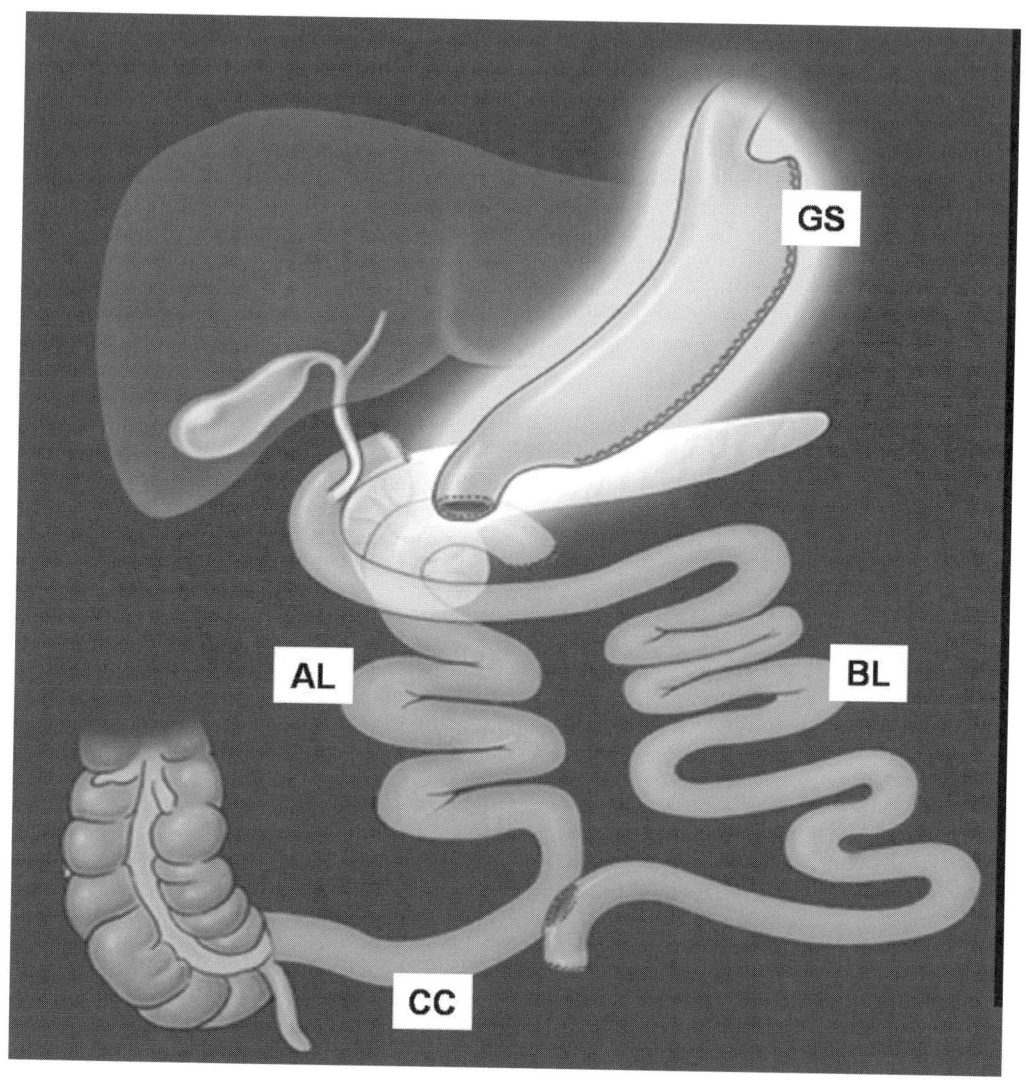

Fig. 3. Biliopancreatic diversion with duodenal switch.

3.2. Cancer

Because the relationship of obesity to some forms of cancer has been known for a long time *(34)*, it is not surprising that researchers have looked closely at the possible role of adipocytikines in the regulation of carcinogenesis *(35)*. Current studies show that obesity and diabetes, closely related diseases, are both risk factors for cancer *(36–37)*. Jee et al. reported that higher blood glucose levels (140 mg/dL and above) increase the risk of all types of cancer by up to 1.29 times. Specifically they found an increased relative risk (RR) for cervical carcinoma (RR = 2.5), breast cancer (RR = 2.3), leukemia (RR = 1.5), pancreatic cancer (RR = 1.7), liver cancer (RR = 1.6), colorectal cancer (RR = 1.3), stomach cancer (RR = 1.2), and esophageal cancer (RR = 1.4) *(38)*. It is still not clear whether disturbances of adipocytokines are directly linked with cancer or whether weight loss can substantially decrease the risk for cancer. However further investigations would help to elucidate whether weight loss can prevent not only the current top cause of death, cardiovascular diseases, but also the second most common cause, cancer.

3.3. Asthma and Obstructive Sleep Apnea

In a study of 33 patients with asthma who were followed for 1 yr after weight-loss surgery, all patients showed improvement in symptoms (severity and frequency) and a third of them had no episodes of asthma and were off all therapy during follow-up *(44)*.

Weigh loss also improves symptoms associated with obstructive sleep apnea (OSA). The seriousness of this problem is emphasized by the fact that the cost of medical care of obese patients with OSA has been estimated to be twice the cost of similar patients without OSA *(45)*. Recently, Guardiano et al. reported on the effect of significant weight loss in eight subjects showing a mean reduction in respiratory disturbance index (RDI ,the sum of all apneas and hypopneas per hour of sleep) by 75%. The same study reported a 100% reduction in the need for continuous positive airway pressure (CPAP) treatment and the mean oxygen saturation improved by 2 ± 2% (95 to 97%). The mean nadir oxygen saturation also improved from 74 to 87% *(45)*.

In addition to the profound reduction in OSA events after weight loss, the risks of sudden death and cardiac arrhythmias are significantly reduced. Perhaps the greatest patient benefit following an improvement in respiratory function after weight loss, is the tremendous improvement in quality of life they experience, with better sleep, higher daytime energy level, and overall satisfaction.

3.4. Cardiovascular Disease

Obesity is associated with an increased risk of coronary artery disease (CAD) and mortality from cardiac events including myocardial infarction, arrhythmias, and sudden cardiac death *(46)*. Morbidity and mortality rates rise proportionally to the degree of obesity in men and women, and the impact of excess body fat is more significant in younger subjects than in older ones *(47)*. Heart function is directly influenced by excess body fat *(52)*. In addition to higher cardiac output in obese patients, left ventricular volume and filling pressures are higher than normal. This frequently results in the development of left ventricular strain, which leads to hypertrophy, often of the eccentric type *(53)*. Left ventricular diastolic function is frequently impaired by this effect. Weight loss has a beneficial impact on the functional and the structural cardiac status. In a study of obese patients with a mean BMI of 32.7, weight loss of 8 kg over a period of 25 wk was associated with a significant decrease in left ventricular mass *(54)*. Weight loss also lowers blood pressure. In a 3-yr follow-up of nonmorbidly obese patients with a mean BMI of 31, patients who maintained a 4.5-kg weight loss had a relative risk of hypertension of 0.35, or a reduction of 0.45 mmHg in systolic blood pressure and 0.35 mmHg in diastolic pressure per kilogram of weight lost (51). Finally, as compared with conventional therapies for weight loss (diet, drugs, lifestyle modification), bariatric surgery appears to be the best option for the treatment of the morbidly obese, resulting in long-term weight loss, improved lifestyle, and amelioration in cardiovascular risk factors that were elevated at baseline.

3.5. Diabetes

Type 2 diabetes affects around 8% of adults in the United States and is substantially related to obesity *(48)*. In a 10-yr follow-up study, men and women with a BMI ≥35.0

had a relative increased risk of developing diabetes of 23- and 17-fold respectively, compared with a control group with a BMI between 18.5 and 24.9 *(49)*. Independent of BMI, the relative risk of developing diabetes mellitus increases with weight gain, as shown in the Nurses' Health Study *(50)*. Moreover, in that study, women who voluntarily lost more than 5.0 kg reduced their risk of diabetes by 50%.

In a review of 440 obese patients (mean weight of 183 kg) who underwent BPD with duodenal switch, all of the 36 patients with type 2 diabetes discontinued their medication over a 7-yr follow-up period. It should be kept in mind, however, that the operative mortality for this particular weight loss surgery is between 0.5 and 2% *(14)*. Weber et al. reported on 103 consecutive patients with laparoscopic gastric bypass who were randomly matched to 103 patients with laparoscopic gastric banding according to age, BMI, and gender. The prevalence of comorbidities, such as arterial hypertension, type 2 diabetes mellitus, and dyslipidemia, was comparable in the two groups before surgery. The frequency of all these comorbidities decreased in the follow-up period, with one exception of dyslipidemia in the banding patients. The prevalence of hypertension dropped from 52 to 13% in the bypass group, and from 60 to 18% in the banding group. Diabetes declined from 37 to 6% and 44 to 18%, respectively, leading to a significantly lower frequency in the bypass patients ($p = 0.007$) on follow-up. Dyslipidemia was also significantly lower in the bypass group than in the banding group after follow-up ($p = 0.001$) and in fact was essentially unchanged in the band patients. This observation might be explained again through the different mode of action, as the gastric bypass procedure leads to a moderate malabsorption, where increased fat consumption can result in uncomfortable steatorrhea. The gastric banding procedure is, overall, less effective in controlling cardiopulmonary comorbidities *(55)*.

Redmon et al. suggested that weight loss at 2 yr of 4 to 5 kg for people with type 2 diabetes can produce improvements in diabetes control that are likely to be clinically significant *(56)*. The Diabetes Prevention Program showed that intensive lifestyle intervention reduced the risk of diabetes by 58%. Diet and regular exercise reduce the incidence of diabetes even the patient is still overweight *(48)*.

The prospective controlled Swedish Obese Subjects Study involved obese subjects who underwent weight-loss surgery and contemporaneously matched, conventionally treated obese control group who have been follow for at least 2 yr (4047 patients) or 10 yr (1703 patients). Two- and 10-yr rates of recovery from diabetes, hypertriglyceridemia, low levels of high density lipoprotein cholesterol, hypertension, and hyperuricemia were more favorable in the surgery group than in the control group, whereas recovery from hypercholesterolemia did not differ between groups *(57)*.

3.6. *Osteoarthritis*

Obesity is strongly associated with an increased risk of osteoarthritis. This has been particularly studied and confirmed for knee osteoarthritis. Analyses of the National Health and Nutrition Examination Survey data show that adults in the United States with a BMI ≥ 30 have a more than fourfold higher prevalence of radiographic knee osteoarthritis than those with a BMI lower than 30 *(58)*. In one study *(59)*, the risk of developing both radiographic and symptomatic knee osteoarthritis was increased in elderly men and women who were obese for an average of 37 yr. The risk of developing disabling knee osteoarthritis over a 10-yr period is strongly and directly correlated with initial BMI *(60)*. Weight loss is recommended for overweight persons with knee osteoarthritis under the American College of Rheumatology treatment guidelines *(61)*. Several studies suggest that weight loss may prevent the development or worsening of knee and hip osteoarthritis. In an observational study, Felson et al. *(62)* found that older women who had lost 5 kg during a 10-yr period had a 50% reduction in the risk of new symptomatic knee osteoarthritis. McGooey et al. *(63)* reported a four- to fivefold reduction in the prevalence of knee and hip pain in morbidly obese subjects who lost an average of 45 kg after gastric stapling. Another study of obese women who lost an average of 5 kg during a 6-mo treatment with phentermine or placebo found that reductions in clinical symptoms of hip and knee ostexoarthritis were correlated with the amount of weight loss *(64)*. Together, these studies demonstrate that overweight and obese older persons with painful knee osteoarthritis can lose 4.5 to 9 kg over 6 mo with a combined diet and exercise program, and that this amount of weight loss is associated with improvements in knee pain and disability.

3.7. Gastroesophageal Reflux Disease

There is a widespread notion that obese persons are more likely to develop gastroesophageal reflux disease (GERD) than leaner subjects, and obese patients who seek medical care for symptoms suggestive of reflux are often recommended by clinicians to reduce their body weight to relieve the symptoms. However, the scientific search for a positive association between BMI and intensity of symptoms has failed to show any correlation.

Lagergren et al. *(65)* have reported that weight reduction may not be justifiable as an antireflux therapy. Even if obesity is a poor predictor of reflux symptoms, this does not necessarily imply that weight reduction will not be of benefit in providing symptom relief. A significant beneficial effect of weight loss on symptoms of gastroesophageal reflux in overweight patients has recently been reported in a small study involving 34 patients *(66)*. In this study, the degree of weight loss was directly correlated with improvement in GERD symptom scores. Elsewhere, strong and independent associations have been reported between obesity, reflux symptoms, and esophageal adenocarcinoma *(67)*. The evidence suggests that an overweight individual with reflux symptoms is at significantly increased risk of esophageal adenocarcinoma.

Although weight loss through lifestyle modification improves GERD symptoms, there are no conclusive studies that document how much weight loss is necessary to alleviate symptoms. The amount of weight loss that will alter intra-abdominal pressure and possibly reduce GERD symptoms in patients with clinically significant obesity (BMI > 35) may be unattainable or unsustainable by dietary changes alone.

The mainstay of operative treatment for medically refractory GERD has been laparoscopic fundoplication, yet the suitability of laparoscopic fundoplication in obese patients has been controversial because of possible higher failure rates. Gastric bypass reduces acid reflux into the esophagus and also induces durable and sustainable weight loss; therefore, Roux-*en*-Y gastric bypass (RYGB) may have a potential application as a primary treatment for GERD in obese patients.

Nelson et al. *(68)* reported that 89% of patients showed improvement of GERD symptoms after 18% excess weight loss, whereas 7% had no change and 4% worsened. Postoperatively, overall medication usage decreased from 30 to 3% ($p < 0.001$ vs preoperative) at the initial 3-mo follow-up. This reduction in medication use was sustained at long-term follow-up, with only 5% using antireflux medications. The most dramatic decrease was in the proton pump inhibitor class, with usage decreasing from 17% preoperatively to 2% at 3 mo after bariatric surgery *(68)*.

RYGB is effective for control of GERD symptoms with a significant decrease in the use of antireflux medications. Both symptomatic improvement and decreased use of antireflux medications occur independent of weight loss. These data suggest that RYGB may be an alternative to fundoplication in obese individuals seeking operative treatment for GERD.

Dixon and O'Brien showed that 76% of patients had complete relief of symptoms after laparoscopic gastric banding with previous moderate or severe reflux, and there was an improvement in symptoms in an additional 14%. The need for medications was markedly reduced, and in 76% of cases, no treatment was required. This degree of effectiveness equates with that achieved with proton pump inhibitors, although it falls short of reported outcomes after antireflux surgery *(69)*.

A recent study by Perez et al. demonstrated a 31% recurrence rate of GERD symptoms following laparoscopic Nissen fundoplication in obese patients *(70)*. Results of this study showed that laparoscopic Nissen fundoplication and laparoscopic gastric bypass are effective in treating both objective acid reflux and heartburn. Heartburn symptoms improved significantly in both groups of patients and completely resolved in most patients after surgery. Patterson et al. *(71)* reported that laparoscopic Nissen fundoplication has proven to be an effective treatment of GERD. However, results have been less successful in the obese population, especially in morbidly obese patients. Laparoscopic Nissen patients did have significantly higher reflux scores preoperatively compared with laparoscopic gastric bypass patients.

3.8. Nonalcoholic Fatty Liver Disease

Several studies have documented that obesity *per se* is a risk factor for liver diseases, including steatosis, steatohepatitis, fibrosis, and cirrhosis. The hepatic alterations described in obesity and morbid obesity are included in the spectrum of lesions considered in the clinicopathological entity called nonalcoholic fatty liver disease (NAFLD). NAFLD is characterized by steatosis-associated hepatic injury that affects individuals who consume little or no alcohol and in whom other causes of liver disease can be excluded. The histological patterns of NAFLD are steatosis alone, nonalcoholic steatohepatitis (NASH)—in which, in addition to steatosis, lobular necroinflammatory alterations are observed—and fibrosis, which can progress to cirrhosis *(72)*.

As early as 1970, Drenick et al. *(73)* reported that extensive weight reduction by dieting or fasting in obese patients was accompanied by a marked decrease in the amount of fatty infiltration in the liver. Other authors have also noted regression of liver steatosis in obese persons after gastroplasty *(74)* or after diet and exercise *(75)*.

Stratopoulos et al. *(76)* reported that NASH is a potentially reversible condition in morbidly obese patients (86% regressed) after weight loss induced by vertical banded gastroplasty. On the other hand, liver fibrosis improved in only 47% of these patients.

Keshishian et al. *(77)* recently published a report on 78 patients who had a NASH grade between moderate and severe at the time of biliopancreatic diversion with duodenal switch (BPD/DS), whereas by the second year postoperatively, most patients had dropped their NASH grade to mild to moderate levels. By the third year, some patients had complete resolution of NASH. Similarly, most patients had hepatic steatosis of 50 to 70% before BPD/DS, which improved to 20 to 30% by 2 yr. Rapid and progressive improvement was observed at and after 12 mo after the BPD/DS operation, and continued beyond 3 yr.

Finally, in view of the frequency of hepatic alterations, a liver biopsy should be part of the surgical protocol for bariatric operations, because it is the only procedure able to determine the presence of potentially progressive lesions and to stage fibrosis.

3.9. Depression

Dixon et al. reported on 262 patients with depression who were studied before and up to 4 yr after weight-loss surgery. Before the procedure, patients fell evenly into each of four categories (major depressive illness, moderate depression, mild depression, and normal status) as measured by the Beck Depression Inventory. At 1 yr, there were major reductions in the proportion of patients in each of the depression groups, with 75% of patients assessed to be normal. Four years after the procedure, the beneficial effect persisted *(40)*.

3.10. Quality of Life

Quality of life is a major reason that people attempt to lose weight. There are currently only a few validated instruments available for bariatric surgery *(41)*. The Bariatric Analysis and Reporting Outcome System (BAROS), the Impact of Weight on Quality of Life (IWQoL), and secondarily the Short form 36 (SF-36), are commonly used by bariatric surgeons.

Tolonen et al. reported on quality-of-life scores among 125 patients, 60 after surgery and 65 before surgery. Significantly better scores were recorded for the group that underwent weight loss surgery in terms of self-esteem and physical, social, labor, and sexual aspects, with an overall better score of quality of life *(42)*.

Recently, Dixon et al. reported on a comparative study between weight-loss surgery patients and controls. Ten out of 13 items in the SF-36 domain score showed significant better scores than those of control patients. Physical function, role limitations from physical problems, pain, general health, energy or vitality, social function, mental health, physical component summary, Beck depression inventory, and appearance evaluation had the best scores, whereas the nonimproved scores were role limitation from emotional problems, mental component summary, and appearance orientation *(4)*.

The Medical Outcomes Trust SF-36 health questionnaire was used to evaluate quality of life among 459 patients before laparoscopic adjustable gastric banding placement and annually thereafter. Scores on all eight subscales of the SF-36 were abnormally low before surgery. At 1 yr, scores on all subscales had returned to normal and the beneficial effect was seen to persist for at least 4 yr *(43)*.

4. CONCLUSIONS

Weight loss has a tremendous beneficial impact on the health of obese patients. Moderate weight loss may improve the severity of obesity-related comorbidities, but more extensive weight loss could resolve many of them. Unfortunately, losing weight remains a difficult goal for the majority of obese patients. There are a variety of weight-reduction methods available for the health practitioner to recommend to patients. Depending on the patient's health status and BMI, different treatments should be chosen, from diet modification, drugs, and lifestyle changes to bariatric surgery.

REFERENCES

1. Flegal KM, Carroll M, Ogden C, et al. JAMA 2002;288:1723.
2. Wing RR, Venditti E, Jakicic JM, et al. Diabetes Care 1998;21:350–359.
3. Poobalan A, Aucott L, Smith WC, et al. Obes Rev 2004;5:43–50.
4. Dixon JB, Anderson M, Cameron-Smith D, et al Obes Res 2004;12(11):1895–1902.
5. Mason EE, Ito C. Surg Clin North Am 1967;47:1345–1351.
6. Wittgrove AC, Clark GW, Tremblay LJ. Obes Surg 1994;4:353–357.
7. Oria HE. Eur J Gastroenterol Hepatol 1999;11:105–114.
8. Kuzmak LI, Yap IS, McGuire L, et al. AORN J 1990;51:1307–1324.
9. Jan J, Hong D, Pereira N, et al. J Gastrointest Surg 2005;9:30–41.
10. Griffen WO Jr, Bivins BA, Bell RM, et al. Surg Gynecol Obstet 1983;157:301–308.
11. Scopinaro N, Gianetta E, Civalleri D, et al. Br J Surg 1979;66(9):618–620.
12. Scopinaro N, Gianetta E, Civalleri D, et al. Am J Clin Nutr 1980;33(2):506–514.
13. Grimms IS, Schindler W, Halluszka O. Am J Gastroenterol 1992;87(6):775–779.
14. Hess DS, Hess DW. Obes Surg 1998;8(3):267–282.
15. Bloomberg RD, Fleishman A, Nalle JE, et al. Obes Surg 2005;15(2):145–154.
16. National Heart, Lung, and Blood Institute. Clinical guidelines on the identification, evaluation, and treatment of overweight and obesity in adults: the evidence report. Obes Res 1998;6:51S–210S.
17. Wadden TA, Foster GD. Med Clin N Am 2000;84:441–462.
18. Kramer FM, Jeffery RW, Forster JL, et al. Int J Obes Relat Metab Disord. 1989;13:123–136.
19. Williamson DF. Ann Intern Med 1993;119:646–649.
20. O'Brien PE, Brown WA, Dixon JB. MJA 2005;183:310–314.
21. Li Z, Maglione M, Tu W, et al. Ann Intern Med 2005;142:532–546.
22. Van Gaal LF, Rissanen AM, Scheen AJ, et al. Lancet 2005;365:1389–1397.
23. Anderson JW, Konz EC, Jenkins DJA. J Am Coll Nutr 2000;19(5):578–590.
24. Chong AP, Rafael RW, Forte CC. Fertil Steril 1986;46:599–603.
25. Hamilton-Fairley D, Kiddy D, Watson H, et al. Br J Obstet Gynaecol 1992;99:128–131.
26. Zaadstra BM, Seidell JC, Van Noord PAH, et al. Br Med J 1993;306:484–487.
27. Crosignani PG, Ragni G, Parazzini F, et al. Hum Reprod 1994;9:420–423.
28. Rogers J, Mitchell GW Jr. N Engl J Med 1952;247:53–55.
29. Mitchell GW Jr, Rogers J. N Engl J Med 1953;249:835–837.
30. Wickelgren I. Science 1998;280:1364–1367.
31. Clark AM, Ledger W, Galletly C, et al. Hum Reprod 1995;10(10):2705–2712.
32. Hollmann M, Runnebaum B, Gerhard I. Hum Reprod 1996;11(9):1884–1891.
33. Clark AM, Thornley B, Tomlinson L, et al. Hum Reprod 1998;13(6):1502–1505.
34. Bray GA. J Nutr 2002;132(11):3451S–3455S.
35. Garofalo C, Surmacz E. J Cell Physiol 2006;7:12–22.
36. Calle EE, Kaaks R. Nat Rev Cancer 2004;4:579–591.
37. Komninou D, Ayonote A, Richie JP, et al. Exp Biol Med (Maywood) 2003;228:396–405.
38. Jee SH, Ohrr H, Sull JW, et al. JAMA 2005;293:194–202.
39. Jee SH, Kim HJ, Lee J. Yonsei Med J 2005;46(4):449–455.
40. Dixon JB, Dixon ME, O'Brien PE. Arch Intern Med 2003;163:2058–2065.
41. Weiner S, Sauerland S, Fein M, et al. Obes Surg 2005;15:538–545.
42. Tolonen P, Victorzon M. Obes Surg 2003;13:424–426.
43. Dixon JB, Dixon ME, O'Brien PE. Obes Res 2001;9:713–721.
44. Dixon JB, Chapman L, O'Brien P. Obes Surg 1999;9:385–389.
45. Guardiano SA, Scott JA, Ware JC, et al. Chest 2003;124:1615–1619.
46. Poirier P, Eckel RH. Curr Atheroscler Rep 2002;4:448–453.

47. Must A, Spadano J, Coakley EH, et al. JAMA 1999;282:1523–1529.
48. The Diabetes Prevention Program Research Group. Obes Res 2004;12(9).
49. Field AE, Coakley EH, Must A, et al. Arch Intern Med 2001;161:1581–1586.
50. Colditz GA, Willett WC, Rotnitzky A, et al. Ann Intern Med 1995;122:481–486.
51. Stevens VJ, Obarzanek E, Cook NR, et al. Ann Intern Med 2001;134:1–11.
52. Alpert MA, Hashimi MW. Am J Med Sci 1993;306:117–123.
53. Poirier P, Eckel RH. In: Fuster V, Alexander RW, King S, et al., eds. *Hurst's The Heart* Vol. 83. 10th ed. McGraw-Hill, New York: 2000, pp. 2289–2303.
54. MacMahon SW, Wilcken DE, Macdonald GJ. N Engl J Med 1986;314:334–339.
55. Weber M, Muller MK, Bucher T, et al. Ann Surg 2004;240(6):975–983.
56. Redmon JB, Reck KP, Raatz SK, et al. Diabetes Care 2005;28:1311–1315.
57. Sjöström L, Lindroos AK, Peltonen M, et al., for the Swedish Obese Subjects Study Scientific Group. N Engl J Med 2004;351:2683–2693.
58. Anderson JJ, Felson DT. Am J Epidemiol 1988;128:179–189.
59. Felson DT, Anderson JJ, Mainmark A, et al. Ann Intern Med 1988;109:18–24.
60. Manninen P, Riihimaki H, Heliovaara M, et al. Int J Obes Relat Metab Dis 1996;20:595–597.
61. ACR Subcommittee on Osteoarthritis Guidelines. Arthritis Rheum 2000;43:1905–2015.
62. Felson DT, Zhang Y, Anthony JM, et al. Ann Intern Med 1992;116:535–539.
63. McGooey B, Deital M, Saplys R, et al. J Bone Joint Surg 1990;72B:322–323.
64. Williams RA, Foulsham BM. Practitioner 1981;225:231–232.
65. Lagergren J, Bergström R, Nyrén O. Gut 2000;47:26–29.
66. Fraser-Moodie CA, Norton B, Gomal C, et al. Scand J Gastroenterol 1999;34:337–340.
67. Lagergren J, Bergstrom R, Nyren O. Ann Intern Med 1999;130:883–890.
68. Nelson LG, Gonzalez R, Haines K, et al. Am Surg 2005;71:950–954.
69. Dixon JB, O'Brien PE. Obes Surg 1999;9:527–531.
70. Perez AR, Moncure AC, Rattner DW. Surg Endosc 2001;15:986–989.
71. Patterson EJ, Davis DG, Khajanchee Y, et al. Surg Endosc 2003;17:1561–1565.
72. Rodrigues Pereira Lima ML, Oliveira Mourão SC, Costa Diniz MT, et al. Obes Surg 2005;15:661–669.
73. Drenick EJ, Simmons F, Murphy JF. N Engl J Med 1970;282:829–834.
74. Ranlov I, Hardt F. Digestion 1990;47:208–214.
75. Ueno T, Sugawara H, Sujaku K, et al. J Hepatol 1997;27:103–107.
76. Stratopoulos C, Papakonstantinou A, Terzis I, et al. Obes Surg 2005;15:1154–1160.
77. Keshishian A, Zahriya K, Willes EB. Obes Surg 2005;15:1418–1423.

V

Adipose Tissue and Disease

21 Adipose Tissue and Insulin Resistance

Stephen E. Borst

Abstract

Adiposity, especially visceral adiposity, is an important risk factor for the development of insulin resistance and type 2 diabetes. In addition to its role in storing energy, adipose tissue also secretes into the circulation a number of hormones and other factors that can alter the response to insulin in distant tissues, such as liver and muscle. Many of these factors are cytokines, which have been associated with the immune system.

Fat-derived hormones that can enhance insulin signaling include leptin, adiponectin, and possibly visfatin. Those impairing insulin signaling include tumor necrosis factor-α, resistin, and several of the interleukins. Obesity has also been identified as a low-grade inflammatory state. Several possible mechanisms are discussed whereby rapid growth of adipose tissue might trigger a local inflammatory response. It is suggested that this inflammatory response and associated release of cytokines may constitute the link between obesity and insulin resistance.

Key Words: Tumor necrosis factor; resistin; adiponectin; interleukins; free fatty acids; subcutaneous fat; visceral fat; insulin signaling.

1. INTRODUCTION

Obesity and type 2 diabetes are the most common metabolic diseases in Western society, together affecting as much as half of the adult population *(1)*. Not only is the prevalence of these conditions high, but it also continues to increase. Insulin resistance is a prediabetic condition, characterized by a failure of target organs to respond normally to insulin. Insulin resistance includes a central component (incomplete suppression of hepatic glucose output) and a peripheral component (impaired insulin-mediated glucose uptake in skeletal muscle and adipose tissue) *(2)*. When increased insulin secretion is no longer sufficient to prevent hyperglycemia, the subject progresses from insulin resistance to type 2 diabetes. Insulin resistance is associated with other conditions such as central obesity, hypertension, and dyslipidemia, all risk factors for cardiovascular disease. The constellation of these metabolic abnormalities has been termed metabolic syndrome.

Obesity is a well-recognized risk factor for the development of insulin resistance and metabolic syndrome. In addition to the total amount of fat, distribution of adipose tissue is also important, with most studies concluding that visceral fat contributes considerably more to insulin resistance than does subcutaneous fat *(3)*. However, one report, by Misra et al., documented a robust correlation between posterior abdominal subcutaneous fat

From: *Nutrition and Health: Adipose Tissue and Adipokines in Health and Disease*
Edited by: G. Fantuzzi and T. Mazzone © Humana Press Inc., Totowa, NJ

and insulin resistance *(4)*. Underscoring the importance of visceral fat is the report by Klein et al. that liposuction, resulting in a substantial reduction of subcutaneous fat, did not enhance insulin responsiveness in insulin-resistant subjects *(5)*. In rats, we *(6)* and others *(7)* have found that surgical removal of visceral fat reverses insulin resistance.

Traditionally, adipose tissue has been regarded largely as a depot for stored fat. More recently, it has become clear that adipose tissue plays an active role in energy metabolism and is the source of hormones, cytokines, and metabolites that play an important role in whole-body metabolism *(8)*. The role of these substances may be either autocrine or endocrine. Adipose tissue, especially visceral fat, is the source of a number of substances that might play a role in the development of insulin resistance. Among the latter are tumor necrosis factor (TNF)-α, adiponectin, interleukin (IL)-6, resistin, and free fatty acids. The difference in the metabolic effects of visceral versus subcutaneous fat may be attributed both to differences in the hormones secreted by the two types of fat and to the fact that hormones secreted by visceral fat reach the liver in high concentration. The latter is due to the fact that visceral fat drains into the portal circulation, whereas subcutaneous fat drains into the systemic circulation *(8)*. This review will focus on the regulation of insulin responsiveness by adipokines and on evidence supporting the hypothesis that these hormones play a role in the pathophysiology of insulin resistance.

2. OBESITY AS AN INFLAMMATORY CONDITION

Obesity produces a state of low-grade inflammation, characterized by elevated circulating concentrations of acute-phase proteins, such as C-reactive protein (CRP), plasminogen activator inhibitor (PAI)-1, and fibrinogen *(9,10)*. CRP is elevated moderately in obese men and highly in obese women *(11)*. Hak et al. reported that in a group of healthy middle-aged women, CRP was approximately threefold higher in the group whose body mass index (BMI) was in the upper half *(9)*. The mean CRP concentration in the upper-BMI group was 1.15 mg/L, around the threshold of a clinically elevated level *(11)*.

There are at least two theories as to why obesity should cause a state of inflammation. First, the growth—especially the rapid growth—of adipose tissue may cause local hypoxia due to inadequate perfusion and result in the generation of angiogenesis factors such as vascular endothelial growth factor (VEGF) and 11β-hydroxysteroid dehydrogenase type 1. Overexpression of the latter enzyme in rodents has been shown to cause adipose production of the inflammatory cytokines resistin and TNF-α, both cytokines being implicated in the development of insulin resistance *(12)*. A second possibility is that rapid expansion of adipose tissue results in the differentiation of preadipocytes into macrophage-like cells *(13)*. It is well-known that obesity is accompanied by an overexpression of TNF-α in adipose tissue. Weisberg et al. have recently shown that obesity is accompanied by an increased number of macrophages in adipose tissue, but not in liver or muscle *(14)*. The overexpression of TNF-α that occurs in adipose tissue of obese humans occurs exclusively in macrophages.

3. ROLE OF LEPTIN RESISTANCE IN OBESITY AND INSULIN RESISTANCE

Leptin is produced by adipocytes and secreted into the blood. In the healthy state, the circulating leptin concentration varies in proportion to adipose mass. Activation of leptin receptors in the hypothalamus decreases food intake and increases energy expenditure via uncoupling proteins (UCPs) in fat and muscle. UCPs are mitochondrial proteins that allow for oxidation of substrates with the production of heat, rather than storage of energy in the form of ATP. Thus leptin serves as part of an "adipostat" mechanism, whereby increased adiposity sets in motion responses that will eventually reduce adiposity. In 1994, Friedman et al. discovered that *ob/ob* mice are leptin-deficient and lose weight following leptin treatment *(15)*. However, obese humans are typically leptin-resistant and have higher-than-normal circulating concentrations of leptin. Leptin resistance in humans has two components: impaired transport of leptin across the blood–brain barrier and impaired signaling via hypothalamic leptin receptors *(16)*.

In addition to its function as a direct regulator of adiposity, leptin is also an insulin-sensitizing hormone *(17,18)*. Thus, the reduced responsiveness to leptin that accompanies obesity and may play a role in causing obesity also plays a role in causing insulin resistance in the brain. Insulin receptors in the hypothalamus play an important role in glucose homeostasis *(18)*. Primate studies have shown that direct delivery of insulin to the brain reduces feeding *(18)*; rodent studies have shown that intracerebroventricular administration of insulin not only reduces food intake, but also suppresses hepatic glucose output *(19,20)*. There is growing evidence that the link between leptin resistance and insulin resistance is suppressor of cytokine signaling (SOCS)3, a molecule that impairs signaling of both leptin and insulin *(21)*, and one that is suppressed by leptin *(22)*. Mice with reduced neuronal expression of SOCS have enhanced sensitivity to leptin and insulin and are protected against diet-induced obesity *(23)*.

4. ROLE OF TNF-α IN LINKING OBESITY TO INSULIN RESISTANCE

TNF-α was first shown by Hotamisligil et al. to be overexpressed in adipose tissue from several strains of obese rodents *(24)*. Weisberg et al. have shown that macrophages are the main source of TNF-α in adipose tissue *(14)*. TNF-α expression is higher in visceral fat (VF) of rodents than in subcutaneous (sc) fat *(25)*. In addition, TNF-α has been shown to impair insulin signaling in cultured cells by three separate molecular mechanisms. TNF-α activates serine/threonine kinases that phosphorylate and impair the function of key elements in the insulin signaling pathway *(26)*. First, TNF-α mediates a serine phosphorylation of IRS-1 *(27)*. This alteration impairs insulin signaling by making IRS-1 resistant to subsequent insulin-stimulated tyrosine phosphorylation. Second, TNF-α phosphorylates and activates a protein tyrosine phosphatase that normally terminates insulin action, thus playing a role in the self-limiting nature of insulin signaling *(28)*. Third, TNF-α phosphorylates and inactivates the protein phosphate PP-1 at site 2, resulting in its inactivation *(29)*. This action of TNF-α opposes the action of insulin, whereby glucose storage is promoted by phosphorylating PP-1 at site 1 and activating it.

The above findings led to the popular theory that TNF-α of adipose origin is secreted into the circulation, from where it reaches targets such as muscle and liver and causes insulin resistance. However, circulating levels of TNF-α are very low compared with the concentrations required to induce insulin resistance when infused into rats *(30)* and tissue levels of TNF-α are several orders of magnitude higher than circulating levels *(31)*. Whereas some studies have shown that circulating TNF-α is elevated in obese and insulin-resistant subjects *(32)*, others have not *(33)*.

We hypothesize that locally produced TNF-α may contribute to insulin resistance in one of two ways. First, obesity may cause insulin resistance in by increasing TNF-α expression in targets such as muscle. Support for this concept is our report that diet-induced obesity in rats is accompanied by reduced insulin-stimulated glucose transport in skeletal muscle, together with an increase in muscle expression of TNF-α *(34)*. Alternatively, obesity may increase TNF-α expression in adipose tissue, leading to the release into the circulation of other cytokines that are capable of causing systemic insulin resistance.

5. ROLE OF ADIPONECTIN IN INSULIN RESISTANCE

Adiponectin is an insulin-sensitizing hormone produced exclusively by adipocytes *(35)*. There is a strong and positive correlation between serum adiponectin and insulin responsiveness. In a group of normal and obese subjects, Weyer et al. studied the relationship between serum adiponectin and the glucose disposal rate, measured during hyperinsulinemic euglycemic glucose clamp studies *(36)*. Serum adiponectin varied over a fivefold range, correlated positively with glucose disposal, and accounted for approx 35% of the variance in insulin responsiveness.

Adiponectin increases oxidation of free fatty acids and reduces postprandial elevation of nonesterified fatty acids (NEFAs) in mice *(37)*. Adiponectin circulates at concentrations in the low μg/mL range, whereas most cytokines produced by adipose tissue are in the pg/mL range *(35)*. In rodents, adiponectin expression is considerably higher in VF than in sc fat *(38)*, and adiponectin secretion by VF has an inverse relationship with VF mass—i.e., the larger the mass of VF, the less adiponectin is secreted. Serum adiponectin is low in obese humans *(39)* and increases following weight loss *(40)*. Yamauchi et al. have shown that administration of adiponectin to obese rats on a high-fat diet reduces weight gain and prevents the development of hepatic insulin resistance *(41)*. The mechanism by which adiponectin enhances insulin responses has not been well-studied, but may involve opposing of TNF-α-induced activation of the transcription factor NFκB *(42)*.

6. ROLE OF RESISTIN IN INSULIN RESISTANCE

Resistin is a recently discovered polypeptide that is secreted by mouse adipocytes and has been implicated in the development of insulin resistance. Resistin was first described in 2001, when a search for genes that are induced during adipocyte differentiation but downregulated in mature adipocytes during exposure to thiazolidinediones led to the discovery of a protein the investigators named resistin, for "resistance to insulin" *(43)*. Administration of resistin in normal mice impairs glucose tolerance and insulin action. Furthermore, immunoneutralization of resistin improved blood glucose and insulin action in animal models of obesity-induced insulin resistance. In rodents, administration of thiazolidinedione drugs reverses insulin resistance. These drugs also reduce gene and protein expression of resistin in some studies *(44)* but not in others *(45)*. These initial data suggested that resistin, at least in part, may explain how adiposity leads to insulin resistance and may also explain the antidiabetic effects of thiazolidinedione drugs. The molecular mechanism for the action of resistin is unknown. A recent study in mice suggested that resistin selectively impairs the inhibitory action of insulin on hepatic glucose production *(44)*. However, the role of resistin in obesity-associated insulin resistance has become controversial because the biology of resistin is different in humans than in rodents and additional evidence has suggested that obesity and insulin resistance are associated with decreased resistin expression *(45–47)*.

Whether resistin is expressed in human adipose tissue is not clear. McTernan et al. found such expression and reported that resistin is relatively highly expressed in the omental visceral fat and the abdominal subcutaneous fat, with lower expression in subcutaneous fat on the thigh *(48)*. This finding is in contrast to the findings of Savage et al. and Nagaev et al., who did not detect resistin in human adipose tissue *(49,50)*. Whereas resistin is expressed mainly in adipocytes in mice *(43)*, Fain et al. reported that most of the resistin secreted by human fat explants is derived from nonadipocytes *(51)*. The reason for the differences in these studies is unclear. Human resistin is only 59% similar to the mouse protein, and this may portend important differences in the endocrine functions of adipocytes and resistin between rodents and humans *(52)*. Furthermore, insulin and TNF-α, both elevated in obesity, have been found to inhibit resistin expression, which may explain the low levels of resistin found in the recent studies of obesity diabetes.

The initial suggestion that resistin may be the link between obesity and insulin resistance is being challenged. The role of resistin in normal and abnormal physiology remains elusive. Studies from knockout mice and better characterization of resistin changes in humans should help determine whether this adipokine is a cause of insulin resistance or simply a bystander. Also, it will important to understand the similarities and differences between mouse and human resistin and mechanisms of obesity-related insulin resistance.

7. ROLE OF IL-6 IN INSULIN RESISTANCE

IL-6 is a pleiotropic circulating cytokine that has important roles in inflammation, host defense, and response to tissue injury *(53)*. It is one of several proinflammatory cytokines with a proposed role in the development of insulin resistance. IL-6 is secreted by many cell types, including immune cells, fibroblasts, endothelial cells, skeletal muscle, and nonadipocyte cells in adipose tissue, and circulates as a variably glycosylated 22- to 27-kDa protein *(2)*.

IL-6 is released from contracting skeletal muscle, causing the serum concentration to increase as much as 100-fold *(54)*. IL-6 increases hepatic glucose production when administered to human subjects, and there is evidence to suggest that the release of IL-6 from exercising muscle mediates the early phase on exercise-induced hepatic glucose output. The fact that IL-6 opposes insulin action in the liver has led to speculation that its oversecretion may play a role in insulin resistance. In the liver, IL-6 causes release of NEFAs and is the primary stimulator of for production of acute phase proteins *(55)*. Administration of IL-6 in healthy volunteers induced dose-dependent increases in blood glucose *(56)*, probably by inducing resistance to insulin action. In vitro, IL-6 has been shown to impair insulin signaling by several distinct molecular mechanisms *(57)*. Weight loss significantly decreases IL-6 levels in both adipose tissue and serum *(58)*. Genetic studies have also demonstrated a high level of correlation between insulin resistance and IL-6 gene polymorphism *(59)*. Besides its glucoregulatory effect, IL-6 increases circulating free fatty acids (FFA) from adipose tissue with their well-described adverse effects on insulin sensitivity *(60)*. Because visceral depots drain into the portal circulation, the metabolic effects of IL-6 on the liver become important. Indeed, there is evidence to suggest that IL-6 inhibits insulin receptor signal transduction in hepatocytes that is mediated, at least in part, by induction of SOCS3 *(61)*. IL-6 may also exert its adverse effects, at least in part, by decreasing adipose secretion adiponectin *(62)*.

Although much evidence implicates IL-6 in insulin resistance, there is some conflicting evidence. In a recent study, acute IL-6 administration did not impair glucose homeostasis in healthy individuals *(63)*. Moreover, IL-6-deficient mice were not protected from development of obesity and glucose intolerance *(64)*. Circulating IL-6 is elevated approximately twofold in obese, insulin-resistant subjects, but although the association is statistically significant, the relationship is not a strong one, accounting only for approx 7% of the variance in insulin responsiveness *(33,65,66)*. Although VF produces two- to threefold more IL-6 than does sc fat, adipose tissue is the source for only about 30% of circulating IL-6 in humans *(38)*, with the majority of adipose tissue-derived IL-6 coming from stromal immune cells and not adipocytes *(67)*.

In summary, a body of evidence indicates that IL-6 of adipose origin may play a role in systemic insulin resistance, although there is also some evidence to the contrary.

8. OTHER POSSIBLE MEDIATORS OF INSULIN RESISTANCE

Visfatin is a recently discovered adipokine that is produced by adipocytes *(68)*, and is expressed in visceral fat at much higher levels than in subcutaneous fat *(69)*. Serum visfatin increases with VF, but not sc fat, in humans and mice. Visfatin has been shown to have insulin-sensitizing properties when administered to insulin-resistant mice. Mice that are heterozygous for a targeted mutation in the visfatin gene display a small impairment in glucose tolerance, whereas homozygous mice die in utero. Visfatin binds to the insulin receptor and activates downstream signaling, but does not compete for binding with insulin. Visfatin has the ability to stimulate glucose transport in cultured muscle and adipose cells and to inhibit glucose output in cultured hepatocytes.

Apart from cytokines, the most important candidate for linking obesity to insulin resistance is circulating nonesterified (or free) fatty acids. Reaven et al. reported that type 2 diabetes is associated with elevations in both fasting and postprandial NEFAs *(70)*. Boden et al. have shown that both acute infusion and chronic elevations of NEFAs can decrease insulin-stimulated glucose disposal in humans *(71)*. It is proposed that increased hydrolysis of NEFAs leads to increased diacyl glycerol, which in turn activates isoforms of protein kinase C (PKC). Although activation of PKC-ζ is a part of normal insulin signaling, activation of PKC-θ has been shown to impair insulin signaling *(72)*. Muscle biopsies from insulin-resistant subjects display serine/threonine phosphorylation of the insulin receptor, an impairment that TNF-α does not cause in cultured cells *(73)*. Current evidence suggests that serine/threonine phosphorylation of IRs is mediated by the theta isoform of protein kinase C, an enzyme that may be activated by an increase in serum FFA *(74)*. Additionally, Bjorntorp has hypothesized that increased visceral fat causes hepatic insulin resistance by a "portal" mechanism, where a higher concentration of NEFAs reach the liver from omental fat *(75)*.

9. CONCLUSIONS

Table 1 lists candidate adipokines that may potentially link obesity with insulin resistance. Such candidates should meet several criteria. The adipokine should have a

Table 1
Circulating Proteins Secreted by Adipose Tissue and That May Link Obesity to Insulin Resistance and Diabetes

Adipokine	*Effect on insulin response*	*Change in blood concentration*	*Tissue expression*	*Expression in VF vs SCF*
Resistin	Impairment	Elevated in rodent models of genetic and diet-induced obesity	Mouse adipocytes *(43)*, human non-adipocyte fat cells *(51)*	Markedly higher in VF of rodents *(25)*. Equal in VF and SCF of humans *(48)*
TNF-α	Impairment	May be elevated in human obesity and insulin resistance	Adipose tissue macrophages *(14)*, liver and muscle *(26)*	Higher in VF of rodents *(25)*
IL-6	Impairment	Dramatic elevation in obese humans *(54)*	Non-adipocyte fat cells, immune cells, skeletal muscle *(2)*	Higher in VF of humans *(38)*
Adiponectin	Enhancement	Serum adiponectin is low in obese humans *(37)* and increases following weight loss *(39)*	Adipocytes *(37)*	Higher in VF of rodents *(38)*
Visfatin	Enhancement	Increased in human visceral adiposity *(64)*	Adipocytes *(68)*	Higher in VF of humans *(66)*
Leptin	Enhancement	High in obese humans, because of leptin resistance	Adipocytes *(2)*	Higher in SCF of humans *(8)*

[a]Candidate proteins should meet the following criteria: (1) the hormone should have an effect on insulin responses, (2) blood levels should be appropriately elevated of reduced in obesity and insulin resistance, (3) a significant fraction of circulating hormone should derive from adipose tissue and (4) production of the hormone should be higher in visceral fat (VF) than in subcutaneous fat (SCF).

major effect on insulin responsiveness; the circulating level should correlate to insulin responsiveness and increase or decrease appropriately in animal models and experimental conditions. A significant fraction of the circulating cytokine should be of adipose origin and expression should be higher in visceral fat than in subcutaneous fat.

Based on these criteria, there is strong evidence in favor of adiponectin as a link between obesity and insulin resistance. Adiponectin is an insulin-sensitizing hormone that is produced almost exclusively in fat, with higher expression in visceral fat. Serum adiponectin is strongly and inversely correlated to insulin resistance and fat mass and increases after weight reduction.

There is evidence both for and against the role of IL-6 in linking obesity to insulin resistance. IL-6 has the ability to impair insulin responses and is elevated in obesity, but circulating levels correlate only weakly to insulin responsiveness.

The case for resistin is strong in rodents, but weaker in humans. Resistin is preferentially expressed in visceral fat of mice and is elevated in rodent models of genetic and diet-induced obesity. However, resistin biology may be different in humans, and some studies suggest that it is not expressed in human adipose tissue. Leptin is an insulin-sensitizing hormone that is elevated in serum of obese subjects owing to leptin resistance. This phenomenon is similar to the elevation of insulin in insulin resistance. Leptin plays a prominent role in the development of obesity. Leptin resistance may also underlie insulin resistance in the brain.

TNF-α is overexpressed in tissues of obese and insulin-resistant animals. TNF-α circulates at low levels; some studies have found serum levels to be elevated in insulin-resistant subjects, while others have not. TNF-α impairs insulin responses in muscle fat and liver by well-established molecular mechanisms. Most evidence suggests that TNF-α plays a paracrine or autocrine role in linking obesity to insulin resistance.

Visfatin is a recently discovered insulin-sensitizing hormone that increases with visceral fat mass. Although less is known about visfatin than other adipokines, the latter finding argues against its role as a link between obesity and insulin resistance.

REFERENCES

1. Must A, Spadano J, Coakley EH, et al. JAMA 1999;282:1523–1529.
2. Pittas AG, Joseph NA, Greenberg AS. J Clin Endocrinol Metab 2004;89:447–452.
3. Pieris AN, Struve MF, Mueller RA, et al. J Clin Endocrinol Metab 1988;67:760–767.
4. Misra A, Garg A, Abate N, et al. Obes Res 1997;5:93–99.
5. Klein S, Fontana L, Young VL, et al. N Engl J Med 2004;350:2549–2557.
6. Borst SE, Conover CF, Bagby GJ. Cytokine 2005;32:39–44.
7. Gabriely I, Barzilai N. Curr Diab Rep 2003;3:201–206.
8. Kershaw EE, Flier JS. J Clin Endocrinol Metab 2005;89:2548–2556.
9. Hak EA, Stenhouer CD, Bo KH, et al. Aterioscler Thromb Vasc Biol 1999;19:1986–1991.
10. Tataranni PA, Ortega E. Diabetes 2005;54:917–927.
11. Visser M, Bouter LM, McQuillan GM, et al. JAMA 1999;282:2131–2135.
12. Mazusaki H, Peterson J, Shinyama H, et al. Science 2001;294:2166–2170.
13. Charriere G, Cousin B, Arnaud E, et al. J Biol Chem 2003;278:9850–9855.
14. Weisberg SP, McCann D, Desai M, et al. J Clin Invest 2003;112:1796–1808.
15. Zhang Y, Proenca R, Maffei M, et al. Nature 1994;372:425–432. Erratum in Nature 1995;374:479.
16. Scarpace PJ, Tumer N. Physiol Behav 2001;74:721–727.
17. Lazar MA. Science 2005;307:373–375.

18. Schwartz MW, Porte D Jr. Science 2005;307:375–379.
19. Zimmet P, Thomas CR. J Intern Med 2003;254:114–125.
20. O'Rahilly S, Farooqi IS, Yeo GS, et al. Endocrinology 2003;144:3757–3764.
21. Frisch RE, McArthur JW. Science 1974;185:949–951.
22. Welt CK, Chan JL, Bullpen J, et al. N Engl J Med 2004;351:987–997.
23. Hales CN, Barker DJ. Diabetologia 1992;35:595–601.
24. Hotamisligil GS, Shargill NS, Spiegelman BM. Science 1993;259:87–91.
25. Das M, Gabriely I, Barzilai N. Obes Rev 2004;5:13–19.
26. Borst SE. Endocrine 2004;23:1771–1782.
27. Hotamisligil GS, Peraldi P, Budavari A, et al. Science 1996;271:665–668.
28. Ahmad F, Goldstein BJ. J Cell Biochem 1997;64:117–127.
29. Ragolia L, Begum N Mol Cell Biochem 1998;182:49–58.
30. Lang CH, Dobrescu C, Bagby GJ. Endocrinology 1992;130:43–52.
31. Borst SE, Bagby GJ. Cytokine 2004;26:217–222.
32. Tsigos C, Papanicolaou DA, Kyrou I, et al. J Clin Endocrinol Metab 1997;82:4167–4170.
33. Kern PA, Ranganathan S, Li C, et al. Am J Physiol Endocrinol Metab 2001;280:E745–E751.
34. Borst SE Conover CF. Life Sci 2005;77:2156–2165.
35. Gil-Campos M, Canete RR, Gil A. Clin Nutr 2004;23:963–974.
36. Weyer C, Funahashi T, Tanaka S, et al. J Clin Endocrinol Metab 2005;86:1930–1935.
37. Fruebis J, Tsao TS, Javorschi S, et al. Proc Natl Acad Sci USA 2001;13:2005–2010.
38. Aldhahi W, Hamdy O. Curr Diab Rep 2003;3:293–298.
39. Argiles JM, Lopez-Soriano J, Almendro V, et al. Med Res Rev 2005;25:49–65.
40. Yang WS, Lee WJ, Funahashi T, et al. J Clin Endocrinol Metab 2001;86:3815–3819.
41. Yamauchi T, Kamon J, Waki H, et al. Nat Med 2001;7:941–946.
42. Ouchi N, Kihara S, Arita Y, et al. Circulation 2000;102:1296–1301.
43. Steppan CM, Bailey ST, Bhat S, et al. Nature 2001;409:307–312.
44. Rajala MW, Obici S, Scherer PE, et al. J Clin Invest 2003;111:225–230.
45. Way JM, Gorgun CZ, Tong Q, et al. J Biol Chem 2001;276:25,651–25,653.
46. Milan G, Granzotto M, Scarda A, et al. Obes Res 2002;10:1095–1103.
47. Juan CC, Au LC, Fang VS, et al. Biochem Biophys Res Commun 2001;289:1328–1333.
48. McTernan PG, McTernan CL, Chetty R, et al. J Clin Endocrinol Metab 2002;87:2407–2410.
49. Savage DB, Sewter CP, Klenk ES, et al. Diabetes 2001;50:2199–2202.
50. Nagaev I, Smith U. Biochem Biophys Res Commun 2001;285:561–564.
51. Fain JN, Cheema PS, Bahouth SW, et al. Biochem Biophys Res Commun 2003;300:674–678.
52. Steppan CM, Lazar MA. Trends Endocrinol Metab 2002;13:18–23.
53. Papanicolaou DA, Wilder RL, Manolagas SC, et al. Ann Intern Med 1998;128:127–137.
54. Febraio MA, Pedersen BK. FASEB J 2002;16:1335–1347.
55. Gabay C, Kushner I. N Engl J Med 1999;340:448–454.
56. Fernandez-Real JM, Ricart W. Endocr Rev 2003;24:278–301.
57. Senn JJ, Klover PJ, Nowalk IA, Mooney RA. Diabetes 2002;51:3391–3399.
58. Bastard JP, Jardel C, Bruckert E, et al. J Clin Endocrinol Metab 2000;85:3338–3342.
59. Vozarova B, Fernandez-Real JM, Knowler WC, et al. Hum Genet 2003;112:409–413.
60. Boden G, Shulman GI. Eur J Clin Invest 2002;32:14–23.
61. Senn JJ, Klover PJ, Nowak IA, et al. J Biol Chem 2003;278:13,740–13,746.
62. Fasshauer M, Kralisch S, Klier M, et al. Biochem Biophys Res Commun 2003;301:1045–1050.
63. Steensberg A, Fischer CP, Sacchetti M, et al. J Physiol 2003;548:631–638.
64. Wallenius V, Wallenius K, Ahren B, et al. Nat Med 2002;8:75–79.
65. Vozarova B, Weyer C, Hanson K, et al. Obes Res 2001;9:414–417.
66. Pradhan AD, Manson JE, Rifai N, et al. JAMA 2001;286:327–334.
67. Fried SK, Bunkin DA, Greenberg AS. J Clin Endocrinol Metab 1998;83:847–850.
68. Hammarstedt A, Pihlajamaki J, Sopasakis VR, et al. J Clin Endocrinol Metab 2006;91:1181–1184.
69. Hug C, Lodish HF. Science 2005;307:366–367.
70. Reaven GM, Hollenbeck C, Jeng CY, et al. Diabetes 1988;37:1020–1024.
71. Boden G. Diabetes Care 1996;19:394–395.

72. Itani SI, Pories WJ, Macdonald KG, et al. Metabolism 2001;50:553–557.
73. Kanety H, Feinstein R, Papa MZ, et al. J Biol Chem 1995;270:23,780–23,784.
74. Dohm GL, Tapscott EB, Pories WJ, et al. J Clin Invest 1988;8:486–494.
75. Bjorntorp P. Arteriosclerosis 1990;10:493–496.

22 Adipokines in Non-Alcoholic Fatty Liver Disease

Ancha Baranova and Zobair M. Younossi

Abstract

Non-alcoholic fatty liver disease (NAFLD) represents a spectrum of clinicopathological conditions in patients who do not consume excessive amounts of alcohol; these conditions are characterized by hepatic steatosis with or without other pathological changes observed in liver biopsy. The pathogenesis of NAFLD and its progressive form (non-alcoholic steatohepatitis [NASH]) appears to be multifactorial and is the subject of intense investigation. Increasing evidence indicates that the pathogenesis of NAFLD and NASH is hastened by a disturbance in adipokine production. Decreased serum adiponectin and increased tumor necrosis factor-α, which are characteristic of obesity, appear to contribute to the development and progression of NASH. The role of leptin in the pathogenesis of NASH remains controversial and the involvement of serum resistin is primarily documented only in animal models, which may or may not be applicable to the human form of NAFLD. Finally, other adipokines such as vaspin, visfatin, and apelin may play important roles in the pathogenesis of NASH and require further investigation.

Key Words: Non-alcoholic fatty liver disease (NAFLD); non-alcoholic steatohepatitis (NASH); oxidative stress; insulin resistance; adiponectin; leptin; resistin; TNF-α; vaspin; visfatin; apelin.

1. INTRODUCTION

Non-alcoholic fatty liver disease (NAFLD) represents a spectrum of clinicopathological conditions characterized by significant lipid deposition in the liver parenchyma of patients who do not consume excessive amounts of alcohol *(1,2)*. At one end of the NAFLD spectrum is steatosis alone ("simple steatosis"), and at the other end are non-alcoholic steatohepatitis (NASH), NASH-related cirrhosis, and hepatocellular carcinoma. The distinction between steatosis alone and NASH can be made only by liver biopsy. NASH is characterized by hepatic steatosis and by evidence for hepatocyte ballooning degeneration, lobular inflammation, and occasionally, Mallory hyaline or sinusoidal fibrosis *(3)*. NASH and steatosis alone have differential risk for progression *(3)*.

Estimates of the prevalence of NAFLD are high and are expected to increase with the global epidemic of obesity. Recent studies suggest that up to 10 to 24% of the general population and 50 to 90% of obese individuals are affected by NAFLD *(2)*. The prevalence of histologically confirmed NASH is estimated as 1.2 to 4%; in morbidly obese patients it is much more common, with estimates ranging from 20 to 47% *(4)*. NAFLD patients have higher-than-average mortality rates (standardized mortality ratio = 1.34) *(5)*.

From: *Nutrition and Health: Adipose Tissue and Adipokines in Health and Disease*
Edited by: G. Fantuzzi and T. Mazzone © Humana Press Inc., Totowa, NJ

Patients with steatosis alone rarely progress to cirrhosis, whereas 10 to 25% of those with biopsy-proven NASH can progress to cirrhosis *(1,3,4)*. In fact, most patients with cryptogenic cirrhosis seem to have "burned-out NASH" that might also cause hepatocellular carcinoma (HCC) *(6)*. The major risk factors for progression in NASH are the presence of type 2 diabetes, obesity, metabolic syndrome, and elevated aminotransferase, and histological features of ballooning degeneration of hepatocytes and Mallory's hyalines *(4,7,8)*. Ultrasound and other noninvasive modalities can only detect steatosis, and are unable to distinguish NASH from steatosis alone or detect hepatic fibrosis *(9)*.

Of the many treatment strategies currently in use, none is proven to be effective for NASH *(10)*. Treatment strategies include modifying the clinical conditions associated with NASH, such as type 2 diabetes mellitus, hyperlipidemia, and obesity *(2)*. Pharmacological interventions for NASH include the use of ursodeoxycholic acid (UDCA), clofibrate, betaine, N-acetylcysteine, gemfibrozil, atorvastatin, thiazolidinedione, pentoxyfillin, and vitamin E *(2,11,12)*. None of these treatments is capable of preventing NASH progression.

2. PATHOGENESIS OF NASH

The pathogenesis of NASH appears to be multifactorial and is the subject of intense investigation. Suggested theories include the influences of abnormal lipid metabolism and the production of reactive oxygen species (ROS), increased hepatic lipid peroxidation, stellate cell activation, and abnormal patterns of cytokine production, promoting liver injury and fibrosis *(13)*.

The "two-hit hypothesis" of NASH pathogenesis suggests that the first "hit" is the accumulation of excessive fat in the hepatic parenchyma *(4,14)*. This first step has been linked to insulin resistance (IR), which is consistently observed in patients with NAFLD *(13)*. Clinical features of metabolic syndrome (obesity, diabetes mellitus, or hypertrigyceridemia) are commonly observed in patients with NAFLD *(1,17)*. Furthermore, unexplained elevations in alanine aminotransferase (ALT) levels in individuals with metabolic syndrome suggest that NAFLD is the hepatic manifestation of this syndrome *(18)*. Additionally, patients with NAFLD with more "severe" forms of IR are at even greater risk of progressive liver disease *(4,19)*. Animal models of NAFLD also have IR, and the use of the insulin-sensitizing agent, metformin, reverses hepatic steatosis *(16)*.

The second "hit" leading to the development of the progressive form of NAFLD involves oxidative stress. In steatotic livers, an imbalance between pro-oxidant and anti-oxidant processes results from the induction of microsomal CYP2E1, peroxisomal β-oxidation of fatty acids (FA), release of cytokines from activated inflammatory cells, changes in adipokine levels, or other unknown factors *(4,13,20)*. NAFLD-related oxidative stress may be linked to mitochondrial dysfunction, as mitochondria are the major source of ROS in living cells *(19)*. ROS, in turn, increases the peroxidation of membrane lipids that induce the production of proinflammatory cytokines and activate stellate cells, leading to hepatic fibrogenesis *(4,13)*. Both the first and second hits may involve changes in circulating levels of various pro- and anti-inflammatory cytokines and adipokines.

3. ADIPOSE TISSUE, ADIPOKINES, AND NAFLD

White adipose tissue produces and releases a variety of proinflammatory and anti-inflammatory factors, including adipokines (leptin, adiponectin, resistin, apelin, vaspin, visfatin, and zinc-α2-glycoprotein), cytokines (such as tumor necrosis factor [TNF]-α and interleukin [IL]-6), and chemokines *(21)*. In addition to adipocytes, white adipose tissue contains several other cell types, including macrophages and monocytes. It is likely that macrophages are retained within adipose tissue in response to both monocyte chemoattrative protein (MCP)-1 and macrophage migration inhibitory factors released by adipocytes in amounts proportional to body mass index (BMI) *(22)*. Cytokines produced by adipose tissue contribute to the increased systemic inflammation associated with obesity *(23)*. The exact contribution of each component of white adipose tissue in the "proinflammatory" state of obesity is not entirely clear. Some studies indicate that more than 90% of the adipokines released from adipose tissue (except for adiponectin and leptin) originate from the nonfat cells embedded in the extracellular matrix *(24)*. In addition, some adipokines (e.g., resistin and adiponectin) are also produced elsewhere in the body *(25)*. Together, these findings suggest that serum adipokine concentrations represent secretions by various cells, including adipocytes. It is increasingly clear that adipokines play an important role in the pathogenesis of NAFLD.

4. ADIPOKINES IN THE EXPERIMENTAL MODELS OF NAFLD

Common experimental models of NAFLD include mice or rats fed high-fat or high-carbohydrate diets, or mice that exhibit a genetic deficiency in leptin, a satiety factor *(15)*. These animal models spontaneously develop steatosis, and some progress to steatohepatitis.

Animal models of NAFLD point to adipokine and cytokine abnormalities in the pathogenesis of NAFLD. For example, leptin-deficient *ob/ob* mice are important animal models of NAFLD because they are obese, insulin-resistant, hyperglycemic, and hyperlipidemic. Similarly, leptin receptor-deficient *fa/fa* rats and *db/db* mice are phenotypically similar to *ob/ob* mice, with the addition of hyperleptinemia. It is noteworthy that NAFLD occurs in both leptin-deficient and hyperleptinemic animals with impaired leptin signaling. However, leptin restoration leads to NAFLD reversal in leptin-deficient animals *(15)*.

TNF-α is another important cytokine involved in the pathogenesis of NAFLD; serum levels are high in all animal models of NAFLD. Nonetheless, its origin (i.e., adipocytes themselves or monocytes and macrophages) is not entirely clear. In NAFLD, the lipotoxic effects of excess fat may enhance TNF-α release. Once initiated, this vicious cycle of NFκB/TNF-α becomes self-perpetuating *(15)*. It seems that chronic exposure to TNF-α promotes the accumulation of inflammatory cells in the liver, thereby exposing hepotocytes to damaging factors released by activated monocytes *(15)*. Indeed, anti-TNF-α treatment in *ob/ob* mice can improve liver histology and reduce total hepatic FA content *(16,26)*. On the contrary, two recent investigations have suggested a "hepatoprotective" role for TNF-α. Leptin-deficient mice with elevated basal TNF-α expression are protected against acute liver damage, as their ability to induce IL-18 is diminished, and T-cell-mediated hepatotoxicity is reduced *(27)*. Because most other studies indicate that TNF-α enhances liver injury in NAFLD, this work has generated some controversy.

Resistin is another important adipokine, but our understanding of its role in NAFLD is complicated by substantial differences between resistin-encoding genes in humans and animal models. The spectrum of resistin-like molecules in humans and mice is different because the resistin-α encoding gene is absent in humans. Resistin serum content is also substantially lower in humans (1/250) than in the rodent model. This is a major consideration when interpreting the applicability of animal studies to humans *(28)*. At present, only one study has described the relationship between resistin and NAFLD *(29)*. This study focused on RELM-β, a resistin-like molecule expressed by intestinal goblet cells. Strictly speaking, RELM-β cannot be considered an adipokine, but its effects might be similar to white-adipose-specific RELM-α. Among other changes, fatty liver results from the overexpression of the RELM-β encoding gene in the liver of transgenic mice maintained on a high-fat diet *(29)*. This intriguing finding requires further investigation.

Finally, adiponectin's potential role in the pathogenesis of NAFLD has generated much interest. Adiponectin reduces IR by decreasing triglyceride (TG) content in the muscle and liver tissue of obese mice. It also increases the ability of subphysiological levels of insulin to suppress glucose production by inhibiting hepatic gluconeogenic enzymes *(30)*. Moreover, in lipoatrophic mice, leptin and adiponectin act synergistically *(31)*. Obese mice produce diminished amounts of adiponectin. When replenished, adiponectin dramatically alleviates steatosis in these animals, and attenuates inflammation by suppressing the hepatic production of TNF-α *(32)*. Adiponectin also attenuates CCl_4-induced hepatic fibrosis *(33)*. Together this work points to the multilevel involvement of this adipokine in the pathogenesis of NAFLD and its progression.

5. ADIPOKINES IN PATIENTS WITH NAFLD

The following sections review current clinical work on the potential role of specific adipokines and cytokines in the development of NAFLD. These include adiponectin, resistin, leptin, TNF-α, and other cytokines.

5.1. Adiponectin

Adiponectin is the most frequently studied adipokine in patients with NASH. Over the past few years, several authors have suggested that hypoadiponectinemia may contribute to the development of NASH in obese individuals *(32,34–37)*. Plasma adiponectin levels are significantly lower in patients with NAFLD than in their matched controls. However, there are no differences in adiponectin levels in patients with simple steatosis versus those with NASH *(35)*. Another study of 68 obese patients shows an independent association of hypoadiponectinaemia with steatosis and markers of liver injury *(34)*. Similar studies confirmed the protective effect of adiponectin against the development of radiologically proven steatosis in adult *(36)* and pediatric *(37)* populations.

The role of adiponectin in distinguishing NASH from simple steatosis remains controversial. One study has shown that a reduction in circulating adiponectin levels in NAFLD is related to hepatic insulin sensitivity and the amount of the hepatic fat, but not to the severity of necroinflammation and fibrosis *(38)*. On the contrary, other studies report that hypoadiponectinemia is associated with increased grades of hepatic necroinflammation, independent of IR *(39)*. Musso and coauthors have also suggested that adiponectin could be protective against NASH *(40)*, as its levels correlate negatively with the presence of necroinflammation and fibrosis *(39,40)*. The same study demonstrates that the changes in adiponectin levels probably precede overt manifestation of diabetes *(40,41)*.

Circulating levels of adiponectin reflect a strong genetic component with an additive genetic heritability of 46% *(42)* that is linked to regions on chromosomes 5p, 14q, and

9p. Individuals homozygous for the +276T allele of the adiponectin-encoding APM1 locus have higher adiponectin levels than other subjects *(43)*. Individuals with an allelic combination of +45T and +276G ("TG" haplotype) are more likely to have various components of metabolic syndrome *(44,45)*. These findings are suggestive, but a correlation between these genetic findings and NAFLD has not yet been reported.

It is also important to mention two recent publications focusing on the role of the adipokine receptors in NAFLD and NASH that report contradictory results. Kaser and colleagues report a significant reduction in the immunostaining of the adiponectin receptor AdipoRII as well as its mRNA expression levels in liver biopsies of patients with NASH as compared with patients with simple steatosis *(46)*. On the other hand, Vuppalanchi and colleagues report an increase in the mRNA expression levels of the same receptor in NASH livers *(47)*. These investigators report several other contradictory findings regarding endogenous adiponectin production in the hepatic sinusoids *(46,47)*. Further clarification of the adiponectin receptors status in NASH is warranted because some common haplotypes of Adip-R1 alleles are associated with hepatic steatosis *(48)*.

5.2. Resistin

Resistin has been the focus of much attention because it is implicated in the pathogenesis of obesity-mediated IR and type 2 diabetes mellitus. In addition, resistin appears to be a proinflammatory cytokine stimulating NFκB-dependent macrophage secretion of TNF-α and IL-12 to the same extent as lipopolysaccharide *(49)*. One recent study shows that plasma resistin concentrations are positively correlated with hepatic fat content *(50)*. Others show that plasma resistin concentrations are similar in NASH patients compared with BMI-matched, non-NASH controls *(37,40)*. Despite these suggestive findings, the role of resistin in the pathogenesis of NAFLD remains speculative and requires further clarification.

5.3. Leptin

It is unclear whether serum leptin elevation is associated with the development of steatosis or NASH. Higher-than-normal leptin concentrations are found in various types of NAFLD and NASH, but not in chronic viral hepatitis without cirrhosis *(36,51)*. Most advanced stages of NASH usually correspond to higher leptin levels *(36,51)*. One recent study shows a correlation between serum leptin and serum ALT *(52)*. Because NAFLD is the most common cause of elevated ALT, this study provides indirect evidence connecting leptin with NAFLD.

One mechanism by which leptin may contribute to the development of NASH is to influence IR as well as FA influx into hepatocytes *(53)*. In the later stages of NASH, leptin may also augment systemic, low-grade inflammation, thus providing the "second hit" responsible for advancing simple steatosis to steatohepatitis *(53)*. Additionally, leptin acts as a profibrogenic adipokine, acting both on endothelial cells and Kupffer cells *(54,55)*. However, despite early enthusiasm, the role of leptin in NAFLD remains controversial. Leptin levels have been associated with NAFLD, but this association becomes insignificant after controlling for important confounders *(56)*. Furthermore, a longitudinal study showed no differences in leptin levels between patients with NAFLD who had fibrosis progression and those who did not *(56)*.

In this context, it is important to remember that NAFLD is commonly seen in conjunction with lipodystrophy, a condition characterized by the partial or complete absence of adipose tissue and hypoleptinemia. In such patients, leptin administration improves IR and corrects hepatic steatosis and hepatocellular ballooning injury, whereas the degree of liver fibrosis remains unchanged *(57)*. Clearly, the exact nature of leptin involvement in the development of NASH and its progression requires further investigation.

5.4. TNF-α

TNF-α is of interest because its levels increase with obesity and NAFLD, and pharmacological interventions decreasing TNF-α appear to be therapeutic in patients with NASH. TNF-α is a proinflammatory cytokine capable of orchestrating the synthesis, secretion, and activity of other proinflammatory molecules. In humans, the majority of TNF-α is produced by macrophages. Other tissues also produce TNF-α in response to infection, ischemia, and trauma *(58)*. TNF-α mRNA is found in very low quantities compared with other proteins in human white adipocytes, but an overall increase in the adipose mass usually leads to substantial, cumulative production of this cytokine *(59)*, potentially contributing to the development of obesity-related NAFLD. Several studies have demonstrated that serum TNF-α levels are significantly higher in patients with NASH than in healthy controls *(39,60)*.

The most comprehensive study of TNF-α in patients with NASH shows remarkable increases in the expression of mRNA encoding TNF-α in both hepatic and adipose tissues *(78)*. Similar mRNA increases have been observed for the p55 receptor, but not for the p75 receptor of TNF-α *(61)*. Additional indirect evidence of TNF-α involvement comes from a 12-mo trial of pentoxifylline (1600 mg/d) in patients with NASH *(62)*. Pentoxifylline is a methylxanthine that can suppress both the accumulation of TNF-α mRNA and the activity of its secreted form. In patients with NASH, both alanine aminotransferase and aspartate aminotransferase levels were significantly lower after 12 mo of therapy compared to the baseline ($p = 0.003$), indicating significant improvement in treated patients *(62)*. The treatment of NASH with 400 mg pentoxifylline three times per day had similar results *(63)*. These findings warrant further investigation.

5.5. Other Cytokines

Two potent proinflammatory cytokines, IL-6 and IL-8, are released by both visceral and subcutaneous adipose tissues of obese subjects *(24)*. In fact, human adipose tissue can release more IL-6 and IL-8 than adiponectin, especially in morbidly obese individuals *(64)*. Two studies show that serum IL-8 and IL-6 levels in patients with NASH are significantly higher than in healthy controls *(60)*. On the other hand, IL-6 seems to induce hepatoprotection both in normal and steatotic liver grafts after liver transplantation *(65)*. These contradictory findings emphasize our incomplete understanding of the role of these cytokines in NAFLD.

6. ROLE OF ADIPOKINES IN PROMOTING HEPATIC STEATOSIS, IR, OXIDATIVE STRESS, AND HEPATIC FIBROSIS IN NAFLD

6.1. Adipokines and Steatosis

As previously noted, hepatic steatosis may result from an increase in the delivery of free FA to the liver, increased FA synthesis, decreased FA degradation, impaired TG release from the liver, or a combination of these factors. Adiponectin exerts a beneficial effect on the accumulation of TGs and on the concentration of FA in skeletal muscle *(68)*. It also enhances FA oxidation both in liver and muscle tissue through activation of acetyl CoA oxidase, carnitine palmitoyltransferase-1 (CPT1), and 5′-AMP activated protein kinase (AMPK) *(69)*, and stimulates lipoprotein lipase activity in animal models *(70)*. Decreased serum adiponectin is associated with lipoprotein lipase (LPL) deficiency in humans, independent of the effects of systemic inflammation and/or IR *(71)*. Therefore, hypoadiponectinemia may stimulate the accumulation of fat in the liver by promoting LPL deficiency, leading to an influx of free FA.

Alternatively, adiponectin may promote hepatic steatosis by increasing FA synthesis or decreasing FA degradation within the liver, or both. For instance, adiponectin treatment normalizes hepatic lipid content in steatotic mice by restoring the activity of CPT1, a rate-limiting enzyme involved in the transport of long-chain FA into mitochondrial matrix *(32)*. Thus, high adiponectin concentrations stimulate β-oxidation of FA in the liver and therefore decreases the intrahepatic lipid load. At the same time, adiponectin downregulates the hepatic lipogenesis pathway *(45)*.

A third mechanism potentially linking hypoadiponectinemia to the development of NAFLD is an increase in hepatic lipid retention owing to adiponectin-dependent suppression of very-low-density lipoprotein (VLDL) synthesis, the chief route of hepatic lipid export *(67)*. The effects of adiponectin on VLDL metabolism are independent of both IR and the size of the adipose tissue compartments *(68)*. Unfortunately, patients included in these studies were not assessed for the presence of NAFLD. Therefore, an important link among adiponectin, VLDL, and the pathogenesis of NAFLD remains uncertain.

Leptin protects against lipotoxicity in nonadipose tissues *(68)*, possibly by a peripheral mechanism. Studies of pair-fed controls receiving the exact amount of food ingested by leptin-treated animals show that controls remain steatotic despite caloric restriction *(73,74)*. In cultured pancreatic islets, leptin lowers TG content by increasing FA oxidation and preventing its esterification *(73)*. A similar mechanism may be at work in the liver, because liver tissue expresses leptin receptors. Indeed, tissue-specific overexpression of wild-type leptin receptors in steatotic livers reduces TG accumulation in the liver but nowhere else *(75)*.

Furthermore, leptin dramatically suppresses the expression of the hepatic stearoyl-CoA desaturase (SCD)-1, the rate-limiting enzyme in the biosynthesis of monounsaturated fats *(74)*. SCD-1 suppression, in turn, supports resistance to both hepatic steatosis and obesity owing to a marked increase in energy expenditure. Two proposed mechanisms for these leptin effects include blocking TG synthesis and exporting VLDL *(74,75)*. These mechanisms lead to a concomitant increase in the pool of saturated fatty acyl CoAs, which allosterically inhibits ACC and reduces the amount of malonyl CoA. Inhibition of the mitochondrial carnityl palmitoyl shuttle system is relieved as a consequence, stimulating the import and oxidation of FA in mitochondria. Thus, leptin administration de-represses FA oxidation, leading to increased fat burning *(74)*. Other proposed mechanisms of antisteatotic effects of the leptin involve increases in a peroxisome proliferator-activated receptor (PPAR) α signaling *(76)* or AMPK activation, or both *(77)*.

Leptin also seems to promote the elimination of the plasma cholesterol through stimulation of its catabolism to bile salts in the setting of decreased cholesterol biosynthesis. Cholesterol elimination is achieved by suppressing the hepatic activity of HMG-CoA

Table 1
Summary of Positive and Negative Effects of Adipokines on Cellular Processes Contributing to Pathogenesis of NASH

	Cellular processes contributing to NASH				
Adipokine	*Lipid accumulation in liver (steatosis)*	*IR*	*Oxidative damage*	*Fibrotic responses*	*Role in NASH progression*
Adiponectin	Suppressed	Suppressed	Suppressed	Suppressed	Prevents NASH
Leptin	Suppression effects are low due to leptin resistance	Suppression effects are low due to leptin resistance	Pro-oxidant	Fibrogenic action	Suppresses initiation of steatosis; stimulates progression of existing steatosis to NASH
Resistin	Possibly steatogenic	Possibly involved in IR; difficult to study in humans	Possibly pro-oxidant	Possibly fibrogenic	Unclear
TNF-α	Steatogenic	Impairs insulin signaling	Pro-oxidant	Fibrogenic action	Augments NASH
Visfatin	Unknown	Mimicking insulin	Unknown	Unknown	Unclear
Vaspin	Unknown	Suppressed	Unknown	Unknown	Unclear
Apelin	Unknown	Inhibits insulin production	Unknown	Unknown	Unclear

reductase, upregulating the activities of both sterol 27-hydroxylase and cholesterol 7α-hydroxylase, and diminishing the cholesterol fraction bound to VLDL by limiting TG supply *(78)*. Lowered leptin signaling might be responsible for the increase in the prevalence of cholesterol gallstones in obese patients compared with the general population *(79)*.

It is important to remember that obesity is associated with leptin resistance and hyperleptinemia. Therefore, exogenous leptin administration does not alleviate lipid accumulation in the liver or improve NAFLD. On the other hand, the development of central and peripheral leptin resistance critically depends on the liver. In animal models, chronic leptin treatment in leptin-naïve animals induces shedding in the soluble leptin receptor protein (SLR). SLR sequesters leptin and prevents productive interactions with its signaling receptor *(80)*, making peripheral leptin activity self-limiting.

It would be interesting to know whether the rate of the SLR synthesis is altered in steatotic livers, or whether the manipulation of SLR production could alter leptin resistance in obesity.

Our understanding of the role of resistin in the development of steatosis is quite preliminary. Resistin is capable of influencing lipid metabolism in rodents. Resistin overexpression in mice and rats leads to plasma TG increases and to significant dyslipidemia *(81)*. Serum resistin levels correlate negatively with HDL cholesterol levels in healthy men *(82)*, suggesting that higher-than-normal resistin levels typically seen in obesity and type 2 diabetes might contribute to the development of fatty liver through its dyslipidemic effects.

Finally, TNF-α's pleiotropic effects can influence lipid metabolism in the liver, as it stimulates *de novo* synthesis of FA, suppresses FA oxidation, and enhances the turnover of VLDL *(83)*. Mice that express T-cell-targeted human TNF-α transgenes provide an animal model for persistent low-grade exposure to TNF-α typical of morbid obesity *(84)*. These mice are dyslipidemic *(84)*. Both mitochondrial and peroxisomal β-oxidations are inhibited in their livers *(84)* with no concomitant increase in the *de novo* FA synthesis *(84)*. Therefore, TNF-α-dependent steatogenesis in the liver is predominantly caused by the suppression of FA decomposition. TNF-α also stimulates VLDL production in the liver and inhibits the activity of lipoprotein lipase in adipocytes *(85)*; these processes favor lipolysis in fat depots and contribute to the development of the TNF-α-dependent hypertriglyceridemia and associated NAFLD.

6.2. Adipokines and IR

Because of the striking association between NASH and IR *(2)*, any factor promoting a vicious cycle of insulin signaling can be steatogenic, and factors counteracting IR can be protective against the development of NAFLD.

Hyperinsulinemia caused by IR increases FA synthesis and impairs both mitochondrial β-oxidation and the export of TGs in multiple ways. Early studies have indicated that adiponectin decreases IR by increasing FA oxidation, which reduces the TG content in nonadipocytes *(31)*, suppresses glucose production in the liver *(69)*, and enhances the hepatic action of insulin *(30)*. These glucose-lowering effects of adiponectin require liver-specific AMPK activation *(69)* and play a key role in the regulation of energy control. AMPK is activated in response to a variety of external signals, including adipokines *(86)*. It is tempting to speculate that AMPK-mediated antiglycemic effects may play a role in the prevention of NAFLD, but this seems unlikely. Recent work indicates that short-term overexpression of a constutively active form of AMPK in the liver can lead to the development of fatty liver in the presence of lowered hepatic glycogen synthesis and circulating lipid levels *(86)*. Most likely, the NAFLD-like disorder in animal models develops from the hepatic accumulation of lipids released from adipose tissue in response to the relative scarcity of glucose. Therefore, additional stimulation of AMPK provided by a sudden increase of adiponectin (e.g., owing to thiazolidinedione [TZD] treatment) may aggravate early stages of the hepatic steatosis. This may also explain the infrequent but potentially serious hepatotoxic side effects of chronic TZD administration *(87)* and the pronounced exacerbation of hepatic steatosis in mice with polygenic obesity treated by rosiglitazone *(88)*.

Leptin exerts a systemic insulin-sensitizing effect *(89)*. An interaction between the insulin and leptin signaling cascades has been studied both in vitro and in vivo *(90)*, but the complete mechanism remains unclear and the results are inconsistent. Most likely, cross-cascade interactions involve insulin receptor substrate (IRS) molecules, PI 3-kinase, Akt, and GSK3 *(90)*. The liver is probably central to the adiposity-independent role of leptin in controlling IR, as some studies have suggested that leptin selectively improves insulin receptor activation only in the liver, but not in skeletal muscle or fat *(91)*. Unfortunately, the insulin-related branch of the leptin-dependent signaling pathway in obese livers is profoundly suppressed *(92)*. Therefore, it is unlikely that therapeutic administration of leptin would alleviate liver steatosis through improved insulin sensitivity.

Resistin reduces glucose tolerance and insulin action, thereby inducing IR. Hyperresistinemia certainly contributes to IR in obese rodents because of decreased gluconeogenic enzyme expression in the liver and to the activation of AMPK *(93)*. In humans, the situation is much more difficult to trace, because serum resistin levels are related to sex, age, and testosterone and estradiol levels *(94)*. These fluctuations in resistin levels and the relatively low homology between resistin and resistin-like molecules in humans and rodents complicate the study of resistin in the development of IR in the liver and NAFLD.

TNF-α alters systemic energy homeostasis in a way that closely resembles the IR phenotype. Mice with a complete knockout of TNF-α signaling show significantly improved insulin sensitivity in both diet-induced and leptin-deficient obesity *(95)*. Long-term exposure to TNF-α completely abolishes insulin-induced glycogen synthesis in hepatocytes *(96)*. Therefore, abnormal production of TNF-α may predispose obese individuals to the development of the IR and NAFLD.

Visfatin is produced both in visceral and subcutaneous adipose tissue and exerts insulin-like effects in various tissues by binding and activation of the insulin receptor *(98)*. Visfatin is upregulated in obesity *(97)* either as a simple reflection of visfatin resistance that parallels the IR in metabolic syndrome, or represents an important compensatory pathway leading to lowered glucose levels. Vaspin, visceral adipose tissue-derived member of the serine protease inhibitor (serpin) family, normalizes serum glucose levels by reversing altered gene expression related to IR, including all other adipokines discussed above *(99)*. In humans, vaspin mRNA expression is not detectable in lean subjects, but is a frequent finding in type 2 diabetes *(100)*. This secreted molecule might be an important insulin sensitizer of adipocytic origin and may play an important role in NAFLD. Finally, apelin is an adipokine that is probably related to peripheral IR. It inhibits glucose-stimulated insulin secretion both in vivo and in vitro by acting on its receptor, which is expressed in β-cells of pancreatic islands *(101)*. Apelin plasma levels are largely increased in all the hyperinsulinemia-associated obese states in mice, independently of diet composition *(102)*, and in obese humans *(102,103)*. In summary, the interplay between insulin-like visfatin, insulin-sensitizing vaspin, and the suppression of insulin production by apelin may represent important avenues for future studies of the pathogenesis of NAFLD and NASH.

6.3. Adipokines and Oxidative Stress

Changes in serum adipokine concentrations augment oxidative stress in patients with NASH. Most studies converge on CYP2E1, peroxisomal release of ROS, and mitochondrial dysfunction. ROS and reactive nitrogen species (RNS) are generated by the parenchymal cells of the liver, Kupffer cells, and inflammatory cells, which further mobilize cellular defense mechanisms and contribute to liver injury and necrosis.

The potential role of adiponectin as an antioxidant is mostly indirect. One study suggests that serum adiponectin levels negatively correlate with urinary levels of isoprostane, an oxidative stress marker *(104)*. A different line of evidence suggests that the production of adiponectin may be suppressed in the pro-oxidative conditions. For example, inhibited adiponectin mRNA expression is observed in differentiated murine adipocytes after exposure to increasing concentrations of glucose oxidase, H_2O_2, and byproducts of lipid peroxidation *(105)*. Adiponectin synthesis may also be suppressed by an excess of angiotensin II (AngII), a vasoactive peptide *(106)*. AngII indirectly activates NAD(P)H oxidase, which favors the production of ROS. It is noteworthy that administration of the angII type 1 receptor antagonist losartan significantly improves liver biochemical indices as well as hepatic necroinflammation in patients with NASH *(52)*.

Leptin increases markers of lipoperoxidation in the liver while decreasing antioxidant GSH levels and the activities of glutathione-S-transferases (GSTs), superoxide dismutase (SOD), and catalase *(107)*. Similar observations have been made in non-NAFLD patients with other chronic liver diseases *(108)*. Intravenous leptin injections induce the release of nitric oxide (NO) *(109)* by both endothelial and inducible nitric oxide synthases (eNOS and iNOS). As uncoupled eNOS changes from a protective enzyme to a contributor to oxidative stress, leptin-induced stimulation of eNOS and iNOS is a pro-oxidative event *(110)*. Leptin also stimulates cytochrome CYP2E1 expression, responsible for the oxidation of alcohol and the production of ROS. Paradoxically, CYP2E1-dependent production of ROS inhibits apoptosis but accelerates necrosis stimulated by polyunsaturated FA *(111)*. This latter observation is consistent with the necroinflammatory features seen in patients with NASH. Finally, CYP2E1 activity is elevated in patients with NASH as assessed by the rates of oral clearance of chlorzoxazone *(112)*.

Additional observations supporting the role of resistin in promoting oxidative stress include a study of resistin's effects in porcine coronary arteries, which shows increased superoxide radical production, and decreased eNOS activity *(113)*. In humans with normal body weight, serum resistin concentrations are negatively correlated with the concentrations of a marker of oxidative stress, nitrotyrosine *(61)*. Conversely, oxidative stress itself can suppress resistin production in adipocytes, similar to suppressed production of adiponectin. The efficiency of such suppression might depend on a particular genotype at a resistin locus *(115)*.

TNF-α certainly plays an important role in the enhancement of ROS production observed in steatotic livers. Key components of TNF-α signaling include ceramide, which influences the mitochondrial electron transport chain and evokes hydrogen peroxide overproduction *(116)*. In addition, ceramide induces mitochondrial membrane permeability transition (MMPT) and subsequent necrosis *(117)*. Another potential sensitizer to TNF-α-induced cell death is the uncoupling of mitochondrial respiration *(118)*. TNF-α enhances the expression of UCP2, a mitochondrial regulator that increases a proton leak across the inner membrane to dissociate respiration from ATP synthesis and reduce ROS generation. TNF-α-dependent UCP2 stimulation is especially pronounced in steatotic *(119)* and regenerating *(120)* livers. Upregulated UCP2 may compromise cellular ATP levels and worsen liver damage by augmenting cell death, or it may be protective by reducing ROS levels. It is also possible that these two effects cancel each other *(121)*. It is important to note that the state of UCP2 activity in patients with NAFLD and patients with NASH is not entirely clear. As mitochondrial uncoupling sensitizes the cells to TNF-α-induced death, this effect might outweigh the simultaneous decrease in the ROS production in human subjects.

6.4. Adipokines in Hepatic Fibrosis

Hepatic fibrosis is a wound-healing response characterized by inflammation, activation of matrix-producing cells, extracellular matrix (ECM) deposition and remodeling, and epithelial cell regeneration *(122)*. Major matrix-producing cells in the liver are hepatic stellate cells (HSCs) that may undergo a phenotypic transition to myofibroblast-like cells that synthesize various ECM components and contribute to fibrogenesis.

Adiponectin suppresses the proliferation and migration of HSCs *(123)* and attenuates CH_4-dependent liver fibrosis through suppression of platelet-derived growth factor (PDGF) and transforming growth factor (TGF)-β1-induced migration and proliferation *(33)*. Adiponectin can also induce apoptosis in activated HSCs, but not in quiescent HSCs *(124)*. Both AdipoR1 and AdipoR2 receptors are present in both quiescent and activated HSCs; however, AdipoR1 mRNA expression is reduced by 50% in activated HSCs *(124)*. These findings indicate that adiponectin is either essential to maintaining the quiescent phenotype of HSCs or it is capable of reversing hepatic fibrosis by hampering the proliferation of activated HSCs and by inducing HSC apoptosis.

Leptin enhances liver inflammation and fibrogenesis, in part, by upregulating TGF-β. Leptin has a profound positive influence on α(2)(I) collagen mRNA expression in HSCs *(125)*. In addition, leptin augments PDGF-dependent HSC proliferation. Taken together, these studies indicate that leptin is a potent promoter of hepatic fibrosis. Observations in lipodystrophic patients treated with recombinant leptin support this conclusion *(57)*.

Resistin has no known connection with hepatic fibrosis, but this molecule has been implicated in pulmonary fibrosis induced by bleomycin. Cocultures of RELM-α-expressing epithelial cells and fibroblasts stimulate α-smooth muscle actin and type I collagen expression independently of TGF-β *(126)*. Similar resistin-dependent responses might be produced in HSCs, if resistin contributes to the development of NASH.

Recent experiments provide direct evidence of the involvement of TNF-α in fibrogenic responses. When double knockout mice lacking both TNF receptors (TNFRDKO mice) are fed methionine- and choline-deficient (MCD) diets, they develop less pronounced liver steatosis than their wild-type counterparts *(127)*. Similar findings in TNFRp55 knockout mice indicate that even partial suppression of TNF-α signaling can alleviate hepatic fibrosis *(128)*. It seems that TNF-α increases the recruitment of Kupffer cells that can, in turn, produce extra TNF-α and hasten fibrosis in either an autocrine or a paracrine manner. Both these processes can contribute to the development of NASH progression to cirrhosis.

7. CONCLUSIONS

We have learned a great deal about the epidemiology and pathogenesis of NAFLD and NASH over the past decade (Table 1). It is increasingly clear that the development of NASH is a complex process involving multiple mechanisms including IR, oxidative stress, abnormal FA metabolism, and disturbances in the production of inflammatory cytokines and adipokines. Decreased production of adiponectin and increased production of TNF-α, which are characteristic of obesity, seem to contribute to all major NASH-related cellular processes. Leptin, on the other hand, behaves as a "wolf in sheep's clothing." Its NASH-suppressive effects are diminished by the widespread effects of leptin resistance, and it becomes potentially pro-oxidant and fibrogenic. Resistin's involvement in NASH is documented in rodent models, but may not be applicable to NAFLD in humans. In addition, other adipokines, such as vaspin, visfatin, and apelin require further study in patients with NAFLD and NASH.

REFERENCES

1. Falck-Ytter Y, Younossi ZM, Marchesini G, et al. Semin Liver Dis 2001;21:17–26.
2. Younossi ZM, Diehl AM, Ong JP. Hepatology 2002;35:746–752.
3. Matteoni CA, Younossi Z, Gramlich T, et al. Gastroenterology 1999;116:1413–1419.
4. Ong JP, Elariny H, Collantes R, et al. Obes Surg 2005;15:310–315.
5. Adams LA, Lymp JF, St Sauver J, et al. Gastroenterology 2005;129:113–121.
6. Caldwell SH, Oelsner DH, Iezzoni JC, et al. Hepatology 1999;29:664–669.
7. Bugianesi E, Leone N, Vanni E, et al. Gastroenterology 2002;123:134–140.
8. Gramlich T, Kleiner DE, McCullough AJ, et al. Hum Pathol 2004;35:196–199.
9. Saadeh S, Younossi ZM, Remer EM, et al. Gastroenterology 2002;123:745–750.
10. Angulo P, Lindor KD. Semin Liver Dis 2001;21:81–88.
11. Mulhall BP, Ong JP, Younossi ZM. J Gastroenterol Hepatol 2002;17:1136–1343.
12. McClain CJ, Mokshagundam SP, Barve SS, et al. Alcohol 2004;34:67–79.
13. Chitturi S, Farrell GC. Semin Liver Dis 2001;21:27–41.
14. Day CP, James OF. Gastroenterology 1998;114:842–845.
15. Koteish A, Diehl AM. Best Pract Res Clin Gastroenterol 2002;16: 679–690.
16. Lin HZ, Yang SQ, Chuckaree C, et al. Nat Med 2000;6:998–1003.
17. Pagano G, Pacini G, Musso G, et al. Hepatology 2002;35:367–372.
18. Liangpunsakul S, Chalasani N. Am J Med Sci 2005;329:111–116.
19. Choudhuri J, Sanyal AJ. Clin Liver Dis 2004;8:575–594.
20. Day CP. Best Pract Res Clin Gastroenterol 2002;16:663–678.
21. Fantuzzi G. J Allergy Clin Immunol 2005;115:911–919.
22. Skurk T, Herder C, Kraft I, et al. Endocrinology 2005;146:1006–1011.
23. Wellen KE, Hotamisligil GS. J Clin Invest 2003;112:1785–1788.
24. Fain JN, Madan AK, Hiler ML, et al. Endocrinology 2004;145:2273–2282.
25. Minn AH, Patterson NB, Pack S, et al. Biochem Biophys Res Commun 2003;310:641–645.
26. Li Z, Yang S, Lin H, et al. Hepatology 2003;37:343–350.
27. Faggioni R, Jones-Carson J, Reed DA, et al. Proc Natl Acad Sci USA 2000;97:2367–2372.
28. Yang RZ, Huang Q, Xu A, et al. Biochem Biophys Res Commun 2003;310:927–935.
29. Kushiyama A, Shojima N, Ogihara T, et al. J Biol Chem 2005;280:42,016–42,025.
30. Berg AH, Combs TP, Du X, et al. Nat Med 2001;7:947–953.
31. Yamauchi T, Kamon J, Waki H, et al. Nat Med 2001;7:941–946.
32. Xu A, Wang Y, Keshaw H, et al. J Clin Invest 2003;112:91–100.
33. Kamada Y, Tamura S, Kiso S, et al. Gastroenterology 2003;125:1796–1807.
34. Targher G, Bertolini L, Scala L, et al. Clin Endocrinol (Oxf) 2004;61:700–703.
35. Pagano C, Soardo G, Esposito W, et al. Eur J Endocrinol 2005;152:113–118.
36. Mendez-Sanchez N, Chavez-Tapia NC, Villa AR, et al. World J Gastroenterol 2005;11:1737–1741.
37. Zou CC, Liang L, Hong F, et al. Endocr J 2005;52:519–524.
38. Bugianesi E, Pagotto U, Manini R, et al. J Clin Endocrinol Metab 2005;90:3498–3504.
39. Hui JM, Hodge A, Farrell GC, et al. Hepatology 2004;40:46–54.
40. Musso G, Gambino R, Biroli G, et al. Am J Gastroenterol 2005;100:2438–2446.
41. Musso G, Gambino R, Durazzo M, et al. Hepatology 2005;42:1175–1183.
42. Comuzzie AG, Funahashi T, Sonnenberg G, et al. J Clin Endocrinol Metab 2001;86:4321–4325.
43. Menzaghi C, Ercolino T, Salvemini L, et al. Physiol Genomics 2004;19:170–174.

44. Zacharova J, Chiasson JL, Laakso M. STOP-NIDDM Study Group. Diabetes 2005;54:893–899.
45. Ukkola O, Santaniemi M, Rankinen T, et al. Ann Med 2005;37:141–150.
46. Kaser S, Moschen A, Cayon A, et al. Gut 2005;54:117–121.
47. Vuppalanchi R, Marri S, Kolwankar D, et al. J Clin Gastroenterol 2005;39:237–242.
48. Stefan N, Machicao F, Staiger H, et al. Diabetologia 2005;48:2282–2291.
49. Silswal N, Singh AK, Aruna B, et al. Biochem Biophys Res Commun 2005;334:1092–1101.
50. Bajaj M, Suraamornkul S, Hardies LJ. Int J Obes Relat Metab Disord 2004;28:783–789.
51. Uygun A, Kadayifci A, Yesilova Z, et al. Am J Gastroenterol 2000;95:3584–3589.
52. Yokoyama H, Hirose H, Ohgo H, et al. Alcohol Clin Exp Res 2004;28:159S–163S.
53. Kaplan LM. Gastroenterology1998;115:997–1001.
54. Piche T, Vandenbos F, Abakar-Mahamat A, et al. J Viral Hepat 2004;11:91–96.
55. Crespo J, Rivero M, Fabrega E, et al. Dig Dis Sci 2002;47:1604–1610.
56. Angulo P, Alba LM, Petrovic LM, et al. J Hepatol 2004;41:943–949.
57. Javor ED, Ghany MG, Cochran EK, et al. Hepatology 2005;41:753–760.
58. Feuerstein GZ, Liu T, Barone FC. Cerebrovasc Brain Metab Rev 1994;6:341–360.
59. Fain JN, Bahouth SW, Madan AK. Int J Obes Relat Metab Disord 2004;28:616–622.
60. Kugelmas M, Hill DB, Vivian B, et al. Hepatology 2003;38:413–419.
61. Crespo J, Cayon A, Fernandez-Gil P, et al. Hepatology 2001;34:1158–1163.
62. Adams LA, Zein CO, Angulo P, et al. Am J Gastroenterol 2004;99:2365–2368.
63. Satapathy SK, Garg S, Chauhan R, et al. Am J Gastroenterol 2004;99:1946–1952.
64. Fain JN, Bahouth SW, Madan AK. Biochem Pharmacol 2005;69:1315–1324.
65. Gao B. Alcohol 2004;34:59–65.
66. Mehta K, Van Thiel DH, Shah N, et al. Nutr Rev 2002;60:289–293.
67. Charlton M, Sreekumar R, Rasmussen D, et al. Hepatology 2002;35:898–904.
68. Ng TW, Watts GF, Farvid MS, et al. Diabetes 2005;54:795–802.
69. Yamauchi T, Kamon J, Minokoshi Y, et al. Nat Med 2002;8:1288–1295.
70. Combs TP, Pajvani UB, Berg AH, et al. Endocrinology 2004;145:367–383.
71. Unger RH, Zhou YT, Orci L. Proc Natl Acad Sci USA 1999;96:2327–2332.
72. Shimabukuro M, Koyama K, Chen G, et al. Proc Natl Acad Sci USA 1997;94:4637–4641.
73. Levin N, Nelson C, Gurney A, et al. Proc Natl Acad Sci USA 1996;93:1726–1730.
74. Cohen P, Friedman JM. J Nutr 2004;134:2455S–2463S.
75. Miyazaki M, Kim YC, Gray-Keller MP, et al. J Biol Chem 2000;275:30,132–30,138.
76. Lee Y, Yu X, Gonzales F, Mangelsdorf DJ, et al. Proc Natl Acad Sci USA 2002;99:11,848–11,853.
77. Minokoshi Y, Kim Y-B, Peroni OD, et al. Nature 2002;415:339–343.
78. VanPatten S, Ranginani N, Shefer S, et al. Am J Physiol Gastrointest Liver Physiol 2001;281: G393–G404.
79. Loria P, Lonardo A, Lombardini S, et al. J Gastroenterol Hepatol 2004;20:1176–1184.
80. Yang G, Ge H, Boucher A, et al. Mol Endocrinol 2004;18:1354–1362.
81. Sato N, Kobayashi K, Inoguchi T, et al. Endocrinology 2005;146:273–279.
82. Chen CC, Li TC, Li CI, et al. Metabolism 2005;54:471–475.
83. Nachiappan V, Curtiss D, Corkey BE, et al. Shock 1994;1:123–129.
84. Glosli H, Gudbrandsen OA, Mullen AJ, et al. Biochim Biophys Acta 2005;1734:235–246.
85. Ruan H, Miles PD, Ladd CM, et al. Diabetes 2002;51:3176–3188.
86. Carling D. 150 Biochimie 2005;87:87–91.
87. Parulkar AA, Pendergrass ML, Granda-Ayala R, et al. Ann Intern Med 2001;134:61–71.
88. Watkins SM, Reifsnyder PR, Pan HJ, et al. J Lipid Res 2002;43:1809–1817.
89. Pocai A, Morgan K, Buettner C, et al. Diabetes 2005;54:3182–3189.
90. Hegyi K, Fulop K, Kovacs K, et al. Cell Biol Int 2004;28:159–169.
91. Lam NT, Lewis JT, Cheung AT, et al. Mol Endocrinol 2004;18:1333–1345.
92. Brabant G, Muller G, Horn R, et al. FASEB J 2005;19:1048–1050.
93. Banerjee RR, Rangwala SM, Shapiro JS, et al. Science 2004;303:1195–1198.
94. Gerber M, Boettner A, Seidel B, et al. J Clin Endocrinol Metab 2005;90:4503–4509.
95. Uysal KT, Wiesbrock SM, Marino MW, et al. Nature 1997;389:610–614.
96. Gupta D, Khandelwal RL. Biochim Biophys Acta 2004;1671:51–58.

97. Berndt J, Kloting N, Kralisch S, et al. Diabetes 2005;54:2911–2916.
98. Fukuhara A, Matsuda M, Nishizawa M, et al. Science 2005;307:426–430.
99. Hida K, Wada J, Eguchi J, et al. Proc Natl Acad Sci USA 2005;102:10,610–10,615.
100. Kloting N, Berndt J, Kralisch S, et al. Biochem Biophys Res Commun. 2006;339:430–436.
101. Sorhede Winzell M, Magnusson C, Ahren B. Regul Pept 2005;131:12–17.
102. Boucher J, Masri B, Daviaud D, et al. Endocrinology 2005;146:1764–1771.
103. Heinonen MV, Purhonen AK, Miettinen P, et al. Regul Pept 2005;130:7–13.
104. Nakanishi S, Yamane K, Kamei N, et al. Metabolism 2005;54:194–199.
105. Soares AF, Guichardant M, Cozzone D, et al. Free Radic Biol Med 2005;38:882–889.
106. Hattori Y, Akimoto K, Gross SS, et al. Diabetologia 2005;48:1066–1074.
107. Balasubramaniyan V, Kalaivani Sailaja J, Nalini N. Pharmacol Res 2003;47:211–216.
108. Uzun H, Zengin K, Taskin M, et al. Obes Surg 2004;14:659–665.
109. Fruhbeck G. Diabetes 1999;48:903–908.
110. Forstermann U. Eur J Clin Pharmacol 2005;26:1–8.
111. Gonzalez FJ. Mutat Res 2005;569:101–110.
112. Chalasani N, Gorski JC, Asghar MS, et al. Hepatology 2003;37:544–550.
113. Kougias P, Chai H, Lin PH, et al. J Vasc Surg 2005;4:691–698.
114. Bo S, Gambino R, Pagani A, et al. Int J Obes (Lond) 2005;29:1315–1320.
115. Smith SR, Bai F, Charbonneau C, et al. Diabetes 2003;52:1611–1618.
116. Garcia-Ruiz C, Colell A, Mari M, et al. J Biol Chem 1997;272:11,369–11,377.
117. Arora AS, Jones BJ, Patel TC, et al. Hepatology 1997;25: 958–963.
118. Hao JH, Yu M, Liu FT, et al. Cancer Res 2004;64:3607–3616.
119. Uchino S, Yamaguchi Y, Furuhashi T, et al. J Surg Res 2004;120:73–82.
120. Lee FY, Li Y, Zhu H, et al. Hepatology 1999;29:677–687.
121. Baffy G. Front Biosci 2005;10:2082–2096.
122. Marra F, Aleffi S, Bertolani C, et al. Eur Rev Med Pharmacol Sci 2005;9:279–284.
123. Goetze S, Bungenstock A, Czupalla C, et al. Hypertension 2002;40:748–754.
124. Ding X, Saxena NK, Lin S, et al. Am J Pathol 2005;166:1655–1669.
125. Saxena NK, Ikeda K, Rockey DC, et al. Hepatology 2002;35:762–771.
126. Otte C, Otte JM, Strodthoff D, et al. Exp Clin Endocrinol Diabetes 2004;112:10–17.
127. Tomita K, Tamiya G, Ando S, et al. Gut 2005;55:415–424.
128. Kitamura K, Nakamoto Y, Akiyama M, et al. Lab Invest 2002;82:571–583.

23 Adiposity and Cancer

Eugenia E. Calle

Abstract

The International Agency for Research on Cancer has classified the evidence of a causal link between adiposity and human cancer as "sufficient" for cancers of the colon, female breast (postmenopausal), endometrium, kidney (renal cell), and esophagus (adenocarcinoma). In addition to these cancers, more recent epidemiological evidence suggests that cancers of the liver and pancreas are obesity related, and that adiposity may also increase the risk for hematopoietic cancers and aggressive prostate cancer. The mechanisms by which obesity is postulated to induce or promote tumorigenesis vary by site. These include insulin resistance and resultant chronic hyperinsulinemia, increased bioavailability of steroid hormones, and localized inflammation. The role that leptin, adiponectin, and other proteins secreted by adipocytes may have in the development and progression of tumors is an active area of research.

Key Words: Anthropometry; body mass index; cancer; epidemiology; obesity; overweight.

1. INTRODUCTION

Obesity has long been recognized to be an important cause of type 2 diabetes mellitus, hypertension, and dyslipidemia *(1,2)*. The adverse metabolic effects of excess body fat are known to accelerate atherogenesis and increase the risk of coronary heart disease, stroke, and early death. The relationship of obesity to cancer has received less attention than its cardiovascular effects. Results from epidemiological studies largely begun in the 1970s indicate that adiposity contributes to the increased incidence and/or death from cancers of the colon, female breast (in postmenopausal women), endometrium, kidney (renal cell), esophagus (adenocarcinoma), gastric cardia, pancreas, gallbladder, liver, and possibly others. It has been estimated that 15 to 20% of all cancer deaths in the United States can be attributed to overweight and obesity *(3)*. At present, the strongest empirical support for mechanisms to link obesity and cancer risk involves the metabolic and endocrine effects of obesity, and the alterations they induce in production of peptide and steroid hormones *(4)*. As the worldwide obesity epidemic has shown no signs of abating, insight into the mechanisms by which obesity contributes to tumor formation and progression is urgently needed, as are new approaches to intervene in this process.

From: *Nutrition and Health: Adipose Tissue and Adipokines in Health and Disease*
Edited by: G. Fantuzzi and T. Mazzone © Humana Press Inc., Totowa, NJ

2. ASSESSMENT OF OVERWEIGHT AND OBESITY IN EPIDEMIOLOGICAL STUDIES OF DISEASE

Definitions for classifying and reporting healthy weight, overweight, and obesity in populations have historically been based on measures of weight and height rather than clinical measures of adiposity *(5,6)*. Although weight is the simplest anthropometric index of excess body fat, it does not distinguish between lean body mass (comprised primarily of muscle, bone, and extracellular water) and adipose tissue *(5)*. Thus, measures of weight adjusted for height provide a better approximation of the proportion or total amount of adipose tissue in the body than does weight alone.

Since the 1980s, indices of weight adjusted for height have gained favor because they provide a single estimate of adiposity regardless of height and can be easily compared across studies and across populations. By far the most widely used weight-for-height measure is the body mass index (BMI, also called Quetelet's index), which is defined as weight (in kilograms) divided by height (in meters squared) *(5)*. The assumption underlying the BMI (and all other such indices) is that true adiposity is unrelated to height. Indeed, among the many indices of weight-for-height that have been proposed, the correlation with height has generally been lowest for BMI *(5)*.

Standards defining healthy weight, overweight, and obesity have evolved over time and reflect existing knowledge of and assumptions about the relation of weight to disease outcomes. Historically, weight-for-height standards prepared by the Metropolitan Life Insurance Company provided "ideal" and "desirable" gender-specific weight ranges for each inch of height based on actuarial data *(7)*. Standards based on BMI have been reported for the US adult population since 1980 in the *Dietary Guidelines for Americans (8)*. Widely accepted current standards based on BMI criteria for overweight and obesity are recommended by the World Health Organization (WHO) *(9)* and supported by other advisory committees and expert panels to federal agencies *(1,8)*. The WHO cut-points for BMI and their corresponding interpretations are shown in Table 1. Although the exact cut-points are somewhat arbitrary, this BMI classification scheme was derived largely from observational and epidemiological studies of BMI and disease end points and thus reflects the relationship of BMI to morbidity and mortality *(1,9)*. The cut-point for the underweight category is based on adverse health consequences of malnutrition in developing countries *(9)*.

Weight and height can be self-reported and thus are more easily determined in epidemiological studies of morbidity and mortality than measured weight and height or clinical measures of adiposity. Even though some systematic error exists in self-reported weight and height (weight tends to be underestimated and height overestimated), self-reported data are highly correlated with measured weight and height ($r = 0.8$ to >0.9) *(5)* and are sufficiently accurate to establish associations with diseases known to be related to obesity in epidemiological studies *(5,10,11)*. However, prevalence estimates of overweight and obesity based on self-reported data tend to be lower than those based on measured values *(12)*.

Many studies have found moderate to strong correlations ($r = 0.6$ to 0.9) between BMI and densitometry estimates of body fat composition in adult populations *(5)*. The validity of BMI as a measure of adiposity is further supported by its association with obesity-related risk factors such as blood triglycerides, total cholesterol, blood pressure,

Table 1
Cut-Points of BMI for Classification of Weight *(9)*

BMI (kg/m2)	*World Health Organization*
<18.5 kg/m^2	Underweight
18.5–24.9 kg/m^2	Normal range
25.0–29.9 kg/m^2	Grade 1 overweight
30.0–39.9 kg/m^2	Grade 2 overweight
≥40.0 kg/m^2	Grade 3 overweight

BMI, body mass index.

and fasting glucose *(5)*. BMI may be a less valid indicator of adiposity among the elderly, who tend to have a shift of fat from peripheral to central sites with a concomitant increase in waist-to-hip ratio (WHR) at the same level of BMI *(13)*. For such populations, and with increasing evidence of health risks associated with abdominal (visceral) fat, two measures of central adiposity, the WHR and, more recently, waist circumference, have been commonly used in epidemiological studies.

3. MECHANISMS RELATING ADIPOSITY TO CANCER RISK

Body weight has strong effects on metabolic factors that may subsequently affect cancer risk—in particular, on circulating levels of peptide and steroid hormones and their binding factors. Specific effects vary somewhat by gender and by menopausal status in women and are summarized in Table 2.

A linear increase in circulating levels of insulin occurs with increasing BMI in both men and women. Insulin acts to control the uptake and use of glucose in peripheral tissues. With excessive calorie consumption and weight gain, tissues become insensitive (resistant) to insulin and the body compensates by producing more insulin, resulting in a chronic state of hyperinsulinemia. Obesity—particularly abdominal obesity—is a major determinant of insulin resistance and hyperinsulinemia.

Insulin-like growth factors (IGFs) are mitogens that regulate energy-dependent growth processes *(14)*. IGF-I stimulates cell proliferation and inhibits apoptosis and has been shown to have strong mitogenic affects in a wide variety of cancer cell lines. The synthesis of IGF-I and its main binding protein, IGFBP-3, are regulated primarily by pituitary growth hormone (GH). In the circulation, more than 90% of IGF-I is bound to IGFBP-3. Obesity and other conditions related to chronic hyperinsulinemia result in elevated blood glucose levels, decreased levels of IGF-binding proteins (IGFBP-1 and IGFBP-2) and higher levels of free plasma IGF-I, the small fraction of IGF-I unbound to any binding protein. Obesity does not increase absolute plasma IGF-I levels, and the mild decrease in IGF-I levels observed in obese and hyperinsulinemic individuals can be explained by the negative feedback of free IGF-I on GH secretion, which is also lower in obese individuals.

Insulin and free IGF-I interact with and regulate the synthesis and bioavailability of sex steroids that affect the development and progression of certain cancers *(15)*. Chronic hyperinsulinemia inhibits hepatic synthesis of sex hormone-binding globulin (SHBG), thus increasing bioavailable androgens and estrogens unbound to SHBG. The unbound fraction determines the actual biological activity of androgens and estrogens, hormones

Table 2
Associations of Obesity With Selected Proteins

Hormone or binding globulin	*Obesity vs normal weight*
Insulin	↑
Insulin-like growth factor-I (IGF-I)	Nonlinear, peak around BMI 24–27 kg/m^2
Free IGF-I	↑
IGF binding protein-1	↓
IGF binding protein-3	↑ or NE
Sex hormone-binding globulin	↓
Total testosterone	↓ (M), NE (F); ↑ (PCOS);
Free testosterone	NE or ↓(M), ↑ (F)
Total estradiol	↑ (M, postmenopausal F), NE (premenopausal F)
Free estradiol	↑ (M, postmenopausal F), NE (premenopausal F)
Progesterone (premenopausal F)	NE, or ↓ in women with suspectibility to develop ovarian hyperandrogenism

↑, increased levels; ↓, decreased levels; NE, no observed effect; M, males; F, females; PCOS, premenopausal women with polycystic ovary syndrome; BMI, body mass index.

Reproduced with permission from ref. *4*.

essential for the growth, differentiation and function of many tissues in both men and women. There is a strong inverse association between the amount and distribution of body fat and circulating levels of SHBG *(16,17)*.

In addition to the effects of insulin on the bioavailability of sex hormones, adipose tissue itself increases the concentration of circulating estrogens in men and postmenopausal women through the aromatization of androstenedione to estrone *(16)*. In postmenopausal women, ovarian production of estrogen falls to very low levels and the adipose is the primary source of circulating estrogen.

Weight loss has been shown to reduce insulin resistance and circulating levels of insulin, glucose, and estradiol, and to increase levels of SHBG *(18)*. In most studies, a 5 to 10% weight reduction is sufficient to improve these metabolic parameters. The influence of weight loss on levels of IGFs and associated binding proteins is less clear. Unfortunately, few studies of the metabolic changes associated with weight loss have measured long-term effects, and there is little empirical evidence regarding the clinical manifestations of weight loss sustained over long periods of time *(2,18,19)*.

4. EPIDEMIOLOGICAL STUDIES OF OBESITY AND CANCER

4.1. Historical Perspective of Epidemiological Studies of Weight

The relationships between excess body weight and mortality from all causes and from cardiovascular disease have been well-established in epidemiological studies *(20–24)*. Excess weight is also known to be associated with an increased risk of morbidity, including cardiovascular diseases, type 2 diabetes mellitus, hypertension, dyslipidemia, glucose intolerance, and osteoarthritis *(1,2)*.

The association of overweight and obesity with most of these noncancer outcomes is generally stronger than the association with all cancer or specific cancer sites. In populations experiencing temporal increases in the prevalence of obesity, increases in hypertension, hyperlipidemia, and diabetes emerge earlier than increases in cancer outcomes. Because the incidence and mortality of specific types of cancer are less common than these noncancer outcomes, the relation of obesity to particular cancer sites has been more difficult to study. Moreover, a biological mechanism that clearly links obesity to forms of cancer without an endocrine component has not been established.

For these reasons, understanding the associations between overweight, obesity, and a wide variety of cancers, as well as the biological mechanisms contributing to these associations, remains an evolving and currently very active area of research. Accumulating research on obesity and cancer suggests that this relationship is not confined to just a few forms of cancers.

4.2. Evaluation by the International Agency for Research on Cancer

An International Agency for Research on Cancer (IARC) Working Group on the Evaluation of Cancer-Preventive Strategies published a comprehensive evaluation of the available literature on weight and cancer that considered epidemiological, clinical, and experimental data *(18)*. Their 2002 report concluded that there is "sufficient evidence" in humans for a cancer-preventive effect of avoidance of weight gain for cancers of the endometrium, female breast (postmenopausal), colon, kidney (renal cell), and esophagus (adenocarcinoma) *(18)*. Regarding premenopausal breast cancer, the report concluded that available evidence on the avoidance of weight gain "suggests lack of a cancer-preventive effect." For all other sites, IARC characterized the evidence for a cancer-preventive effect of avoidance of weight as "inadequate" in humans.

The conclusions regarding the evidence in humans are based on epidemiological studies of overweight and/or obese individuals compared with leaner individuals, not on studies of individuals who have lost weight. Unfortunately, few individuals lose and maintain significant amounts of weight, making it extremely difficult to examine cancer outcomes in large populations of weight losers. Consequently, the IARC report concluded that there is "inadequate evidence" in humans for a cancer-preventive effect of intentional weight loss for any cancer site. Of note, though, three very recent studies of the impact of weight loss on breast cancer *(25,26)* and endometrial cancer *(27)* suggest that weight loss over the course of adult life may substantially reduce the risk for these cancers.

The IARC also concluded that in experimental animals, there is "sufficient or limited evidence" for a cancer-preventive effect of avoidance of weight gain by calorie restriction, based on studies of spontaneous and chemically induced cancers of the mammary gland, liver, pituitary gland (adenoma), and pancreas, for chemically induced cancers of the colon, skin (nonmelanoma), and prostate, and for spontaneous and genetically induced lymphoma. An association between overweight and obesity and cancer at many sites is consistent with animal studies showing that caloric restriction dramatically decreases spontaneous and carcinogen-induced tumor incidence, multiplicity, size, and growth *(28–32)*. Possible mechanisms for these observations include altered carcinogen metabolism, decreased oxidative DNA damage, greater DNA repair capacity *(18)*, and a reduction of IGF-I levels in calorie-restricted animals *(30)*.

5. ADIPOSITY AND INDIVIDUAL CANCER SITES

5.1. Endometrial Cancer

Endometrial cancer (cancer of the uterine lining) was the first cancer to be recognized as being obesity-related. There is convincing and consistent evidence from both case-control and cohort studies that overweight and obesity are associated strongly with endometrial cancer *(18,33)*. A linear increase in the risk of endometrial cancer with increasing weight or BMI has been observed in most studies *(3,18,34–36)*. The increase in risk generally ranges from 2- to 3.5-fold in overweight and/or obese women (Table 3), and might be somewhat higher in studies of mortality than incidence *(3,18,37)*.

The probable mechanism for the increase in risk of endometrial cancer associated with obesity in postmenopausal women is the obesity-related increase in circulating estrogens *(38)*. In premenopausal women, endometrial cancer risk is also increased among women with polycystic ovary syndrome, which is characterized by chronic hyperinsulinemia and progesterone deficiency *(39)*. Thus, in both pre- and postmenopausal women, endometrial cancer is increased by the mitogenic effects of estrogens on the endometrium when these effects are not counterbalanced by sufficient levels of progesterone. Many studies have shown large increases in endometrial cancer risk among postmenopausal women who take unopposed estrogen replacement therapy (i.e., estrogen in the absence of progesterone) *(40)*, as well as increases in risk among women with higher circulating levels of total and bioavailable estrogens *(18)*.

Studies on BMI and prognosis among women with endometrial cancer suggest that heavier women may have a better prognostic profile, as indicated by more favorable pathological features, and longer survival *(41–44)*. This finding lends additional support to the mechanistic hypothesis of unopposed estrogens, as it mirrors the better prognostic profile seen in women whose endometrial cancer was induced by estrogen replacement therapy *(45,46)*.

5.2. Female Breast Cancer

Many epidemiological studies since the 1970s have assessed the association between anthropometric measures and female breast cancer occurrence and/or prognosis *(18,47)*. Early studies established that the association between body size and risk of breast cancer varied based on menopausal status—that heavier women were at increased risk of postmenopausal, but not premenopausal, breast cancer *(18)*. In fact, among premenopausal women, there is consistent evidence of a modest reduction in risk among women with high (≥28) BMI. This reduction in risk could be due to the increased tendency for young obese women to have anovulatory menstrual cycles and lower levels of circulating steroid hormones, notably progesterone and estradiol *(17)*.

Obesity has been shown consistently to increase rates of breast cancer in postmenopausal women by 30 to 50% (Table 3) *(48–53)*. Some studies have found central adiposity to be an independent predictor of postmenopausal breast cancer risk beyond the risk attributed to overweight alone, but a recent systematic review has indicated that this is not the case *(54)*. In addition, adult weight gain has generally been associated with a larger increase in risk of postmenopausal breast cancer than has BMI, in studies that examined both *(55–59)*. Both BMI and weight gain are more strongly related to risk of breast cancer among postmenopausal women who have never used hormone replacement

Table 3
Relative Risks Associated With Overweight and Obesity, and Percentage of Cases Attributable to Overweight and Obesity in the United States

	Relative risk[a]		
Type of cancer	*BMI ≥25 – <30*	*BMI 30+*	*PAF% for US adults, 2000*[b]
Colorectal (men)	1.5	2.0	35.4
Colorectal (women)	1.2	1.5	20.8
Female breast (post)	1.3	1.5	22.6
Endometrium	2.0	3.5	56.8
Kidney (renal cell)	1.5	2.5	42.5
Esophagus (adeno)	2.0	3.0	52.4
Pancreas	1.3	1.7	26.9
Liver	?	1.5–4.0	–[c]
Gallbladder	1.5	2.0	35.5
Gastric cardia (adeno)	1.5	2.0	35.5

[a]Relative risk estimates are summarized from the literature cited.
[b]Data on prevalence of overweight and obesity are from the National Health and Nutrition Examination Survey (NHANES) (1999–2000) *(224)* for US men and women 50 to 69 yr of age.
[c]PAFs were not estimated because the magnitude of the relative risks across studies are not sufficiently consistent.
BMI, body mass index in kg/m^2; RR, relative risk; PAF, population attributable fraction.
Adapted with permission from ref. *4*.

therapy, compared with women who have used hormones *(52,56,58,60,61)*. This finding lends support to the hypothesis that adiposity increases breast cancer risk through its estrogenic effects. It is likely that the uniformly high levels of circulating estrogens among women who use exogenous hormones, regardless of weight, obscure much or all of the association between BMI and breast cancer.

High levels of circulating estrogens and low levels of SHBG have been shown to be associated with increased risk of breast cancer in postmenopausal women *(62)*. Another mechanism by which obesity may affect the risk of breast cancer involves insulin and/or IGFs. IGF-I is a potent mitogen for normal and transformed breast epithelial cells *(63)*, and is associated with mammary gland hyperplasia *(64)* and mammary cancer *(65)* in animals. In addition, IGF-I receptors are present in most human breast tumors and in normal breast tissue *(66,67)*. Two case–control studies *(67,68)* and two prospective cohort studies *(69,70)* found positive associations between serum or plasma IGF-I concentrations and breast cancer in premenopausal, but not postmenopausal, women. The magnitude of the association increased when both IGF-I and IGFBP-3 were considered *(68,70)*. That the association with IGF-I is stronger in studies of premenopausal than postmenopausal breast cancer has been interpreted as suggesting that IGF-I may increase risk only in the presence of high levels of endogenous estrogens *(18)*.

Studies of circulating insulin or C-peptide (a marker of insulin secretion) levels and breast cancer have had inconsistent results *(69,71,72)*. A recent prospective cohort study found that postmenopausal, but not premenopausal, women with type 2 diabetes were at greater risk for breast cancer than those without diabetes; the association was particularly evident among women with estrogen receptor-positive breast cancer *(73)*. Prediagnosic levels of plasma leptin were not found to be associated with postmenopausal breast cancer in a recent study *(74)*, whereas an inverse association has been seen with serum adiponectin and postmenopausal breast cancer *(75)*.

Studies of breast cancer mortality and survival among breast cancer cases illustrate that adiposity is associated with both increased likelihood of recurrence and reduced likelihood of survival among those with the disease, regardless of menopausal status and after adjustment for stage and treatment *(47,76–84)*. Very obese women (BMI $\geq$ 40.0) have breast cancer death rates that are three times higher than very lean (BMI < 20.5) women *(85)*.

The greater risk of death among heavier women likely reflects both a true biological effect of adiposity on survival and delayed diagnosis in heavier women. There are substantial data to suggest that adiposity is associated with a more aggressive tumor; obese women are more likely than lean women to have increased tumor size, lymph node involvement, and later stage disease at diagnosis *(78–80,82,84,86)*. Among a large prospective cohort of postmenopausal women, adult weight gain was associated more strongly with risk for breast cancer diagnosis at advanced stage than risk for localized disease *(86)*. Studies of prognosis among breast cancer patients show that weight gain, both before and after diagnosis, adversely affects outcome *(81,87)*.

In addition to the direct effects of adiposity, there is evidence that heavier women are less likely to receive mammography screening *(88–90)*, and among women who self-detect their tumors, high BMI increases the likelihood of nonlocalized disease *(91)*. Recent studies, however, suggest that mammography screening is not a confounder of the relationship between adiposity and poor prognosis *(81,83,92)*, nor is mammography any less sensitive among heavier women *(93)*. Thus, the impact of obesity on prognosis is unlikely to be solely, or even largely, an artifact of delayed diagnosis.

5.3. Colorectal Cancer

Obesity has also been consistently associated with higher risk of colorectal cancer in men (relative risks of approx 1.5 to 2.0) and women (relative risks of approx 1.2 to 1.5) in both case–control and cohort studies (Table 3) *(18,34,35,94,95)*. In studies that were able to examine the colon and rectum separately, relative risks have been generally higher for the colon *(18,36,96)*. Similar relationships are seen for colon adenomas, with stronger associations observed between obesity and advanced adenomas *(97–100)*.

A gender difference, in which obese men are more likely to develop colorectal cancer than obese women, has been observed consistently across studies and populations. The reasons for this gender difference are speculative. One hypothesis is that central adiposity, which occurs more frequently in men, is a stronger predictor of colon cancer risk than peripheral adiposity or general overweight. Support for the role of central obesity in colorectal cancer comes from studies reporting that waist circumference and WHR are related strongly to risk of colorectal cancer and large adenomas in men *(101)*. Two recent prospective cohort studies specifically examining the predictive value of anthropometric measurements for risk of colon cancer found waist circumference *(102)* or WHR *(103)* to be associated with colon cancer risk, independent of, and with greater magnitude than, BMI, and this result was seen in women as well as in men. However, the association between WHR and colorectal cancer in women was not stronger than the association between BMI and colorectal cancer in other studies that examined both measures *(104,105)*, making it less likely that body fat distribution completely explains the gender differences.

Another explanation is that there might be an offsetting beneficial effect of obesity on colorectal cancer risk in women based on evidence that exogenous estrogens (in the form of postmenopausal hormone therapy) reduce the risk of colorectal cancer in women *(106,107)*. However, this hypothesis also is quite speculative, as circulating levels of endogenous estrogens are higher in obese men as well as obese women, compared with lean subjects *(108)* and oral intake of exogenous estrogens could have different effects than endogenous estrogens on the risk of colon cancer.

Giovannucci was the first to propose the mechanistic hypothesis that high body mass, and central obesity in particular, increased colon cancer risk through their effect on insulin production *(101,109,110)*. Insulin and IGFs have been shown to promote the growth of colonic mucosal cells and colonic carcinoma cells in in vitro studies *(111,112)*. This hypothesis has received recent support from many epidemiological studies. Higher risk of colorectal cancer has been associated with elevated fasting plasma glucose and insulin levels following a standard dose of oral glucose challenge *(101,113)* and with elevated serum insulin or C-peptide levels *(114–116)*. Several prospective cohort *(114,117,118)* and case–control studies *(115,119,120)* have found increased risk of colorectal cancer and large adenomas with increasing absolute levels of IGF-I and decreasing levels of IGFBP-3. A recent meta-analysis of six case-control and nine cohort studies found diabetes mellitus, which is preceded by many years of hyperinsulinemia, to be associated with a 30% increase in risk of colorectal cancer *(121)*. Elevated levels of serum leptin recently have been found to be associated with increased risk of colon cancer, independent of circulating insulin levels *(122,123)*. Low levels of plasma adiponectin have also been found to be associated with increased risk of colorectal cancer *(124)* and colorectal adenoma *(125)*.

There are very few studies examining the impact of adiposity on prognosis among colorectal cancer patients. Available studies suggest that adiposity may have an adverse effect on prognosis *(126–128)*, although results are somewhat inconsistent by gender and for colon rectal cancer.

5.4. Kidney Cancer

The risk of kidney cancer (specifically, renal cell cancer) is 1.5- to 3-fold higher in overweight and obese persons than in normal weight men and women in study populations worldwide (Table 3); most studies have found a dose–response relationship with increasing weight or BMI *(18,35,96,129–133)*. In several studies, the increase in risk with increasing BMI was greater in women than in men *(3,134–141)*, although at present this finding remains unexplained and was not confirmed in a review of published studies *(129)*, nor in a recent prospective cohort study *(35)*. Importantly, the obesity-associated risk of renal cell cancer appears to be independent of blood pressure, indicating that hypertension and obesity might influence renal cell cancer through different mechanisms *(142)*. The hypothesis that chronic hyperinsulinemia contributes to the association of BMI and renal cell cancer is supported indirectly by the increased risk of kidney cancer seen in diabetics *(143)*.

5.5. Adenocarcinoma of Esophagus

The incidence of adenocarcinoma of the esophagus has been rapidly increasing in Westernized countries in recent decades *(144,145)*, whereas rates for the other main histological subtype, squamous cell carcinoma, have remained stable or have decreased. Thus, an increasing proportion of all esophageal cancers in Western countries are adenocarcinomas. Obesity is consistently associated with a two- to threefold increase in risk for adenocarcinoma of the esophagus (Table 3) *(18,146)*, with stronger associations seen in nonsmokers *(3,147)*. Obesity is not associated with an increased risk of squamous cell carcinoma of the esophagus.

Independent of obesity, gastroesophageal reflux disease (GERD) has been associated with esophageal adenocarcinoma and with its metaplastic precursor, Barrett's esophagus *(145,148,149)*. Obesity has been hypothesized to increase the risk of adenocarcinoma of the espophagus indirectly, by increasing the risk of GERD and Barrett's esophagus *(150,151)*. The association between obesity and esophageal adenocarcinoma has been shown in some studies to be independent of reflux *(147,152)*. Thus, obesity might increase the risk of esophageal adenocarcinoma through mechanisms other than, or in addition to, reflux.

5.6. Adenocarcinoma of Gastric Cardia

Risk for adenocarcinoma of the gastric cardia has been found to be obesity-related *(147,153,154)*, but the magnitude of the association is not as great as for adenocarcinoma of the esophagus. Relative risks are in the range of 1.5 to 2.0. It is unclear at present why risks associated with obesity are greater for esophageal adenocarcinoma than for gastric cardia adenocarcinoma. It is possible that reflux mechanisms are more closely related to adenocarinoma of the esophagus than to adenocarcinoma of the gastric cardia. Data are limited for noncardia cancers of the stomach, but there is no suggestion of increased risk with obesity *(147,153)*.

5.7. Gallbladder Cancer

There have been a limited number of studies of gallbladder cancer and obesity; most have been relatively small, as gallbladder cancer is quite rare, especially in men. However, these few studies have consistently found elevated risks of about twofold (Table 3) *(3,37,96,132,155–158)*. One study found a greater than fourfold increase in risk for the highest category of BMI (≥30) in a Japanese cohort, but only among women *(36)*. Obesity is thought to operate indirectly to increase the risk of gallbladder cancer by increasing the risk of gallstones, which in turn, causes chronic inflammation and increased risk of biliary tract cancer *(134)*.

5.8. Liver Cancer

Seven studies that have examined obesity and liver cancer or hepatocellular carcinoma (HCC) found excess relative risk in both men and women in the range of 1.5 to 4.0 *(3,34,95,96,132,155,158)*; however, two studies did not find any suggestion of an increased risk *(36,131)*. Taken together, these studies suggest that obesity increases the risk of liver cancer, but the magnitude of the observed relative risk from existing studies is not consistent.

Obesity, and especially visceral adiposity, is strongly associated with nonalcoholic fatty liver disease (NAFLD), a chronic liver disease that occurs in nondrinkers but that is histologically similar to alcohol-induced liver disease *(159)*. NAFLD is an emerging clinical problem among obese patients and is now recognized as the most common cause of abnormal liver tests *(160)*. Disorders of glucose regulation are significantly associated with NAFLD, indicating that insulin resistance is the link between NAFLD and metabolic diseases *(160)*. NAFLD is characterized by a spectrum of liver tissue changes ranging from accumulation of fat in the liver to nonalcoholic steatohepatitis (NASH), cirrhosis, and HCC at the most extreme end of the spectrum. Progression to NASH appears to represent the turning point from a seemingly nonprogressive condition to fibrosis, necrosis, and inflammation, and multiple cellular adaptations to the resulting oxidative stress *(159)*. Visceral adiposity likely contributes to the risk of HCC by promoting NAFLD and NASH.

5.9. Pancreatic Cancer

Several recent studies suggest that high body mass is associated with increased risk for pancreatic cancer in men and women, with relative risk estimates for obesity generally in the range of 1.5 to 2.0 (Table 3) *(3,34,131,155,158,161–166)*. However, other studies found smaller positive associations *(36,132,167)* or, in some cases, no association *(35,95,96,168,169)*. Further research is needed to refine the magnitude of the risk in both men and women and to explain the inconsistency in current estimates of risk. Many of these studies are based on small numbers of cases, and retrospective studies of adiposity are hampered by weight loss that accompanies pancreatic cancer and that often begins prior to diagnosis. In addition, smoking is an important potential confounder of the relationship between adiposity and pancreatic cancer, and the smoking habits of the various study populations and differential adequacy of control for smoking may partly explain differences across studies. It is thought that chronic hyperinsulinemia and glucose intolerance may contribute to an increased risk of pancreatic cancer, as suggested by the well-established positive association between diabetes and pancreatic cancer in prospective studies *(170,171)*. A recent study suggests that individuals with the highest vs lowest quartiles of fasting serum levels of glucose and insulin, and insulin resistance, have more than a twofold increased risk of pancreatic cancer *(172)*.

5.10. Prostate Cancer

There are many studies that do not support an association between body mass and incident prostate cancer *(18,34–36,173)*, although four large recent studies found a small but statistically significant increased risk of prostate cancer among obese men *(131,132,174)* or a significant trend toward increasing risk with increasing BMI *(96)*. In addition, there is accumulating evidence that obesity is associated with an increase in risk of advanced prostate cancer or death from prostate cancer *(175–181)*. Recent studies consistently indicate that obese men with prostate cancer are more likely to have aggressive disease that recurs after radical prostatectomy than nonobese men *(182–184)*.

As with breast cancer, "nonbiological" issues of screening, detection, and treatment are important to the evaluation of the impact of adiposity on prostate cancer prognosis. It can be harder to perform a digital rectal examination in obese men because of their general adiposity in combination with larger prostate size *(181)*. Additionally, despite larger prostate sizes, obese men may have lower serum levels of prostate-specific antigen (PSA) *(185–187)*, potentially biasing them toward later stage at diagnosis even in the presence of PSA screening. Surgery is more difficult to perform in obese men, with a greater risk of positive surgical margins *(181)*. Still, mortality from prostate cancer was found to be increased in obese men in a prospective cohort study conducted from 1950 to 1972 *(177)*, long before PSA screening, and at a time when surgery was rarely done, suggesting that the adverse impact of adiposity on prognosis is rooted in biological characteristics of the tumor.

Several studies suggest that insulin, which is strongly correlated with obesity, acts to promote prostate cancer development and growth *(188–191)*. Increased levels of circulating leptin, also correlated with adiposity, have been found in two studies to be associated with larger, more advanced, higher-grade tumors *(192,193)*, although a third study did not confirm this finding *(194)*. Decreased adiponectin levels have also been linked to high-grade, advanced prostate cancer *(195)*.

5.11. Cervical Cancer

Studies on BMI and cervical cancer are limited and inconclusive *(18)*. Two prospective studies of mortality from cervical cancer found that it was associated with high BMI (two- to threefold increased risk) *(3,37)*, whereas much lower relative risks were observed in two cohorts of hospitalized patients diagnosed with obesity, compared with general populations *(155,158)*, and no association was observed in three cohort studies *(34,36,196)*. A recent case-control study that controlled for human papillomavirus infection found about a twofold increased risk of cervical adenocarcinoma among overweight and obese women; smaller increases in risk were seen for cervical squamous cell carcinoma *(197)*. However, differential screening behavior (obese women may be less likely to be screened on a regular basis than women of normal weight) could also explain some of these observations.

5.12. Ovarian Cancer

Although endogenous hormones are believed to be involved in the etiology of ovarian cancer *(198)*, and obesity is a well-established risk factor for other hormone-related cancers in women (e.g., breast and endometrial cancers), ovarian cancer has not been linked consistently to obesity *(18,34,36,158,199–206)*. Some studies have reported an association between the two, and relative risks in these studies have been in the range of 1.5 to 2.0 for the highest categories of BMI studied *(35,37,131,207–212)*. However, several large studies *(155,196,202,213,214)* have not found an association between ovarian cancer and obesity, so no solid conclusions should be drawn at this time. It is not clear what factors might explain the divergent results among studies. Weight loss several years prior to the time of cancer diagnosis would bias the relative risk downward in case–control studies, but such a bias would not be operative in several prospective cohort studies that found no association. It is possible that obesity increases the risk of specific histological subtypes of ovarian cancer (e.g., endometroid) but not others. Most studies have not examined risk by histological subtype of ovarian cancer, and this may contribute to the inconsistent findings.

5.13. Hematopoietic Cancers

Several studies have examined the relationship between hematopoietic cancers and BMI, but results from most of these studies are based on relatively small numbers of events. Still, most of the available studies have observed modest obesity-associated

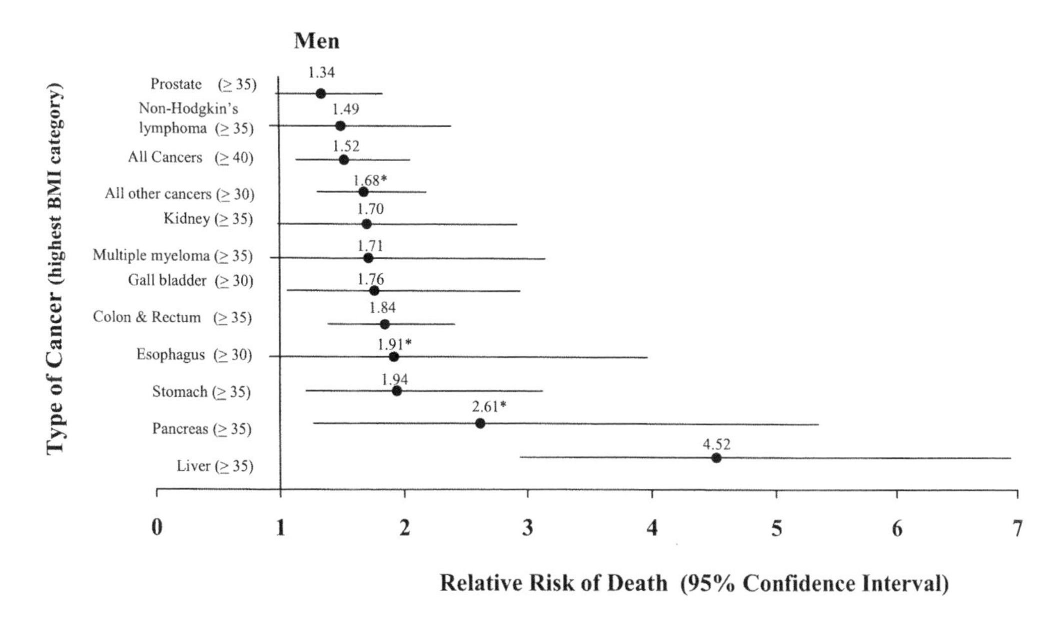

Fig. 1. Summary of mortality from cancer according to body mass index for US men in the Cancer Prevention Study II, 1982 through 1998. For each relative risk, the comparison was between men in the highest BMI category (indicated in parentheses) and men in the reference category (BMI 18.5 to 24.9). Asterisks indicate relative risks for men who never smoked. Results of the linear test for trend were significant ($p \leq 0.05$) for all cancer sites. Reproduced with permission from ref. *3*.

increases in the risk of non-Hodgkin's lymphoma *(3,34,35,37,95,96,131,155,158,215–217)*, multiple myeloma *(3,131,132,218–220)*, and leukemia *(3,37,96,131,132,155,158,221,222)*. Relative risks from these studies have been generally in the range of 1.2 to 2.0.

5.14. Lung Cancer

BMI has been reported to be inversely associated with lung cancer in several study populations that did not exclude smokers from the analysis *(18)*. This finding is explained by the confounding effects of smoking; smoking is the primary cause of lung cancer, and is inversely associated with BMI *(222)*. Studies that do not exclude smokers cannot separate the effects of BMI on the risk of death from the effects of smoking, namely, decreased BMI and increased risk of death. No association is seen between BMI and lung cancer in nonsmoking populations *(3,131)*.

6. OVERWEIGHT, OBESITY, AND CANCER MORTALITY

Recent results from a large American Cancer Society prospective mortality study illustrate that increased body weight is associated with increased death rates from all cancers combined and for cancers at multiple specific sites in both men and women (Figs. 1,2). These results were based on a population of more than 900,000 US adults who were followed from 1982 through 1998, and on more than 57,000 deaths from cancer that occurred during the 16-yr follow-up period *(3)*. The results are based on cancer mortality and thus may reflect the influence of BMI on either cancer incidence or survival, or both.

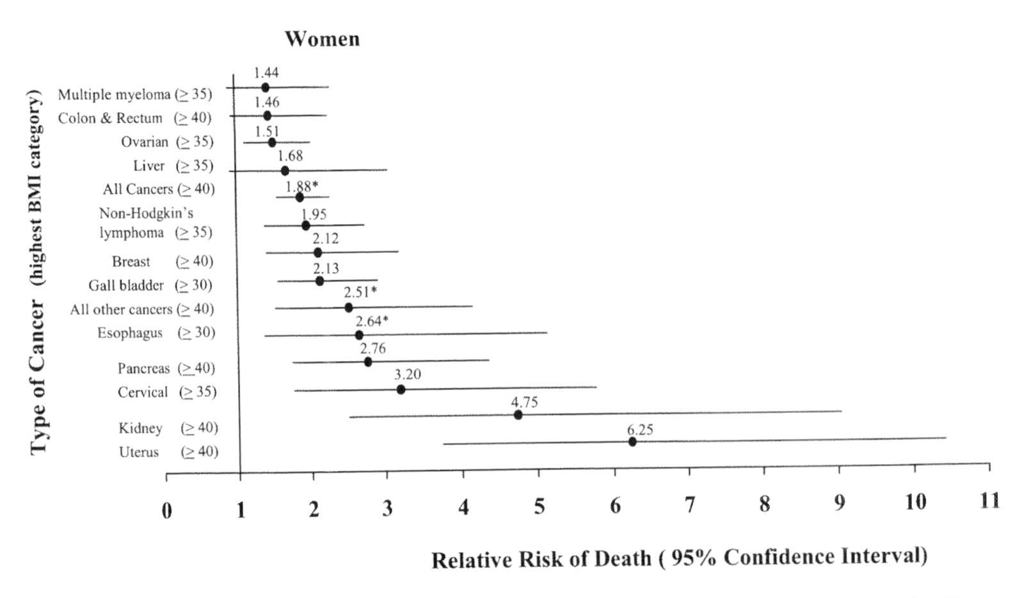

Fig. 2. Summary of mortality from cancer according to body mass index for US women in the Cancer Prevention Study II, 1982 through 1998. For each relative risk, the comparison was between women in the highest BMI category (indicated in parentheses) and women in the reference category (BMI 18.5 to 24.9). Asterisks indicate relative risks for women who never smoked. Results of the linear test for trend were significant ($p \leq 0.05$) for all cancer sites. Reproduced with permission from ref. *3*.

Survival may be influenced by adiposity-related differences in diagnosis or treatment of cancer, as well as by true biological effects of adiposity on cancer progression.

7. POPULATION ATTRIBUTABLE FRACTION

The proportion or percentage of disease in a population that can be attributed to a risk factor is termed the "population attributable fraction" (PAF) and is used as a measure of the public health impact of the risk factor. The PAF is sometimes referred to as the population attributable risk, population attributable risk percent, and excess fraction. The magnitude of the PAF depends on both the magnitude of the association between the risk factor and the disease (i.e., the size of the relative risk) and the prevalence of the risk factor in the population of interest. The PAF will increase as either one of its components increases. Because the PAF is very sensitive to the population prevalence of the risk factor, in this case overweight and obesity, it is not generalizable to populations with different distributions of the risk factor. The PAFs presented in Table 3 are estimates of the percentage of cancer cases at each indicated site that could be attributed to excess adiposity, defined as a BMI higher than 25.0. These estimates were based on summary relative risks estimated from the existing published literature for each cancer site and on the distribution of BMI in adults in the United States in 2000.

8. CONCLUSIONS

Further research defining the causal role of obesity and cancers of specific sites is needed, including mechanistic research, and studies that are able to separate the effects of obesity and several highly correlated factors such as physical activity and dietary composition. At present, the biological mechanisms clearly linking overweight and obesity to many forms of cancer, other than those with an endocrine component, are poorly understood. In addition to causing changes in hormone metabolism (insulin, IGF-I, sex steroids), proteins secreted by adipose tissue also contribute to the regulation of immune response (leptin), inflammatory response (TNF-α, IL-6, serum amyloid A), vasculature and stromal interactions and angiogenesis (vascular endothelial growth factor-1), and extracellular matrix components (type VI collagen) *(224)*. Obesity-associated dysregulation of multiple adipokines are likely to be of great significance for the occurrence, promotion, and metastatic potential of human cancers.

ACKNOWLEDGMENT

I would like to thank Janet Hildebrand for her assistance in carefully reviewing the literature.

REFERENCES

1. National Institutes of Health and National Heart Lung and Blood Institute. Obes Res 1998;6:51S–209S.
2. National Task Force on the Prevention and Treatment of Obesity. Arch Intern Med 2000;160:898–904.
3. Calle E, Rodriguez C, Walker-Thurmond K, et al. N Engl J Med 2003;348:1625–1638.
4. Calle EE, Kaaks R. Nat Rev Cancer 2004;4:579–591.
5. Willett W. In: Willett W, ed. *Nutritional Epidemiology*. Oxford University Press, New York:1998.
6. Kuczmarski R, Flegal KM. Am J Clin Nutr 2000;72:1074–1081.
7. Metropolitan Life Insurance Company. Stat Bull Metropol Life Insur Co 1959;40:1–4.
8. US Department of Agriculture and US Department of Health and Human Services. US Government Printing Office, Washington, DC: 2000.
9. WHO Expert Committee on Physical Status. World Health Organization, Geneva: 1995.
10. Stevens J, Keil JE, Waid R, et al. Am J Epidemiol 1990;132:1156–1163.
11. Willett W, Dietz W, Colditz G. N Eng J Med 1999;341:427–434.
12. Mokdad A, Serdula M, Dietz W, et al. JAMA 1999;282:1519–1522.
13. Borkan G, Hultz D, Gerzof S, et al. J Gerontol 1983;38: 673–677.
14. Le Roith D. Insulin-like growth factors. N Engl J Med 1997;336: 633–640.
15. Kaaks R. Cancer Causes Control 1996;7:605–625.
16. Madigan MP, Troisi R, Potischman N, et al. Cancer Causes Control 1998;9:199–207.
17. Potischman N, Swanson C, Siiteri P, et al. J Natl Cancer Inst 1996;88:756–758.
18. IARC. IARC Handbooks of Cancer Prevention. Weight Control and Physical Activity. International Agency for Research on Cancer, Lyon: 2002.
19. Mann J. Br J Nutr 2000;83 Suppl 1:S169–S172.
20. Manson JE, Willett WC, Stampfer MJ, et al. N Engl J Med 1995;333:677–685.
21. Willett,WC, Manson JE, Stampfer MJ, et al. JAMA 1995;273:461–465.
22. Stevens J, Plankey MW, Williamson DF, et al. Obesity. 1998;6:268–277.
23. Lindsted KD, Singh PD. Int J Obes 1998;22:544–548.
24. Calle E, Thun M, Petrelli J, et al. N Engl J Med 1999;341:1097–1105.
25. Harvie M, Howell A, Vierkant RA, et al. Cancer Epidemiol Biomarkers Prev 2005;14: 656–661.
26. Eng SM, Gammon MD, Terry MB, et al. Am J Epidemiol 2005;162:229–237.
27. Schouten LJ, Goldbohm RA, van den Brandt PA. J Natl Cancer Inst 2004;96:1635–1638.
28. Das M, Gabriely I, Barzilai N. Obesity Rev 2004;5:13–19.
29. Dirx M, Zeegers M, Dagnelie P, et al. Int J Cancer 2003;106:766–770.
30. Dunn SE, Kari FW, French J, et al. Cancer Res 1997;57:4667–4672.
31. Hursting S, Perkins S, Brown C, et al. Cancer Res 1997;57:2843–2846.
32. Kritchevsky D. Toxicol Sci 1999;52:13–16.

33. Kaaks R, Lukanova A, Kurzer M. Cancer Epidemiol Biomark Prev 2002;11:1531–1543.
34. Rapp K, Schroeder J, Klenk J, et al. Br J Cancer 2005;93:1062–1067.
35. Lukanova A, Bjor O, Kaaks R, et al. Int J Cancer 2006;118:458–466.
36. Kuriyama S, Tsubono Y, Hozawa A, et al. Int J Cancer 2005;113:148–157.
37. Lew EA, Garfinkel L. J Chron Dis 1979;32:563–576.
38. Lukanova A, Lundin E, Micheli A, et al. Int J Cancer 2004;108:425–432.
39. Kaaks R, Lukanova A, Kurzer MA. Cancer Epidemiol Biomark Prev 2002;11:1531–1543.
40. Key T, Pike M. Br J Cancer 1988;57:205–212.
41. Everett E, Tamimi H, Greer B, et al. Gynecol Oncol 2003;90:150–157.
42. Pavelka JC, Ben-Shachar I, Fowler JM, et al. Gynecol Oncol 2004;95:588–592.
43. Duska LR, Garrett A, Rueda BR, et al. Gynecol Oncol 2001;83:388–393.
44. Anderson B, Connor JP, Andrews JI, et al. Am J Obstet Gynecol 1996;174:1171–1178; discussion 1178–1179.
45. Nyholm H, Nielsen A, Norup P. Oncology 1993;49:229–235.
46. Schwartzbaum J, Hulka B, Fowler W, et al. Am J Epidemiol 1987;126:851–860.
47. Stephenson G, Rose D. Nutr Cancer 2003;45:1–16.
48. Ballard-Barbash R, Swanson C. Am J Clin Nutr 1996;63(Suppl):437S–441S.
49. Galanis D, Kolonel L, Lee J, et al. Cancer Causes Control 1998;9:217–224.
50. Trentham-Dietz A, Newcomb PA, Storer BE, et al. Am J Epidemiol 1997;145:1011–1019.
51. Hunter DJ, Willett WC. Epidemiol Rev 1993;15:110–132.
52. Lahmann PH, Hoffmann K, Allen N, et al. Int J Cancer 2004;111:762–771.
53. Tehard B, Clavel-Chapelon F. Int J Obes (Lond) 2006;30:156–163.
54. Harvie M, Hooper L, Howell A. Obes Rev 2003;4:157–173.
55. Folsom AR, Kaye SA, Prineas RJ, et al. Am J Epidemiol 1990;131:794–803.
56. Huang Z, Hankinson SE, Colditz GA, et al. JAMA 1997;278:1407–1411.
57. Barnes-Josiah D, Potter JD, Sellers TA, et al. Cancer Causes Control 1995;6:112–118.
58. Feigelson H, Jonas C, Teras L,et al. Cancer Epidemiol Biomark Prev 2004;13:220–224.
59. Sweeney C, Blair CK, Anderson KE, et al. Am J Epidemiol 2004;160:868–875.
60. Collaborative Group on Hormonal Factors in Breast Cancer. Lancet 1997;350:1047–1059.
61. Schairer C, Lubin J, Troisi R, et al. JAMA 2000;283:485–491.
62. The Endogenous Hormones and Breast Cancer Collaborative Group. J Natl Canc Inst 2002;94: 606–616.
63. Lee A, Lee D. In: Manni A, ed. *Endocrinology of Breast Cancer*. Humana Press, Totowa, NJ, 1999: pp. 187–203.
64. Ng ST, Zhou J, Adesanya OO, et al. Nat Med 1997;3:1141–1144.
65. Bates P, Fisher R, Ward A, et al. Br J Cancer 1995;72:1189–1193.
66. Papa V, Gliozzo B, Clark GM, et al. Cancer Res 1993;53:3736–3740.
67. Peyrat JP, Bonneterre J, Hecquet B, et al. Eur J Cancer 1993;29A:492–497.
68. Bruning PF, Van Doorn J, Bonfrer JMG, et al. Int J Cancer 1995;62:266–270.
69. Toniolo P, Bruning PF, Akhmedkhanov A, et al. Int J Cancer 2000;88:828–832.
70. Hankinson SE, Willett WC, Colditz GA, et al. Lancet 1998;351:1393–1396.
71. Del Giudice M, Fantus I, Ezzat S, et al. Breast Cancer Res Treat 1998;47:111–120.
72. Bruning PF, Bonfrer JMG, van Noord PAH, et al. Int J Cancer 1992;52:511–516.
73. Michels KB, Solomon CG, Hu FB, et al. Diabetes Care 2003;26:1752–1758.
74. Stattin P, Soderberg S, Biessy C, et al. Br Cancer Res Treat 2004;86:191–196.
75. Mantzoros C, Petridou E, Dessypris N, et al. J Clin Endocrinol Metab 2004;89:1102–1107.
76. Rock C, Demark-Wahnefried W. J Clin Oncol 2002;20:3302–3316.
77. Chlebowski R, Aiello E, McTiernan A. J Clin Oncol 2002;20:1128–1143.
78. Dignam JJ, Wieand K, Johnson KA, et al. J Natl Cancer Inst 2003;95:1467–1476.
79. Berclaz G, Li S, Price KN, et al. Ann Oncol 2004;15:875–884.
80. Maehle BO, Tretli S, Thorsen T. Apmis 2004;112:349–357.
81. Baumgartner KB, Hunt WC, Baumgartner RN, et al. Am J Epidemiol 2004;160:1087–1097.
82. Carmichael AR, Bendall S, Lockerbie L, et al. Breast 2004;13:93–96.
83. Whiteman MK, Hillis SD, Curtis KM, et al. Cancer Epidemiol Biomarkers Prev 2005;14:2009–2014.
84. Loi S, Milne RL, Friedlander ML, et al. Cancer Epidemiol Biomarkers Prev 2005;14:1686–1691.

85. Petrelli JM, Calle EE, Rodriguez C, et al. Cancer Causes Control 2002;13:325–32.
86. Feigelson H, Patel AV, Teras LR, Gansler T, Thun MJ, Calle EE. Cancer 2006;107:12–21.
87. Kroenke CH, Chen WY, Rosner B, et al. J Clin Oncol 2005;23:1370–1378.
88. Wee C, McCarthy E, Davis R, et al. Ann Intern Med 2000;132:697–704.
89. Ostbye T, Taylor DH Jr, Yancy WS Jr, et al. Am J Public Health 2005;95:1623–1630.
90. Fontaine KR, Heo M, Allison DB. J Womens Health Gend Based Med 2001;10:463–470.
91. Reeves M, Newcomb P, Remington P, et al. Cancer 1996;77:301–307.
92. Daling JR, Malone KE, Doody DR, et al. Cancer 2001;92:720–729.
93. Elmore JG, Carney PA, Abraham LA, et al. Arch Intern Med 2004;164:1140–1147.
94. Lin J, Zhang SM, Cook NR, et al. Cancer Causes Control 2004;15:581–589.
95. Batty GD, Shipley MJ, Jarrett RJ, et al. Int J Obes (Lond) 2005;29:1267–1274.
96. Oh SW, Yoon YS, Shin SA. J Clin Oncol 2005;23:4742–4754.
97. Giovannucci E, Colditz GA, Stampfer MJ, et al. Cancer Causes Control 1996;7:253–263.
98. Almendingen K, Hofstad B, Vatn M. Am J Gastroenterol 2001;96:2238–2246.
99. Bird C, Frankl H, Lee E, et al. Am J Epidemiol 1998;147:670–680.
100. Boutron-Ruault M, Senesse P, Meance S, et al. Nutr Cancer 2001;39:50–57.
101. Giovannucci E, Ascherio A, Rimm E, et al. Ann Intern Med 1995;122:327–334.
102. Moore LL, Bradlee ML, Singer MR, et al. Int J Obes Relat Metab Disord 2004;28:559–567.
103. Macinnis RJ, English DR, Hopper JL, Giles GG. Int J Cancer 2006;118:2628–2631.
104. Bostick R, Potter J, Kushi L, et al. Cancer Causes Control 1994;5:38–52.
105. Martinez M, Giovannucci E, Spiegelman D, et al. J Natl Cancer Inst 1997;89:948–955.
106. Calle EE, Miracle-McMahill HL, Thun MJ, et al. J Natl Cancer Inst 1995;87:517–523.
107. Writing Group for the Women's Health Initiative Investigators. JAMA 2002;288:321–333.
108. Tchernof A, Despres JP. Horm Metab Res 2000;32:526–536.
109. Sandhu M, Dunger D, Giovannucci E. J Natl Canc Inst 2002;94:972–980.
110. Komninou D, Ayonote A, Richie JJ, et al. Exp Biol Med 2003;228:396–405.
111. Macaulay VM. Br J Cancer 1992;65:311–320.
112. LeRoith D, Baserga R, Helman L, et al. Ann Intern Med 1995;122:54–59.
113. Schoen R, Tangen C, Kuller L, et al. J Natl Canc Inst 1999;91:1147–1154.
114. Kaaks R, Toniolo P, Akhmedkhanov A, et al. J Natl Canc Inst 2000;92:1592–1600.
115. Wei EK, Ma J, Pollak MN, et al. Cancer Epidemiol Biomarkers Prev 2005;14:850–855.
116. Schoen R, Tangen C, Kuller L, et al. J Natl Cancer Inst 1999;91:1147–1154.
117. Ma J, Pollak M, Giovannucci E, et al. J Natl Cancer Inst 1999;91:620–625.
118. Giovannucci E, Pollak MN, Platz E, et al. Cancer Epidemiol Biomark Prev 2000;9:345–349.
119. Manousos O, Souglakos J, Bosetti C, et al. Int J Cancer 1999;83:15–17.
120. Renehan A, Jones J, Potten C, et al. Br J Cancer 2000;83:1344–1350.
121. Larsson SC, Orsini N, Wolk A. J Natl Cancer Inst 2005;97:1679–1687.
122. Stattin P, Lukanova A, Biessy C, et al. Int J Cancer 2004;109:149–152.
123. Stattin P, Palmqvist R, Soderberg S, et al. Oncol Rep 2003;6:2015–2021.
124. Wei E, Giovannucci E, Fuchs CS, et al. J Natl Cancer Inst 2005;97:1688–1694.
125. Otake S, Takeda H, Suzuki Y, et al. Clin Cancer Res 2005;11:3642–3646.
126. Meyerhardt JA, Catalano PJ, Haller DG, et al. Cancer 2003;98:484–495.
127. Meyerhardt JA, Tepper JE, Niedzwiecki D, et al. J Clin Oncol 2004;22:648–657.
128. Haydon AM, Macinnis RJ, English DR, et al. Gut 2006;55:62–67.
129. Bergstrom A, Hsieh C-C, Lindblad P, et al. Br J Cancer 2001;85:984–990.
130. Hu J, Mao Y, White K. The Canadian Cancer Registries Epidemiology Research Group. Soz-Praventivmed 2003;48:178–185.
131. Pan S, Johnson K, Ugnat A-M, et al. The Canadian Cancer Registries Epidemiology Research Group. Am J Epidemiol 2004;159:259–268.
132. Samanic C, Gridley G, Chow W-H, et al. Cancer Causes Control 2004;15:35–43.
133. van Dijk BA, Schouten LJ, Kiemeney LA, et al. Am J Epidemiol 2004;160:1159–67.
134. World Cancer Research Fund and American Institute for Cancer Research. In: *Food, Nutrition and the Prevention of Cancer: A Global Perspective*. American Institute for Cancer Research, Washington, DC: 1997, pp. 371–373.

135. Hill H, Austin H. Cancer Causes Control 1996;7:19–32.
136. Wolk A, Lindblad P, Adami H-O. Cancer Causes Control 1996;7:5–18.
137. Chow W-H, McLaughlin J, Mandel J, et al. Cancer Epidemiol Biomarkers Prev 1996;5:17–21.
138. Mellemgaard A, Engholm G, McLaughlin J, et al. Int J Cancer 1994;56:66–71.
139. McLaughlin J, Mandel J, Blot W, et al. J Natl Canc Inst 1984;72:275–284.
140. McLaughlin J, Gao Y, Gao R, et al. Int J Cancer 1992;52:562–565.
141. Pischon T, Lahmann PH, Boeing H, et al. Int J Cancer 2006;118:728–738.
142. Chow W, Gridley G, Fraumeni J, et al. N Engl J Med 2000;343:1305–1311.
143. Lindblad P, Chow W, Chan JM, et al. Diabetologia 1999;42:107–112.
144. Devesa S, Blot W, Fraumeni J. Cancer 1998;83:2049–2053.
145. Wild C, Hardie L. Nat Rev Cancer 2003;3:676–685.
146. Wu A, Bernstein L. Cancer Causes Control 2001;12:721–732.
147. Chow W-H, Blot W, Vaughan T, et al. J Natl Cancer Inst 1998;90:150–155.
148. Lagergren J, Bergstrom R, Lindgren A, et al. N Engl J Med 1999;340:825–831.
149. Chow W, Finkle W, McLaughlin J, et al. JAMA 1995;274:474–477.
150. Nilsson M, Johnsen R, Ye W, et al. JAMA 2003;290:66–72.
151. Lagergren J, Bergstrom R, Nyren O. Gut 2000;47:26–29.
152. Lagergren J, Bergstrom R, Adami HO, Nyren O. Ann Intern Med 2000;133:165–175.
153. Ji B-T, Chow W-H, Yang G, et al. Cancer Epidemiol Biomarkers Prev 1997;6:481–485.
154. Vaughan T, Davis S, Kristal A, et al. Cancer Epidemiol Biomarkers Prev 1995;4:85–92.
155. Moller H, Mellemgaard A, Lindvig K, et al. Eur J Cancer 1994;30A:344–350.
156. Strom B, Soloway R, Rios-Dalenz J, et al. Cancer 1995;76:1747–1756.
157. Zatonski W, Lowenfels A, Boyle P, et al. J Natl Cancer Inst 1997;89:1132–1138.
158. Wolk A, Gridley G, Svensson M, et al. Cancer Causes Control 2001;12:13-21.
159. Harrison S, Diehl A. Semin Gastroint Dis 2002;13:3–16.
160. Festi D, Colecchia A, Sacco T, et al. Obes Rev 2004;5:27–42.
161. Fryzek JP, Schenk M, Kinnard M, et al. Am J Epidemiol 2005;162:222–228.
162. Gapstur S, Gann P, Lowe W, et al. JAMA 2000;283:2552–2558.
163. Michaud D, Giovannucci E, Willett W, et al. JAMA 2001;286:921–929.
164. Silverman D, Swanson C, Gridley G, et al. J Natl Cancer Inst 1998;90:1710–1719.
165. Hanley A, Johnson K, Villeneuve P, et al. Int J Cancer 2001;94:140–147.
166. Patel A, Rodriguez C, Bernstein L, Chao A, Thun MJ, Calle EE. Cancer Epidemiol Biomarkers Prev 2005;14:459–466.
167. Sinner PJ, Schmitz KH, Anderson KE, et al. Cancer Epidemiol Biomarkers Prev 2005;14:1571–1573.
168. Lee I-M, Sesso H, Oguma Y, et al. Br J Cancer 2003;88:679–683.
169. Stolzenberg-Solomon RZ, Pietinen P, Taylor PR, et al. Cancer Causes Control 2002;13:417–426.
170. Calle E, Murphy T, Rodriguez C, et al. Cancer Causes Control 1998;9:403–410.
171. Everhart J, Wright D. JAMA 1995;273:1605–1609.
172. Stolzenberg-Solomon R, Graubard B, Chari S, et al. JAMA 2005;294:2872–2878.
173. Calle E. Am J Epidemiol 2000;151:550–553.
174. Engeland A, Tretli S, Bjorge T. Br J Cancer 2003;89:1237–1242.
175. Giovannucci E, Rimm E, Stampfer M, et al. Cancer Epidemiol Biomarkers Prev 1997;6:557–663.
176. Andersson S, Wolk A, Bergstrom R, et al. J Natl Cancer Inst 1997;89:385–389.
177. Rodriguez C, Patel A, Calle E, et al. Cancer Epidemiol Biomarkers Prev 2001;10:345–353.
178. Amling CL, Kane CJ, Riffenburgh RH, et al. Urology 2001;58:723–728.
179. Rohrmann S, Roberts WW, Walsh PC, et al. Prostate 2003;55:140–146.
180. Neugut AI, Chen AC, Petrylak DP. J Clin Oncol 2004;22:395–398.
181. Freedland SJ. Clin Cancer Res 2005;11: 6763–6766.
182. Amling C, Riffenburgh R, Sun L, et al. J Clin Oncol 2004;22:439–445.
183. Freedland S, Aronson W, Kane C, et al. J Clin Oncol 2004;22:446–453.
184. Strom SS, Wang X, Pettaway CA, et al. Clin Cancer Res 2005;11:6889–6894.
185. Baillargeon J, Pollock BH, Kristal AR, et al. Cancer 2005;103:1092–1095.
186. Barqawi AB, Golden BK, O'Donnell C, et al. Urology 2005;65:708–712.
187. Kristal AR, Chi C, Tangen CM, Goodman PJ, Etzioni R, Thompson IM. Cancer 2006;106:320–328.

188. Hammarsten J, Hogstedt B. Eur J Cancer 2005;41:2887–2895.
189. Hsing AW, Gao YT, Chua S Jr, et al. J Natl Cancer Inst 2003;95:67–71.
190. Hsing AW, Chua S Jr, Gao YT, et al. J Natl Cancer Inst 2001;93:783–789.
191. Tulinius H, Sigfusson N, Sigvaldason H, et al. Cancer Epidemiol Biomarkers Prev 1997;6:863–873.
192. Ho E, Boileau TW, Bray TM. Arch Biochem Biophys 2004;428:109–117.
193. Chang S, Hursting SD, Contois JH, et al. Prostate 2001;46:62–67.
194. Freedland SJ, Sokoll LJ, Mangold LA, et al. J Urol 2005;173:773–776.
195. Goktas S, Yilmaz MI, Caglar K, et al. Urology 2005;65:1168–1172.
196. Tornberg S, Carstensen J. Br J Cancer 1994;69:358–361.
197. Lacey J, Swanson C, Brinton L, et al. Cancer 2003;98:814–821.
198. Risch H. J Natl Canc Inst 1998;90:1774–1786.
199. Rodriguez C, Calle EE, Fakhrabadi-Shokoohi D, et al. Cancer Epidemiol Biomarkers Prev 2002; 11:822–828.
200. Fairfield K, Willett W, Rosner B, et al. Obstet Gynecol 2002;100:288–296.
201. Kuper H, Cramer D, Titus-Ernstoff L. Cancer Causes Control 2002;13:455–463.
202. Lukanova A, Toniolo P, Lundin E, et al. Int J Cancer 2002;99:603–608.
203. Byers T, Marshall J, Graham S, et al. J Natl Cancer Inst 1983;71:681–686.
204. Hartge P, Schiffman MH, Hoover R, et al. Am J Obstet Gynecol 1989;161:10–16.
205. Parazzini F, Moroni S, La Vecchia C, et al. Eur J Cancer 1997;33:1634–1637.
206. Polychronopoulou A, Tzonou A, Hsieh CC, et al. Int J Cancer 1993;55:402–407.
207. Anderson JP, Ross JA, Folsom AR. Cancer 2004;100:1515–1521.
208. Hoyo C, Berchuck A, Halabi S, et al. Cancer Causes Control 2005;16:955–963.
209. Zhang M, Xie X, Holman CD. Gynecol Oncol 2005;98:228–234.
210. Purdie D, Bain C, Whiteman P, et al. Cancer Causes Control 2001;12:855–863.
211. Schouten L, Goldbohm A, van den Brandt P. Am J Epidemiol 2003;157:424–433.
212. Lubin F, Chetrit A, Freedman L, et al. Am J Epidemiol 2003;157:113–120.
213. Engeland A, Tretli S, Bjorge T. J Natl Cancer Inst 2003;95:1244–1248.
214. Mink P, Folsom A, Sellers T, et al. Epidemiol 1995;7:38–45.
215. Willett EV, Skibola CF, Adamson P, et al. Br J Cancer 2005;93:811–816.
216. Cerhan JR, Bernstein L, Severson RK, et al. Cancer Causes Control 2005;16:1203–1214.
217. Holly E, Lele C, Bracci P, et al. Am J Epidemiol 1999;150:375–389.
218. Blair CK, Cerhan JR, Folsom AR, et al. Epidemiology 2005;16:691–694.
219. Friedman GD, Herrinton LJ. Cancer Causes Control 1994;5:479–483.
220. Brown LM, Gridley G, Pottern LM, et al. Cancer Causes Control 2001;12:117–125.
221. MacInnis RJ, English DR, Hopper JL, et al. J Natl Cancer Inst 2005;97:1154–1157.
222. Ross JA, Parker E, Blair CK, et al. Cancer Epidemiol Biomarkers Prev 2004;13:1810–1813.
223. Henley SJ, Flanders WD, Manatunga A, et al. Epidemiol 2002;13:268–276.
224. Rajala MW, Scherer PE. Endocrinology 2003;144:3765–3773.

24 Obesity and the Heart

Alison M. Morris, Paul Poirier,
and Robert H. Eckel

Abstract

Obesity is a major contributor to the prevalence of cardiovascular disease (CVD) in the developed world. As adipose tissue stores increase, the structure and function of the cardiovascular system changes to enable circulation requirements to be met. Adipose tissue is not simply a passive organ to store excess energy in the form of fat, but an endocrine organ that is capable of synthesizing and releasing into the bloodstream a variety of molecules that may have unfavourable impact on CVD risk. Obesity may increase risk factors for coronary artery disease and congestive heart failure through parameters such as dyslipidemia, high blood pressure, glucose intolerance, inflammatory markers, and the prothrombotic state. By favorably modifying lipids, decreasing blood pressure, and decreasing levels of glycemia, and proinflammatory cytokines, a reduction in adipose tissue may prevent the progression of atherosclerosis or the occurrence of acute coronary syndrome events in the obese high-risk population. This chapter will discuss the relationship between obesity and CVD and will explore the differences amongst locations of adipose stores and the extent of adiposity.

Key Words: Cardiovascular disease; metabolic dyndrome; dyslipedmia; mortality; obstructive sleep apnea syndrome; hypertension.

1. INTRODUCTION

Obesity is a multifactorial condition that develops via interaction between genetics and environment *(1)*. In 1998, the American Heart Association added obesity to its list of *major* modifiable risk factors for coronary heart disease (CHD) *(2)*, with both a high body mass index (BMI) and a high waist-to-hip ratio (WHR) being independent risk factors for CHD and mortality irrespective of the presence of other coronary risk factors *(3)*. With excess body weight comes the risk of long-term health consequences. Adipose tissue increases when energy intake is greater than energy expenditure. Weight reduction is difficult to achieve, and even harder to maintain, in part due to homeostatic mechanisms protecting against loss of nutrient stores *(4,5)*. Following weight loss, resumption of the obese state typically occurs, with fewer than 5% of subjects remaining lean for more than 4 yr *(6)*.

2. OBESITY AND MORTALITY

Currently, all overweight and obese adults (aged > 18 yr with a BMI ≥ 25 kg/m^2) are considered at risk for developing one or more of the many comorbidities associated with obesity, such as hypertension, dyslipidemia, type 2 diabetes mellitus, CHD, gallbladder

From: *Nutrition and Health: Adipose Tissue and Adipokines in Health and Disease*
Edited by: G. Fantuzzi and T. Mazzone © Humana Press Inc., Totowa, NJ

disease, ischemic stroke, osteoporosis, obstructive sleep apnea, and some types of cancers *(7)*. Although the focus of this chapter is on the relationship between overweight/obesity and cardiovascular disease (CVD), the list of associated conditions also highlights the impact of excess body weight on overall morbidity and mortality *(8)*.

As the incidence of obesity has expanded dramatically in the past two decades, so has the number of deaths associated with the condition. Data from five large prospective cohort studies have been used to calculate the annual number of deaths attributable to obesity in adults in 1991 *(9)*. The authors found that 80% of deaths attributable to obesity occurred in people with a BMI higher than 30. However, a number of limitations are encountered in this meta-analysis: data were collected from a predominantly Caucasian population; half the studies included self-reported body weight data; the statistical analysis model considered only age, smoking, and sex; and the outcomes were not assessed after 1991. Flegal et al. *(10)* reported that the relative risk of mortality associated with obesity with a BMI higher than 35 was twice that of a healthy-weight individual, with influence diminishing in the elderly (age > 70 yr). Of note, mortality related to obesity has declined since the first National Health and Nutrition Examination Survey (NHANES) study in 1971 *(10)*, a result most likely due to a more aggressive approach to the treatment of associated CVD risk factors in obese patients *(11)*. In the most recent analysis by Yan et al., baseline measurements of BMI were recorded before the age of 65 yr in 17,643 subjects from the Chicago Heart Study, free of CHD and diabetes *(12)*. The outcome of death from CHD, CVD, or diabetes after the age of 65 yr was adjusted for hypertension, hypercholesterolemia, and cigarette smoking. At a low level of risk, obesity had very little impact on CVD outcomes but did have an impact on the relative risk of diabetes; however, for obese subjects at moderate risk, there was a twofold increased risk of CHD mortality and a fivefold increase risk for diabetes. Overall, these data suggest that obese subjects during midlife have a greater likelihood of mortality after the age of 65 yr from CHD independent of the presence of hypertension, hypercholesterolemia, and cigarette smoking.

An important parameter often overlooked in outcome studies in overweight and obese subjects is the level of physical activity. In a recent report from the Harvard School of Public Health, the level of physical activity was an important predictor of CHD, an effect independent of the presence of obesity *(13)*. Previous findings suggest that even light-to-moderate activity is associated with lower CHD rates in women, with as little as 1 h of walking per week lowering the risk of CHD. It was encouraging to note from this study that the intensity of exercise did not dictate the benefit gained *(14)*. In women with established CHD, higher self-reported physical fitness scores were independently associated with fewer CAD risk factors, less angiographic CAD, and lower risk for adverse CV events *(15)*. These studies promote an increase in physical activity for cardiovascular health, not just to assist with weight loss.

3. STRUCTURAL AND METABOLIC CHANGES IN OBESITY AND IMPACT ON CVD RISK

Excess adipose tissue results in a number of structural and functional adaptations by the cardiovascular system. With progressive and central accumulations of body fat, many cardiac complications often follow *(16)*. The mechanisms involved are discussed in more detail in the next section.

3.1. Impact of Obesity on Metabolism

For many years, adipose tissue was considered a passive storage organ, but it is now clear that adipose tissue plays an active role in controlling energy balance. The metabolic alterations of adipose tissue that occur in obesity are numerous. These include an increased release of secretory molecules such as fatty acids, hormones, and proinflammatory cytokines. Relative decreases in insulin receptor number and function lead to reduced insulin sensitivity, a condition that may contribute to altered fuel partitioning. It has been hypothesized that the insulin resistance associated with obesity may in fact be a protective/adaptive mechanism against further weight gain *(17)*.

Adipose tissue produces metabolically active proinflammatory molecules called adipocytokines such as tumour necrosis factor (TNF)-α, interleukins (IL), leptin, adiponectin *(18–20)*, and some newly identified molecules such as visfatin *(21)* and omentin *(22)*. The cytokines within adipose tissue originate predominantly from *in situ* macrophages *(23–25)* but also from adipocytes and preadipocytes *(26)*. The altered production of these molecules has characterized obesity as a state of chronic, low-grade inflammation *(27)*, which may contribute to the development of insulin resistance and endothelial dysfunction *(28–30)*. It is postulated that paracrine and endocrine communication between macrophages and adipocytes, mediated by cytokines and fatty acids, creates a positive feedback loop to aggravate inflammatory changes in adipose tissue *(31)*. Although the adipocytokines are reviewed in detail in previous chapters in this book, it is important to briefly highlight these compounds here, as they appear to play a role in the pathophysiology of CVD *(32)*.

The regulation of adipocytokines in adipose tissue is complex and many of these molecules work together in autocrine regulation. For example, the increase of transforming growth factor (TGF)-β1 production in adipose tissue can induce plasminogen activator inhibitor (PAI)-1 and TGF-β1 can also inhibit leptin production *(33)*. An increase in TNF-α can trigger the previously mentioned cascade of events and can also inhibit adiponectin expression *(34)*, which together reduce insulin sensitivity and unfavourably modify vascular function and increase lipolysis *(35)*. Persistent inflammation also seems to play a role in congestive heart failure (CHF) by reducing cardiac contractility, inducing cardiac hypertrophy, and promoting apoptosis, a process that contributes to undesirable myocardial remodeling *(36)*.

3.2. Impact of Obesity on Cardiovascular Structure and Function

Obesity is associated with abnormalities in cardiac structure and function *(37–39)*, which can often be alleviated by weight loss (*see* Table 1). As there is an increased energy requirement to move excess body weight at any given level of activity, the cardiac workload is greater for obese subjects than for nonobese individuals *(39)*. Thus, obese subjects are known to have higher cardiac output (CO) and a lower total peripheral resistance in the absence of hypertension *(23)*. The high CO is attributable to increased stroke volume, whereas heart rate (HR) is usually unchanged *(24)*. The increase in blood volume and CO in obesity is in proportion to the amount of excess body weight *(40)*. Recent evidence from the HyperGen study shows that both increased total fat mass and fat-free mass are able to cause these physiological changes although centrally located adipose tissue is particularly strongly associated with increased CO *(41)*. In moderate to severe cases of obesity, an increased CO may lead to left ventricular

Table 1
Prevalence of Cardiovascular Complications Associated With Obesity and Changes With Weight Loss

	Obesity	*Weight loss with increased physical activity*
Blood volume	↑	↓
Cardiac output	↑	↓
Endothelial function	↓	↑
Stroke volume	↑	↓
Resting heart rate	Unchanged	↓
Blood pressure	↑	↓
Left ventricular mass	↑	↓

Table 2
Extent of Weight Loss Needed to See a Benefit in CVD Risk Factors

	Extent of weight loss (%)[a]
Hypertension	5
Left ventricular function	5
OSA	5
Dyslipidemia	10
Hyperglycemia	10

[a]These effects are not certain and may depend on the length and severity of disease.
OSA, obstructive sleep apnea.

dilation, increased left ventricular wall stress, compensatory (eccentric) left ventricular hypertrophy *(42–44)*, and left ventricular diastolic dysfunction *(45)*. It is important to emphasize that left ventricular hypertrophy is an important risk factor for CHF.

These complications from obesity occur irrespective of age. It has been reported that in children as young as 12 yr, obesity impairs the ability to exercise, elevates blood pressure (BP), and increases left ventricular mass (LVM), indicating the development of early cardiovascular adaptation/damage in young individuals *(46)*. In fact, the P-DAY Study in young men aged 15 to 34 yr demonstrated an accelerated progression of atherosclerosis at autopsy in obese individuals *(47)*. Higher LVM and left ventricular dysfunction have been documented with longer durations of obesity *(45)*. As previously mentioned, weight loss is able to diminish some of these anatomical and pathophysiological adaptations, including increased LVM *(48)* and abnormal ventricular filling *(49,50)* (Table 2).

The overproduction of adipocytokines in obesity also contributes to physiological changes to cardiac function. Changes in adipose tissue production of TGF-β1 may be a potential pathophysiological mechanism for the development of left ventricular filling abnormalities in obesity-associated hypertension *(51)*. A relative deficiency of adiponectin may promote inflammation and vascular dysfunction by a reduced ability to inhibit local proinflammatory signals and prevent plaque formation *(52)*. Proatherogenic chemokines, such as monocyte chemoattractant protein (MCP)-1, are also elevated in obesity. Such molecules may modulate the migration of granulocytes and monocytes into the arterial wall *(53)*. Increased MCP-1 is associated with a number of alterations in the cardiac system, including increased LVM and altered diastolic filling *(54)*.

3.3. Degree of Adiposity

Data from 115,886 women in the Nurses' Health Study showed that even mild-to-moderate overweight (BMI = 25.0–28.9) increased the risk of nonfatal CHD in middle-aged women after adjustment for age and smoking (RR: 1.8; 95% CI: 1.2–2.5). Among those with a BMI ≥ 29, the risk increased more than threefold (RR: 3.3; 95% CI: 2.3–4.5). The effect was substantially reduced after adjusting for other CVD risk factors but remained significant among those with a BMI ≥ 29 (RR: 1.9; 95% CI: 1.3–2.6) *(55)*. Willett et al. *(56)* concluded that higher levels of body weight within the "normal" range, as well as modest weight gain (more than 5 kg) after 18 yr of age, appear to increase risks of CHD in middle-aged women. After controlling for age, smoking, menopausal status, hormone replacement therapy, and parental history of CHD, significant increases in risk were still observed among those with a BMI 23 compared with those with a BMI less than 21. The RRs for CHD were 1.5 (95% CI: 1.2–1.8) for a BMI = 23.0 to 24.9, 2.1 (95% CI: 1.7–2.5) for a BMI = 25.0 to 28.9, and 3.6 (95% CI: 3.0–4.3) for a BMI ≥ 29. However, significant weight gain during adulthood (range: 20–34.9 kg) approximately doubled the coronary risk after controlling for initial relative weight level at age 18 yr (RR: 2.5; 95% CI: 1.7–3.7). In contrast to weight gain throughout life, "morbid" obesity (defined as BMI ≥ 40) early in adult life is emerging as a significant risk factor for CHD mortality, the duration of morbid obesity being the strongest predictor of CHF *(57)*.

3.4. Location of Body Fat

In humans and most animal models, the development of obesity leads not only to increased fat depots in classical adipose tissue locations but also to significant lipid deposits within and around other tissues and organs, a phenomenon known as ectopic fat storage *(58)*. Cardiac fat depots within and around the heart impair both systolic and diastolic functions, and may promote CHF in the long term *(59)*. Accumulation of fat around blood vessels (perivascular fat) may affect vascular function in a paracrine manner, as perivascular fat cells secrete vascular modulating factors, proatherogenic cytokines, and smooth muscle cell growth factors *(53)*. Furthermore, high amounts of perivascular fat could mechanically contribute to the increased vascular stiffness seen after years of obesity *(58)*.

The relative excess of fat in the abdomen aids in the development of diabetes and atherosclerosis *(60)*. The distribution of fat depots in the body is a strong independent predictor of CHD *(61–63)*. Indeed, disturbances in lipoprotein metabolism, glucose homeostasis, and hypertension have been reported in subjects with an excessive deposition of adipose tissue in the abdomen *(32,61,64,65)*. In addition, abdominal distribution of body fat is associated with increased plasma levels of fibrinogen, and other factors that modulate coagulation (e.g., PAI-I antigen). These same molecules may also contribute to left ventricular dysfunction *(66)*.

There is strong debate as to which anthropometric measure is the best technique to assess the risk of CVD *(4,67–69)*. Waist circumference (WC) is strongly correlated with abdominal fat content and is the easiest way to assess a patient's abdominal fat *(70)*. The Health Professionals Follow-up Study found that WC, but not BMI, predicted risk of death from CVD *(71)*. However, the WC cutoffs lose their incremental predictive power in patients with a BMI $\geq$ 35 *(72)*. Evidence from the Heart Outcomes Prevention Evaluation (HOPE) study of more than 8000 patients with stable CVD suggests that elevated WC was significantly associated with an increased risk of myocardial infarction (RR 1.23, $p < 0.01$), heart failure (RR 1.38, $p < 0.03$), and total mortality (RR 1.17, $p < 0.05$) *(63)*. This supports previous findings from the Framingham Study *(73)*, which reported that the risk of CVD incidence and mortality increased with the degree of regional, central, or abdominal obesity.

4. CARDIOVASCULAR RISK FACTORS OF OBESITY

4.1. CVD Risk Associated With Obesity

The association between overweight/obesity and CVD risk has been known for many years with evidence from several large cohort studies *(74–76)*. After 44 yr of follow-up of the Framingham Heart Study, Wilson et al. *(77)* showed that CVD risk (including angina, myocardial infarction, CHD, or stroke) was higher among overweight men (RR: 1.24; 95% CI: 1.07–1.44), obese men (RR: 1.38; 95% CI: 1.12–1.69), and obese women (RR: 1.38; 95% CI: 1.14–1.68) after adjustment for age, smoking, high blood pressure, high cholesterol, and diabetes. During a 14-yr follow-up of 1 million adults in the United States, it was found that as BMI increased there was an increase in the risk of death from all causes, CVD, cancer, or other diseases for both men and women in all age groups *(78)*. These findings confirmed the previous report of the Nurses' Health Study *(79)*. In the Nurses' Health Study, weight gain of 5 to 8 kg increased CHD risk (nonfatal myocardial infarction and CHD death) by 25%, and weight gain of $\geq$ 20 kg increased risk more than 2.5 times in comparison with women whose weight was stable within a range of 5 kg *(56)*. In British men, an increase of 1 BMI unit was associated with a 10% increase in the rate of coronary events *(80)*.

5. METABOLIC SYNDROME

The term "metabolic syndrome" refers to a clustering of specific CVD risk factors whose underlying pathophysiology is likely insulin resistance *(9)*. Although several sets of diagnostic criteria exist (WHO *[81]*, NCEP *[82]*, IDF *[83]*), waist circumference, dyslipidemia, elevated blood pressure, and glucose intolerance are shared by all. Uncertainty about the mechanism of pathogenesis has resulted in a debate to determine whether metabolic syndrome is a syndrome or an independent CVD risk factor *(84)*. This is reviewed in detail in Chapter 20.

5.1. Atherosclerosis and Coronary Artery Disease

Atherosclerosis leads to the development of CHD. Individuals with atherosclerosis can be identified by the ultrasound examination of carotid intimal–medial thickness (IMT). Evidence from the Bogalusa Heart Study found that the carotid IMT at age 35 yr was correlated with BMI measured throughout life *(85)*. Longitudinal evidence from the Muscatine Study suggests that increased adipose tissue in youth is correlated with a greater risk of developing coronary artery calcification in later life and that this association was stronger in males *(86)*. Berenson et al. *(87)* observed a similar association in young adults, particularly among young men, between obesity and the development of atherosclerotic lesions as evidenced by fatty streaks and/or fibrous plaque lesions.

Obesity may be an independent risk factor for ischemic heart disease. However, numerous studies have been unable to confirm this association because of the short time period of observation. Indeed, the association between obesity and ischemic heart disease seems evident only after two decades of follow-up *(76)*. The Manitoba Heart Study reported that a high BMI was significantly associated with development of myocardial infarction, coronary insufficiency, and sudden death *(76)*.

A cross-sectional study by Takami et al. *(88)* of 849 Japanese men aged 20 to 78 yr investigated the relationship between body fatness (particularly abdominal fat) and carotid atherosclerosis. They found that general adiposity (as measured by BMI), WC, waist-to-hip ratio (WHR), abdominal subcutaneous fat, and intra-abdominal fat were all correlated with carotid IMT after adjustment for age and smoking habits. Adjustment for BMI eliminated all other associations except those of WHR with IMT, suggesting that in this population abdominal fat is not as strongly associated with carotid atherosclerosis as is general body fatness. The Progetto ATENA study is a large (more than 5000 participants) ongoing investigation of the causes of CVD and cancer in Italian females aged 30 to 69 yr. Within that study, De Michele et al. *(89)* reported on a subsample of 310 women and concluded that BMI and WHR were significant predictors of carotid wall thickness independent of other cardiovascular risk factors (age, BP, lipid abnormalities, and fasting insulin). As BMI increased, IMT increased along with other coronary risk factors (systolic blood pressure [SBP], diastolic blood pressure [DBP], triglycerides, fasting glucose, insulin, and lower high-density lipoprotein [HDL] cholesterol concentrations).

5.2. Hypertension

The majority of patients with high BP are overweight, and hypertension is about six times more frequent in obese than lean subjects *(90)*. This represents an estimated 12% increased risk for CHD and 24% increased risk for stroke *(72)*. Although the association between obesity and hypertension is well recognized, the underlying pathophysiological mechanisms are still poorly understood. The expansion of extracellular volume and increased CO are characteristic hemodynamic changes that occur with obesity-related hypertension *(28)*. A variety of endocrine, genetic, and metabolic mechanisms have also been linked to the development of obesity hypertension *(1,32,91–93)*. One potential mechanism leading to the development of obesity-induced hypertension may be through leptin-mediated sympathoactivation *(94)*. The association between obesity and hypertension begins in early life. Longitudinal observations of children, adolescents, and young adults enrolled in the Bogalusa Heart Study show that obesity persists over time and is linked to the commonly clustered components of metabolic syndrome, including hypertension, hyperinsulinemia/insulin resistance, and dyslipidemia, which in turn is associated with the processes leading to CVD *(95)*.

The INTERSALT Study *(23,96)* examined the relationship between BMI and BP among more than 10,000 people from 52 centers and 32 countries around the world. They found a significant and independent relationship between high BP and increased BMI in 98% and 90% of all centers among men and women, respectively. Irrespective of age, for every BMI unit increase, there was an associated increase in SBP of 0.91 mmHg for men and a 0.72 mmHg increase for women. For DBP, this increase was 0.75 mmHg for men and 0.5 mmHg for women per BMI unit *(23,96)*. Overall, a 10-kg increase in body weight was associated with an elevation of 3.0 mmHg in SBP and 2.2 mmHg in DBP *(23)*.

The Framingham Heart Study reported that obesity was significantly correlated with increased LVM *(97)*; it has been shown that a 10% reduction in weight of obese hypertensive patients not only reduced blood pressure, but also decreased left ventricular wall thickness and LVM *(48)* (Table 2). There is evidence in both overweight hypertensive and nonhypertensive patients that weight loss produced by lifestyle modifications reduces BP levels *(98)*. Weight reduction is one of the rare antihypertensive strategies that decreases BP in normotensive as well as hypertensive individuals *(72)*. However, this reduction is not always maintained once weight is stable *(99)*, and it has been suggested that the extent to which BP decreases is influenced by several factors including the duration of hypertension *(100)* and the composition of the diet *(101)*.

The reduction in BP could also be attributable to (1) reductions in salt intake concomitant with caloric restriction *(98)* or (2) reductions in total circulating and cardiopulmonary blood volume, as well as (3) reductions in sympathetic nervous system activity *(97)*. The reduction in plasma catecholamines and plasma renin activity, which are associated with decreased sympathetic activity, are also probably playing a role *(98,99)*.

5.3. Dyslipidemia

The presence of dyslipidemia is well established in obesity. "At risk" overweight/obese individuals commonly present with reduced HDL cholesterol, elevated apolipoprotein B levels, and a prevalence of small, dense low-density lipoprotein (LDL) particles *(102)*. Visceral adiposity, in particular, has a negative impact on the lipid profile; it has been suggested that dyslipidemia is the main contributor to the increase in CHD in abdominally obese patients *(74)*. A BMI change of 1 unit is associated with a decrement change in HDL-cholesterol of 1.1 mg/dL for young adult men and 0.69 mg/dL for young adult women *(2)*. There is evidence that weight loss achieved by lifestyle modification in overweight individuals is accompanied by a reduction in serum triglycerides and an increase in HDL cholesterol *(86)*. Weight loss may also contribute to a reduction in serum total cholesterol and LDL cholesterol levels *(87)*. Moreover, in subjects with type 2 diabetes, aerobic exercise may mediate an improvement in the lipid profiles through fat loss *(103–108)*.

5.4. Type 2 Diabetes Mellitus

Several prospective studies in numerous countries have demonstrated an elevated risk of diabetes mellitus as weight increases *(75–77)*. The development of type 2 diabetes is associated with weight gain after age 18 yr in both men and women such that the relative risk of diabetes increases by approx 25% for each additional unit of BMI over 22 *(109)*. Moreover, cross-sectional and longitudinal studies show that abdominal obesity is a major risk factor for type 2 diabetes *(74,82,91)*. There is strong evidence that weight loss reduces blood glucose levels and hemoglobin A_{1c} levels in patients with type 2 diabetes. Moreover, in three European cohorts (>17,000 men) followed for more than 20 yr, nondiabetic men with higher blood glucose had a significantly higher risk of cardiovascular and CHD death *(110)*. In addition, it has been demonstrated in the Framingham Offspring Cohort that metabolic factors associated with obesity (overall and central), including hypertension, low levels of HDL cholesterol, and increased levels of TG and insulin, worsen continuously across the spectrum of glucose tolerance *(111)*. Although BMI increased steadily with increasing glucose intolerance, the association between most other measures of metabolic risk and glycemia were independent of overall obesity, and the gradient of increasing risk was similar for nonobese and obese participants *(111)*. Thus, asymptomatic glucose intolerance is not a benign metabolic condition, and characteristics associated with insulin resistance syndrome should be taken seriously. This is further reinforced by the Quebec Cardiovascular Study, where hyperinsulinemia was reported as an independent risk factor for CHD *(112)*.

5.5. Obstructive Sleep Apnea Syndrome

Obstructive sleep apnea syndrome (OSAS) is one of the many respiratory complications associated with obesity. It is defined as repeated episodes of obstructive apnea and hypopnea during sleep, in association with altered cardiopulmonary function *(113)*. Evidence is emerging that patients with apneic events that occur during sleep have associated acute and chronic hemodynamic changes during waking time, including elevated sympathetic tone, decreased stroke volume (SV) and CO, increased HR, and changes in circulating hormones that regulate BP, fluid volume, vasoconstriction, and vasodilation *(114,115)*. Weight loss is an effective method for reducing the extent of OSAS *(116)* and associated disruptive symptoms, such as habitual snoring and daytime sleepiness *(117)*.

OSAS is thought to be both a systemic and local inflammatory condition *(118)*. Inflammatory processes associated with OSAS may contribute to cardiovascular morbidity in these patients. Indeed, it is the presence of systemic inflammation, characterized by elevated levels of certain proinflammatory mediators, such as C-reactive protein *(119)*, leptin *(120)*, TNF-α, IL-1β, IL-6 *(37)*, reactive oxygen species, and adhesion molecules, that may predispose people to the development of cardiovascular complications observed in patients with OSAS. Interestingly, both TNF-α and IL-6 have been found to be significantly elevated in OSAS independent of obesity *(121)*. To date it is unclear how the cytokines directly mediate OSAS *(122)*.

6. CONCLUSIONS

Obesity is a chronic metabolic disorder associated with a number of CVD risk factors. There are complex paracrine and endocrine communication pathways that promote homeostasis in healthy individuals. However, when challenged in conditions such as obesity, by modulations of genes or environment, these networks can be altered in ways that result in deleterious changes to the CV system that ultimately reduce life span (Fig. 1). It therefore comes as no surprise that CVD is more frequent in subjects with obesity. Moreover, when BMI is $\geq$ 30, mortality rates from all causes, and especially CVD, are increased by 50 to 100% *(39)*. With an increasing incidence of obesity worldwide, it is important that we not only understand the problems associated with excess weight but that we also strive to identify the underlying mechanisms and ways to prevent further increases. There is strong evidence that weight loss in overweight and obese individuals reduces risk factors for diabetes and CVD. This includes a reduction in deleterious

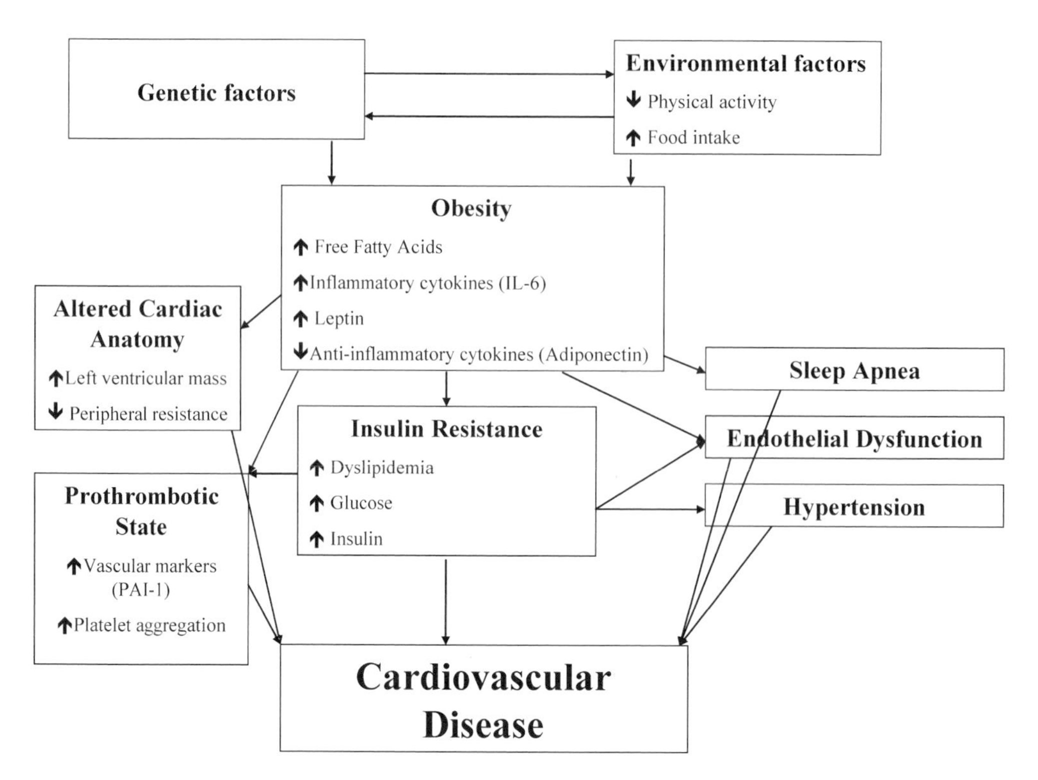

Fig. 1. There is a complex interaction between the conditions that cluster together to influence risk of CVD. Obesity appears to be central to the problem. With excess body weight comes the risk of long-term health consequences. Overweight and obesity are associated with the morbidity and mortality of many health conditions, such as hypertension, sleep apnea, insulin resistance, and endothelial dysfunction. All these conditions ultimately affect CVD.

circulating cytokines. Although there have been no prospective trials to convincingly show changes in mortality with weight loss in obese patients, it is anticipated that a reduction in risk factors would predict a reduced incidence of CVD, and perhaps CVD-related mortality. Thus, although the influence of obesity on all-cause and CVD-related mortality remains controversial, there is a necessary concern for our younger populations with decades of obesity ahead of them.

REFERENCES

1. Bouchard C, Despres JP, Mauriege P. Endocr Rev 1993;14:72–93.
2. Eckel RH, Krauss RM. Circulation 1998;97:2099–2100.
3. Rao SV, Donahue M, Pi-Sunyer FX, et al. Am Heart J 2001;142:1102–1107.
4. Arch JR. Proc Nutr Soc 2005;64:39–46.
5. Eckel RH. In: Eckel RH, ed. *Obesity Mechanisms and Clinical Management*. Lippincott, Williams and Wilkins, Philadelphia: 2003, pp. 3–30.
6. Wing RR, Hill JO. Annu Rev Nutr 2001;21:323–341.
7. World Health Organization. In: WHO Technical Report Series, Vol. 894. Geneva: 2000.
8. Eckel RH, Grundy SM, Zimmet PZ. Lancet 2005;365:1415–1428.
9. Allison DB, Fontaine KR, Manson JE, et al. JAMA 1999;282:1530–1538.

10. Flegal KM, Graubard BI, Williamson DF, et al. JAMA 2005;293:1861–1867.
11. Smith SC Jr, Blair SN, Bonow RO, et al. Circulation 2001;104:1577–1579.
12. Yan LL, Daviglus ML, Liu K, et al. JAMA 2006;295:190–198.
13. Li TY, Rana JS, Manson JE, et al. Circulation 2006;113:499–506.
14. Lee IM, Rexrode KM, Cook NR, et al. JAMA 2001;285:1447–1454.
15. Wessel TR, Arant CB, Olson MB, et al. JAMA 2004;292:1179–1187.
16. Poirier P, Giles TD, Bray GA, et al. Circulation 2006;113:898–918.
17. Eckel RH. Lancet 1992;340:1452–1453.
18. Messerli FH, Nunez BD, Ventura HO, et al. Arch Intern Med 1987;147:1725–1728.
19. Mohamed-Ali V, Goodrick S, Rawesh A, et al. J Clin Endocrinol Metab 1997;82:4196–4200.
20. Lundgren CH, Brown SL, Nordt TK, et al. Circulation 1996;93:106–110.
21. Fukuhara A, Matsuda M, Nishizawa M, et al. Science 2005;307:426–430.
22. Schaffler A, Neumeier M, Herfarth H, et al. Biochim Biophys Acta 2005;1732:96–102.
23. Dyer AR, Elliott P. J Hum Hypertens 1989;3:299–308.
24. Fain JN, Madan AK, Hiler ML, et al. Endocrinology 2004;145:2273–2282.
25. Fain JN, Tichansky DS, Madan AK. Metabolism 2005;54:1546–1551.
26. Sewter CP, Digby JE, Blows F, et al. J Endocrinol 1999;163:33–38.
27. Coppack SW. Proc Nutr Soc 2001;60:349–356.
28. Morse SA, Bravo PE, Morse MC, et al. Expert Rev Cardiovasc Ther 2005;3:647–658.
29. Ritchie SA, Ewart MA, Perry CG, et al. Clin Sci (Lond) 2004;107:519–532.
30. Hotamisligil GS, Shargill NS, Spiegelman BM. Science 1993;259:87–91.
31. Suganami T, Nishida J, Ogawa Y. Arterioscler Thromb Vasc Biol 2005;25:2062–2068.
32. Poirier P, Lemieux I, Mauriege P, et al. Hypertension 2005;45:363–367.
33. Gottschling-Zeller H, Birgel M, Scriba D, et al. Eur J Endocrinol 1999;141:436–442.
34. Fasshauer M, Klein J, Neumann S, et al. Biochem Biophys Res Commun 2002;290:1084–1089.
35. Ruan H, Lodish HF. Cytokine Growth Factor Rev 2003;14:447–455.
36. Aukrust P, Gullestad L, Ueland T, et al. Ann Med 2005;37:74–85.
37. Alberti A, Sarchielli P, Gallinella E, et al. J Sleep Res 2003;12:305–311.
38. Yokoe T, Minoguchi K, Matsuo H, et al. Circulation 2003;107:1129–1134.
39. Poirier P, Despres JP. Cardiol Clin 2001;19:459–470.
40. Alpert MA. Am J Med Sci 2001;321:225–236.
41. de Simone G, Devereux RB, Kizer JR, et al. Am J Clin Nutr 2005;81:757–761.
42. Ku CS, Lin SL, Wang DJ, et al. Am J Cardiol 1994;73:613–615.
43. Hense HW, Gneiting B, Muscholl M, et al. J Am Coll Cardiol 1998;32:451–457.
44. Ferrara AL, Vaccaro O, Cardoni O, et al. Nutr Metab Cardiovasc Dis 2003;13:126–132.
45. Alpert MA, Lambert CR, Panayiotou H, et al. Am J Cardiol 1995;76:1194–1197.
46. Giordano U, Ciampalini P, Turchetta A, et al. Pediatr Cardiol 2003;24:548–552.
47. McGill HC Jr, McMahan CA, Herderick EE, et al. Circulation 2002;105:2712–2718.
48. MacMahon SW, Wilcken DE, Macdonald GJ. N Engl J Med 1986;314:334–339.
49. Caviezel F, Margonato A, Slaviero G, et al. Int J Obes 1986;10:421–426.
50. Alpert MA, Lambert CR, Terry BE, et al. Am J Cardiol 1995;76:1198–1201.
51. Parrinello G, Licata A, Colomba D, et al. J Hum Hypertens 2005;19:543–550.
52. Schondorf T, Maiworm A, Emmison N, et al. Clin Lab 2005;51:489–494.
53. Henrichot E, Juge-Aubry CE, Pernin A, et al. Arterioscler Thromb Vasc Biol 2005;25:2594–2599.
54. Malavazos AE, Cereda E, Morricone L, et al. Eur J Endocrinol 2005;153:871–877.
55. Manson JE, Colditz GA, Stampfer MJ, et al. N Engl J Med 1990;322:882–889.
56. Willett WC, Manson JE, Stampfer MJ, et al. JAMA 1995;273:461–465.
57. Alpert MA, Terry BE, Mulekar M, et al. Am J Cardiol 1997;80:736–740.
58. Montani JP, Carroll JF, Dwyer TM, et al. Int J Obes Relat Metab Disord 2004;28 Suppl 4:S58–S65.
59. Poirier P, Martin J, Marceau P, et al. Expert Rev Cardiovasc Ther 2004;2:193–201.
60. Vague J, Vague P, Tramoni M, et al. Acta Diabetol Lat 1980;17:87–99.
61. Larsson B, Svardsudd K, Welin L, et al. Br Med J (Clin Res Ed) 1984;288:1401–1404.
62. Lapidus L, Bengtsson C, Larsson B, et al. Br Med J (Clin Res Ed) 1984;289:1257–1261.
63. Dagenais GR, Yi Q, Mann JF, et al. Am Heart J 2005;149:54–60.

64. Cigolini M, Targher G, Bergamo Andreis IA, et al. Arterioscler Thromb Vasc Biol 1996;16:368–374.
65. Reaven G. Diab Vasc Dis Res 2005;2:105–112.
66. Licata G, Scaglione R, Avellone G, et al. Metabolism 1995;44:1417–1421.
67. Hsieh SD, Yoshinaga H. Int J Obes Relat Metab Disord 1995;19:585–589.
68. Ohrvall M, Berglund L, Vessby B. Int J Obes Relat Metab Disord 2000;24:497–501.
69. Seidell JC, Perusse L, Despres JP. Am J Clin Nutr 2001;74:315–321.
70. Pouliot MC, Despres JP, Lemieux S, et al. Am J Cardiol 1994;73:460–468.
71. Baik I, Ascherio A, Rimm EB, et al. Am J Epidemiol 2000;152:264–271.
72. National Institutes of Health. Clinical guidelines on the identification, evaluation, and treatment of overweight and obesity in adults—the evidence report. Obes Res 1998;6 Suppl 2:51S–209S
73. Kannel WB, Cupples LA, Ramaswami R, et al. J Clin Epidemiol 1991;44:183–190.
74. Kannel WB, LeBauer EJ, Dawber TR, et al. Circulation 1967;35:734–444.
75. Hubert HB, Feinleib M, McNamara PM, et al. Circulation 1983;67:968–977.
76. Rabkin SW, Mathewson FA, Hsu PH. Am J Cardiol 1977;39:452–458.
77. Wilson PW, D'Agostino RB, Sullivan L, et al. Arch Intern Med 2002;162:1867–1872.
78. Calle EE, Thun MJ, Petrelli JM, et al. N Engl J Med 1999;341:1097–1105.
79. Manson JE, Willett WC, Stampfer MJ, et al. N Engl J Med 1995;333:677–685.
80. Shaper AG, Wannamethee SG, Walker M. BMJ 1997;314:1311–1317.
81. Alberti KG, Zimmet PZ. Diabet Med 1998;15:539–553.
82. Executive Summary of The Third Report of The National Cholesterol Education Program (NCEP) Expert Panel on Detection, Evaluation, And Treatment of High Blood Cholesterol In Adults (Adult Treatment Panel III) JAMA 2001;285:2486–2497.
83. Alberti KG, Zimmet P, Shaw J. Lancet 2005;366:1059–1062.
84. Kahn R, Buse J, Ferrannini E, et al. Diabetes Care 2005;28:2289–2304.
85. Freedman DS, Dietz WH, Tang R, et al. Int J Obes Relat Metab Disord 2004;28:159–166.
86. Mahoney LT, Burns TL, Stanford W, et al. J Am Coll Cardiol 1996;27:277–284.
87. Berenson GS, Wattigney WA, Tracy RE, et al. Am J Cardiol 1992;70:851–858.
88. Takami R, Takeda N, Hayashi M, et al. Diabetes Care 2001;24:1248–1252.
89. De Michele M, Panico S, Iannuzzi A, et al. Stroke 2002;33:2923–2928.
90. Stamler R, Stamler J, Riedlinger WF, et al. JAMA 1978;240:1607–1610.
91. Poirier P, Despres JP. J Cardiopulm Rehab 2003;23:161–169.
92. Ferrannini E. J Hypertens 1992;10:1417–1423.
93. Kozakova M, Fraser AG, Buralli S, et al. Hypertension 2005;45:619–624.
94. Haynes WG. Exp Physiol 2005;90:683–688.
95. Berenson GS. Prev Cardiol 2005;8:234–241.
96. Stamler J, Rose G, Elliott P, et al. Hypertension 1991;17:I9–I15.
97. Lauer MS, Anderson KM, Kannel WB, et al. JAMA 1991;266:231–236.
98. Himeno E, Nishino K, Nakashima Y, et al. Am Heart J 1996;131:313–319.
99. Sjostrom L, Lindroos AK, Peltonen M, et al. N Engl J Med 2004;351:2683–2693.
100. Lawrence VJ, Kopelman PG. Clin Dermatol 2004;22:296–302.
101. Nowson CA, Worsley A, Margerison C, et al. Am J Clin Nutr 2005;81:983–989.
102. Despres JP. Baillieres Clin Endocrinol Metab 1994;8:629–660.
103. Ross R, Freeman JA, Janssen I. Exerc Sport Sci Rev 2000;28:165–170.
104. Poirier P, Catellier C, Tremblay A, et al. Metabolism 1996;45:1383–1387.
105. Lee S, Kuk JL, Davidson LE, et al. J Appl Physiol 2005;99:1220–1225.
106. Lee S, Kuk JL, Katzmarzyk PT, et al. Diabetes Care 2005;28:895–901.
107. Ross R, Janssen I, Dawson J, et al. Obes Res 2004;12:789–998.
108. Katzmarzyk PT, Church TS, Janssen I, et al. Diabetes Care 2005;28:391–397.
109. Colditz GA, Willett WC, Rotnitzky A, et al. Ann Intern Med 1995;122:481–486.
110. Balkau B, Shipley M, Jarrett RJ, et al. Diabetes Care 1998;21:360–367.
111. Meigs JB, Nathan DM, Wilson PW, et al. Ann Intern Med 1998;128:524–533.
112. Despres JP, Lamarche B, Mauriege P, et al. N Engl J Med 1996;334:952–957.
113. Strollo PJ Jr, Rogers RM. N Engl J Med 1996;334:99–104.
114. Partinen M, Jamieson A, Guilleminault C. Chest 1988;94:1200–1204.

115. Merritt SL. Prog Cardiovasc Nurs 2004;19:19–27.
116. Gami AS, Caples SM, Somers VK. Endocrinol Metab Clin North Am 2003;32:869–894.
117. Bearpark H, Elliott L, Grunstein R, et al. Am J Respir Crit Care Med 1995;151:1459–1465.
118. Hatipoglu U, Rubinstein I. Respiration 2003;70:665–671.
119. Kokturk O, Ciftci TU, Mollarecep E, et al. Int Heart J 2005;46:801–809.
120. Ulukavak Ciftci T, Kokturk O, Bukan N, et al. Respiration 2005;72:395–401.
121. Ciftci TU, Kokturk O, Bukan N, et al. Cytokine 2004;28:87–91.
122. Vgontzas AN, Bixler EO, Lin HM, et al. Neuroimmunomodulation 2005;12:131–140.

25 Obesity and Asthma

Elisabeth Luder

Abstract

Obesity and asthma are both common complex traits responsible for substantial morbidity in the developed world. The consistency of the relationship between obesity and asthma, the temporal and dose-response association, and the correlation of obesity with intermediate phenotypes for asthma suggest that the obesity–asthma link is causal. With few exceptions, the existing epidemiological studies show a consistent positive association of obesity with both incidence and prevalence of asthma in children and adults, with the effect being greater in females than in males. Obesity precedes and predicts the development of asthma, not the other way around, and the effect persists after controlling for diet and physical activity. The dose–response relationship is demonstrated by the finding that the greater the obesity the greater the effect on asthma. Studies are showing improvements in asthma in subjects who lose weight. From these observations, the main research issues at present relate to the actual biological mechanisms by which obesity influences the asthma phenotype. The specific areas requiring further investigation are: direct effect of obesity on mechanical functioning of the lung; changes in the immune or inflammatory responses directly or through genetic mechanism; sex-specific influences relating to hormones; and the influence of maternal diet on fetal programming. At present, we still do not have a good understanding of the precise relationship between obesity and asthma.

Key Words: Obesity; asthma; lung; inflammation; genes; hormones; diet.

1. INTRODUCTION

Obesity and asthma are both chronic conditions affecting millions worldwide. Over the past two decades there has been a rapid increase in the prevalence of both these conditions *(1,2)*. In the United States, in 2002, 65% of adults were overweight, with a body mass index (BMI) higher than 25, compared with 46% in 1980; in addition, 16% of children, age 6 to 19 yr had a BMI higher than 25, compared with 5% in 1980 *(3)*. From 1980 to 1996, there was an increase in self-reported asthma prevalence of 74% *(4)*. Given the parallel increase in obesity and asthma, it is not surprising that the prevalence and incidence of asthma and its related symptoms have been associated with BMI and obesity as described in a number of epidemiological studies for adults and children. In this chapter, the definition of asthma, association between obesity and asthma as described in cross-sectional and longitudinal studies, and causal hypotheses of obesity and asthma will be addressed.

From: *Nutrition and Health: Adipose Tissue and Adipokines in Health and Disease*
Edited by: G. Fantuzzi and T. Mazzone © Humana Press Inc., Totowa, NJ

2. DEFINITION AND EPIDEMIOLOGY OF ASTHMA

Asthma is defined by episodic airflow obstruction, increased airway responsiveness, and airway inflammation characterized by infiltration with eosinophils and T-lymphocytes, particularly $CD4^+$ T-lymphocytes that express T-helper (Th) cell type 2 cytokines such as interleukin (IL)-4, IL-5, and IL-13. The histopathological appearance of the airways includes denudation of the airway epithelium, thickening of the basement membrane, mucus production, and airway smooth muscle hypertrophy *(5)*. Although asthma is a chronic—often lifelong—disease that affects humans of all ages, the onset of the disease occurs primarily in early childhood. Fifty percent of all male asthma cases are diagnosed by age 3, and 50% of all female cases are diagnosed by age 8 *(6)*. This increase of asthma in early childhood has been most marked in minority populations, particularly African Americans and Puerto Rican Hispanics *(7)*.

3. DEFINITION OF OBESITY

A number of methods have been proposed to describe increases in body weight. The most widely used measurement is the BMI, calculated as weight/height^2 (kg/m^2). Among adult subjects overweight is defined as a BMI of 25 to 29.9 kg/m^2, and obesity as a BMI of at least 30 kg/m^2. Different ethnicities, such as Caucasians and Chinese, have highly contrasting distributions of body weight and height. Males and females differ in their BMI distribution. Age is an obvious modifier of body weight and height. Therefore, age- and sex-specific definitions for different ethnic groups must be applied when comparing effects in various study populations. Growth charts from the Centers for Disease Control and Prevention (CDC) include age- and sex-specific BMI reference values for children and adolescents aged 2 to 20 yr *(8)*. However, BMI does not account for body frame and proportion of muscle mass. This limitation is particularly relevant in pediatric studies because of the effects of maturation and growth of lean muscle mass, fat mass, and hydration status. Furthermore, muscle mass increases with higher activity level, and fat mass values are higher among females than males and vary across ethnic groups *(9)*. Therefore, other measures of body weight have been applied, such as assessment of body fat by skinfold thickness, dual-energy X-ray absorptiometry (DEXA) evaluation or bioelectrical impedance. However, large comparative studies on skinfold, DEXA, and bioelectrical impedance variables with BMI are missing to date.

4. ASSOCIATION BETWEEN OBESITY AND ASTHMA

Conflicting results of various studies investigating the potential association between obesity and asthma may be attributable to different study design—cross-sectional or longitudinal—or effect modifiers.

4.1. Cross-Sectional Studies

The cross-sectional diagnosis of asthma has been associated with obesity in both children *(10–13)* and adults *(14–16)*. In most studies, the definition of asthma was based on a doctor diagnosis of asthma, partly including airway hyper-responsiveness (AHR), but in some studies only symptoms such as wheeze or asthma attacks were assessed. Several of these studies noted a relationship only in women, but not in men *(17,18)*.

However, findings are not consistent, and associations between obesity and AHR were even stronger in males than females *(19)*. AHR may reflect airway inflammation but not asthma itself, as AHR is a feature rather loosely associated with a doctor's diagnosis of asthma in population-based studies. Effect modification by sex, however, may also be based on differences in the shape of the relationship. This notion is supported by a survey demonstrating that the association between BMI and asthma differed only in the lowest weight category. Among women a monotonic association was seen, whereas in men a U-shaped relation was found, but both extremes of weight were associated with higher prevalence of asthma *(20)*.

Cross-sectional studies may be prone to bias because they do not allow an assessment of the timing of the exposure in relation to the occurrence of asthma. Thus the relationship between BMI and asthma may reflect an asthmatic patient's predisposition to gain weight because of reduced exercise tolerance rather than causal association between a high BMI and the inception of asthma. Therefore results from prospective studies may help to better interpret the findings *(9)*.

4.2. Prospective Studies

Nearly all prospective studies demonstrate a positive association between BMI and the development of asthma and AHR, respectively *(19,21,22)*. In these studies weight gain occurred before the new onset of asthma or asthma symptoms, suggesting a true relation between both conditions. In some studies stratification by sex either by study design *(21,22)* or in the analyses revealed sex-specific effects. Camargo et al. *(21)* reported a significant association between overweight development and new-onset asthma in the US Nurses' Health Study, whereas Litonjua et al. *(22)* detected a U-shaped relation with AHR at high and low BMI in the Normative Aging Study, including only males. In a study of 135,000 Norwegians aged 14 to 60 yr who were followed on average for 21 yr, the risk of asthma increased steadily with an increase in BMI. For men the risk of asthma increased starting with a BMI of 20 and in women with a BMI of 22. In men, the risk of asthma increased by 10% and for women by 7% with each unit of increased BMI between 25 and 30 *(23)*. A population of 10,597 adult twins, initially free of asthma, was followed for 9 yr. Obese men with a BMI ≥ 30 had a significantly increased risk of developing asthma when compared to those with a BMI of 20 to 24.99 (OR = 3.47). More men were obese than women, and the association between BMI and asthma was not significant in women *(24)*.

In a longitudinal population-based birth cohort study of 781 children, in boys and girls the presence of obesity during the prepubertal period and early onset of puberty were significant and independent risk factors for persistent asthma after puberty *(25)*. Among 3,792 participants aged 7 to 18 yr in the Children's Health Study who were asthma-free at enrollment and were followed for 5 yr, the risk of asthma development increased among overweight and obese boys but not girls (OR = 2.06 vs OR = 1.25) *(26)*. Among 9828 children aged 6 to 14 yr examined annually over a follow-up time of 5 yr in six US cities, an increased risk of new asthma diagnosis in girls was associated with higher BMI at entry into the study and greater increase in BMI during follow-up. Boys with the largest and smallest annual change in BMI also had an increased risk of asthma. For boys and girls, extremes of annual BMI growth rates increased the risk of asthma *(27)*. As addressed in the following section, in these children, overweight or extreme leanness may represent a combination of *in utero* and postnatal influences on growth and development that increase the risk of airway mechanical dysfunction or inflammation *(28,29)*.

5. CAUSAL HYPOTHESES

In the association between obesity and asthma, potential misclassifications of wheeze and asthma must be considered. Obstructive sleep apnea or hypoventilation are frequent among obese patients *(5,9)*. However, it seems unlikely that these symptoms are misclassified as asthma in longitudinal surveys. As previously stated, all the existing prospective epidemiological studies show a consistent positive association of obesity with both incidence and prevalence of asthma in children and adults. Obesity precedes asthma and predicts the development of asthma and the effect persists after controlling for diet and physical activity. The dose–response relationship is demonstrated by the finding that the greater the obesity, the greater the observed effect on asthma *(10–12,21)*. The effects of obesity seem greater for asthma and airway responsiveness than they do for other allergy phenotypes, although these effects have not been assessed as frequently. From these observations, studies suggest that obesity has the potential to affect airway function through a variety of agencies, including mechanical functioning of the lung, changes in immune or inflammatory responses directly or through genetic mechanisms, sex-specific influences relating to hormones, lung development, and the influence of maternal diet on fetal programming *(28,29)*. These are many possibilities, but they are not mutually exclusive, and the dominant mechanism among them is yet to be identified.

5.1. Effects of Lung Mechanics

Obesity leads to decreased lung tidal volume as well as decreased functional residual capacity. These volume changes result in reduced smooth muscle stretch or latching. Consequently, the ability to respond to a physiological stress such as exercise is hampered by small tidal breaths, which alters smooth muscle contraction, worsening the respiratory condition. Normal smooth muscle has an intrinsic rate of excitation and contraction called the cycling rate. In obese people, lower cycling rates of the airway smooth muscles and thus decreased functional capacity result from the conversion of rapidly cycling actin–myosin crossbridges to slowly cycling latch bridges *(28,30)*. The exact dose–effect relationship between the amount and distribution of body fat and the mechanical changes remains unknown and is an area for further research.

5.2. Comorbidities

Obesity may lead to asthma not directly, but through its role in other disease processes. For example, obesity increases the risk of both gastroesophageal reflux disease (GERD) and sleep-disordered breathing (SDR). An increased prevalence of asthma has been observed in subjects with each of these conditions; furthermore, subjects undergoing surgical induced weight loss showed improvements not only in asthma but also in GERD and sleep apnea. Consequently, there has been speculation that obesity leads to asthma through its effects on these other conditions. Two recent studies have examined the interrelationships between these conditions. Multivariate logistic regression in data from more than 16,000 participants in the European Community Respiratory Health Survey demonstrated that the relationship between obesity and the onset of asthma was unaffected by adjustment for GERD or habitual snoring *(31)*. Similarly, Sulit et al. demonstrated that adjustment for SDB and asthma did not substantially alter the association between obesity and asthma *(32)*. Taken together, these data indicate that the increased risk of asthma in the obese is independent of GERD and SDB.

5.3. Chronic Systemic Inflammation

There is increasing evidence that obesity is a proinflammatory state *(33)*. Initial studies have focused mainly on the association of obesity and tumor necrosis factor (TNF), IL-6, IL-1β, and C-reactive protein. IL-6 and TNF are constitutively expressed by adipocytes and correlate with total fat mass. TNF is increased in asthma, and it increases further with allergen exposure. Thus, the TNF inflammatory pathway is common to both obesity and asthma, and it is plausible that it is upregulated by the presence of both conditions *(28,29)*.

Recent research shows that in obese humans, even in the absence of any overt inflammatory insult, there is chronic, low-grade systemic inflammation characterized by increased circulating leukocytes and increased serum concentration of cytokines, cytokine receptors, chemokines, and acute-phase proteins *(34)*. Similar results are obtained in obese mice. The origin of this inflammation appears to be, at least in part, the adipose tissue itself, because expression of a variety of inflammatory genes is upregulated in adipose tissue from obese humans or mice. The cellular source of some of these factors appear to be macrophages that infiltrate adipose tissue *(35,36)*. Systemic inflammatory markers in humans correlate with the presence of diseases common to obesity, including type 2 diabetes and atherosclerosis, suggesting that the inflammation is functionally important. Obese *Cpe^fat^* mice display innate airway hyper-responsiveness, as well as increased airway responsiveness and inflammation following ozone (O_3) exposure. These increased effects of O_3 appear to be independent of changes in lung volume or lung mass, suggesting that obesity augments the airway response to O_3 in mice *(34)*.

Adiponectin is one of the most abundant gene products in adipose tissue. In contrast with many of the other adipokines, the levels of which rise in obesity, plasma adiponectin levels are decreased in obesity, and levels increase following weight loss. The predominant metabolic effects of adiponectin are in the liver and in skeletal muscle and include increased glucose uptake, inhibition of gluconeogenesis, and increased fatty acid oxidation *(37)*. Adiponectin also has anti-inflammatory properties. Pertaining to asthma, adiponectin inhibits proliferation and migration of cultured vascular smooth muscle cells induced by mitogens *(38)*. It will be important to determine whether adiponectin has similar effects on airway smooth muscles (ASM), especially because both the AdipoR1 and AdipoR2 receptors are expressed in cultured human ASM cells. In this context, it should be noted that increased ASM mass is a feature of human asthma, and modeling studies have shown that increased muscle mass alone can account for a large part of the AHR of asthma. Taken together, the anti-inflammatory effects of adiponectin and the possibility that adiponectin may have antimitogenic effects on ASM suggest that the decreased serum concentration of adiponectin observed in the obese may contribute to the propensity toward AHR in this population *(29)*.

There are also limited data showing greater systemic inflammation in obese vs lean asthmatics, as measured by serum amyloid A, fibrinogen, and C-reactive protein *(39)*. Such changes are to be expected because levels of these acute-phase proteins are also elevated in nonasthmatic obese versus lean subjects. However, in some cases, obese asthmatics had higher serum acute-phase proteins than lean asthmatics even after correction for BMI, suggesting that systemic as well as airway inflammation exists in asthma *(29)*.

5.4. Genetics

The possibility that genes known to be important in asthma may also be important in obesity is among the most interesting areas of research in this field. Because genes tend to be pleiotropic, it is biologically plausible that genes important in one complex trait could be important in another. Linkage analysis has identified several linkage peaks with chromosomal regions that are shared for obesity and asthma phenotypes *(28)*. Chromosomal areas of 5q, 6p, 11q, and 12q all contain regions with loci common to both complex phenotypes. Chromosome 5q contains the β_2-adrenergic receptor gene *ADRB2* and the glucocorticoid receptor gene *NR3C1*. Furthermore, *ADRB2* encodes a receptor that influences sympathetic nervous system activity, which is important in controlling both airway tone and metabolic rate. The glucocorticoid receptor is involved in modulating inflammation important in both diseases. Chromosome 6p, which contains the HLA gene cluster and TNF, influences the immune and inflammatory response important in both these conditions. Chromosome 11q13 contains *UCP2, UCP3,* and the gene encoding the low-affinity immunoglobulin E receptor FCεRB. The uncoupling proteins (encoded by *UCP2* and *UCP3*) influence metabolic rate but have no known function in asthma. The low-affinity immunoglobulin E receptor is part of the T-helper type 2 inflammatory response, which is increased in asthma and has not been assessed for modification by obesity. Chromosome 12q contains the inflammatory cytokine genes *STAT6, IGF1, IL1A,* and *LTA4H.* As already noted, inflammation is a feature common to both obesity and asthma. Research needs to be conducted to relate specific genetic polymorphisms in these and other loci to the effects of the obesity phenotype on asthma *(28,40,41)*.

5.5. Female Sex Steroid Hormones

The association between obesity and asthma has been particularly strong in adult women and postpubertal girls, suggesting that female sex hormones may be contributing to the increased risk of asthma in obesity *(5,28,41)*. Aromatase, the enzyme responsible for converting androgens to estrogens, is found in adipose tissue. Therefore, it is reasonable to hypothesize that obesity increases estrogen and is associated with early menarche *(42)*, and the risk of developing asthma is particularly strong in girls with early menarche *(25)*. The two different estrogen receptors (ERs), ERα and ERβ, are expressed in adipose tissue. In general, estrogen leads to an increase in basal metabolic rate as well as increased ambulatory activity and decreased activity of lipoprotein lipase in laboratory animals. Mice genetically deficient in ERα, as well as those lacking in aromatase, are obese; however, oophorectomized ERα mice have much less fat than intact ERα mice, suggesting that the estrogen signaling through ERβ may promote fat deposition. Although there is some literature on the effect of estrogen on airway responsiveness in animal models, it is unclear how estrogen might impact the development of asthma, but both estrogen and progesterone have been shown to increase IL-4 and IL-13 in peripheral blood mononuclear cells *(5,28,41)*, and there may be other effects on immune or inflammatory cells. Pertaining to sex differences, in mice, O_3 induced injury and inflammation of the lungs was enhanced in both male and female *Cpe^fat^* mice when compared to their respective gender-matched controls, indicating, at least in obese mice, that gender had no impact augmenting response to O_3 *(34)*.

5.6. Developmental Effects, Physical Activity, and Diet

Asthma is primarily a disease of early childhood, with 90% of all cases being diagnosed by age 6. There is increasing evidence that prenatal, neonatal, and early childhood events affect the subsequent development of both asthma and obesity *(28,41)*. Although physical activity has not been shown to diminish the relationship between obesity and asthma in prospective epidemiological studies, physical activity of the mother during pregnancy may be important to the development of the sympathetic nervous system (SNS) *in utero (43)*. For example, activation of brown adipose tissue, which is regulated by the SNS, is important in increasing thermogenesis and basal metabolism through activation of uncoupling proteins. All three types of β-adrenergic receptors are expressed in adipose tissue and hence are relevant to this physiological effect *(41,43)*. Other environmental factors acting during pregnancy, such as maternal diet, maternal stress, and ambient temperature, may also affect fetal SNS development. Asthma and obesity share the possibility of decreased or defective SNS activity that may contribute to both disease phenotypes *(28,41,43)*.

A variety of dietary factors have been linked to asthma prevalence in adults and children. Specifically, antioxidant vitamins C and E, carotene, riboflavin, and pyridoxine may have important effects, with greater intake being associated with enhanced immune function, reduced asthma symptoms, less eczema, and higher lung function *(44)*. Cross-sectional studies have demonstrated a reduced risk of asthma in relation to a high intake of fruits, vegetables, whole-grain products, and fish *(44,45)*. A cross-sectional prevalence study of 1312 children (mean age = 11.4 yr) showed that frequent consumption of "fast food type of meals" had a dose-dependent association with asthma symptoms *(46)*. The benefit of diet for asthma and obesity may be achieved from the combined nutritional value in particular foods, from the interaction of foods, or the combined effect of foods in a balanced diet throughout life *(44,45)*. Weight loss achieved through diet intervention has been shown to improve lung function and asthma symptoms *(47)*.

Another dietary factor worth greater attention is the omega-3 fatty acids, for which emerging data suggest a protective effect on asthma development in childhood *(41,44, 45,48)*. Unfortunately, most of the work on diet and its effect on asthma has been done in adults or in children after the diagnosis of asthma has already been made. The ideal time in the lifecycle to assess the effects of diet is in the pregnant mother, in whom the effects of total caloric intake and dietary constituents can be measured and their effects on birthweight, obesity, and asthma can be assessed. Barker et al. *(49)* have proposed that many chronic diseases arise from adaptations the fetus makes when it is undernourished. The prototypical example of the relationship of fetal development to both asthma and obesity is the Dutch winter famine of 1944–1945. Women exposed during early and mid-pregnancy to the severe nutritional limitations imposed by the famine had offspring of reduced birth size *(28,49)*. Lower lung function and increased risk of death from obstructive airways disease as adults was increased in those exposed to famine in early and mid-gestation, but not in late gestation *(49)*. Interestingly, in follow-up studies, the prevalence of obesity was higher in 19-yr-old men exposed to famine during early to mid-gestation *(50)*, and maternal malnutrition during early gestation was associated with higher BMI and waist circumference in 50-yr-old women but not in men *(51)*. Pembrey et al. *(52)* described in a recent study that food and tobacco consumption may have sex-specific, male line transgenerational effects on health and that these transmissions are mediated by the sex chromosomes, X and Y.

With respect to asthma, Raby et al. *(53)* reported a strong relationship between low-normal gestational age and asthma symptoms at age 6 yr. Shaheen and coworkers *(54)* reported that impaired fetal growth is a risk facture for adult asthma. Low birth weight is associated with lower adult lung function *(49)* and small lung size is a known risk factor for asthma, likely because small lung size results in small airway caliber *(27)*. Animal models of obesity may provide clues about whether or how obesity affects lung development. *Cpe^fat^* mice, which become obese more slowly than lean mice, have normal lung mass at 14 to 16 wk of age, at which time they weigh about 50% more than lean controls. However, there are changes in the pressure–volume curve of their lungs, suggesting that lung development has been affected by the obesity. It is possible that obesity affects lung anatomy, airway branching structure, the nature or distribution of connective tissue, the production of surfactant, or the innervation of the tracheobronchial tree *(34)*. Ultimately, all fetal programming phenomena must have their basis in the altered expression of genes or epigenetic states. Interactions of the *in utero* environment with fetal genes may thus also contribute to the development of obesity and asthma *(28)*.

6. CONCLUSIONS

There is a significant temporal relationship between alterations in body mass and asthma. The relationship is probably multifactorial and the potential independent influences of biomechanics, inflammation, genetics, and sex-specific effects demonstrate the complex interactions in the complex traits of obesity and asthma. The likelihood of additional direct, interactive, or otherwise related contributions of physical activity, diet, and *in utero* development to the relationship between obesity and asthma further strengthens this notion. Although complex, this relationship has much to teach about how the environment and genes interact to produce disease phenotypes, and great insight can be gained by considering this potential interrelationship in a developmental context. That there are so many theoretical hypotheses underlying this relationship only enhances the intrigue related to suspected causality.

REFERENCES

1. World Health Organization. Obesity: prevention and managing the global epidemic. WHO Technical Report Series 894. Geneva, Switzerland: 2000.
2. Masoli M, Fabian D, Holt S, et al. Allergy 2004;59;469–478.
3. Hedley AA, Ogden CL, Johnson CL, et al. JAMA 2004;291:2847–2850.
4. Mannino DM, Homa DM, Akinbami L, et al. MMWR CDC Surveill Summ 2000;51:1–13.
5. Weiss ST, Shore S. Am J Respir Crit Care Med 2004;169:963–968.
6. Yunginger JW, Reed CE, O'Connell EJ, et al. Am Rev Respir Dis 1992;146:888–894.

7. The National Heart, Lung, and Blood Institute Working Group. Chest 1995;108:1380–1392.
8. Kuczmarsiki RJ, Ogden CL, Guo SS, et al. Vital Health Stat 2002;11:1–190.
9. Schaub B, von Mutius E. Curr Opin Allergy Clin Immunol 2005;5:185–193.
10. Luder E, Melnik TA, DiMaio M. J Pediatr 1998;132:699–703.
11. von Mutius E, Schwartz J, Neas LM, et al. Thorax 2001;56:835–838.
12. Belamarich PF, Luder E, Kattan M, et al. Pediatrics 2000;106:1436–1441.
13. Romieu I, Mannino DM, Redd SC, et al. Pediatr Pulmonol 2004;38:31–42.
14. Celedon JC, Palmer LJ, Litonjua AA, et al. Am J Respir Crit Care Med 2001;164:1835–1840.
15. Schachter LM, Salome CM, Peat JK, et al. Thorax 2001;56:4–8.
16. Jarvis D, Chinn S, Potts J, et al. Clin Exp Allergy 2002;32:831–837.
17. Chen Y, Dales R, Krewski D, et al. Am J Epidemiol 1999;150:255–262.
18. Del-Rio-Navarro BE, Fanghanel G, Berber A, et al. J Investig Allergol Clin Immunol 2003;13: 118–123.
19. Chinn S, Jarvis D, Burney P. Thorax 2002;57:1028–1033.
20. Luder E, Ehrlich RI, Lou WY, et al. Respir Med 2004;98:29–37.
21. Camargo CA Jr, Weiss ST, Zhang S, et al. Arch Intern Med 1999;159:2582–2588.
22. Litonjua AA, Sparrow D, Celedon JC, et al. Thorax 2002;57:581–585.
23. Nystad W, Meyer HE, Nafstad P, et al. Am J Epidemiol 2004;160:969–976.
24. Huovinen E, Kaprio J, Koskenvuo M. Respir Med 2003;97:273–280.
25. Guerra S, Wright AL, Morgan WJ, et al. Am J Respir Crit Care Med 2004;170:78– 85.
26. Gilliland FD, Berhane K, Islam T, et al. Am J Epidemiol 2003;158:406–415.
27. Gold DR, Damokosh AI, Dockery DW, et al. Pediatr Pulmonol 2003;36:514–521.
28. Tantisira KG, Weiss ST. Thorax 2001;56(Suppl.2):ii64–ii73.
29. Shore SA, Johnston RA. Pharmacol Ther 2006;110:83–110.
30. Fredberg JJ, Inouye D, Miller B, et al. Am J Respir Crit Care Med 1997;156:1752–1759.
31. Hampel H, Abraham NS, El-Serag HB. Ann Intern Med 2005;143:199–211.
32. Sulit LG, Storfer-Isser A, Rosen CL, et al. Am J Respir Crit Care Med 2005;171:659–664.
33. Visser M, Bouter LM, McQuillan GM, et al. JAMA 1999;282:2131–2135.
34. Johnston RA, Theman TA, Shore SA. Am J Physiol Regul Integr Comp Physiol 2006;290:R126–R133.
35. Weisberg SP, McCann D, Desai M, et al. J Clin Invest 2003;112:1796–1808.
36. Xu H, Barnes GT, Yang Q, et al. J Clin Invest 2003;112:1821–1830.
37. Berg AH, Combs TP, Du X, et al. Nat Med 2001;7:947–953.
38. Arita Y, Kihara S, Ouchi N, et al. Circulation 2002;105:2893–2898.
39. Jousilahti P, Salomaa V, Hakala K, et al. Ann Allergy Asthma Immunol 2002;89:381–385.
40. Weiss ST, Raby BA. Hum Mol Genet 2004;13(Spec No 1):R83–R89.
41. Weiss ST. Nat Immunol 2005;6:537–539.
42. Castro-Rodriguez JA, Holberg CJ, et al. Am J Respir Crit Care Med 2001;163:1344–1349.
43. Young JB, Morrison SF. Diabetes Care 1998;21:B156–B160.
44. McKeever TM, Britton J. Am J Respir Crit Care Med 2004;170:725–729.
45. Tabak C, Wijga AH, de Meer G, et al. Thorax 2005 in press.
46. Wickens K, Barry D, Friezema A, et al. Allergy 2005;60:1537–1541.
47. Stenius-Aarniala B, Poussa T, Kvarnstrom J, et al. BMJ 2000;320:827–832.
48. Oddy WH, de Klerk NH, Kendall GE, et al. J Asthma 2004;41:319–326.
49. Barker DJ, Godfrey KM, Fall C, et al. BMJ 1991;303:671–675.
50. Ravelli GP, Stein ZA, Susser MW. N Engl J Med 1976;295:349–353.
51. Ravelli ACJ, von der Meulen JHP, Osmond C, et al. J Clin Nutr 1999;70:811–816.
52. Pembrey ME, Bygren LO, Kaati G, and The ALSPAC Study Team. Eur J Hum Genet, 2006;14:159–166.
53. Raby BA, Celedon JC, Litonjua AA, et al. Pediatrics 2004;114:e327–e332.
54. Shaheen SO, Sterne JA, Montgomery SM, et al. Thorax 1999;54:396–402.

26 Adiposity and Kidney Disease

Srinivasan Beddhu
and Bonnie Ching-Ha Kwan

Abstract

The prevalence of chronic kidney disease is growing worldwide. Epidemiological data suggest there is a causal relationship with adiposity. There are also laboratory research and clinical studies showing that adiposity is involved in the development and the progression of kidney disease itself. Mechanisms include adaptation to increased body mass, activation of sympathetic nervous and renin–angiotensin systems, effects of insulin resistance, lipid overload, and release of adipokines. Kidney disease may also affect the association of adiposity with cardiovascular outcomes. In this chapter, the interactions of adiposity and kidney disease and their effects on clinical outcomes are examined.

Key Words: Adipose tissue; obesity; kidney disease; insulin resistance; cardiovascular outcomes.

1. INTRODUCTION

Chronic kidney disease (CKD) is increasingly common. From the Third National Health and Nutrition Examination Survey (NHANES III), it was found that the prevalence of CKD in the US adult population was 11% (19.2 million), of which 7.6 million had stage 3 (glomerular filtration rate [GFR] 30 to 59 mL/min/1.73 m^2) and 400,000 had stage 4 (GFR 15 to 29 mL/min/1.73 m^2) *(1)*. In 2002, the number of people with stage 5 CKD (GFR < 15 mL/min/1.73 m^2) requiring dialysis in the United States was approx 310,000, representing an 86% increase over the past decade. This number is expected to rise exponentially, to affect more than 650,000 individuals by 2010 *(2)*.

There is evidence that adiposity is involved in the development, and the progression, of kidney disease itself. Further, whereas adiposity is a well-known cardiovascular risk factor in the general population, epidemiological studies have raised uncertainties regarding the impact of adiposity on clinical outcomes in CKD and dialysis patients. In this chapter, we address these two issues: the effects of adiposity on kidney disease, and the effects of kidney disease on the associations of adiposity with cardiovascular risk factors and cardiovascular disease.

From: *Nutrition and Health: Adipose Tissue and Adipokines in Health and Disease*
Edited by: G. Fantuzzi and T. Mazzone © Humana Press Inc., Totowa, NJ

2. EFFECT OF ADIPOSE TISSUE ON PROGRESSION OF KIDNEY DISEASE

2.1. *Epidemiological Data*

There is evidence that in many populations, the rising trend of kidney disease in the population has also mirrored that of obesity *(3–5)*. Obesity, which is mainly caused by an increase in adipose tissue, has a direct relationship to the development and progression of diabetes mellitus, hypertension, and dyslipidemia. Diabetes and hypertension are well known to be the two most common causes of renal impairment. More than 40% of the CKD and dialysis populations have diabetes as a cause of renal failure, whereas hypertension takes up around 28%. Analyses of the data from the Modification of Diet in Renal Disease (MDRD) *(6)* and Atherosclerosis Risk in Communities (ARIC) *(7)* studies showed that high triglyceride and low high-density lipoprotein (HDL) levels are related to the development of CKD.

Kidney damage is clinically manifest as loss of albumin in urine or decline in GFR. Analysis of NHANES data further showed that abdominal obesity was associated with both a decrease in GFR and microalbuminuria (24-h urinary albumin excretion in the range of 150–300 mg/d) *(5)*. This association was also seen with each of the other elements of metabolic syndrome (i.e., insulin resistance, hypertension, hypertriglyceridemia, and low HDL level). Furthermore, there was a graded relationship between the number of components present and the corresponding prevalence of CKD and microalbuminuria.

Thus, the above data raise the question whether the association of adiposity with CKD is a mere reflection of associated diabetes, hypertension, and dyslipidemia or whether adiposity is an independent risk factor for kidney disease. In another analysis of the ARIC data, the odds ratio (OR) of developing CKD during a 9-yr follow-up period in participants with metabolic syndrome was 1.43, and remained 1.24 after adjusting for the subsequent development of diabetes and hypertension. Compared with participants with no traits of metabolic syndrome, those with one, two, three, four, or five traits of the metabolic syndrome had an OR of CKD of 1.13 (95% CI: 0.89–1.45), 1.53 (95% CI: 1.18–1.98), 1.75 (95% CI: 1.32–2.33), 1.84 (95% CI: 1.27–2.67), and 2.45 (95% CI: 1.32–4.54), respectively. Thus, metabolic syndrome is independently associated with an increased risk for incident CKD in nondiabetic adults *(8)*.

2.2. *Renal Pathology in Adiposity*

Histologically, renal biopsies of obese patients with renal failure have shown glomerulomegaly and focal segmental glomerulosclerosis *(9,10)*. To date, it is not known whether obesity-related glomerulomegaly is a cause or just an associated feature of proteinuria. It is also not certain if glomerulomegaly is a precursor of obesity-related focal segmental glomerulosclerosis.

2.3. *Mechanisms of Kidney Damage in Adiposity*

There are several biological mechanisms through which adiposity could lead to kidney damage (Table 1). Pathophysiology of increased microalbuminuria and proteinuria may include glomerular hyperfiltration, increased renal venous pressure, glomerular hypertrophy (due to mesangial cell hypertrophy and matrix production), and increased synthesis of vasoactive and fibrogenic substances (including angiotensin II, insulin,

Table 1
Mechanisms of Kidney Damage

Cardiovascular	Hypertension
Renal	Altered vascular structure and function
	Enhanced renin–angiotensin–aldosterone system
	Enhanced sympathetic nervous system
Metabolic	Hyperinsulinemia/insulin resistance
	Dyslipidemia
	Hypercortisolemia
Inflammatory	Hyperleptinemia
Hematological	Hypercoagulability
	Altered kallikrein–kinin system

leptin, and transforming growth factor [TGF]-β1) *(11)*. The following discussion elaborates on these mechanisms.

2.3.1. Adaptation to Increased Body Mass

Increased body mass leads to increased excretory load of nitrogen and metabolic waste. In the face of stable number of nephrons, this leads to increased workload with hyperperfusion and hyperfiltration of each single nephron. It has been shown that obesity-related glomerular hyperfiltration ameliorates after weight loss *(12)*.

Second, obesity and increased body mass correlate positively with systemic hypertension. This causes increase in glomerular capillary pressure, proteinuria, endothelial dysfunction, vasculopathy, and fibrosis within the kidney, leading to nephronal damage.

2.3.2. Adverse Effects of Adaptations to Obesity-Induced Sodium Retention

Obesity leads to activation of the sympathetic system, possibly caused, in part, by hyperleptinemia that stimulates the hypothalamic pro-opiomelanocortin pathway *(13)*. Angiotensin II, and hence the renin–angiotensin system, is upregulated. These cause volume expansion and increased blood pressure. Further, excess visceral adipose tissue may lead to physical compression of the kidneys, medullary compression owing to accumulation of adipose tissue around the kidney, and increased extracellular matrix within the kidney *(13,14)*, causing increasing intrarenal pressures and tubular reabsorption.

Because of increased tubular reabsorption of sodium, marked renal vasodilation and glomerular hyperfiltration set in to serve as compensatory mechanisms to maintain sodium balance *(15)*. However, chronic renal vasodilation and increased systemic arterial pressure cause increased hydrostatic pressure and glomerular capillary wall stress. Along with increased lipids and glucose intolerance, these may cause glomerular cell proliferation, matrix accumulation, and eventually glomerulosclerosis and loss of nephron function in obese subjects. This creates a slowly developing vicious cycle, with ever-increasing arterial pressure and urinary protein excretion leading to gradual but continual loss of nephron function.

2.3.3. Direct or Indirect Effects of Hyperinsulinemia/Insulin Resistance

Adipose tissue is associated with insulin resistance. This causes systemic hyperinsulinemia, which contributes to renal vascular injury by stimulating smooth muscle cell proliferation *(16)*. It also has both direct and indirect effects on the progression of glomerular damage. Direct effects include irreversible glycosylation of glomerular proteins—e.g., renal mesangial cells and stimulation of expression of inflammatory collagens *(17)*. Hyperinsulinemia augments endothelial-dependent vasodilation (increased nitric oxide production via PI3-k/Akt pathway), thus contributing to preglomerular vasodilation and glomerular hypertension *(15,18)*. Indirect effects include interaction with elevated intrarenal angiotensin II to augment angiotensin II-induced contraction of glomerular mesangial cells *(19)*. In vitro studies have shown that hyperinsulinemia can induce glomerular hypertrophy both directly, or via insulin-like growth factor (IGF)-1 *(20)*.

2.3.4. Renal Lipotoxicity

Overload of intracellular lipid leads to intracellular shunting of excess fatty acids toward synthesis of products that induce cell damage *(21–23)*. This impairs function of the individual cells and causes apoptosis, leading to reduction of total cell mass. Lipotoxicity is associated with progression of metabolic syndrome and can involve multiple organs, including kidney, liver, skeletal, pancreas, and cardiac cells *(21)*. Free fatty acids *per se* also increase oxidative stress *(24)*. In the kidneys, dyslipidemia enhances the amount of lipoproteins being filtered in the Bowman's capsule, damaging glomerular and tubular cells and enhancing endothelial dysfunction and atherosclerosis *(25,26)*.

2.3.5. Adipose Tissue as Secretory Organ

White adipose tissue can be considered as the largest secretory organ in the body. It releases a wide range of protein signals and factors, termed adipokines *(27)*. A number of adipokines—including leptin, adiponectin, tumor necrosis factor (TNF)-α, interleukin (IL)-1β, IL-6, monocyte chemoattractant protein-1, macrophage migration inhibitory factor, nerve growth factor, vascular endothelial growth factor, plasminogen activator inhibitor 1 and haptoglobin—are linked to inflammation and the inflammatory response. This leads to a low-grade inflammatory condition that is important in the causation and progression of hypertension and endothelial dysfunction.

Reduction of renal mass contributes to retention of proinflammatory adipokines, leading to adipokine imbalance *(28)*, and augments the inflammatory state in end-stage renal disease (ESRD).

Leptin is a proinflammatory adipokine. Leptin is cleared by the kidney *(29–31)*, and its concentration increases in renal impairment. In glomerular endothelial cells, leptin stimulates cellular proliferation, TGF-β1 synthesis, and type IV collagen production *(32)*. In mesangial cells, leptin upregulates synthesis of TGF-β2 receptor *(32)* and type I collagen production *(33,34)*. Physiologically, leptin activates sympathetic nervous activity *(35)*, increases sodium reabsorption, and stimulates reactive oxygen species *(36)*. All these contribute to extracellular matrix deposition and glomerulosclerosis, leading to hypertension, proteinuria *(37)*, and progression of kidney disease. There is also evidence to suggest that leptin and TGF-β1 promote mesangial sclerosis by different mechanisms and act synergistically to potentiate mesangial matrix production.

Adiponectin, on the other hand, is anti-inflammatory. Plasma adiponectin level is negatively associated with fat mass *(38,39)*. It has been implicated in "absorbing" fat cells from other organs and also reducing the degree of macrophage-to-foam cell transformation and TNF-α expression in macrophages and adipose tissue *(40)*. This may lead to reduced lipotoxicity and counteract inflammation. In addition, plasma adiponectin level has been found to be negatively associated with insulin sensitivity *(39)*.

3. MANAGEMENT

There is substantial evidence that adipose tissue and obesity are related to the progression of renal disease. If managed properly in the early stages, most of the physiological and structural changes may be reversible. On the other hand, prolonged increase in adipose tissue leads to a cycle of increased hypertension, renal damage, and further maladaptation mechanisms. Early interventions targeted toward hypertension, adiposity, and insulin resistance might minimize renal damage associated with obesity.

4. EFFECTS OF KIDNEY DISEASE ON ASSOCIATIONS OF ADIPOSITY WITH CARDIOVASCULAR RISK FACTORS AND CARDIOVASCULAR DISEASE

In contrast with the data for the general population, dialysis patients with higher body mass index have lower mortality compared with dialysis patients with normal BMI *(41,42)*. Strikingly, these data have been consistent in several studies *(43–45)*. Thus, it has been suggested that obesity is protective in dialysis patients *(41)*. In other words, as the associations of body size with mortality appear to vary depending on the presence or absence of advanced kidney failure, it can be said that kidney disease is an effect modifier of this association.

However, there are three problems with the suggestion that adiposity is protective in dialysis patients. First, the real paradox of the "BMI paradox" in dialysis patients is the possible association of high BMI with inflammation, yet with decreased mortality. Adipocytes are rich sources of proinflammatory cytokines such as IL-6 and TNF-α, which in turn stimulate the production of C-reactive protein (CRP) in the liver *(46)*. It was shown in a cross-sectional study that abdominal adiposity is strongly associated with elevated CRP levels in dialysis patients *(47)*. Further, the cross-sectional associations of high BMI, abdominal adiposity, and other components of metabolic syndrome *(48–50)* with inflammation in stage III CKD have been demonstrated. Therefore, the current evidence suggests that in stages III and V of CKD, obesity is associated with inflammation as in the general population.

Second, high BMI might result from high muscle mass, fat mass, or both. It is possible that high BMI owing to high muscle mass might be more protective than high BMI from high fat mass. In 70,028 patients initiated on hemodialysis in the United States from January 1996 to December 1998 with reported measured creatinine clearances at initiation of dialysis, BMI in conjunction with 24-h urinary creatinine excretion (an indicator of muscle mass) was used to estimate body composition, and the effects of estimated body composition on all-cause and cardiovascular mortality were examined *(51)*. High body size was associated with better survival. However, compared with normal BMI, normal or high-muscle patients, those with high BMI and low muscle mass had increased mortality, whereas those with high BMI and normal or high muscle mass had decreased mortality. These data suggest that high BMI is not uniformly associated with better survival, and body composition is important in high-BMI dialysis patients. In another study of incident peritoneal dialysis patients, similar results were shown *(52)*.

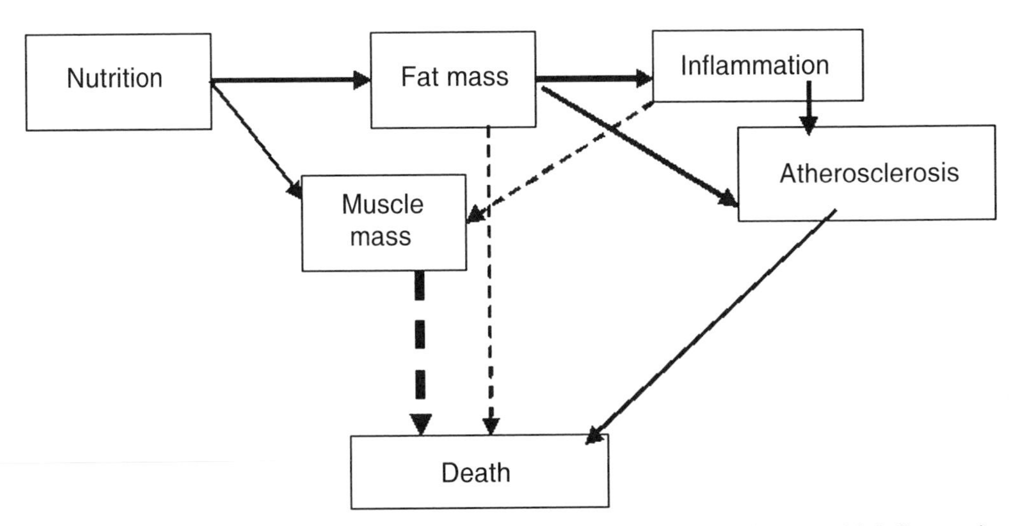

Fig. 1. Directed acylic graph of the hypothesized associations of nutritional status with inflammation, atherosclerosis, and death. Dotted lines represent a negative effect, whereas the unbroken lines represent a positive effect.

Third, previous studies have shown that in dialysis patients, adiposity and high BMI are associated with diabetes *(53),* inflammation *(54)*, coronary calcification *(55,56)*, and carotid atherosclerosis *(57)*. These data raise the question that if adiposity is associated with diabetes, inflammation, coronary calcification, and atherosclerosis in dialysis patients, how is adiposity associated with better survival in dialysis patients?

We propose the following framework (Fig. 1) to integrate these seemingly contradicting data. In Fig. 1, dotted lines represent a negative effect, whereas the unbroken lines represent a positive effect.

When the association of high BMI with survival is examined in dialysis patients, there might actually be two issues that are examined—what is the effect of nutrition on survival, and what is the effect of adiposity on atherosclerotic events and cardiovascular events? We hypothesize that the effects of nutrition on survival are much stronger than the effects of atherosclerotic events on survival in dialysis patients. Further, we also propose that the effects of nutrition on survival might differ based on body composition (muscle vs fat). Better nutrition, as evidenced by higher muscle mass, decreases the hazard of death from concomitant cardiovascular and noncardiovascular events, resulting in the lowest cardiovascular and noncardiovascular deaths. On the other hand, fat mass has dual effects: a negative effect on death as a result of nutrition, and a positive effect on death mediated through its association with inflammation and atherosclerosis. Thus, compared with undernutrition, adiposity decreases the hazard of death from concomitant disease processes but is associated with inflammation, oxidative stress, and atherosclerotic events in dialysis patients as in the general population. In other words, adiposity confers a survival advantage over undernutrition but not compared with higher muscle mass in dialysis patients.

Further, as shown in Fig. 1, the above paradigm could also incorporate the current theories on the association of inflammation with malnutrition, in particular, the observed associations of inflammation with decreased muscle mass in dialysis patients

(58,59). In other words, the association of inflammation with loss of muscle mass does not contradict adipose tissue as a source of inflammation in CKD.

5. CONCLUSIONS

In summary, obesity is a risk factor for kidney damage, as evidenced by albuminuria and loss of GFR. The effects of kidney disease on the associations of adiposity with cardiovascular risk factors and cardiovascular disease remain controversial. It is possible that in dialysis patients, despite the protective effect of adiposity with survival, it is still associated with insulin resistance, inflammation, and atherosclerosis. Further studies will shed light on this complex area.

REFERENCES

1. Coresh J, Astor BC, Greene T, et al. Am J Kidney Dis 2003;41:1–12.
2. Hostetter TH, Lising M. J Am Soc Nephrol 2003;14:S114–S116.
3. Hoehner CM, Greenlund KJ, Rith-Najarian S, et al. J Am Soc Nephrol 2002;13:1626–1634.
4. Palaniappan L, Carnethon M, Fortmann SP. Am J Hypertens 2003;16:952–958.
5. Chen J, Muntner P, Hamm LL, et al. Ann Intern Med 2004;140:167–174.
6. Hunsicker LG, Adler S, Caggiula A, et al. Kidney Int 1997;51:1908–1919.
7. Muntner P, Coresh J, Smith JC, et al. Kidney Int 2000;58:293–301.
8. Kurella M, Lo JC, Chertow GM. J Am Soc Nephrol 2005;16:2134–2140.
9. Kasiske BL, Crosson JT. Arch Intern Med 1986;146:1105–1109.
10. Cohen AH. Am J Pathol 1975;81:117–130.
11. Adelman RD. Curr Opin Nephrol Hypertens 2002;11:331–335.
12. Chagnac A, Weinstein T, Herman M, et al. J Am Soc Nephrol 2003;14:1480–1486.
13. Hall JE, Jones DW, Kuo JJ, et al. Curr Hypertens Rep 2003;5:386–392.
14. Shimotomai T, Kakei M, Narita T, et al. Ren Fail 2005;27:323–328.
15. Hall JE, Henegar JR, Dwyer TM, et al. Adv Ren Replace Ther 2004;11:41–54.
16. Bruemmer D, Law RE. Am J Med 2003;115:87S–92S.
17. Abrass CK, Spicer D, Raugi GJ. Kidney Int 1994;46:613–620.
18. Sowers JR. Am J Physiol Heart Circ Physiol 2004;286:H1597–H1602.
19. Kreisberg JI. Proc Natl Acad Sci USA 1982;79:4190–4192.
20. Abrass CK, Raugi GJ, Gabourel LS, et al. Endocrinology 1988;123:2432–2439.
21. Unger RH, Orci L. Biochim Biophys Acta 2002;1585:202–212.
22. Unger RH. Trends Endocrinol Metab 2003;14:398–403.
23. Bagby SP. J Am Soc Nephrol 2004;15:2775–2791.
24. Unger RH. Endocrinology 2003;144:5159–5165.
25. Kamijo A, Kimura K, Sugaya T, et al. Kidney Int 2002;62:1628–1637.
26. Thomas ME, Schreiner GF. Am J Nephrol 1993;13:385–398.
27. Trayhurn P, Wood IS. Biochem Soc Trans 2005;33:1078–1081.
28. Axelsson J, Heimburger O, Lindholm B, et al. J Ren Nutr 2005;15:131–136.
29. Sharma K, Considine RV, Michael B, et al. Kidney Int 1997;51:1980–1985.
30. Cumin F, Baum HP, Levens N. Int J Obes Relat Metab Disord 1996;20:1120–1126.
31. Merabet E, Dagogo-Jack S, Coyne DW, et al. J Clin Endocrinol Metab 199782:847–850.
32. Wolf G, Hamann A, Han DC, et al. Kidney Int 1999;56:860–872.
33. Wolf G, Chen S, Han DC, et al. Am J Kidney Dis 2002;39:1–11.
34. Han DC, Isono M, Chen S, et al. Kidney Int 2001;59:1315–1323.
35. Haynes WG, Morgan DA, Walsh SA, et al. J Clin Invest 1997;100:270–278.
36. Bouloumie A, Marumo T, Lafontan M, et al. FASEB J 1999;13:1231–1238.
37. Ballermann BJ. Kidney Int 1999;56:1154–1155.
38. Vilarrasa N, Vendrell J, Maravall J, et al. Clin Endocrinol (Oxf) 2005;63:329–335.
39. Farvid MS, Ng TW, Chan DC, et al. Diabetes Obes Metab 2005;7:406–413.

40. Ouchi N, Kihara S, Funahashi T, et al. Curr Opin Lipidol 2003;14:561–566.
41. Kalantar-Zadeh K, Abbott KC, Salahudeen AK, et al. Am J Clin Nutr 2005;81:543–554.
42. Kalantar-Zadeh K, Ikizler TA, Block G, et al. Am J Kidney Dis 2003;42:864–881.
43. Kopple JD, Zhu X, Lew NL, et al. Kidney Int 1999;56:1136–1148.
44. Leavey SF, McCullough K, Hecking E, et al.. Nephrol Dial Transplant 2001;16:2386–2394.
45. Leavey SF, Strawderman RL, Jones CA, et al. Am J Kidney Dis 1998;31:997–1006.
46. Bastard JP, Jardel C, Delattre J, et al. Circulation 1999;99:2221–2222.
47. Axelsson J, Qureshi AR, Heimburger O, et al. Am J Kidney Dis 2005;46:628–634.
48. Beddhu S, Kimmel PL, Ramkumar N, et al. Am J Kidney Dis 2005;46:577–586.
49. Menon V, Wang X, Greene T, et al. Am J Kidney Dis 2003;42:44–52.
50. Ramkumar N, Cheung AK, Pappas LM, et al. J Ren Nutr 2004;14:201–207.
51. Beddhu S, Pappas LM, Ramkumar N, et al. J Am Soc Nephrol 2003;14:2366–2372.
52. Ramkumar N, Beddhu S. Perit Dial Int, In press.
53. Beddhu S. Semin Dial 2004;17:229–232.
54. Stenvinkel P, Marchlewska A, Pecoits-Filho R. Kidney Int 2004;65:274–281.
55. Stompor T, Pasowicz M, Sullowicz W, et al. Am J Kidney Dis 2003;41:203–211.
56. Goodman WG, Goldin J, Kuizon BD, et al. N Engl J Med 2000;342:1478–1483.
57. Yamauchi T, Kuno T, Takada H, et al. Nephrol Dial Transplant 2003;18:1842–1847.
58. Kaizu Y, Ohkawa S, Odamaki M, et al. Am J Kidney Dis 2003;42:295–302.
59. Kaysen GA, Greene T, Daugirdas JT, et al. Am J Kidney Dis 2003;42:1200–1211.

27 Obesity and Joint Disease

Andrew J. Teichtahl, Anita E. Wluka, and Flavia M. Cicuttini

Abstract

Obesity is a risk factor for joint disease, in particular osteoarthritis, at both weight-bearing joints such as the knee, and non-weight bearing joints such as the carpometacarpal joint of the hand. Despite this, how obesity is mechanistically associated with joint disease is unclear. Both metabolic and biomechanical factors are likely to mediate the association between obesity and joint disease, although different joint tissues such as bone and cartilage are likely to differ in their response to adiposity. It may be that biomechanical factors contribute more to joint disease at weight-bearing joints such as the knee, whereas metabolic factors associated with obesity predispose joint changes at relatively non-weight bearing joints, such as those in the hand. This chapter aims to examine the evidence for a role of obesity in joint pathology using osteoarthritis as a disease paradigm.

Key Words: Obesity; cartilage; bone; osteoarthritis; metabolic; biomechanical.

1. INTRODUCTION

Musculoskeletal conditions represent a global socioeconomic problem. In 1998, it was estimated that the direct and indirect costs of musculoskeletal conditions in America were equivalent to 2.5% of the gross national product (GNP), and represented a major cause of work disability *(1)*. Currently, more than half of older American adults report chronic joint problems, and with the projected growth of the older population in the next 25 yr, the number of elders with arthropathies is expected to double, to 41 million *(2)*.

Osteoarthritis (OA) is the most common of the arthropathies; it is the eighth most common cause of disability globally and the second most common form of disability in the United States *(3)*. Obesity is likely to be the most important preventable risk factor for OA at weight-bearing joints, such as the knee. However, obesity is also associated with OA at non-weight-bearing joints, such as those in the hand. Therefore, it is not surprising that the costs in the workplace of musculoskeletal diseases were the highest among people with a body mass index (BMI) of 27.5 or greater *(4)*. Despite these significant socioeconomic issues, the association between obesity and joint disease is not mechanistically well understood. Both metabolic and biomechanical factors are likely to be important in the pathogenesis of OA, although the different variables associated with joint pathology may have a specific predilection for particular anatomical sites.

This chapter aims to examine the evidence for a role of obesity in joint pathology using OA as a disease paradigm. However, it is important to first examine the association

From: *Nutrition and Health: Adipose Tissue and Adipokines in Health and Disease*
Edited by: G. Fantuzzi and T. Mazzone © Humana Press Inc., Totowa, NJ

between obesity and the characteristic features of joint structure, such as the properties of cartilage and bone, whose phenotypic variation, along with a constellation of patient symptoms such as pain, stiffness, and loss of function, represent the hallmark of OA.

2. OBESITY AND JOINT STRUCTURE

Although OA manifests as joint changes that primarily include progressive cartilage loss and bony abnormalities such as subchondral sclerosis, osteophytes, and bone cysts, it is unclear how obesity influences joint morphology. To date, the majority of studies have measured the BMI (kg/m^2) as an indicator of obesity, and only relatively recently have parameters of body composition, such as fat distribution and lean body mass, been examined in the context of joint disease. Nevertheless, there still remains a paucity of data examining the association between joint structure and increased body mass. This lack of specific data may be partly attributable to the difficulty in obtaining reliable, valid, and sensitive measures of cartilage, bone, and body composition properties noninvasively in human subjects. However, magnetic resonance imaging (MRI) and dual-energy X-ray absorptiometry (DEXA) are becoming widely accepted as noninvasive, reliable, valid, and sensitive measures of cartilage and bony properties, as well as body composition in vivo *(5–8)*.

2.1. Association Between Obesity and Cartilage Properties

2.1.1. Cartilage as Measured by Radiology

Few studies have directly examined the association between obesity and cartilage properties. Radiographic joint space width (JSW) had previously been used as an indirect measure of articular cartilage volume. A study demonstrated that the medial and lateral knee JSW was narrower in obese patients (BMI > 30) compared with nonobese patients *(9)*. This may suggest that obesity is associated with a reduction in the amount of cartilage. However, confounding factors such as varus malignment at the knee, which strongly mediates the association between obesity and medial tibiofemoral JSW *(10)*, were not adjusted for in this study, which challenges the generalizability of the association between obesity and JSW. Indeed, other previous work had failed to demonstrate that obesity was a risk factor for joint space narrowing (JSN) *(11)*. Moreover, the assumption that JSW is a surrogate marker of articular cartilage is often misleading, as the radiographic joint space is composed of structures other than articular cartilage. For instance, mensical extrusion has been shown to account for much early joint space loss *(12)*. Additionally, radiographs employ a one-dimensional measure to assess change in a three-dimensional structure such as the joint space. Therefore, the validity of the radiographic JSW as a measure of articular cartilage is challenged, and any association between obesity and JSW does not necessitate an association between cartilage volume and obesity.

2.1.2. Cartilage Measured Directly Using MRI

There has been increasing interest in using MRI to examine joint cartilage as a measure of disease severity of OA. MRI allows direct visualization of all components of the joint simultaneously. Knee cartilage volume measured by MRI has been shown to be valid and reproducible *(6,13,14)*, sensitive to change in both normal subjects *(15)* and those with OA *(16)*, and has also been demonstrated to correlate with radiographic

Table 1
Comparisons Between MRI and Radiographs

	MRI	*Radiograph*
Direct visualization of cartilage	Yes	No
Visualization of other joint structures	Menisci, ligaments, subchondral bone, etc.	Bone only
Ionizing radiation	No	Yes
Three-dimensional	Yes	No
Continuous or ordinal variables	Articular cartilage volume—continuous	ROA grading system—ordinal Joint space width alone—continuous
Cost	Relatively high	Relatively low
Access	Limited	Routinely available
Post-processing times	Relatively high	Relatively low
Reproducibility	High	Moderate–high
Sensitivity to change	High	Low–moderate

grade of OA *(17,18)* and to predict the clinically important outcome of pain *(16,19)* and joint replacement *(20)*. MRI has therefore provided the first opportunity to directly examine articular knee cartilage noninvasively (Table 1).

Studies that have directly examined articular cartilage using MRI have demonstrated inconsistent associations between articular cartilage volume or thickness and BMI. For instance, no association was demonstrated between BMI and either articular cartilage volume or thickness in healthy and arthritic adult knee joints *(21,22)*. Similarly, overweight children have not demonstrated significantly different articular cartilage volumes than children with normal BMIs either cross-sectionally or longitudinally *(23)*. However, BMI has been inversely associated with both patellar cartilage thickness *(24)* and tibial cartilage volume *(25)*. These contrasting results imply that the effect of added body mass on cartilaginous properties in the presence and or absence of joint disease is not yet well understood.

2.1.3. Cartilage Defects and BMI

Although there does not appear to be any consistent evidence to support or refute an association between cartilage volume or thickness and BMI, there has recently been evidence to support an association between cartilage defects (measured from MRI assessment) and BMI. Cartilage defects are typically measured on an ordinal scale (Table 2), and indicate the degree of cartilage abnormality, as opposed to thickness and volume *per se*. Using this classification system, BMI is associated positively with cartilage defects *(24)*. Nevertheless, this area of research is in its infancy, and future work is required to substantiate and further examine the relationship between cartilage defects and BMI.

Table 2
Cartilage Defect Classification System

Grade 0—Normal cartilage
Grade 1—Focal blistering and intracartilaginous low-signal intensity area with an intact surface
Grade 2—Irregularities on the surface or bottom and loss of thickness of less than 50%
Grade 3—Deep ulceration with loss of thickness of more than 50%
Grade 4—Full-thickness chondral wear with exposure of subchondral bone

2.1.4. Cartilage and Body Composition

Although many studies have chosen to use the BMI as a surrogate marker of obesity, there is a paucity of data directly examining specific measures of body composition and cartilage properties. Of the limited studies that have examined the association between measures of body composition and cartilage properties, there has been a general tendency to refute an association between BMI and articular cartilage volume. In particular, studies have failed to demonstrate a significant association between the distribution of body fat and tibial cartilage volume *(25–29)*.

In contrast to body fat distribution, muscle mass in the lower limbs, muscle mass in all limbs, and total body muscle mass are significantly associated with the magnitude of medial tibial cartilage volume. Loss of muscle mass is also associated with longitudinal loss of medial and lateral tibial cartilage volume *(25)*. Whereas fat distribution does not appear to be a significant determinant of cartilage volume at the knee, measures of lean body and muscle mass appear to be important determinants of cartilage volume.

2.2. Association Between Obesity and Bony Properties

2.2.1. Bone Mineral Density and Content and BMI

The literature examining the association between obesity and bony properties has predominantly focused on bone mineral density (BMD) and bone mineral content (BMC). Compared with normal-weight people, overweight individuals (BMI $\geq$ 26) have higher BMD and BMC at both weight-bearing (e.g., femur) and non-weight-bearing (e.g., radius) sites *(30)*. Nevertheless, the association between obesity, defined as an increased BMI, and BMD/BMC is not this simplistic and is dependent on several other factors, including body composition. Abdominal obesity, body weight, and muscle strength have emerged as strong correlates of BMD in older persons *(31)*.

Although certain parameters of body composition, such as abdominal obesity, are strongly associated with BMD, gender is arguably the strongest factor that mediates the obesity–BMD relationship. In both men and women, decreased BMD occurs after the age of 50 *(32)*, although women demonstrate greater variability than men. In particular, lean body mass and total fat mass are significant determinants of BMD among postmenopausal women *(33)*. In premenopausal women, lean mass, but not total fat mass, is a significant determinant of BMD. These findings infer that whereas lean body mass is associated with BMD across the female lifespan, adiposity is most strongly associated with BMD after menopause. This may be partly attributable to the interdependence of the increased mechanical forces that occur across the obese skeleton, as well as the metabolic changes, such as the adipose-derived estrogen, that occurs after menopause. Whatever the mechanism, a positive outcome of postmenopausal obesity is its protective role against the onset and progression of osteoporosis, which is characterized by reduced BMD.

2.2.2. Other Bony Features and BMI

Although the association between an increased BMI and BMD/BMC is important in the pathogenesis of osteoporosis, it is likely that other properties of bone are associated with OA. Osteophytosis in the thoracic and lumbar spine is associated with increased BMI *(34)*. Likewise, longitudinal work has demonstrated that women in the top tertile of obesity (BMI > 26.4) have a significantly increased risk of radiographic knee osteophytes *(11)*. Twin studies have demonstrated that for each kilogram increase in body weight, the likelihood of developing osteophytes is greatly increased at the tibiofemoral, patellofemoral, and carpometacarpal joints *(35)*. Nevertheless, few studies have examined the link between measures of body composition and the bony features of OA. Of the limited studies, it does not appear that fat distribution is as strongly associated with the bony features of OA as it is for osteoporosis *(26–29)*.

Some studies have suggested that bone size alters in response to increasing BMI. Studies using DEXA, have reported a positive correlation between BMI and bone area in the tibia and femoral diaphyses in boys and girls *(36,37)*. Whereas one study found that BMI was not significantly correlated with tibial plateau area measured on radiographs *(38)*, another found that both medial and lateral tibial bone area increased significantly with increasing BMI *(24)*. Given such contrasting results, the association between BMI and bone size is equivocal, and whether a greater BMI can induce subchondral bone growth is unclear. Future work is required to examine such issues, as well as determining whether the features of body composition, such as adiposity and lean body mass, are associated with bony properties. Such work will have important ramifications for the prevention and management of joint diseases, most prominently OA. In particular, if cartilage volume reduces with the passage of time, and bone size is increased, then theoretically the combined result of these two events may result in the exposure of highly innervated articular bone surfaces and the presentation or progression of symptomatic and radiological OA.

3. ASSOCIATION BETWEEN OBESITY AND OA AT DIFFERENT JOINTS

OA is characterized by the degeneration of hyaline articular cartilage and the formation of new bone at synovial joints. The condition most notably affects weight-bearing sites, although non-weight-bearing joints, such as those in the hand, are commonly affected. Although obesity, defined by a larger than normal BMI, is associated with OA at both weight-bearing and non-weight-bearing joints, the strength and consistency of this relationship vary between different anatomical locations. Additionally, little is known about the parameters of body composition, such as fat distribution, and the prevalence of OA at different joints.

3.1. Association Between Obesity and Knee OA

The association between obesity and OA is arguably strongest and most consistent at the knee joint. In 1958, Kellgren and Lawrence found that knee OA was more common in obese people, particularly women. Since then, cross-sectional studies have consistently shown an association between obesity and knee OA, which has been stronger for women than men. Among obese middle-aged females with knee OA, it has been reported that the proportion of the disease attributable to obesity is approx 63% *(26)*.

Longitudinal studies have consistently demonstrated an association between obesity and knee OA. A 35-yr follow-up study demonstrated a strong association between being overweight and the development of OA, particularly in women *(27)*. Likewise, an increased BMI at a young age was a risk factor for knee OA in males *(39)*. Twin studies have also demonstrated that a twin with tibiofemoral and patellofemoral OA is likely to be 3 to 5 kg heavier than the co-twin. Moreover, twin studies have also demonstrated a 14% increased risk of developing tibiofemoral osteophytes and a 32% increased risk of developing patellfemoral osteophytes for every kilogram gain in body weight *(35)*.

Given that the knee is composed of distinct compartments, the association between obesity and knee OA may differ between the different knee compartments. Of the few studies that have examined the association between obesity and compartment OA, one cross-sectional study of middle-aged women demonstrated that obesity was an important risk factor for both medial tibiofemoral and patellofemoral joint disease *(35)*. Another study demonstrated an association between obesity and tibiofemoral OA, but failed to show a relationship at the patellofemoral joint *(40)*. These contrasting results highlight the need to clarify the association between obesity and OA at the different compartments in the knee complex. Nevertheless, the association between obesity and knee OA is unequivocal.

3.2. Association Between Obesity and Hand OA

The evidence examining the association between obesity and hand OA is conflicting. Data from the National Health Examination Survey demonstrated an association between BMI and the presence of hand OA in men after adjustment for age, race, and skin fold thickening *(28)*. However, this relationship was not significant after adjustment for waist girth and seat breadth. A case–control study found that obesity and hand OA were associated *(41)*. Longitudinal data also confirmed an association between radiographic hand OA and BMI in men *(42)*, although the New Haven Survey demonstrated that finger OA and obesity were more strongly associated in women than in men *(43)*.

In the Chingford study, obesity was only moderately associated with distal interphalangeal and carpometacarpal OA, but not with proximal interphalyngeal OA in women *(26)*. Another study found that there was no significant difference in weight within twin pairs discordant for osteophytes at the distal and the proximal interphalyngeal joints, although there was a 9% increased risk for developing carpometacarpal osteophytes for ever kilogram increase in body weight *(35)*. Other studies, such as the National Health Examination Survey, did not find a significant association between BMI and hand OA *(28)*. A lack of association was also identified between indices of obesity and hand OA in men in the Baltimore Longitudinal Study of Aging *(44)*.

3.3. Association Between Obesity and Hip OA

Similar to the hand, the association between obesity and hip OA is equivocal. A case-control study that examined BMI at 10-yr intervals in men who had received a hip prosthesis because of OA demonstrated that a BMI greater than one standard deviation above the mean was associated with the development of severe OA *(45)*. Relative weight was only weakly associated with OA of the hips when examining data from 4225 persons in the National Health and Nutrition Examination Survey *(46)*.

Data from the First National Health and Nutrition Examination Survey (NHANES-I) failed to demonstrate an association between obesity and hip OA *(47)*. Additionally, parameters of body composition, such as fat distribution, were not associated with hip OA. No association was demonstrated between BMI in the three decades prior to the onset of hip OA *(39)*. Given these contrasting results, no definitive conclusions can be drawn from the available data examining the possible association between obesity and hip OA. To date, the data indicate that at best, obesity is only weakly associated with OA of the hip.

4. ROLE OF OBESITY IN THE ONSET AND PROGRESSION OF OA

The natural history of OA is poorly understood, with a paucity of studies examining the factors that may alter the onset and or progression of the disease. The available evidence, which has focused primarily on knee OA, has demonstrated that obesity is an important factor in both the onset and progression of the disease.

4.1. Obesity as Risk for Onset of OA

Longitudinal studies have shown that obesity is a powerful risk factor for the development of knee OA, with one twin study finding a 9 to 13% increased risk for the onset of the disease with every kilogram increase in body weight *(35)*. For every kilogram increase in body weight, a twin had an increased likelihood of developing features of OA at the tibiofemoral, patellofemoral, and carpometacarpal joints. In women with established unilateral knee OA, obesity was the most important factor for the development of OA in the contralateral knee *(48)*. This was based on the finding that 47% of women in the top BMI tertile developed contralateral knee OA, whereas only 10% in the lowest tertile developed contralateral disease, yielding a relative risk of 4.69 for incident disease at the knee in the presence of obesity.

Obesity is also a significant risk factor the development of features associated with OA at the knee in middle-aged and elderly people. Data from the Chingford study demonstrated that middle-aged women in the top tertile of obesity (BMI > 26.4) had significantly increased risk of incident knee osteophytes (OR: 2.38, 95% CI: 1.29–4.39) *(11)*. Likewise, the Framingham study demonstrated that higher baseline BMI increased the risk of radiological knee OA at follow-up assessment in the elderly *(49)*.

Although a larger-than-normal BMI is a risk factor for OA, particularly at the knee, there is a paucity of data examining the specific features of body composition that mediate this risk. Although it is well documented that adipose tissue distribution is a risk factor for a number of metabolic complications, particularly central abdominal fat and increased risk of diabetes, cardiovascular disease, and mortality, independent of degree of obesity *(50)*, central fat does not appear to be a risk for the onset of OA at the hand or knee *(28)*. This finding may infer a stronger biomechanical, rather than metabolic, predisposition toward the onset of OA.

4.2. Obesity as Risk for Progression of OA

Despite the significant limitations associated with the radiological assessment of OA (Table 1), X-rays have been routinely used as the gold standard to assess the progression of OA. In particular, a reduction in the JSW is regarded as the hallmark of disease progression. Whereas cross-sectional studies examining the association between BMI and radiological JSW have reported conflicting results *(10,11)*, obesity has been consistently associated with a longitudinal reduction in the JSW.

A 12-yr follow-up study found that among people with knee OA, larger body mass indices were a risk for a reduction in the JSW, and therefore the radiological progression of OA (OR: 11.1; 95% CI: 3.3–37.3) *(51)*.

Although obesity is a risk factor for longitudinal radiological JSN, the assessment of the JSW as an outcome measure for the progression of OA is often insensitive. Raynauld et al. found that over a 2-yr period, radiological assessment was unable to distinguish significant changes in the JSW in people with knee OA, despite a significant loss of articular cartilage volume *(52)*. In contrast, MRI studies have revealed that as little as 2% change in cartilage volume may be reliably detected when a maximum of 6 individuals (patella), 10 (femur), 28 (medial tibia), and 33 (lateral tibia) are followed longitudinally *(53)*. Although obesity is a risk factor for the progression of radiological JSN, it would appear that the assessment of cartilage volume measured from MRI is a more sensitive indicator of disease progression in OA. Nevertheless, no studies have directly assessed the relationship between obesity and longitudinal loss of articular cartilage volume from MRI assessment. Moreover, no longitudinal study has examined the specific parameters of body composition, such as fat distribution, and the risk of the progression of knee OA. Further work is required in these areas.

5. ASSOCIATION BETWEEN OA AND BODY FAT DISTRIBUTION

Although the preponderance of studies support an association between the onset and progression of OA and BMI, particularly at the knee, little is known about the association between specific parameters of body composition and OA. Of the limited studies, fat distribution does not appear to be a significant determinant of OA.

A study examining 317 women found that the waist-to-hip ratio and percent body fat were not associated with the grade of hand OA after adjustment for age *(54)*. Similarly, the association between the percentage of body fat and the waist-to-hip ratio with knee OA in men and women was not significant after adjustment for BMI *(29)*.

In contrast, lower-limb, but not upper-limb, lean body mass is associated with the severity of knee OA *(55)*. This may suggest that limb-specific lean body mass is more of a determinant than fat distribution in mediating the association between OA and BMI. Indeed, muscle mass is an independent predictor of medial–tibial cartilage volume in healthy people, and is associated with a reduction in the rate of loss of tibial cartilage *(25)*. Although BMI is associated with OA, preliminary results indicate that muscle, rather than fat mass, is the more important feature of body composition mediating this relationship.

6. MECHANISMS FOR OBESITY IN THE PATHOGENESIS OF OA

As discussed previously, obesity is associated with OA in weight-bearing and non-weight-bearing joints and is more common in women than men, particularly after menopause. Although the mechanisms by which obesity influences the pathogenesis of OA are unknown, metabolic and biomechanical hypotheses have been proposed.

6.1. Biomechanical Mechanism for OA

Given that the knee, which is a weight-bearing joint, has the strongest and most consistent evidence for an association between obesity and OA, it is plausible that biomechanical factors may be important in mediating the association between obesity and OA. However, the mechanical hypothesis has received little attention in epidemiological studies.

Even though the knee adduction moment, which concentrates load to the medial tibiofemoral compartment, represents one of the most important biomechanical variables associated with knee OA, no study has examined its relationship with obesity. Nevertheless, it is intuitive to suggest that added weight would increase joint reaction forces, which may adversely affect joint structure. Indeed, the knee adduction moment has been associated with the size of the medial tibial plateau *(56)*, although it is unclear whether obesity may help mediate this relationship. It has also been argued that obesity increases subchondral bony stiffness *(57)*, making bone less adept at coping with impact loads. The increased bony stiffness may subsequently redistribute greater force across the articular cartilage, increasing its vulnerability to degenerative changes.

Although the knee adduction moment has received little attention in the context of obesity, the relationship between obesity and knee malalignment in the pathogenesis of OA is of growing interest. In a study examining 300 adults with knee OA, the severity of radiological OA was related to BMI in people with varus knees *(10)*. A similar association was not demonstrated for those people with valgus knees. A varus aligned knee has been consistently shown to increase medial tibiofemoral load, whereas a valgus aligned knee inconsistently increases lateral tibiofemoral load. Interestingly, it was also shown that the impact of BMI on radiological JSW was greatly reduced when controlling for varus malalignment. This indicated that almost all of the effect of BMI on medial tibiofemoral disease severity was explained by varus malalignment.

Limb alignment is also strongly associated with the risk of radiological progression in people with pre-existing OA *(58)*. In particular, the risk of progression is increased in obese people with moderately malaligned knees, but not in obese people with neutral or severe malalignment. This may be the result of the combined focus of load from moderate malalignment and the excess load from increased weight. Future studies examining the radiological features of knee OA should therefore adjust for malalignment.

6.2. Metabolic Mechanism for OA

The female disparity and increased incidence of postmenopausal onset of generalized OA, as well as the prevalence of disease in non-weight-bearing joints, such as the hand, suggest the likely existence of a metabolic/systemic component in the pathogenesis of OA. Despite this, the majority of studies have not been able to identify a metabolic link between obesity and OA.

Generally, both cross-sectional and longitudinal studies have not revealed factors that that may help explain the association between obesity and OA. Adjusting for blood pressure, body fat distribution, serum lipids, serum uric acid, and blood glucose has generally failed to reduce the association between obesity and OA, implying that these metabolic factors are not significant mediators of the obesity–OA relationship *(28,29,42,59–61)*.

Few studies have found significant associations between osteoarthritis and other metabolic conditions. The Chingford population study demonstrated that "ever treated" hypertension was associated with the development of OA, particularly for bilateral knee disease *(11,26)*. This study also suggested an association between bilateral knee disease and hypercholesterolemia and raised blood glucose. Similarly, after adjusting for age, slightly higher levels of plasma glucose were demonstrated in arthritic women than in normal controls *(62)*. Nevertheless, most studies generally do not support an association between hypertension, raised serum cholesterol, glucose, and OA, and the data therefore remain inconclusive.

Despite these inconsistent findings, it may be that unexamined or unidentified metabolic factors mediate the association between obesity and OA. For example, there is emerging evidence that leptin may be important in the pathogenesis of OA *(63)*. In particular, the discovery that osteoblasts and chondrocytes are capable of leptin synthesis and secretion *(64,65)*, as well as existence of leptin receptors at articular cartilage *(66)*, may have significant implications for future studies examining the metabolic link between obesity and OA. Indeed, significant levels of leptin were observed in the cartilage and osteophytes of people with OA, yet few chondrocytes produced leptin in the cartilage of healthy people *(64)*.

To date, the data have tended to support mechanical rather than metabolic factors to help to account for the association between obesity and OA. However, there is increasing evidence that OA is not a single disorder but a heterogenous group of disorders, with a complex interplay of several factors that may result in a common pathway of joint damage. Whereas obesity may have a mechanical effect on some joints, it is possible that it may have a metabolic effect on other joints. It may be that obesity manifests increased mechanical stress across weight-bearing joints such as the knee, but the effect of obesity on the small non-weight-bearing joints of the hand is through metabolic mechanisms.

7. ROLE OF WEIGHT LOSS IN PREVENTION AND MANAGEMENT OF OA

Weight loss may represent the most modifiable risk factor amenable to conservative treatment in the pathogenesis of OA. Despite this, few studies have examined the effect of weight loss in subjects with established disease.

Although the contribution of obesity to symptom development in individuals with OA is unclear, weight loss is correlated with reduced pain among people with established disease *(67)*. In particular, when a reduction in the total percentage of body fat occurs via physical activity, symptomatic improvement of knee OA can be achieved *(68)*. Similarly, when weight loss is mediated by a significant loss of body fat, significant functional improvement is apparent *(69)*.

As well as assisting with symptom management and functional improvement, weight loss can significantly affect the risk for the development of knee OA *(27)*. When 64 women with confirmed radiographic knee OA who had developed recent-onset symptomatic OA were compared with women without the disease, weight change significantly affected the risk for the development of knee OA. For instance, a decrease in BMI of two or more units over the 10 yr before follow-up examination decreased the odds for developing OA by more than 50% (OR: 0.46; 95% CI: 0.24–0.86). Women with a high risk for OA, defined by elevated baseline BMI (>25), also decreased their risk for the onset of OA by weight loss of two or more units of BMI (OR: 0.41).

Although the limited data suggest that a reduction of body mass—and, in particular, body fat—may be important in the prevention of OA, there is a lack of data examining whether weight loss affects the progression of OA. Given the aging population, as well as the growing number of obese individuals in the developed world, further work must examine the role of weight loss in altering the natural history of OA.

8. CONCLUSIONS

Few studies have examined the relationship between obesity and joint structure in the absence of significant joint disease. Of these studies, obesity has been more strongly and consistently associated with bone mineral density. Studies of the association between added body mass and other joint properties, such as cartilage volume and bone size, have yielded conflicting results. Future studies examining the link between obesity and joint structure will however benefit from direct examination of the joint, such as MRI assessment, rather than indirect and arguably invalid examinations such as measuring the radiological joint space width as a marker of cartilage volume. The limited number of studies that have directly examined the association between cartilage defects and BMI have shown promising results.

Although a larger-than-normal BMI is a risk factor for the onset and progression of OA at the knee, the data associating obesity and OA of the hand and hip are equivocal. Mechanistically, it is likely that the association between obesity and the pathogenesis of OA is mediated by both biomechanical and metabolic factors. To date, no consistently strong associations have been shown between metabolic factors and OA. It may be that unidentified or unexamined metabolic factors are important in mediating the relationship between obesity and OA. Among the few biomechanical studies available, varus malalignment in the presence of obesity is strongly associated with the severity of knee OA, and is also a risk factor for the radiological progression of the disease.

Despite the BMI–OA relationship, it is unclear how features of body composition influence disease. Unlike cardiovascular disease, central adiposity does not appear to be associated with OA. To date, muscle mass appears to be the strongest identifiable feature of body composition that is associated with both normal and arthritic knee joints. Nevertheless, preliminary findings have suggested that weight loss in obese people appears to be the most modifiable factor to reduce the risk for the onset of knee OA. Further work is required to help establish whether weight loss slows the progression of OA in people with established disease.

REFERENCES

1. Yelin EH. Arthritis Care Res 1995;8(4):311–317.
2. Leveille SG. Curr Opin Rheumatol 2004;16(2):114–118.
3. Kee CC. Nurs Clin North Am 2000;35(1):199–208.
4. Burton WN, Chen CY, Schultz AB, et al. J Occup Environ Med 1998;40(9):786–792.
5. Pilch L, Stewart C, Gordon D, et al. J Rheumatol 1994;21(12):2307–2321.
6. Peterfy CG, van Dijke CF, Janzen DL, et al. Radiology 1994;192(2):485–491.
7. Burgkart R, Glaser C, Hyhlik-Durr A, et al. Arthritis Rheum 2001;44(9):2072–2077.
8. Graichen H, von Eisenhart-Rothe R, Vogl T, et al. Arthritis Rheum 2004;50(3):811–816.
9. Cimen OB, Incel NA, Yapici Y, et al. Ups J Med Sci 2004;109(2):159–164.
10. Sharma L, Lou C, Cahue S, et al. Arthritis Rheum 2000;43(3):568–575.

11. Hart DJ, Doyle DV, Spector TD. Arthritis Rheum 1999;42(1):17–24.
12. Adams JG, McAlindon T, Dimasi M, et al. Clin Radiol 1999;54(8):502–506.
13. Marshall KW, Mikulis DJ, Guthrie BM. J Orthop Res 1995;13(6):814–823.
14. Cicuttini F, Forbes A, Morris K, et al. Osteoarthritis Cartilage 1999;7(3):265–271.
15. Wluka AE, Stuckey S, Snaddon J, et al. Arthritis Rheum 2002;46(8):2065–2072.
16. Wluka AE, Wolfe R, Stuckey S, et al. Ann Rheum Dis 2004;63(3):264–268.
17. Cicuttini FM, Wluka AE, Forbes A, et al. Arthritis Rheum 2003;48(3):682–688.
18. Cicuttini FM, Wang YY, Forbes A, et al. Clin Exp Rheumatol 2003;21(3):321–326.
19. Hunter DJ, March L, Sambrook PN. Osteoarthritis Cartilage 2003;11(10):725–729.
20. Cicuttini FM, Jones G, Forbes A, et al. Ann Rheum Dis 2004;63(9):1124–1127.
21. Eckstein F, Winzheimer M, Westhoff J, et al. Anat Embryol (Berl) 1998;197(5):383–390.
22. Karvonen RL, Negendank WG, Teitge RA, et al. J Rheumatol 1994;21(7):1310–1318.
23. Jones G, Ding C, Glisson M, et al. Pediatr Res 2003;54(2):230–236.
24. Ding C, Cicuttini F, Scott F, et al. Obes Res 2005;13(2):350–361.
25. Cicuttini FM, Teichtahl AJ, Wluka AE, et al. Arthritis Rheum 2005;52(2):461–467.
26. Hart DJ, Spector TD. J Rheumatol 1993;20(2):331–335.
27. Felson DT, Zhang Y, Anthony JM, et al. Ann Intern Med 1992;116(7):535–539.
28. Davis MA, Neuhaus JM, Ettinger WH, et al. Am J Epidemiol 1990;132(4):701–707.
29. Hochberg MC, Lethbridge-Cejku M, Scott WW Jr, et al. J Rheumatol 1995;22(3):488–493.
30. Holbrook TL, Barrett-Connor E. Bone Miner 1993;20(2):141–149.
31. Stewart KJ, Deregis JR, Turner KL, et al. J Intern Med 2002;252(5):381–388.
32. Marcus R. Rheum Dis Clin North Am 2001;27(1):131–141, vi.
33. Douchi T, Yamamoto S, Oki T, et al. Maturitas 2000;34(3):261–266.
34. O'Neill TW, McCloskey EV, Kanis JA, et al. J Rheumatol 1999;26(4):842–848.
35. Cicuttini FM, Baker JR, Spector TD. J Rheumatol 1996;23(7):1221–1226.
36. Nordstrom P, Pettersson U, Lorentzon R. J Bone Miner Res 1998;13(7):1141–1148.
37. Pettersson U, Nordstrom P, Alfredson H, et al. Calcif Tissue Int 2000;67(3):207–214.
38. Dacre JE, Scott DL, Da Silva JA, et al. Br J Rheumatol 1991;30(6):426–428.
39. Gelber AC, Hochberg MC, Mead LA, et al. Am J Med 1999;107(6):542–548.
40. Cooper C, McAlindon T, Snow S, et al. J Rheumatol 1994;21(2):307–313.
41. Oliveria SA, Felson DT, Cirillo PA, et al. Epidemiology 1999;10(2):161–166.
42. Bagge E, Bjelle A, Eden S, et al. J Rheumatol 1991;18(8):1218–1222.
43. Acheson RM, Collart AB. Ann Rheum Dis 1975;34(5):379–387.
44. Hochberg MC, Lethbridge-Cejku M, Plato CC, et al. Am J Epidemiol 1991;134(10):1121–1127.
45. Vingard E. Acta Orthop Scand 1991;62(2):106–109.
46. Hartz AJ, Fischer ME, Bril G, et al. J Chronic Dis 1986;39(4):311–319.
47. Tepper S, Hochberg MC. Am J Epidemiol 1993;137(10):1081–1088.
48. Spector TD, Hart DJ, Doyle DV. Ann Rheum Dis 1994;53(9):565–568.
49. Felson DT, Zhang Y, Hannan MT, et al. Arthritis Rheum 1997;40(4):728–733.
50. Joos SK, Mueller WH, Hanis CL, et al. Ann Hum Biol 1984;11(2):167–171.
51. Schouten JS, van den Ouweland FA, Valkenburg HA. Ann Rheum Dis 1992;51(8):932–937.
52. Raynauld JP, Martel-Pelletier J, Berthiaume MJ, et al. Arthritis Rheum 2004;50(2):476–487.
53. Eckstein F, Westhoff J, Sittek H, et al. AJR Am J Roentgenol 1998;170(3):593–597.
54. Hochberg MC, Lethbridge-Cejku M, Scott WW Jr, et al. Osteoarthritis Cartilage 1993;1(2):129–135.
55. Toda Y, Segal N, Toda T, et al. J Rheumatol 2000;27(10):2449–2454.
56. Jackson BD, Teichtahl AJ, Morris ME, et al. Rheumatology (Oxford) 2004;43(3):311–314.
57. Dequeker J, Goris P, Uytterhoeven R. JAMA 1983;249(11):1448–1451.
58. Felson DT, Goggins J, Niu J, et al. Arthritis Rheum 2004;50(12):3904–3909.
59. Davis MA, Ettinger WH, Neuhaus JM. J Rheumatol 1988;15(12):1827–1832.
60. Anderson JJ, Felson DT. Am J Epidemiol 1988;128(1):179–189.
61. Carman WJ, Sowers M, Hawthorne VM, et al. Am J Epidemiol 1994;139(2):119–129.
62. Cimmino MA, Cutolo M. Clin Exp Rheumatol 1990;8(3):251–257.
63. Teichtahl AJ, Wluka AE, Proietto J, et al. Med Hypotheses 2005;65(2):312–315.
64. Dumond H, Presle N, Terlain B, et al. Arthritis Rheum 2003;48(11):3118–3129.

65. Kume K, Satomura K, Nishisho S, et al. J Histochem Cytochem 2002;50(2):159–169.
66. Figenschau Y, Knutsen G, Shahazeydi S, et al. Biochem Biophys Res Commun 2001;287(1):190–197.
67. McGoey BV, Deitel M, Saplys RJ, et al. J Bone Joint Surg Br 1990;72(2):322–323.
68. Toda Y, Toda T, Takemura S, et al. J Rheumatol 1998;25(11):2181–2186.
69. Christensen R, Astrup A, Bliddal H. Osteoarthritis Cartilage 2005;13(1):20–27.

APPENDICES

Appendix I

ABBREVIATIONS

ACC	acetyl-CoA carboxylase
ACTH	adrenocorticotropin
adipoR	adiponectin receptor
AGAPT	1-acylglycerol-3-phosphate acyltransferase
AGL	acquired generalized lipodystrophy
AGRP	agouti-related protein
AHR	airway hyper-responsiveness
AIA	antigen-induced arthritis
AIDS	acquired immunodeficiency syndrome
ALBP	adipocyte lipid binding protein
ALT	alanine aminotransferase
AMPK	AMP-dependent protein kinase
ANP	atrial natriuretic peptide
APL	acquired partial lipodystrophy
AQPap	aquaporin
AR	adrenergic
Arc	arcuate nucleus
ASM	airway smooth muscle
AT	adipose tissue
ATGL	adipose tissue lipase
ATM	adipose tissue macrophage
ATP	adenosine triphosphate
BAT	brown adipose tissue
BBB	blood–brain barrier
BM	bone marrow
BMC	bone mineral content
BMD	bone mineral density
BMI	body mass index
BP	blood pressure
BPD	biliopancreatic diversion

From: *Nutrition and Health: Adipose Tissue and Adipokines in Health and Disease*
Edited by: G. Fantuzzi and T. Mazzone © Humana Press Inc., Totowa, NJ

BSCL2	Berardinelli Seip congenital lipodystrophy 2
CAD	coronary artery disease
cAMP	cyclic adenosine monophosphate
CART	cocaine and amphetamine-regulated transcript
CCl4	carbon tetrachloride
CD	Crohn's disease
CDC	Centers for Disease Control and Prevention
CEBP	CCAAT/enhancer binding protein
CGL	congenital generalized lipodystrophy
cGMP	cyclic guanosine monophosphate
CGRP	calcitonin gene-related peptide
ChCREBP	carbohydrate response element binding protein
CHD	coronary heart disease
CHO	high carbohydrate
CKD	chronic kidney disease
CNS	central nervous system
CO	cardiac output
ConA	concanavalin A
COX	cyclooxygenase
CPAP	continuous positive airway pressure
CRH	corticotropin-releasing hormone
CRP	C-reactive protein
CSF	cerebrospinal fluid
CVD	cardiovascular disease
DAG	diacylglycerol
DBP	diastolic blood pressure
DEXA	dual X-ray absorptiometry
DGAT	diacylglycerol acyltransferase
DM	diabetes mellitus
DMN	dorsomedial nucleus
DNL	*de novo* lipogenesis
DSS	dextran sulfate sodium
EAE	experimental autoimmune encephalomyelitis
EAT	epicardial adipose tissue
ECM	extracellular matrix
EIH	experimentally-induced hepatitis
eNOS	endothelial nitric oxide synthase
ER	estrogen receptor
ERK	extracellular-regulated kinase
ESRD	end-stage renal disease
FA	fatty acid
FABP	fatty acid binding protein
FAS	fatty acid synthase
FAT	fatty acid transporter
FDA	Food and Drug Administration
FFA	free fatty acids

FMLP	*N*-formyl-methionyl-leucyl-phenylalanine
FOX	forkhead box
FPL	familial partial lipodystrophy
FPLD	familial partial lipodystrophy, Dunningan variety
G3P	glycerol 3 phosphate
GalN	D-galactosamine
G-CSF	granulocyte-colony stimulating factor
GERD	gastroesophageal reflux disease
GFR	glomerular filtration rate
GH	growth hormone
GHRH	growth hormone-releasing hormone
Glut	glucose transporter
GM-CSF	granulocyte macrophage colony-stimulating factor
GPAT	G3P acyltransferase
GPCR	G protein-coupled receptor
GST	glutathione-*S*-transferase
HAART	highly active antiretroviral therapy
HARS	HIV-associated adipose redistribution syndrome
HB-EGF	heparin-binding epidermal growth factor-like growth factor
HCC	hepatocellular carcinoma
HDL	high-density lipoprotein
HG	hunter-gatherer
HIV	human immunodeficiency virus
HLA	human leukocyte antigen
HMW	high molecular weight
HPA	hypothalamus-pituitary-adrenal axis
HR	heart rate
HSC	hepatic stellate cells
HSL	hormone-sensitive lipase
HTA	hypertension
IBD	inflammatory bowel disease
IBMX	isobutyl-methylxantine
ICAM	intercellular adhesion molecule
IFN	interferon
Ig	immunoglobulin
IGF	insulin-like growth factor
IGFBP	insulin-like growth factor binding protein
IκB	inhibitor of κB
IL	interleukin
IMT	intimal-medial thickness
iNOS	inducible nitric oxide synthase
IR	insulin resistance
IRS	insulin receptor substrate
JAK	Janus-activated kinase
JSN	joint space narrowing
JSW	joint space width

KO	knockout
LCFA-CoA	long-chain fatty acid Coenzyme A
LDHIV	lipodystrophy in HIV-infected subjects
LDL	low-density lipoprotein
LEPR	leptin receptor
LH	luteinizing hormone
LIF	leukemia inhibitory factor
LPA	lysophosphatidic acid
LPL	lipoprotein lipase
LPS	lipopolysaccharide
LVM	left ventricular mass
LXR	liver X receptor
MAD	mandibuloacral dysplasia
MAG	monoacylglyerol
MAPK	mitogen-activated protein kinase
MBP	myelin basic protein
mBSA	methylated bovine serum albumin
MC4R	melanocortin 4 receptor
MCH	melanin-concentrating hormone
MCP	monocyte chemoattractant protein
MHC	major histocompatibility complex
MIP	macrophage inflammatory protein
MMP	matrix metalloproteinase
MRI	magnetic resonance imaging
MS	multiple sclerosis
MSC	mesenchimal stem cells
MSH	melanin-stimulating hormone
NAFLD	non-alcoholic fatty liver disease
NASH	non-alcoholic steatohepatitis
NEFA	non-esterified fatty acids
NFκB	nuclear factor kappa B
NGF	nerve growth factor
NHANES	National Health and Nutrition Examination Survey
NIH	National Institutes of Health
NK	natural killer
NO	nitric oxide
NOD	non-obese diabetic
NOS	nitric oxide synthase
NPY	neuropeptide Y
NRTI	nucleoside reverse transcriptase inhibitor
OA	osteoarthritis
Ob-R	leptin receptor
OR	odds ratio
OSA	obstructive sleep apnea
OSAS	obstructive sleep apnea syndrome
PAI	plasminogen activator inhibitor

PBEF	pre-B cell colony-enhancing factor
PBMC	peripheral blood mononuclear cells
PDGF	platelet-derived growth factor
PE	pelvic endometriosis
PECAM	platelet-endothelial cell adhesion molecule
PEPCK	Phosphoenolpyruvate carboxykinase $\gamma\gamma$
PGC	peroxisome proliferator-activated receptor γ coactivator
PHA	phytoemagglutinin
PI	protease inhibitor
PI3K	phosphoinositide-3 kinase
PIF	proteolysis-inducing factor
PIR	poverty–income ratio
PKA	protein kinase A
PKB	protein kinase B
PKC	protein kinase C
PMA	phorbol myristate acetate
PMN	polymorphonuclear cells
POMC	proopiomelanocortin
PPAR	peroxisome proliferator-activated receptor
PSA	prostate-specific antigen
PTP	protein tyrosine phosphatase
PUFA	polyunsaturated fatty acids
RA	rheumatoid arthritis
RDI	respiratory disturbance index
RELM	resistin-like molecule
RNS	reactive nitrogen species
ROS	reactive oxygen species
RR	relative risk
RXR	retinoic acid receptor
RYGB	Roux-en-Y gastric bypass
SA	subsistence agriculture
SAA	serum amyloid A
SBP	systolic blood pressure
sc	subcutaneous
SCD	stearoyl-CoA desaturase
SDR	sleep-disordered breathing
SES	socioeconomic status
sFRP	secreted frizzled-related protein
SH2	src homology 2 domain
SHBG	sex hormone binding globulin
SHP-2	SH-2-phosphatase
SLE	systemic lupus erythematosus
SLR	soluble leptin receptor
SNS	sympathetic nervous system
SOCS	suppressor of cytokine signaling
SOD	superoxide dismutase

SP	substance P
SR-BI	scavenger receptor-BI
SREBP	sterol regulatory element binding protein
STAT	signal transducer and activator of transcription
T1D	type 1 diabetes
T2DM	type 2 diabetes mellitus
T3	triiodotyronine
T4	thyroxine
TAG	triacylglycerols
TAO	thyroid-associated ophthalmopathy
TG	triglyceride
TGF	transforming growth factor
Th	T helper
TIMP	tissue inhibitor of metalloproteinase
TNF	tumor necrosis factor
TRAF	TNF-α receptor associated factor
TRH	thyrotropin-releasing hormone
Trk-A	tyrosine kinase A
TSH	thyroid-stimulating hormone
TZD	thiazolidinediones
UC	ulcerative colitis
UCP	uncoupling protein
UDCA	ursodeoxycholic acid
VCAM	vascular cell adhesion molecule
VDR	vitamin D receptor
VEGF	vascular endothelial growth factor
VF	visceral fat
VLDL	very low density lipoprotein
VMN	ventromedial nucleus
VSMC	vascular smooth muscle cells
WAT	white adipose tissue
WC	waist circumference
WHO	World Health Organization
WHR	waist-to-hip ratio
Wnt	wingless-related MMTV integration site
WT	wild-type
ZAG	zinc-α2-glycoprotein
ZIA	zymosan-induced arthritis

Appendix II

USEFUL READING MATERIAL

This appendix contains suggestions about books, journals and web sites that discuss adipose tissue and obesity, with its associated pathologies. This is not intended to be a comprehensive list and is only aimed at providing indications on recently published material on the topic.

BOOKS

1. Aihaud G, ed. Adipose Tissue Protocols. Humana Press, Totowa, NJ: 2001.
2. Bray GA, Bouchard C, eds. Handbook of Obesity, Second Edition ed. Taylor and Francis Group, New York, NY: 2003.
3. Bray GA, ed. Overweight and the Metabolic Syndrome. Humana Press, Totowa, NJ: 2007.
4. Cameron N, Norgan NGE, Ellison GTH, eds. Childhood Obesity: Contemporary Issues. Taylor and Francis, Boca Raton, FL: 2006.
5. Ehud U, ed. Neuroendocrinology of Leptin. Karger, Basel, New York, NY: 2000.
6. Ford J. Diseases and Disabilities Caused by Weight Problems: The Overloaded Body. Mason Crest Publishers, Philadelphia, PA: 2006.
7. Frisch RE, ed. Female Fertility and the Body Fat Connection. University of Chicago Press, Chicago, IL: 2002.
8. Goldstein DJ, ed. Management of Eating Disorders and Obesity, The Second Edition. Humana Press, Totowa, NJ: 2004.
9. Hunter W. How Genetics and the Environment Shape Us: The Destined Body. Mason Crest Publishers, Philadelphia, PA: 2006.
10. Klaus S, ed. Adipose Tissue. Landes Bioscience, Austin, TX: 2001.
11. Kushner RF, Bessesen DH, eds. Treatment of the Obese Patient. Humana Press, Totowa, NJ: 2007.
12. Mantzoros CS, ed. Obesity and Diabetes. Humana Press, Totowa, NJ: 2006.
13. McElroy SL, Allison DB, Bray GA, eds. Obesity and Mental Disorders. Taylor and Francis, New York, NY: 2006.
14. Opara E, ed. Nutrition and Diabetes: Pathophysiology and Management. CRC, Taylor and Francis, Boca Raton, FL: 2006.

From: *Nutrition and Health: Adipose Tissue and Adipokines in Health and Disease*
Edited by: G. Fantuzzi and T. Mazzone

15. Robinson MK, Thomas A, eds. Obesity and Cardiovascular Disease. Taylor and Francis Group, New York, NY: 2006.
16. Shah PK, ed. Risk Factors in Coronary Artery Disease. Taylor and Francis Group, New York, NY: 2006.
17. Shils ME, ed. Modern Nutrition in Health and Disease, 10th ed. Lippincott Williams & Wilkins, Philadelphia, PA: 2006.
18. Sugerman HJ, Nguyen NT, eds. Management of Morbid Obesity. Taylor and Francis, New York, NY: 2006.
19. Vaidya V, ed. Health and Treatment Strategies in Obesity. Karger, Basel, New York, NY: 2006.
20. Wolf G, ed. Obesity and the Kidney. Karger, Basel, New York, NY: 2006.
21. Wood PA. How Fat Works. Harvard University Press, Cambridge, MA: 2006.

JOURNALS

Many scientific journals of general interest to biomedical scientists and clinicians periodically publish research and review articles on the topic of adipose tissue, adipokines and obesity. These publications include, among others, Nature, Nature Medicine, PloS Medicine, Science, The Lancet, The Journal of the American Medical Association, The New England Journal of Medicine, and many others. Below is a selected list of journals specifically covering issues related to adipose tissue and obesity. In addition to these, articles on adipose tissue and its role in pathology can be found in publications specifically devoted to each pathology (asthma, cancer, cardiovascular disease, diabetes, osteoarthritis, etc.)

1. *Cell Metabolism*. Cell Press, Cambridge, MA.
2. *Diabetes*. American Diabetes Association, Alexandria, VA.
3. *Diabetes Care*. American Diabetes Association, Alexandria, VA.
4. *Diabetologia*. Springer, New York, NY.
5. *Endocrine Reviews*. Endocrine Society, Chevy Chase, MD.
6. *Endocrinology*. Endocrine Society, Chevy Chase, MD.
7. *International Journal of Obesity*. Nature Publishing Group, London, UK.
8. *Journal of Clinical Endocrinology and Metabolism*. Endocrine Society, Chevy Chase, MD.
9. *Nature Clinical Practice Endocrinology and Metabolism*. Nature Publishing Group, London, UK.
10. *Obesity*. North American Association for the Study of Obesity, Silver Spring, MD.
11. *Trends in Endocrinology and Metabolism*. Elsevier Science, London, UK.

USEFUL WEB SITES

1. American Diabetes Association: www.diabetes.org.
2. American Heart Association: www.amhrt.org.
3. American Obesity Association: www.obesity.org.
4. Centers for Disease Control and Prevention (CDC) Overweight and Obesity: http://www.cdc.gov/nccdphp/dnpa/obesity/index.htm.
5. European Childhood Obesity Group (ECOG): www.childhoodobesity.net.
6. European Public Health Alliance (EPHA): www.epha.org.

7. The Obesity Society (North American Association for the Study of Obesity, NAASO): www.naaso.org.
8. United Stated Department of Agriculture (USDA) Nutrition information: www.nutrition.gov.
9. World Health Organization (WHO) Global Strategy on Diet, Physical Activity and Health: http://www.who.int/dietphysicalactivity/en.

Index

A

B

D

I

J

K

L

M

P

About the Series Editor

Dr. Adrianne Bendich is Clinical Director of Calcium Research at GlaxoSmithKline Consumer Healthcare, where she is responsible for leading the innovation and medical programs in support of several leading consumer brands including TUMS and Os-Cal. Dr. Bendich has primary responsibility for the coordination of GSK's support for the Women's Health Initiative (WHI) intervention study. Prior to joining GlaxoSmithKline, Dr. Bendich was at Roche Vitamins Inc., and was involved with the groundbreaking clinical studies proving that folic acid-containing multivitamins significantly reduce major classes of birth defects. Dr. Bendich has co-authored more than 100 major clinical research studies in the area of preventive nutrition. Dr. Bendich is recognized as a leading authority on antioxidants, nutrition and bone health, immunity, and pregnancy outcomes, vitamin safety, and the cost-effectiveness of vitamin/mineral supplementation.

In addition to serving as Series Editor for Humana Press and initiating the development of the 20 currently published books in the *Nutrition and Health*™ series, Dr. Bendich is the editor of 11 books, including *Preventive Nutrition: The Comprehensive Guide for Health Professionals.* She also serves as Associate Editor for *Nutrition: The International Journal of Applied and Basic Nutritional Sciences,* and Dr. Bendich is on the Editorial Board of the *Journal of Women's Health and Gender-Based Medicine,* as well as a past member of the Board of Directors of the American College of Nutrition. Dr. Bendich also serves on the Program Advisory Committee for Helen Keller International.

Dr. Bendich was the recipient of the Roche Research Award, was a Tribute to Women and Industry Awardee, and a recipient of the Burroughs Wellcome Visiting Professorship in Basic Medical Sciences, 2000–2001. Dr. Bendich holds academic appointments as Adjunct Professor in the Department of Preventive Medicine and Community Health at UMDNJ, Institute of Nutrition, Columbia University P&S, and Adjunct Research Professor, Rutgers University, Newark Campus. She is listed in *Who's Who in American Women.*

About the Editors

Dr. Giamila Fantuzzi is an Associate Professor in the Department of Human Nutrition at the University of Illinois at Chicago. From 2000 to 2004 she was an Assistant Professor in the Department of Medicine at the University of Colorado Health Sciences Center.

Dr. Fantuzzi is a graduate of the University of Milan, Milan, Italy, where she also obtained her PhD in Experimental Endocrinology. She completed her post-doctoral fellowships in the laboratory of Neuroimmunology at the Mario Negri Institute for Pharmacological Research in Milan, Italy and in the Division of Geographic Medicine and Infectious Diseases at the New England Medical Center of Tufts University in Boston, MA.

Dr. Fantuzzi has investigated and published extensively on the role of cytokines in the regulation of inflammation. Her current research focuses on the role of adipose tissue-derived factors in the modulation of the inflammatory response, with a particular interest on the gastrointestinal system. Her research is currently funded by the National Institutes of Health and she received past funding from the Crohn's and Colitis Foundation of America, the Broad Medical Research Program, and the Cystic Fibrosis Foundation.

Dr. Theodore Mazzone received his MD with distinction from Northwestern University in Chicago, IL. He was an Intern and Resident in Internal Medicine at UCLA Medical Center and a Senior Fellow in Metabolism, Endocrinology and Nutrition at the University of Washington.

Dr. Mazzone has been elected to membership in the American Society for Clinical Investigation. He is a Fellow of the American College of Physicians, the Council on Arteriosclerosis, Thrombosis and Vascular Biology, and the Council on Nutrition, Physical Activity and Metabolism. He is a member of numerous professional organizations, including the American Heart Association, where he is Vice-Chairman of the Diabetes Advisory Committee. He has served on the Steering Committee for the Complications Council of the American Diabetes Association and on the Planning Committee for the annual Scientific Sessions of the American Diabetes Association.

Dr. Mazzone has been a member of multiple peer review panels for the National Institutes of Health and the American Heart Association. He currently serves on the Editorial Board of *Arteriosclerosis, Thrombosis, and Vascular Biology*, and the *Journal of Clinical Endocrinology and Metabolism*. His research interests include the pathogenesis of atherosclerosis, the macrovascular complications of diabetes, and the complications of obesity. He has directed an ongoing research program in these areas for more than 25 years.

Dr. Mazzone has been named in "America's Top Doctors" and as one of "Chicago's Top Doctors." His clinical interests include risk factor for diabetes, metabolic syndrome, and refractory lipid disorders. Dr. Mazzone is currently Professor of Medicine, and Chief of the Section of Endocrinology, Diabetes and Metabolism, at the University of Illinois Medical Center in Chicago.